David E. McAdams

Geometric Nets
Mega Project Book
Tabbed

by David E. McAdams
http://www.demcadams.com

A hands-on introduction to three-dimensional geometry
using geometric nets with instructions.

Index

See Also:
- **Alphabetical Index page 338**
- **Keyword Index page 341**
- **Index by Number of Faces page 348.**

Getting Started

What is a geometric net?

A geometric net is a flat drawing that can be folded into a three dimensional figure. For example, six identical squares can be made into a cube. This is because a cube has six sides, all of which are identical squares. Each of the drawings in this book can be folded into a three dimensional geometric object.

Most geometric nets fold into solids with flat sides. There are a few exceptions. A cylinder can be made from a rectangle and two circles. A cone can be made from a circle and a triangle with a curved bottom.

How hard is it to make a solid from a geometric net?

Some of them are easy, and some are hard. Basically, the more sides a solid has, the harder it is to build from a net. Start with the easy ones, and build up to the hard ones.

How do I build a solid from a geometric net?

Start by making a copy of the page on which the geometric net is drawn. If you want to decorate your net by drawing on it or coloring it, do so before you cut it out.

Then use scissors to carefully cut out the net along the solid lines. Sometimes two adjacent faces share a line in the drawing that must be cut. This line will be a solid line.

Once the shape is cut out, start folding along the dotted lines. Use small pieces of clear tape to attach the edges together. When all the edges are taped together, your shape is finished.

What do all the words in the names mean?

Most of the words used in the names of the three dimensional solid shapes were made up by the Greeks over two thousand years ago. Greek mathematicians put together words to make names for the shapes. Some of the words mean numbers. For example 'Tetra' is used to mean 'four'. Some of the words used are:

antiprism	a solid with polygons for bases and alternating, identical triangles for sides.
antipyramid	a pyramid with sides that are alternating triangles, forming a truncated pyramid.
augmented	having a polygon "pasted" onto one or more faces.
biaugmented	having a polygon "pasted" onto two faces.
bipyramid	a solid that can be made by 'gluing' the bottoms of two identical pyramids together.
bireduced	having two faces 'dug out' to form a concave shape.
cantellated	cut at the edges and vertices.
chamfered	having the edges cut off in a bevel.

cubed	Placing an identical polygon on each face of a cube, then cutting corners so that only the face of the polygon is in the plane of the face of the original cube.
cupola	having a dome.
deca-	ten.
decagonal	based on a polygon with ten sides.
deltohedron	a polyhedron constructed from usually identical deltoids.
deltoid	a kite shaped object with four sides.
deltoidal	being made of kite shaped objects (deltoids) for faces.
digonal	based on a two-sided polygon (in reality, the base is a single line segment).
diminished	changed by having a polygon "cut off" of a polygon to form a single face.
disdyakis	square faces on a polygon are replaced with quadrilateral pyramids.
dodeca-	twelve
elongated	a solid that starts with another shape, but has rectangles added to make it longer.
endo-	bending inwards, concave
frustum	a pyramid or cone with the top cut off.
great	used with 'small' to distinguish between two different polyhedra with the same number of faces.
gyrate	changed by rotating everything on one side of an intersecting plan by a certain angle.
gyro-	different on two sides, alternating.
gyroelongated	made longer adding an antiprism to the base.
-hedron	a solid whose sides are flat.
heptagonal	based on a seven sided polygon.
hexagonal	based on a six sided polygon.
icosi-	having twenty sides.
icosa-	having twenty sides.
metabiaugmented	having polygons "pasted" onto two non-opposite faces.
metabidiminished	having two polygons "cut off" to form a single face, of two non-opposite faces.
oblique	not right-angled, usually referring to the angle of a segment between the apex and the center of the base.
octa-	eight.
octagonal	based on an eight sided polygon.
omnireduced	having all faces 'dug out' to form a concave shape.
ortho-	the same on both sides, not alternating.
parabiaugmented	having a polygon "pasted" onto two opposite faces.

     Geometric Nets Mega Project Book, Tabbed By David E. McAdams

pentagonal	based on a pentagon (five sized polygon)
pentakis	a polyhedron formed by replacing regular pentagonal faces with pentagonal pyramids
prism	a solid with polygons for tops and bottoms and identical rectangles for sides.
pyramid	a solid with a polygon for a bottom and triangular sides that come to a point.
quadrilateral	based on a four sided polygon.
reduced	having one or more faces 'dug out' to form a concave shape
regular	having faces made of identical regular polygons.
rhombi-	having to do with or containing one or more rhombuses.
rhombic	containing rhombuses for one or more faces.
rhombus	a flat figure with four sides that are not perpendicular.
right	a line joining the center of the base and the center of the top is perpendicular to the top and to the base; or a line joining the center of the base to the apex (point) of a figure is perpendicular to the base.
rotunda	having an round dome, taller than a cupola.
snub	a process on a polyhedron, where vertices are cut off to form new faces.
stella-	star
stellated	having faces replaced with a pyramid that has the face as a base.
tetra	four
tetrakis	a polygon formed by replacing square faces of another polygon with quadrilateral pyramids
trapezohedron	
triakis	a polyhedron formed by replacing triangular faces with a triangular pyramid.
triangular	based on a triangle.
triaugmented	having a polygon "pasted" onto three faces
tridiminished	having three polygons "cut off", forming a single face on each.
trigonal	based on a triangle
truncated	having a apex, vertex, or other part cut off by a plane.

Tetrahedron

1. Color the shape before you cut it out.

2. Cut out along the solid lines.

3. Fold along all of the dotted lines.

4. Dab a little glue on each tab and attach it.

5. Once the glue has dried, attach any decoration

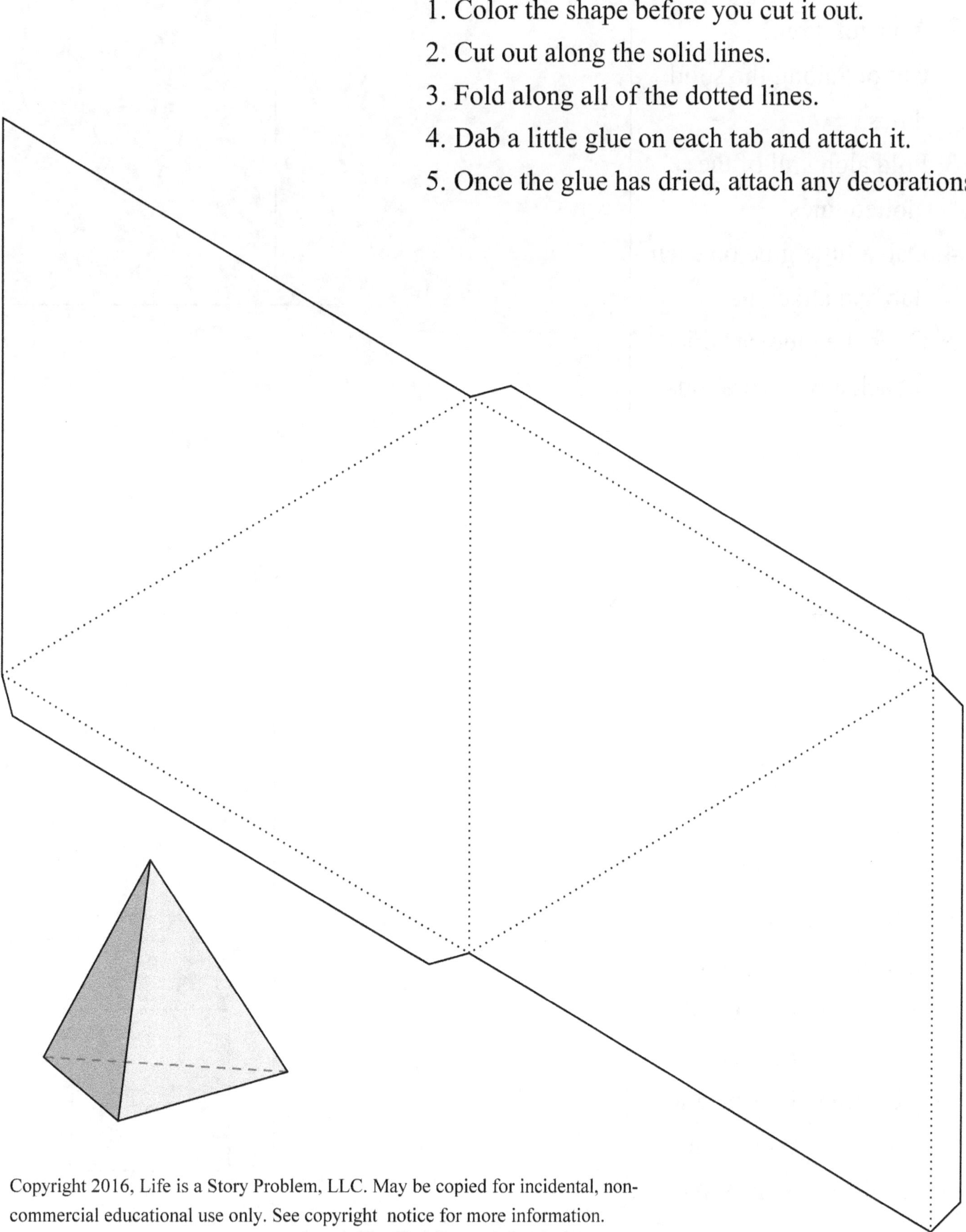

Cube

1. Color the shape before you cut it out.
2. Cut out along the solid lines.
3. Fold along all of the dotted lines.
4. Dab a little glue on each tab and attach it.
5. Once the glue has dried, attach any decorations.

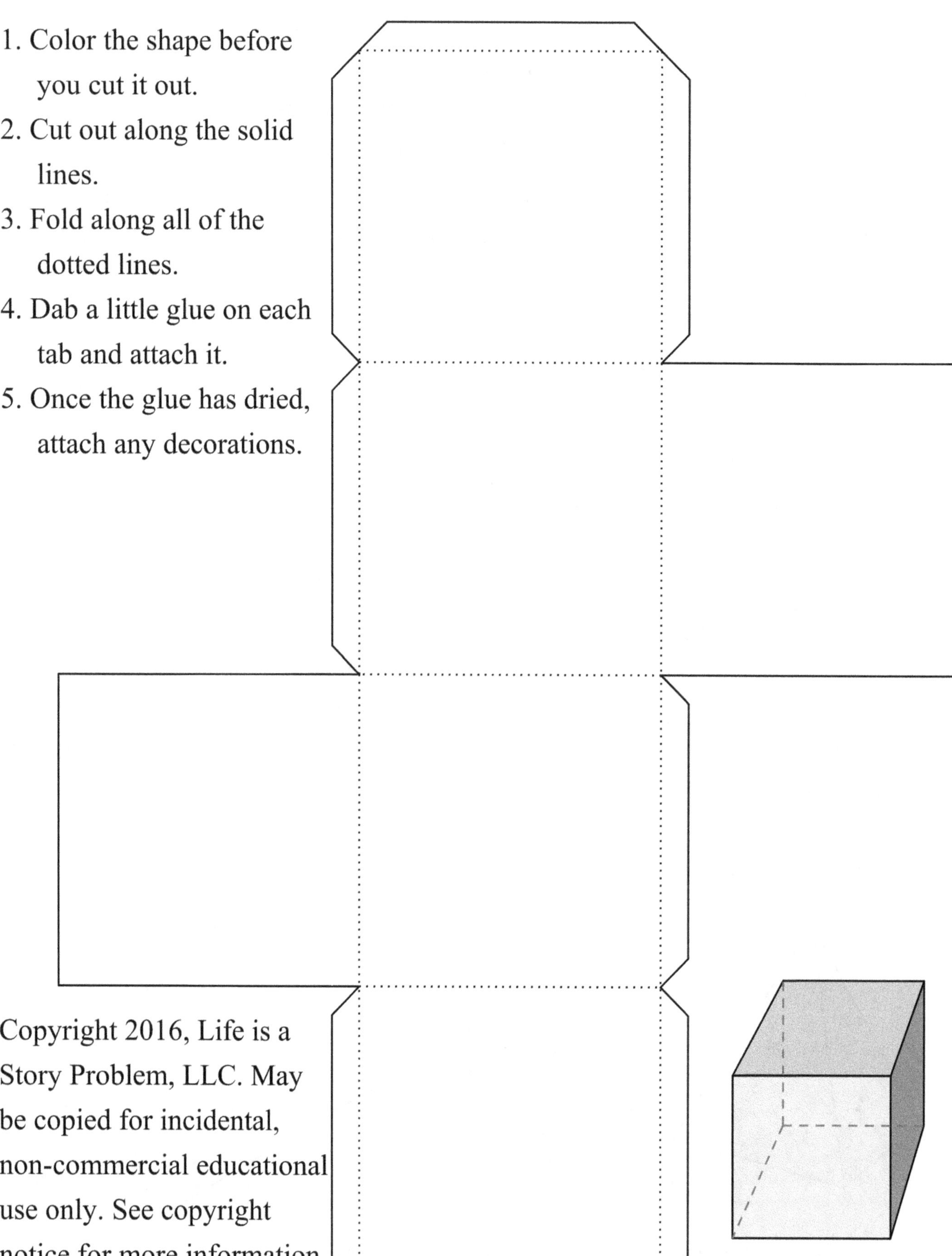

 Geometric Nets Mega Project Book, Tabbed By David E. McAdams

Octahedron

1. Color the shape before you cut it out.
2. Cut out along the solid lines.
3. Fold along all of the dotted lines.
4. Dab a little glue on each tab and attach it.
5. Once the glue has dried, attach any decorations.

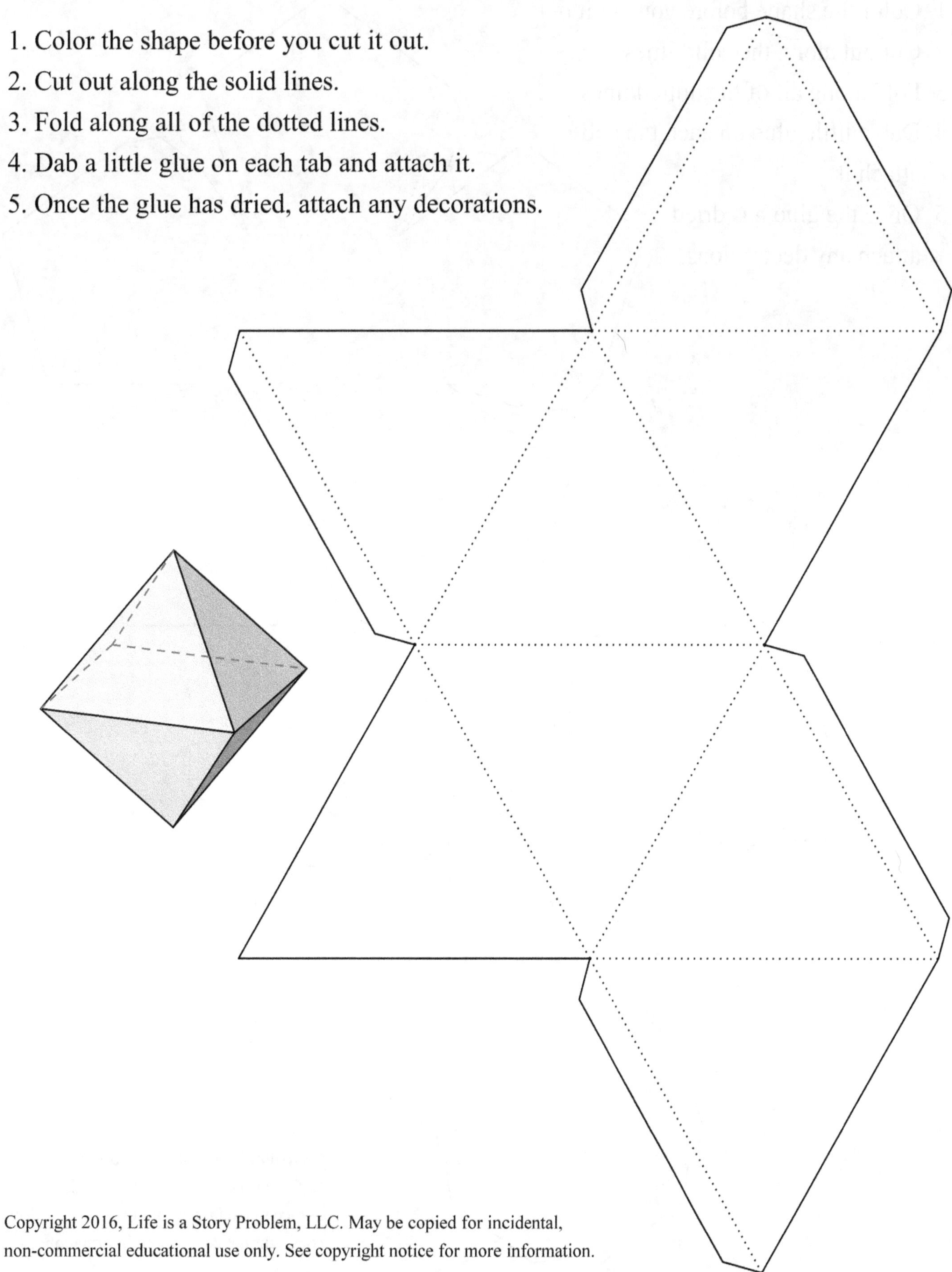

Dodecahedron

1. Color the shape before you cut it out.
2. Cut out along the solid lines.
3. Fold along all of the dotted lines.
4. Dab a little glue on each tab and attach it.
5. Once the glue has dried, attach any decorations.

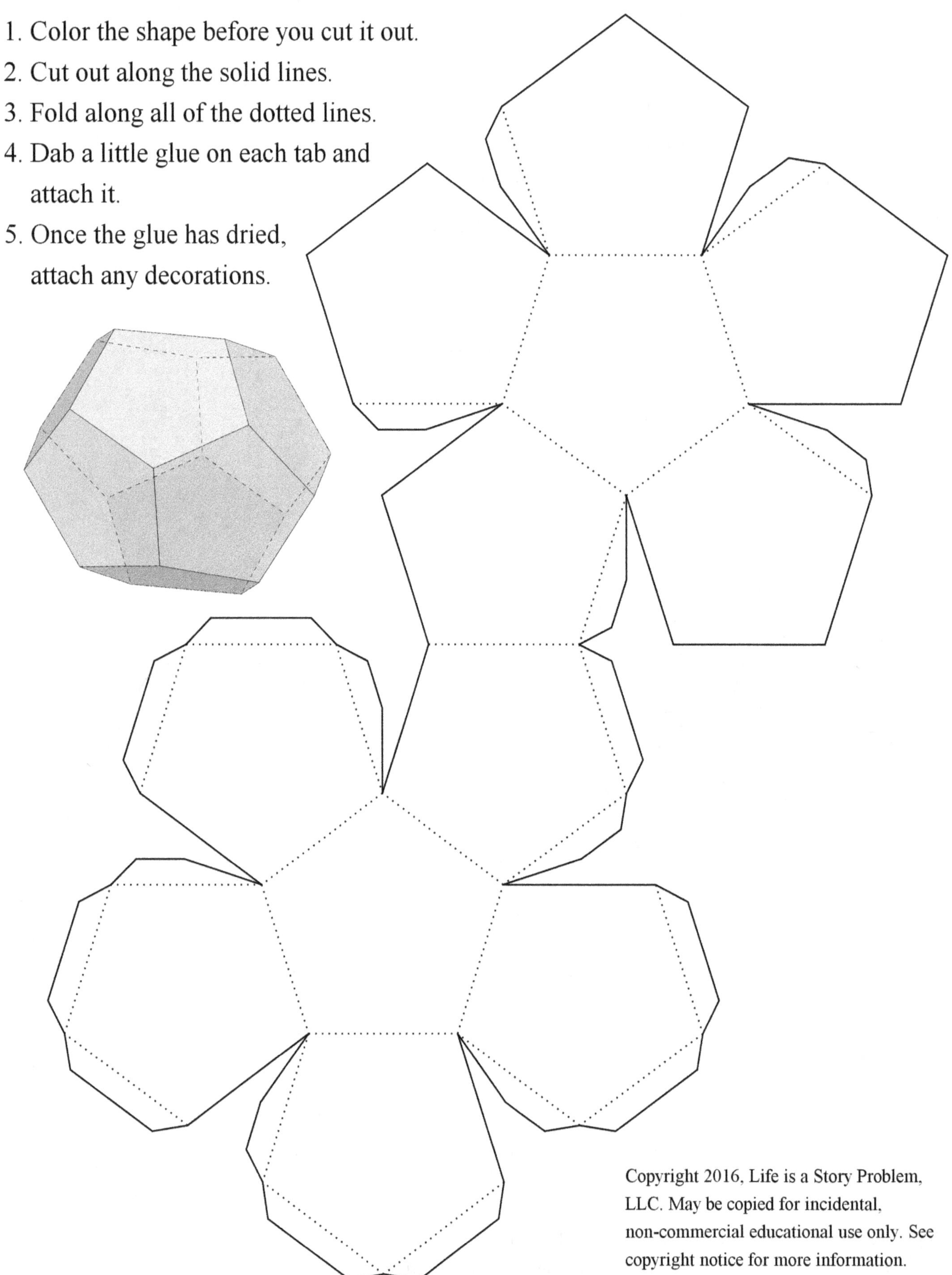

 Geometric Nets Mega Project Book, Tabbed By David E. McAdams

Icosahedron

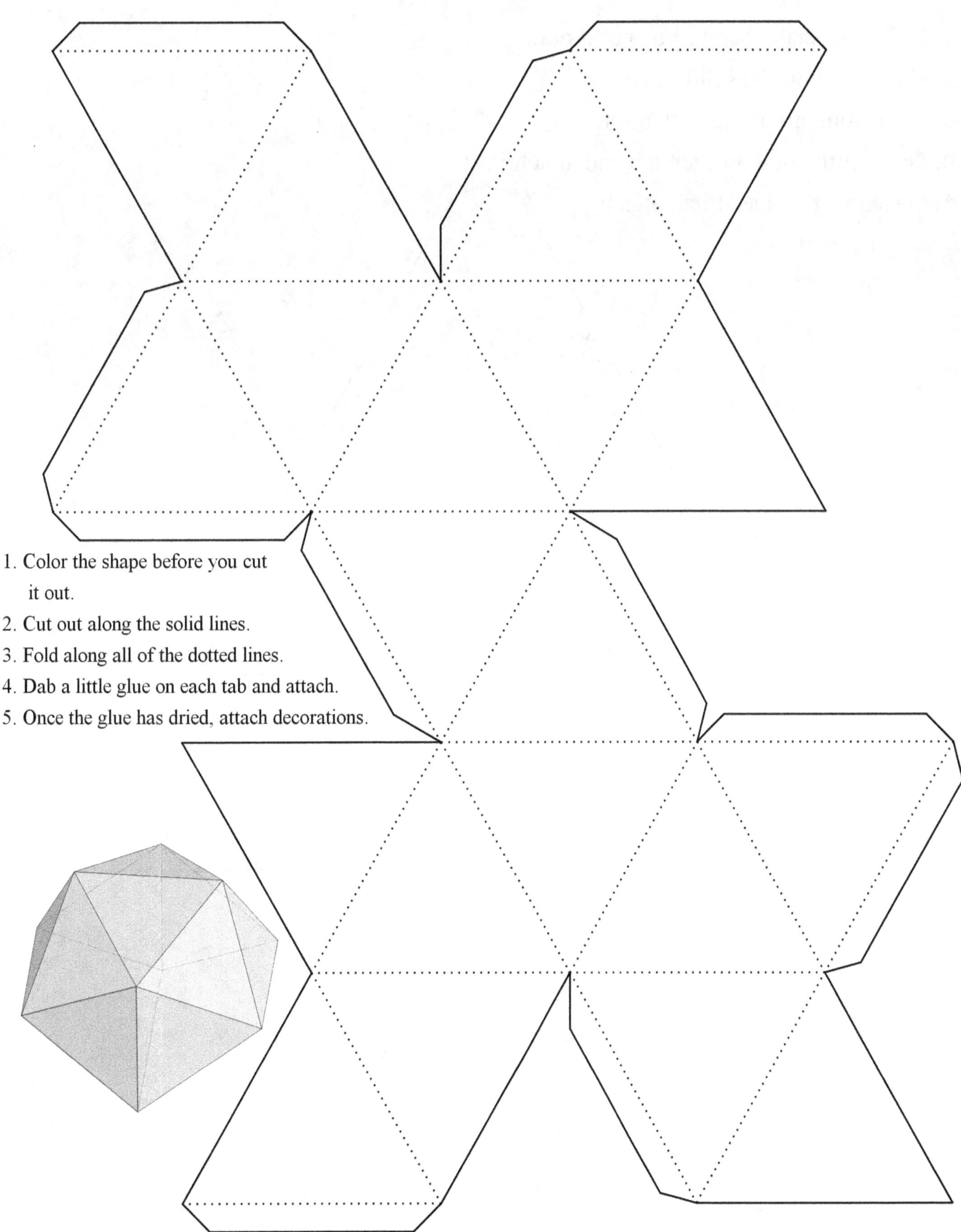

1. Color the shape before you cut
   it out.
2. Cut out along the solid lines.
3. Fold along all of the dotted lines.
4. Dab a little glue on each tab and attach.
5. Once the glue has dried, attach decorations.

Truncated tetrahedron

1. Color the shape before you cut it out.
2. Cut out along the solid lines.
3. Fold along all of the dotted lines.
4. Dab a little glue on each tab and attach it.
5. Once the glue has dried, attach any decorations.

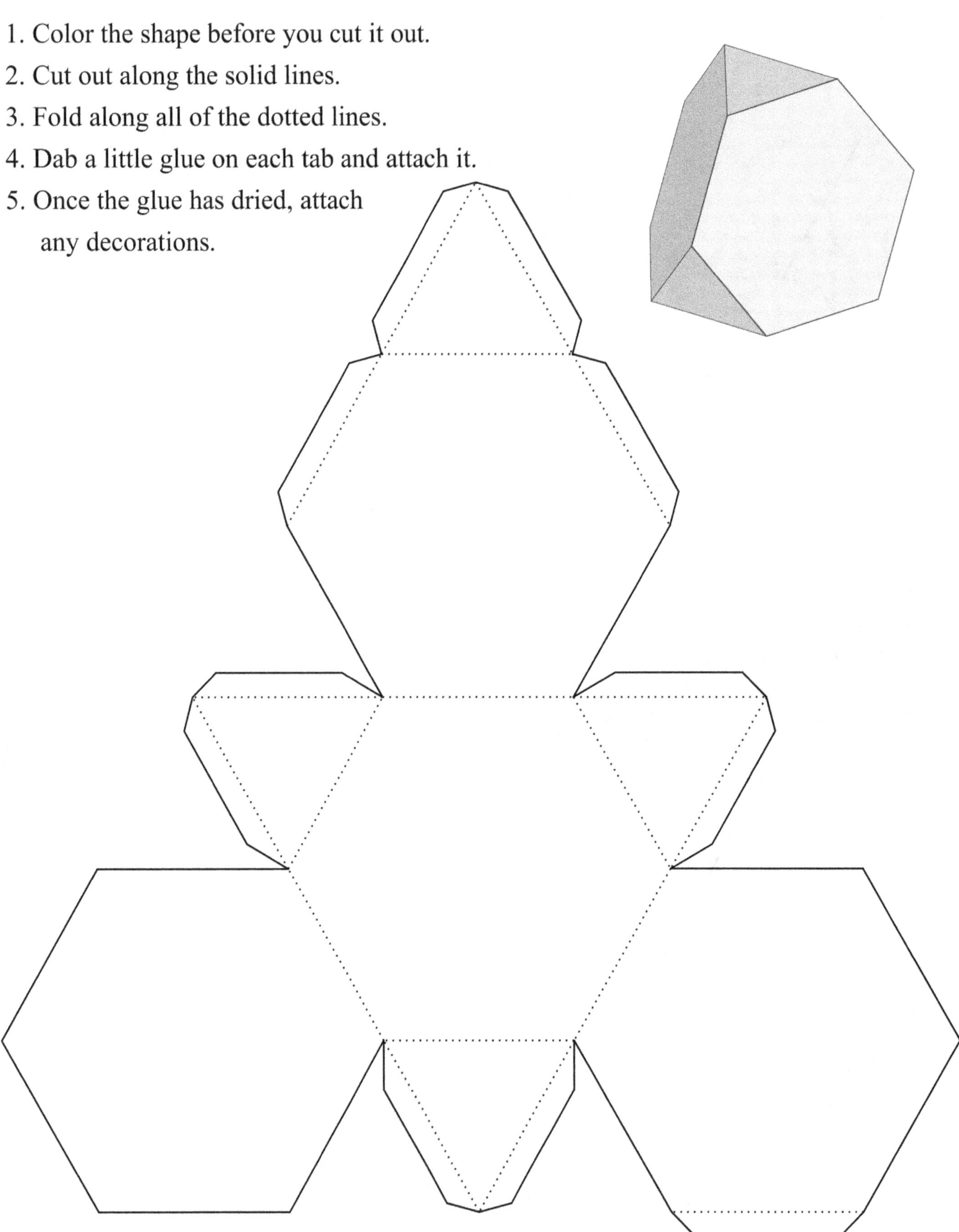

 Geometric Nets Mega Project Book, Tabbed By David E. McAdams

Cuboctahedron

1. Color the shape before you cut it out.
2. Cut out along the solid lines.
3. Fold along all of the dotted lines.
4. Dab a little glue on each tab and attach it.
5. Once the glue has dried, attach any decorations.

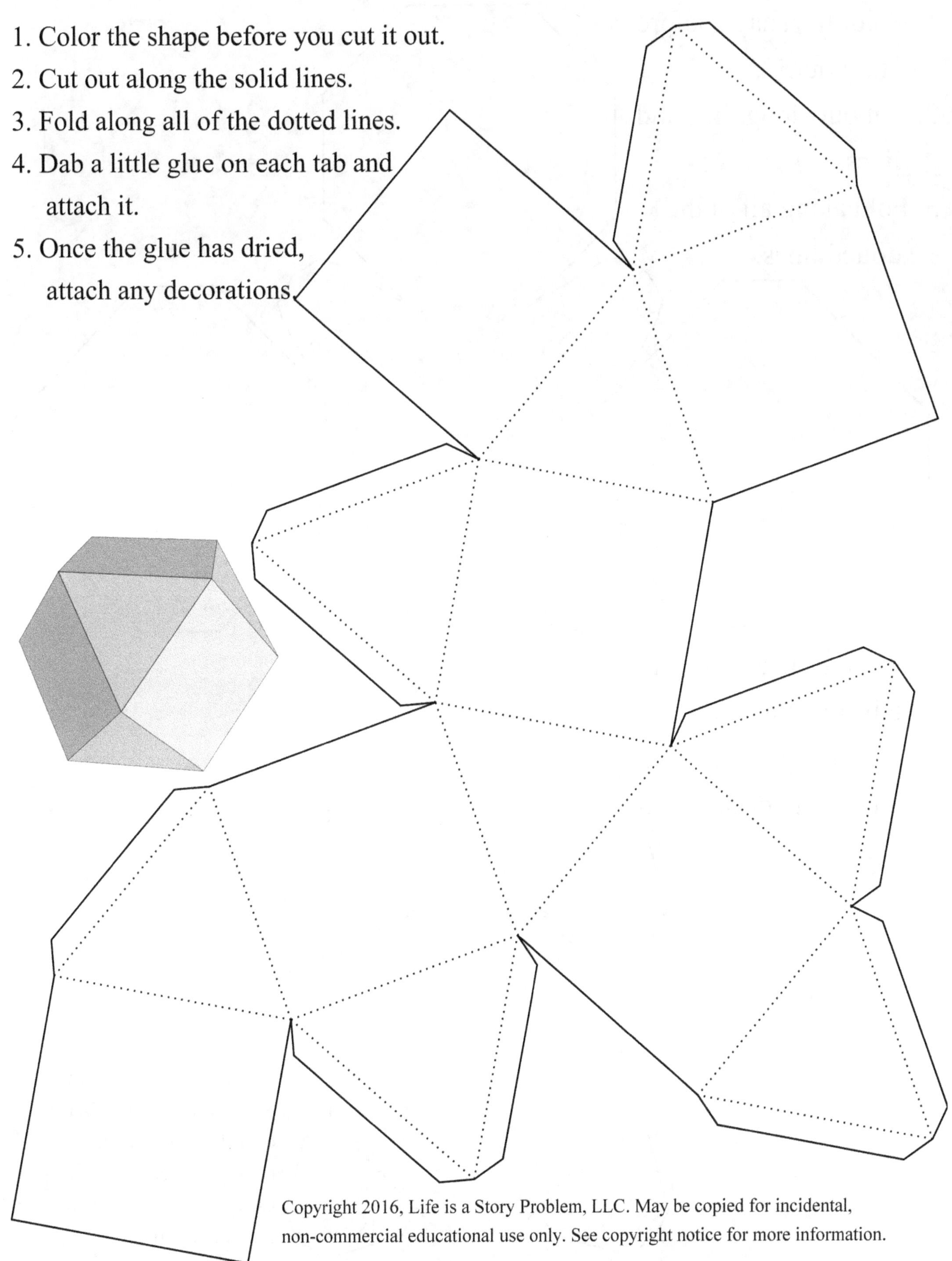

Truncated cube

1. Color the shape before you cut it out.

2. Cut out along the solid lines.

3. Fold along all of the dotted lines.

4. Dab a little glue on each tab and attach it.

5. Once the glue has dried, attach any decorations.

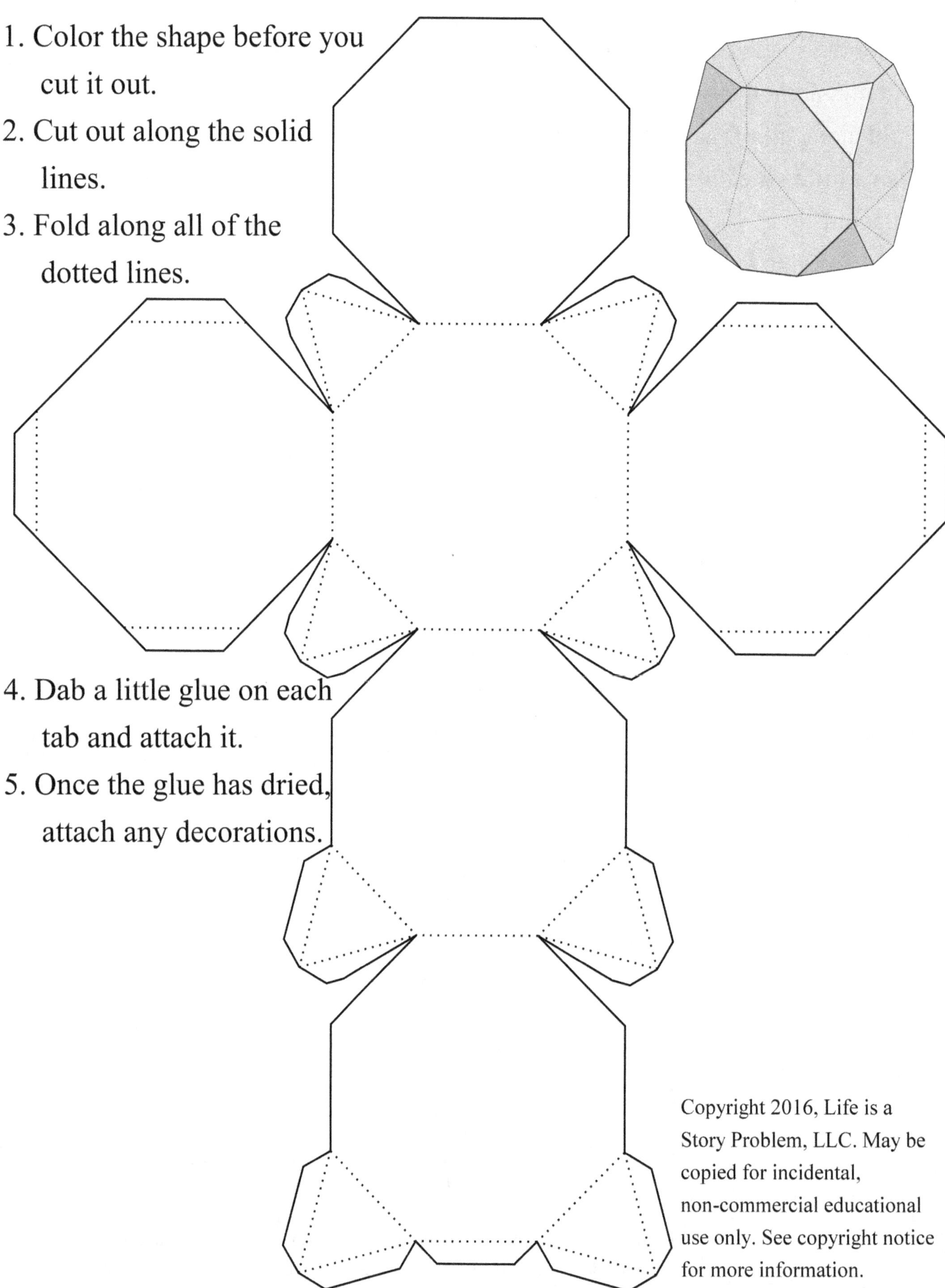

 Geometric Nets Mega Project Book, Tabbed By David E. McAdams

Truncated octahedron

1. Color the shape before you cut it out.
2. Cut out along the solid lines.
3. Fold along all of the dotted lines.
4. Dab a little glue on each tab and attach it.
5. Once the glue has dried, attach any decorations.

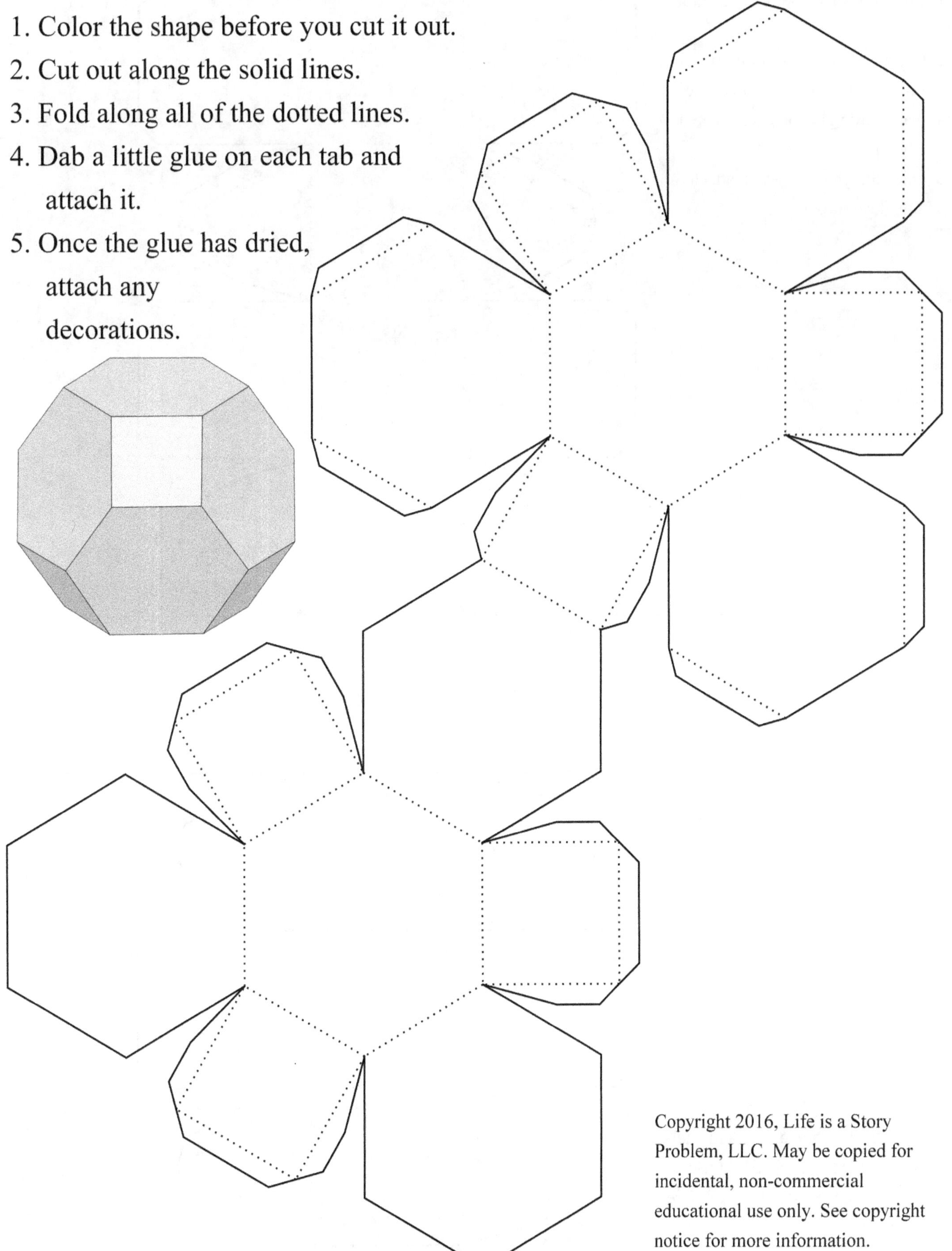

Rhombicuboctahedron

1. Color the shape before you cut it out.
2. Cut out along the solid lines.
3. Fold along all of the dotted lines.
4. Dab a little glue on each tab and attach it.
5. Once the glue has dried, attach any decorations.

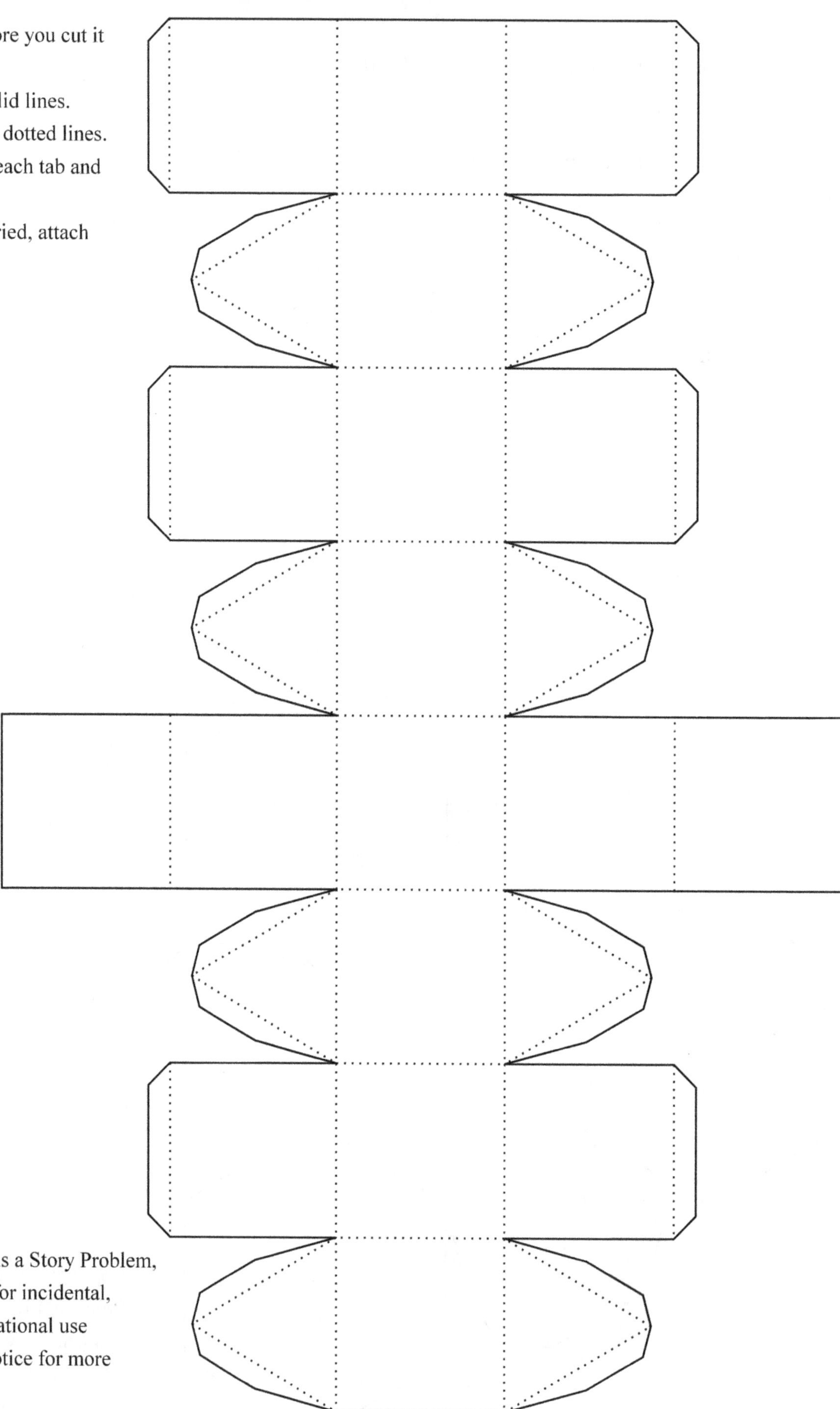

Truncated cuboctahedron

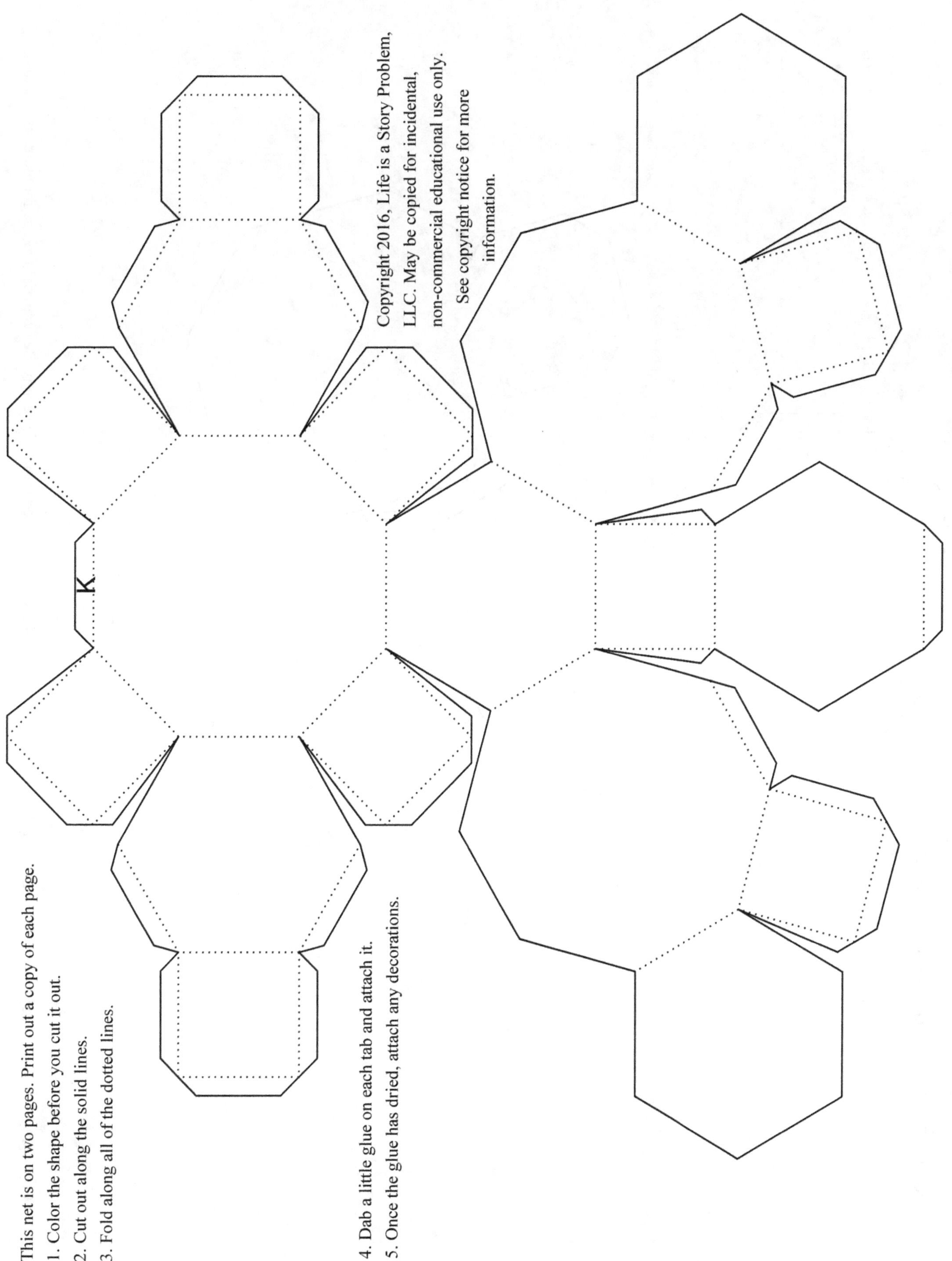

2nd page, truncated cuboctahedron

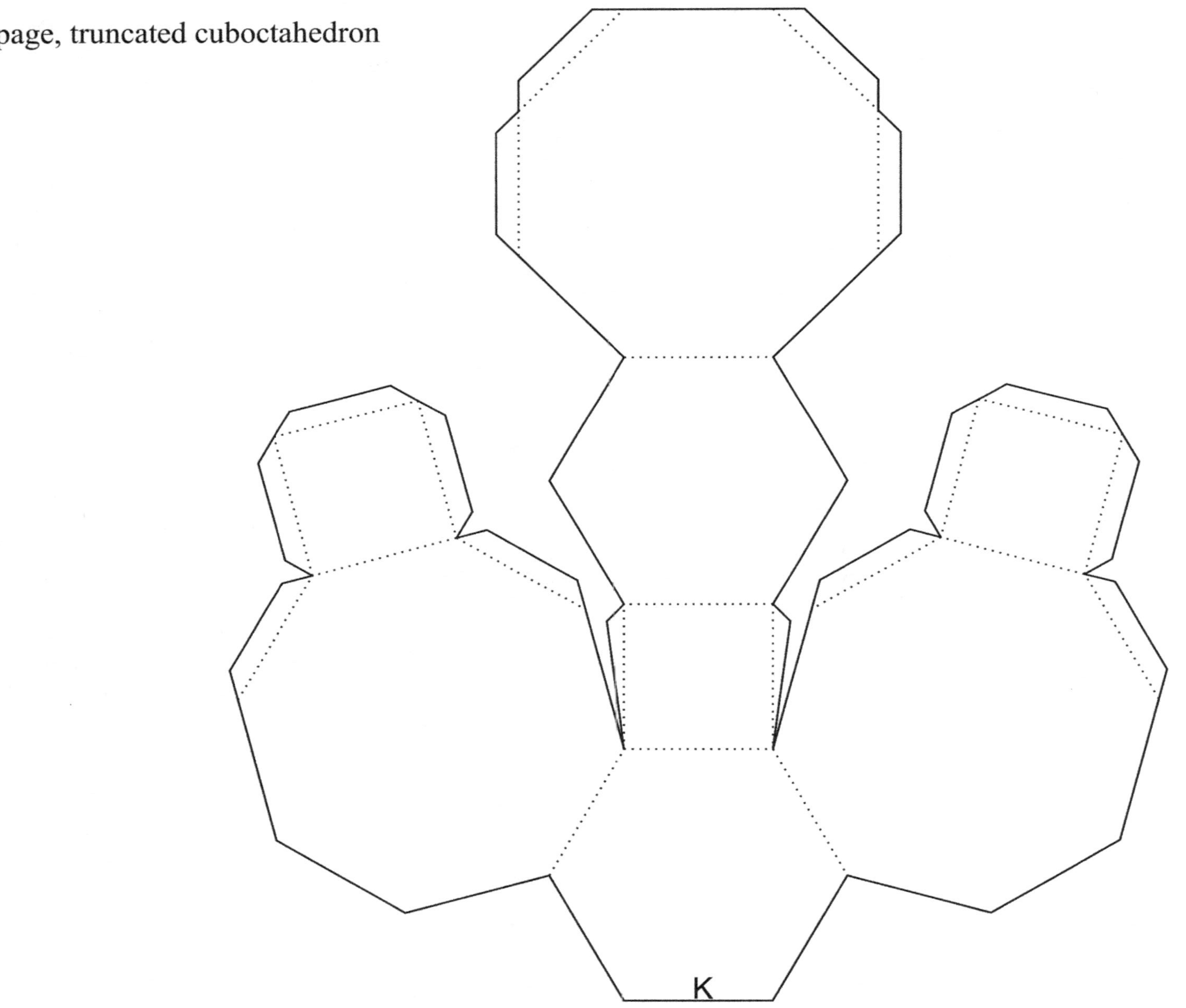

Snub cube

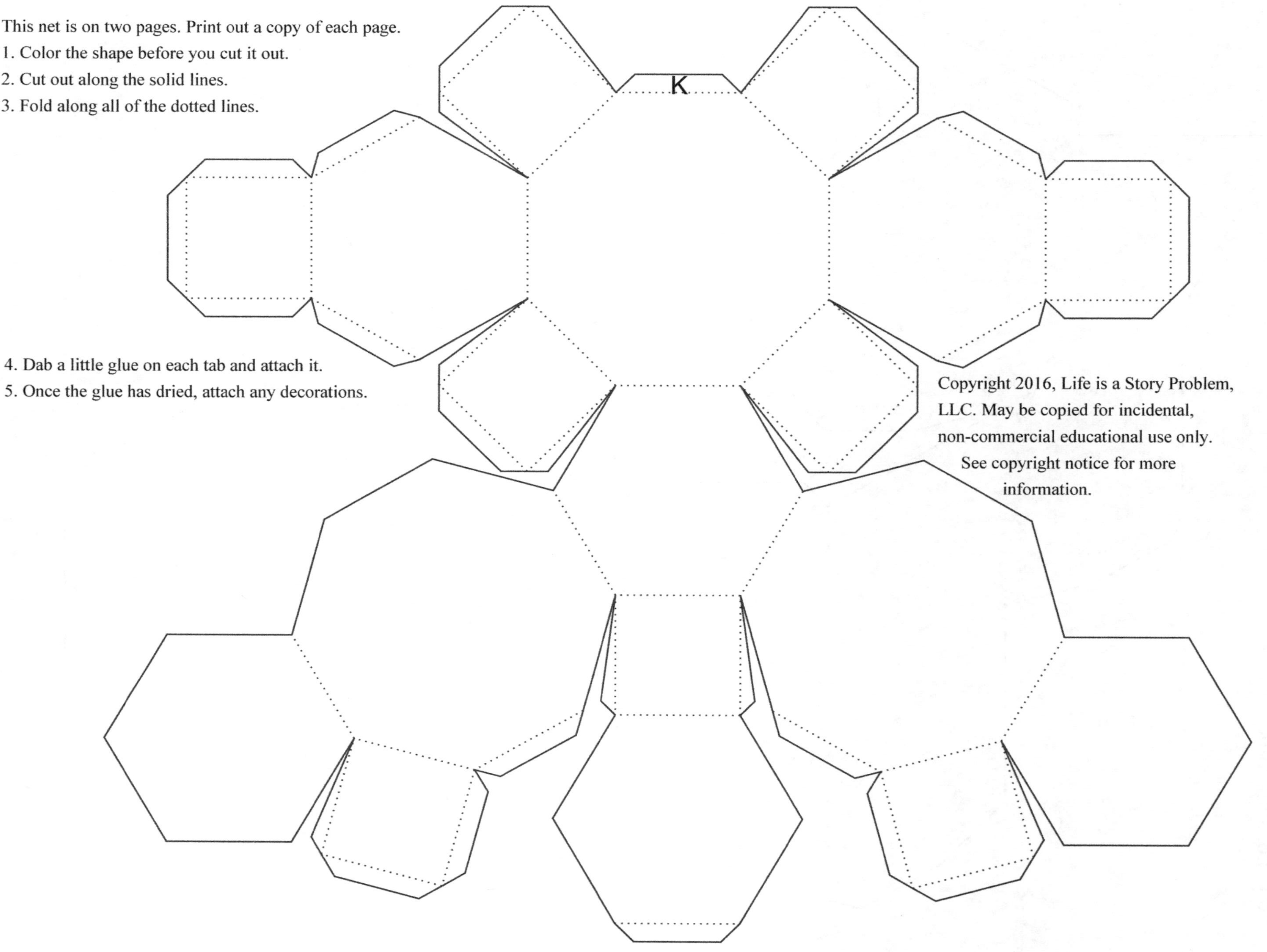

This net is on two pages. Print out a copy of each page.

1. Color the shape before you cut it out.
2. Cut out along the solid lines.
3. Fold along all of the dotted lines.
4. Dab a little glue on each tab and attach it.
5. Once the glue has dried, attach any decorations.

2nd page, snub cube

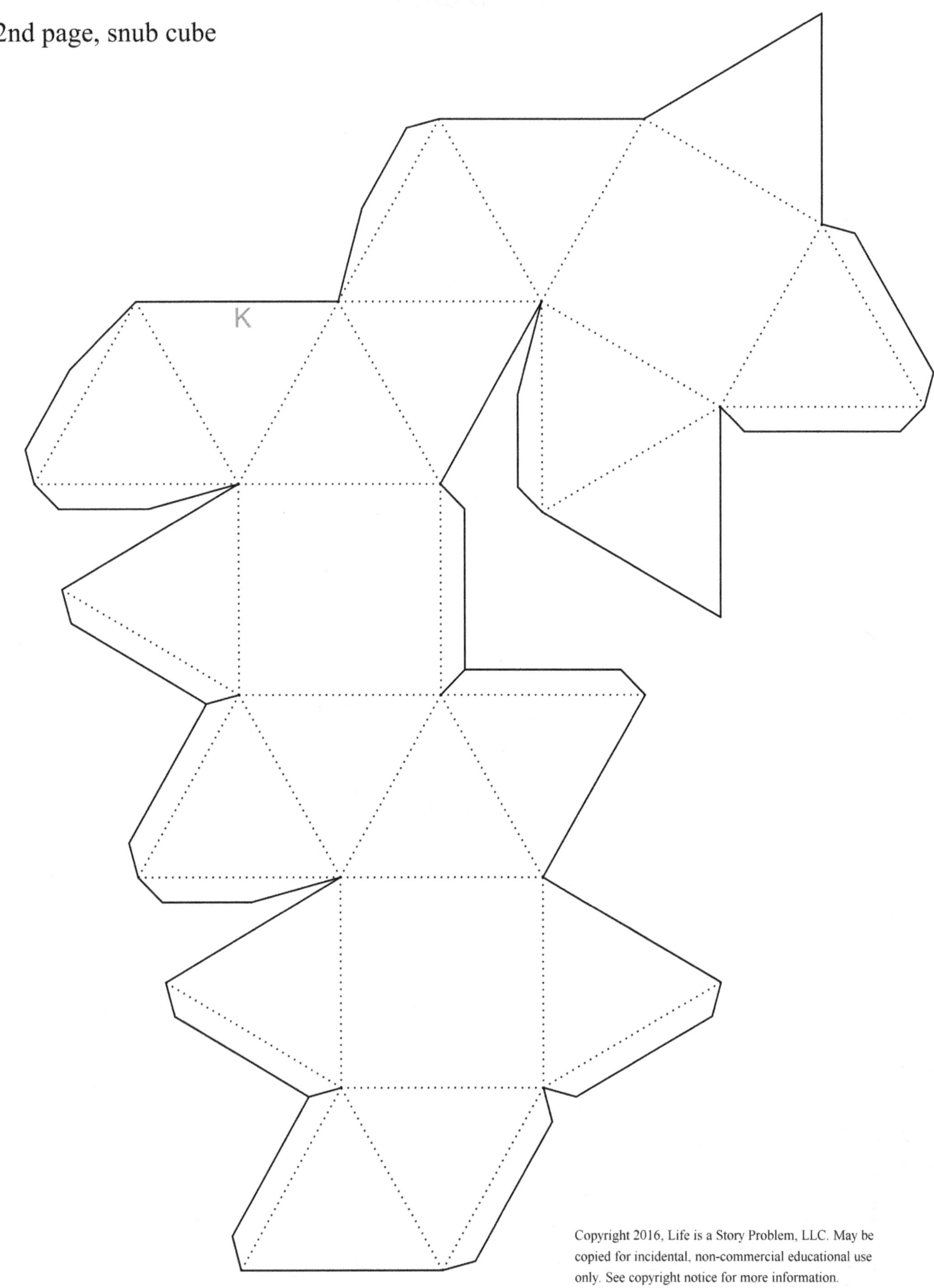

Geometric Nets Mega Project Book, Tabbed By David E. McAdams

Icosidodecahedron

This net is on two pages. Print out a copy of each page.

1. Color the shape before you cut it out.

2. Cut out along the solid lines.

3. Fold along all of the dotted lines.

4. Dab a little glue on each tab and attach it.

5. Once the glue has dried, attach any decorations.

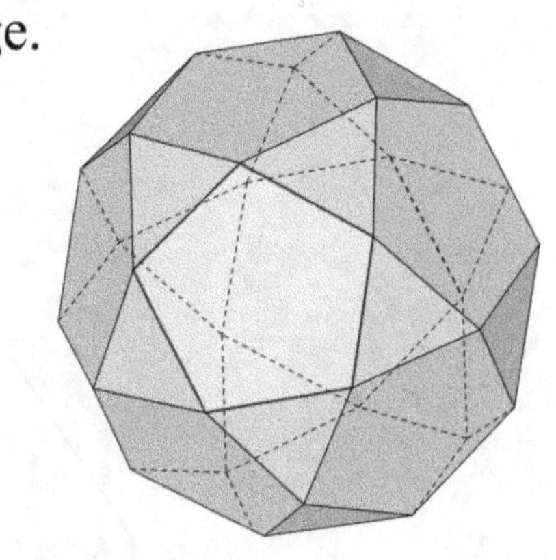

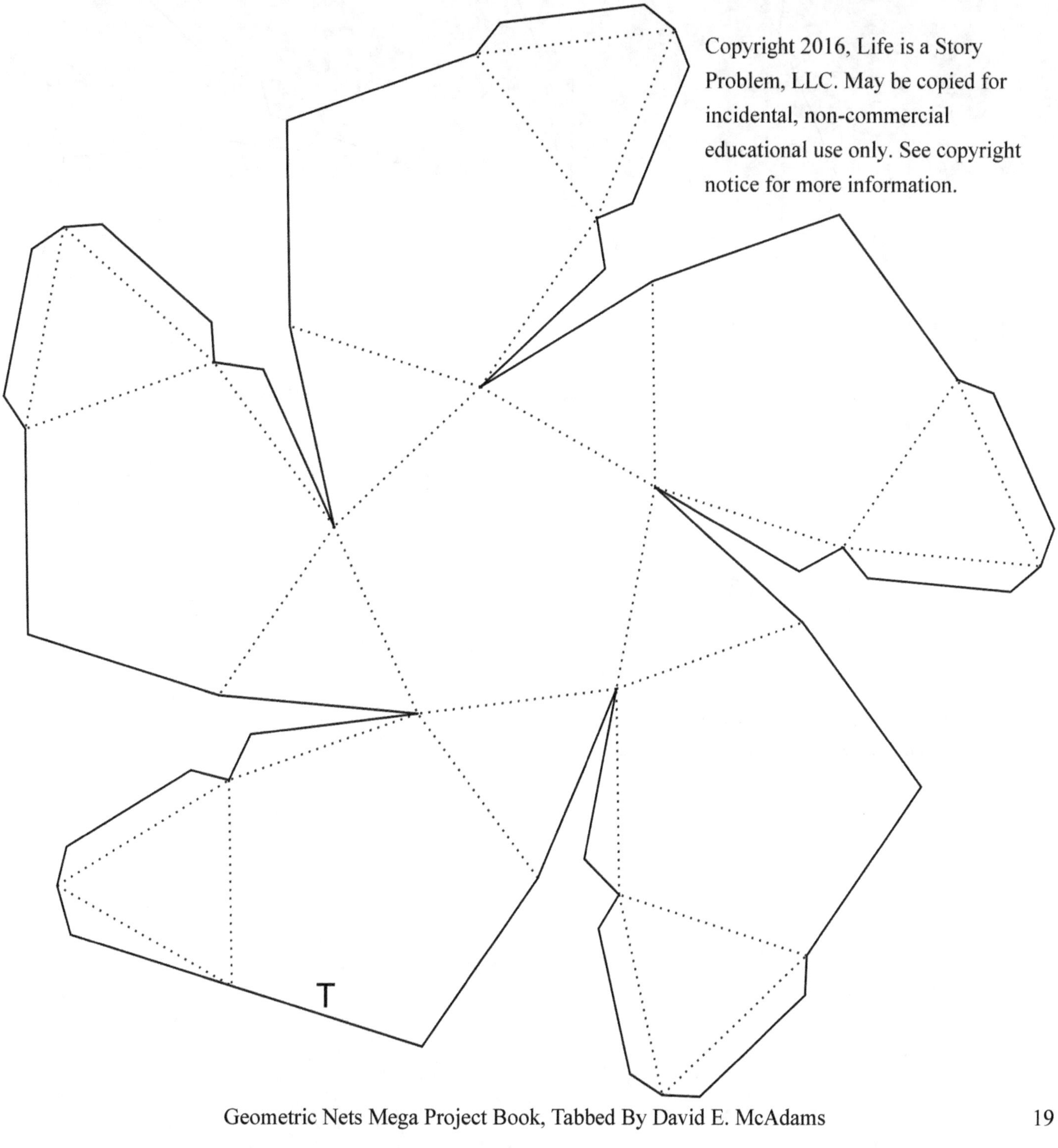

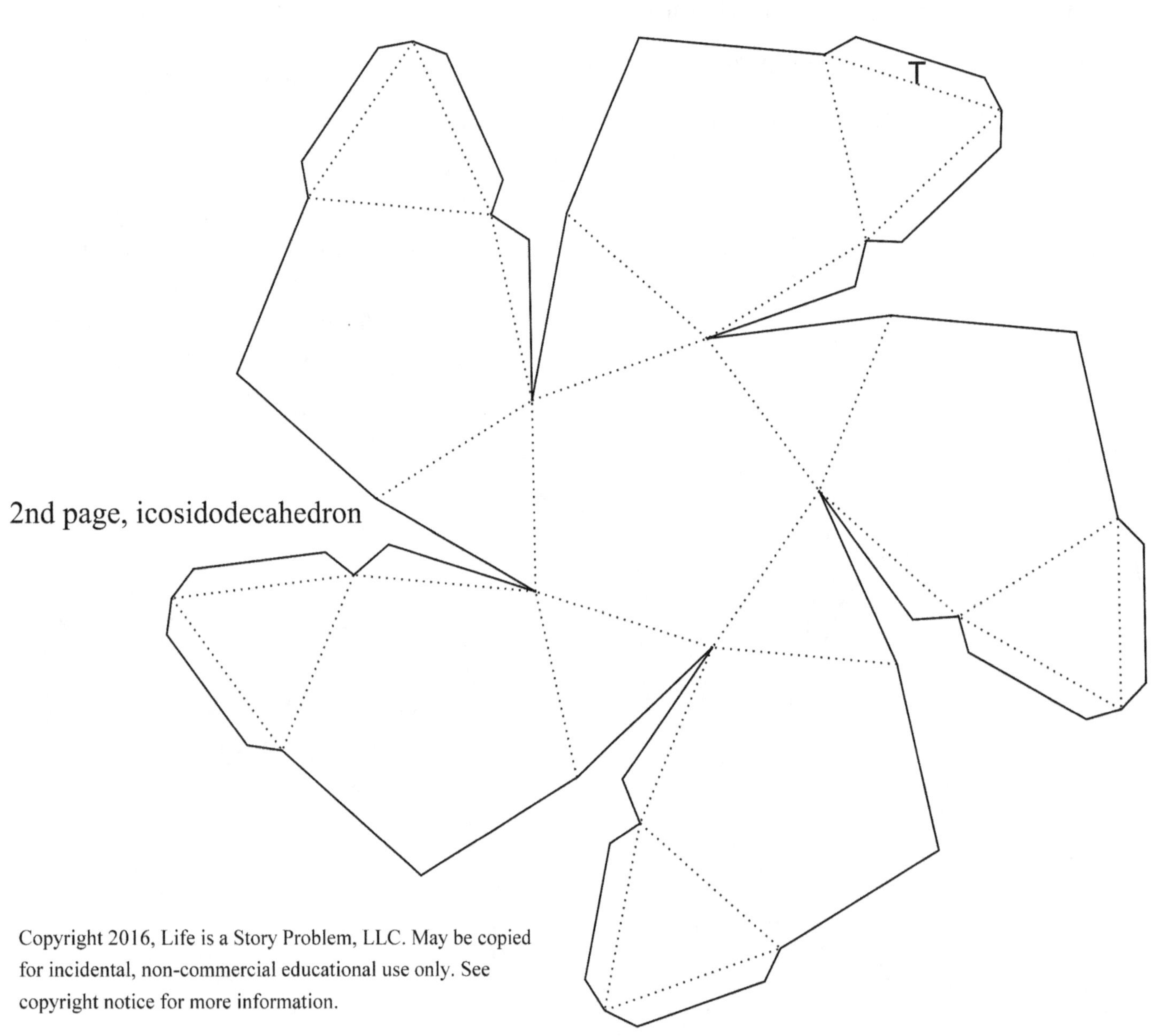

2nd page, icosidodecahedron

Geometric Nets Mega Project Book, Tabbed By David E. McAdams

Truncated dodecahedron

This net is on two pages. Print out a copy of each page.

1. Color the shape before you cut it out.

2. Cut out along the solid lines.

3. Fold along all of the dotted lines.

4. Dab a little glue on each tab and attach it.

5. Once the glue has dried, attach any decorations.

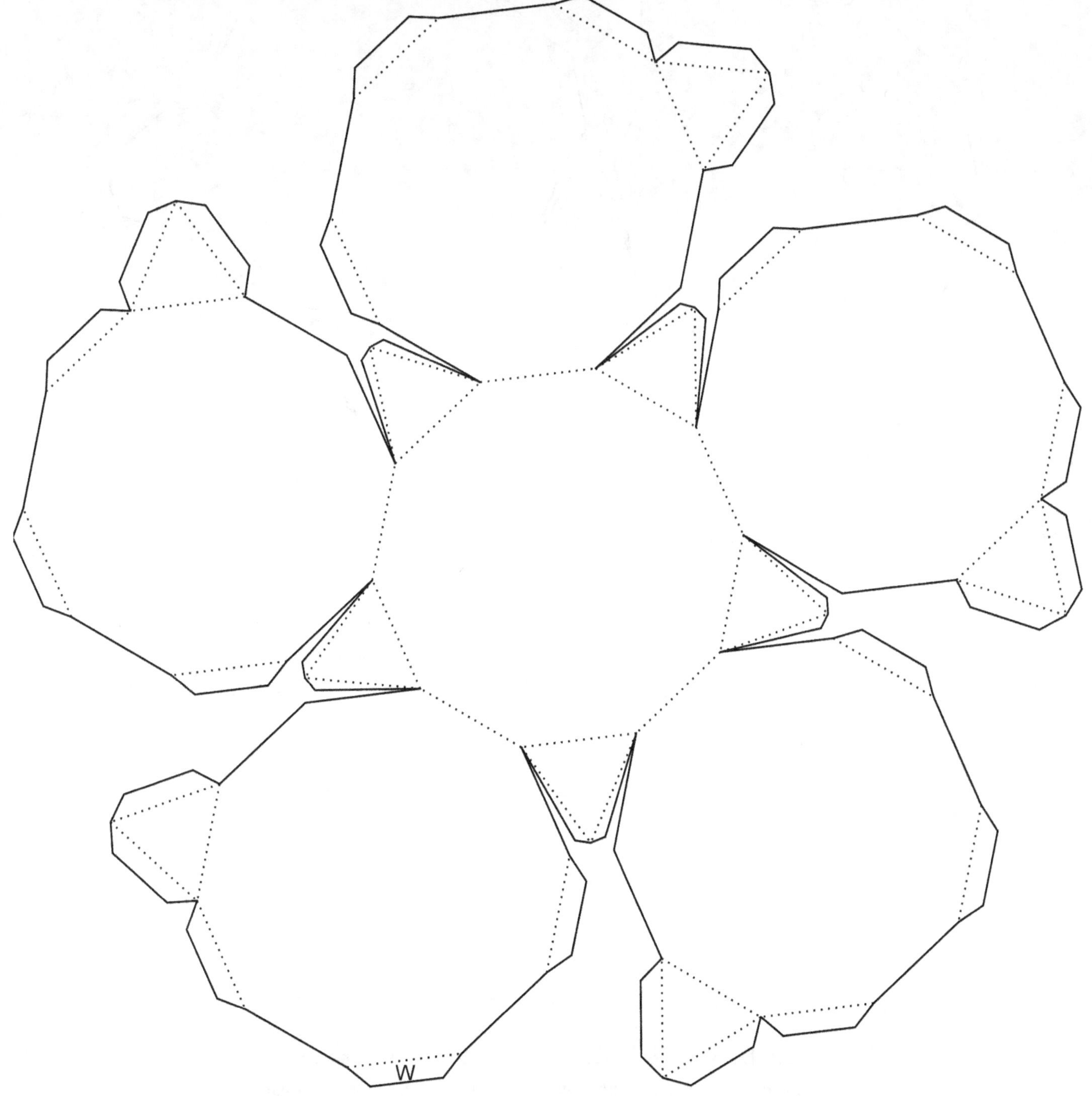

2nd page, truncated dodecahedron

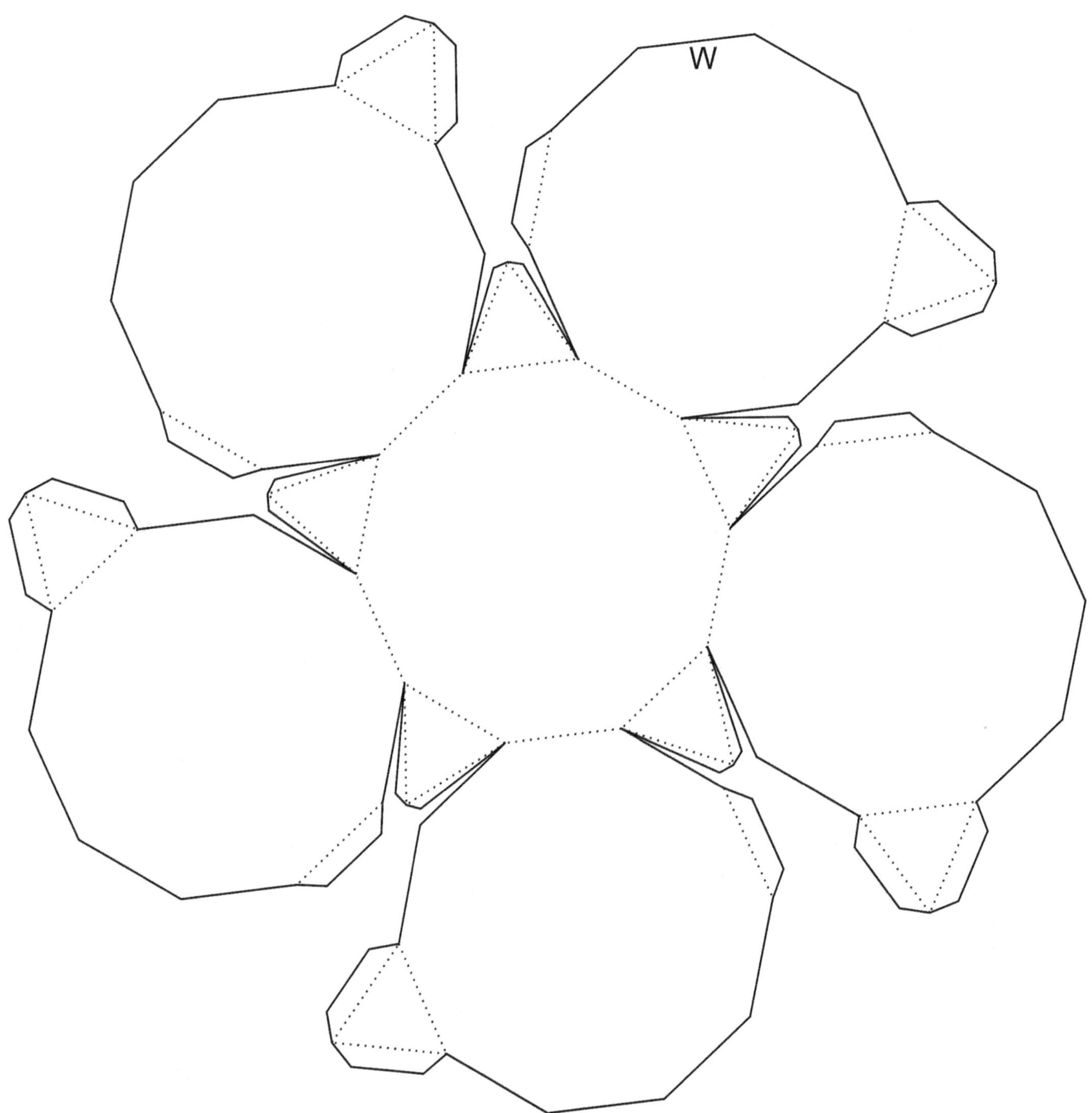

Truncated icosahedron

This net is on three pages. Print out a copy of each page.

1. Color the shape before you cut it out.
2. Cut out along the solid lines.
3. Fold along all of the dotted lines.
4. Dab a little glue on each tab and attach it.
5. Once the glue has dried, attach any decorations.

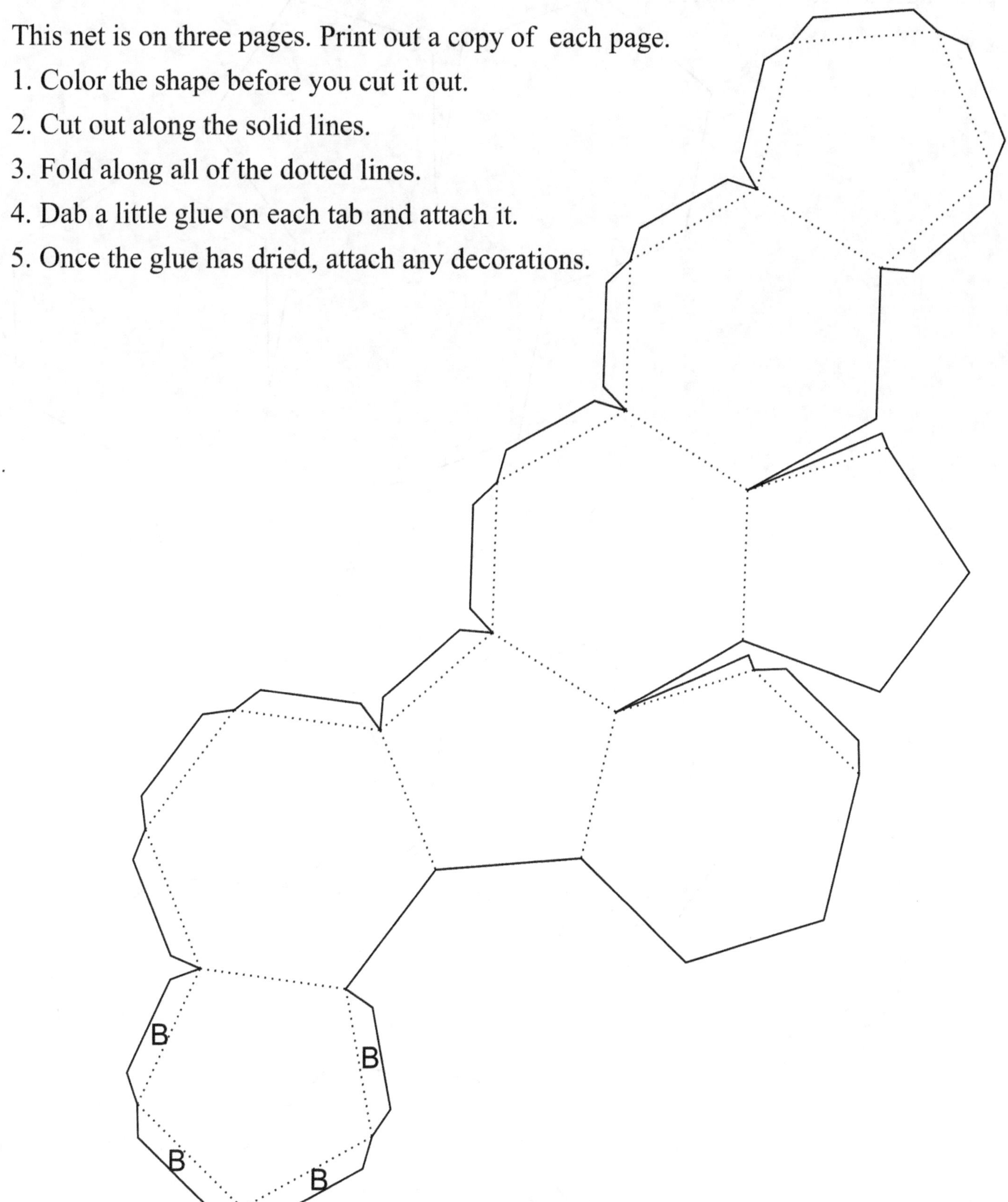

2nd page, truncated icosahedron

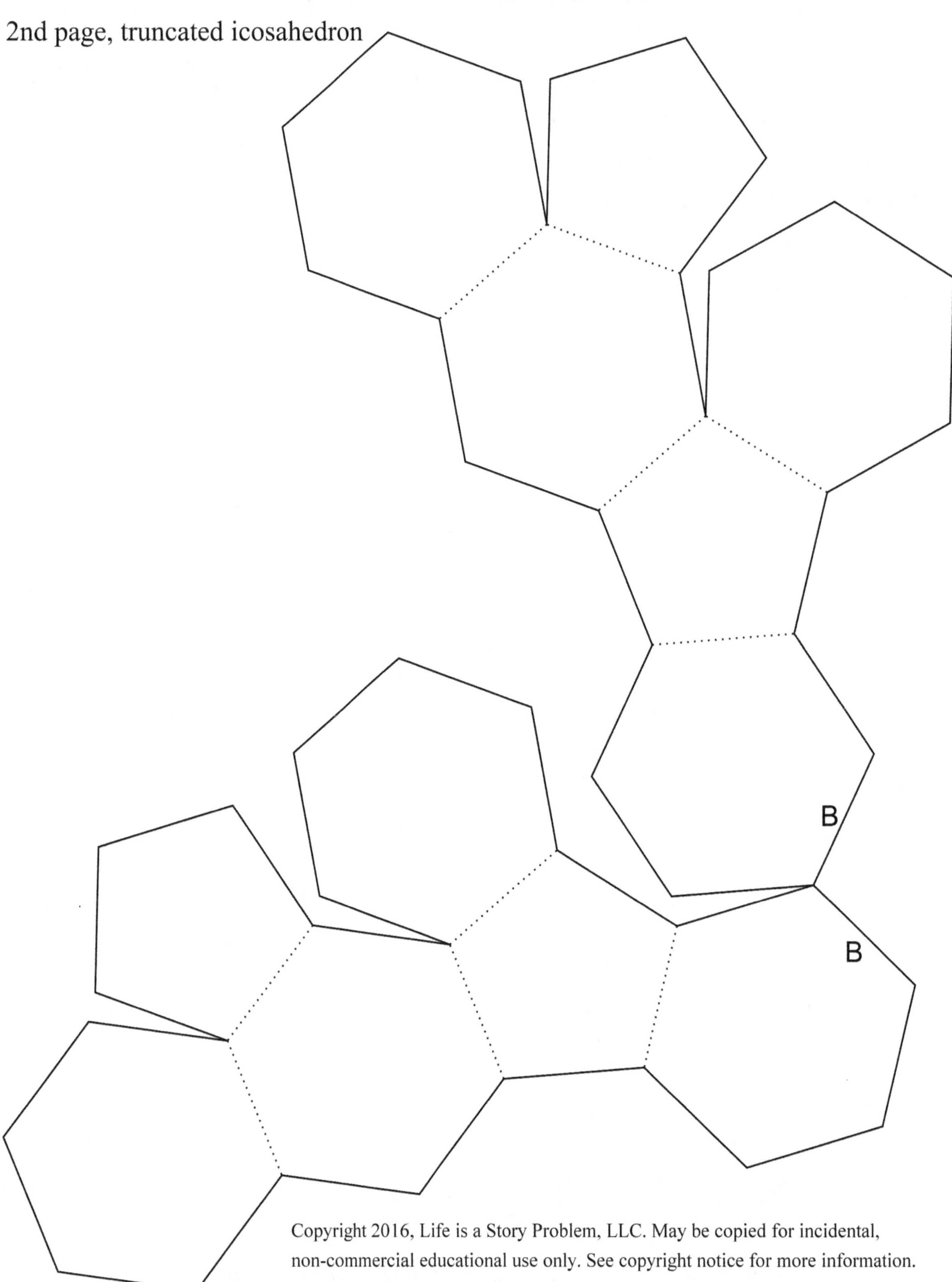

 Geometric Nets Mega Project Book, Tabbed By David E. McAdams

3rd page, truncated icosahedron

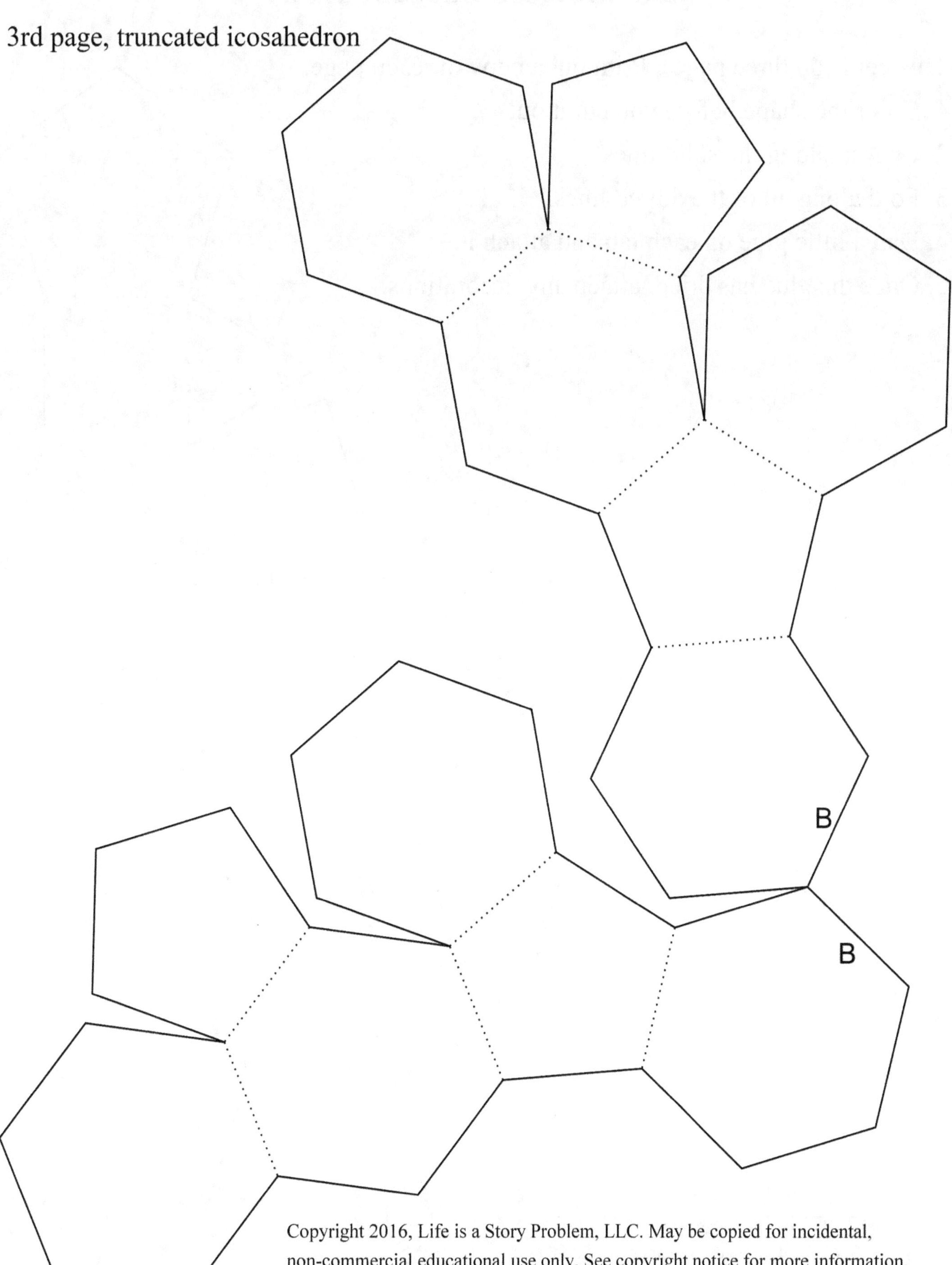

Rhombicosidodecahedron

This net is on three pages. Print out a copy of each page.

1. Color the shape before you cut it out.
2. Cut out along the solid lines.
3. Fold along all of the dotted lines.
4. Dab a little glue on each tab and attach it.
5. Once the glue has dried, attach any decorations.

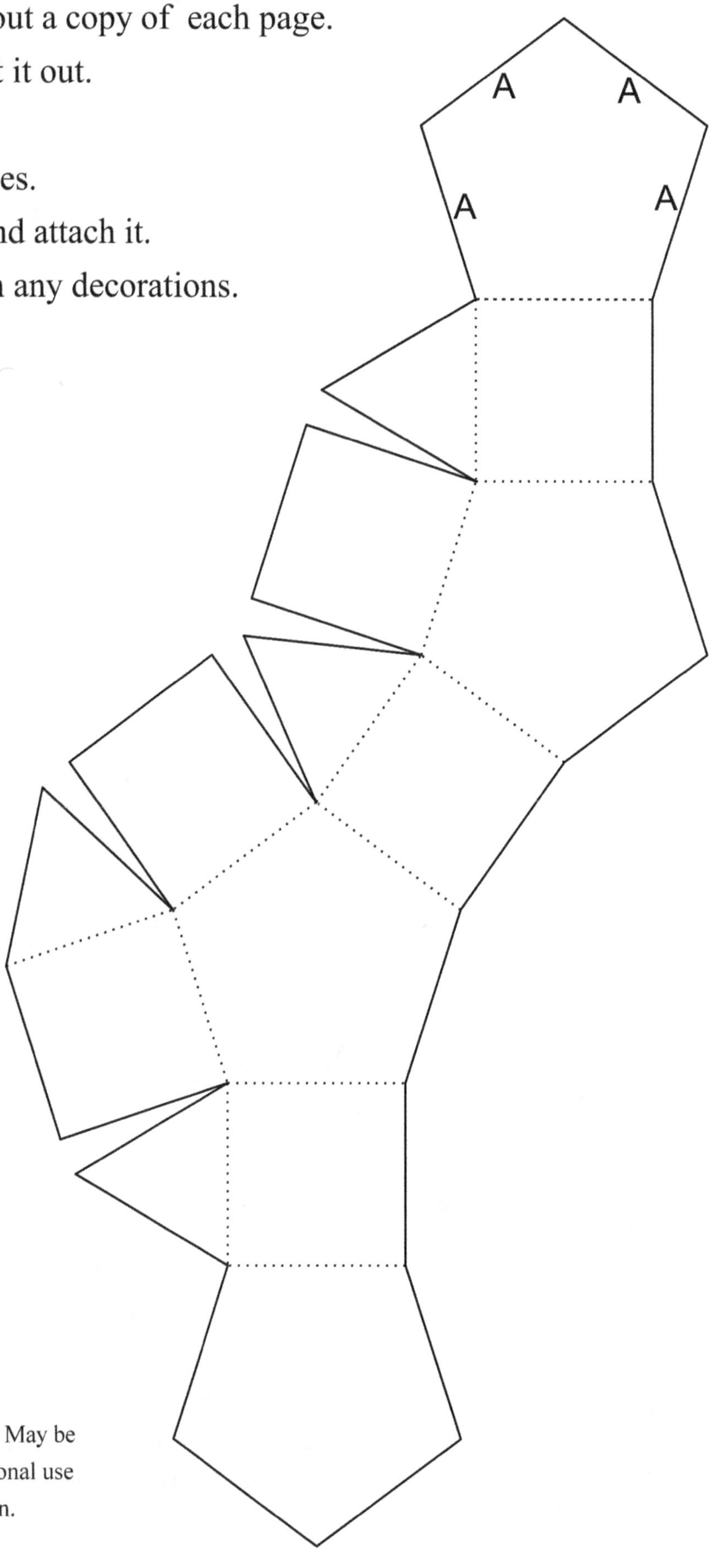

 Geometric Nets Mega Project Book, Tabbed By David E. McAdams

2nd page, rhombicosidodecahedron

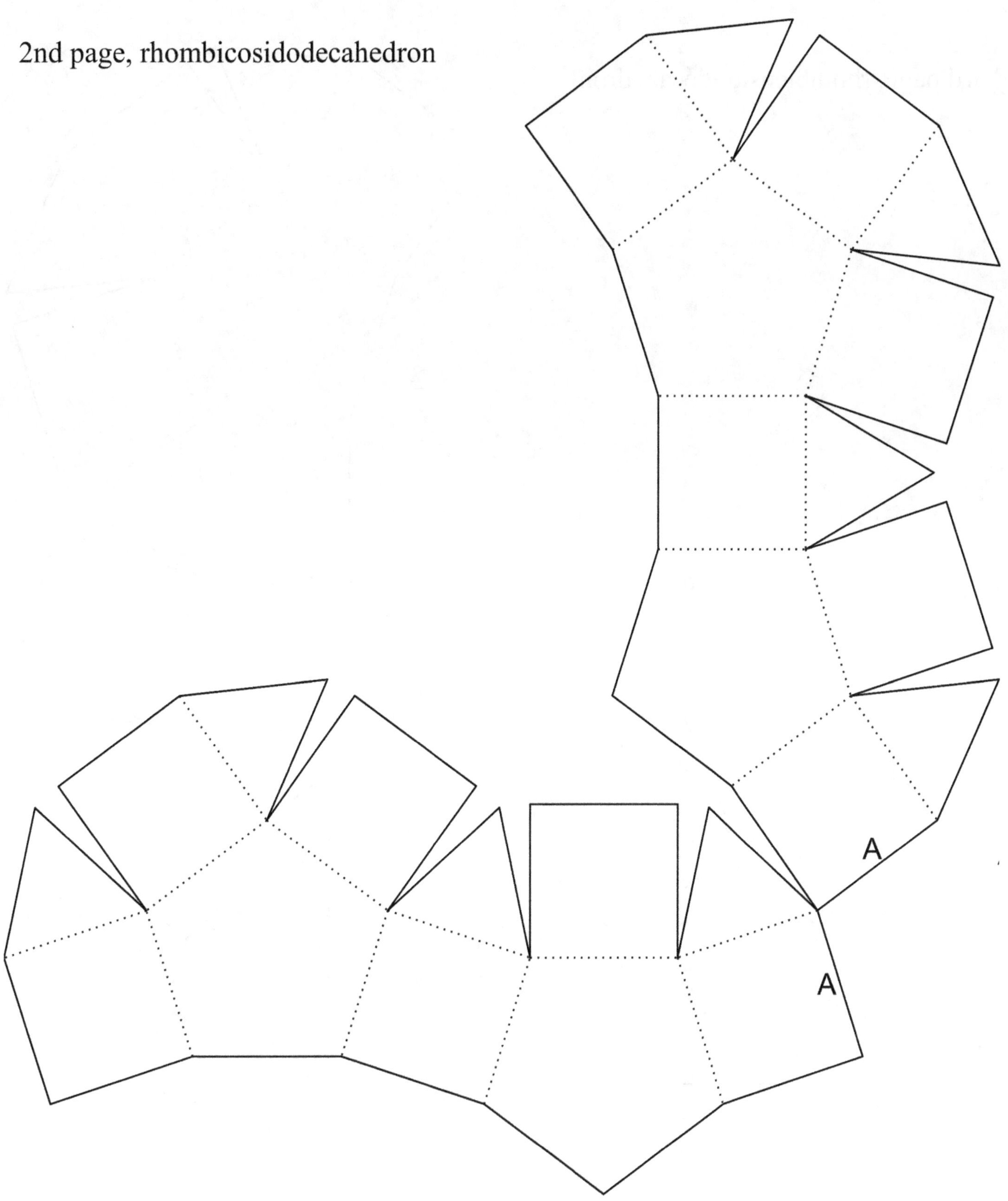

3rd page, rhombicosidodecahedron

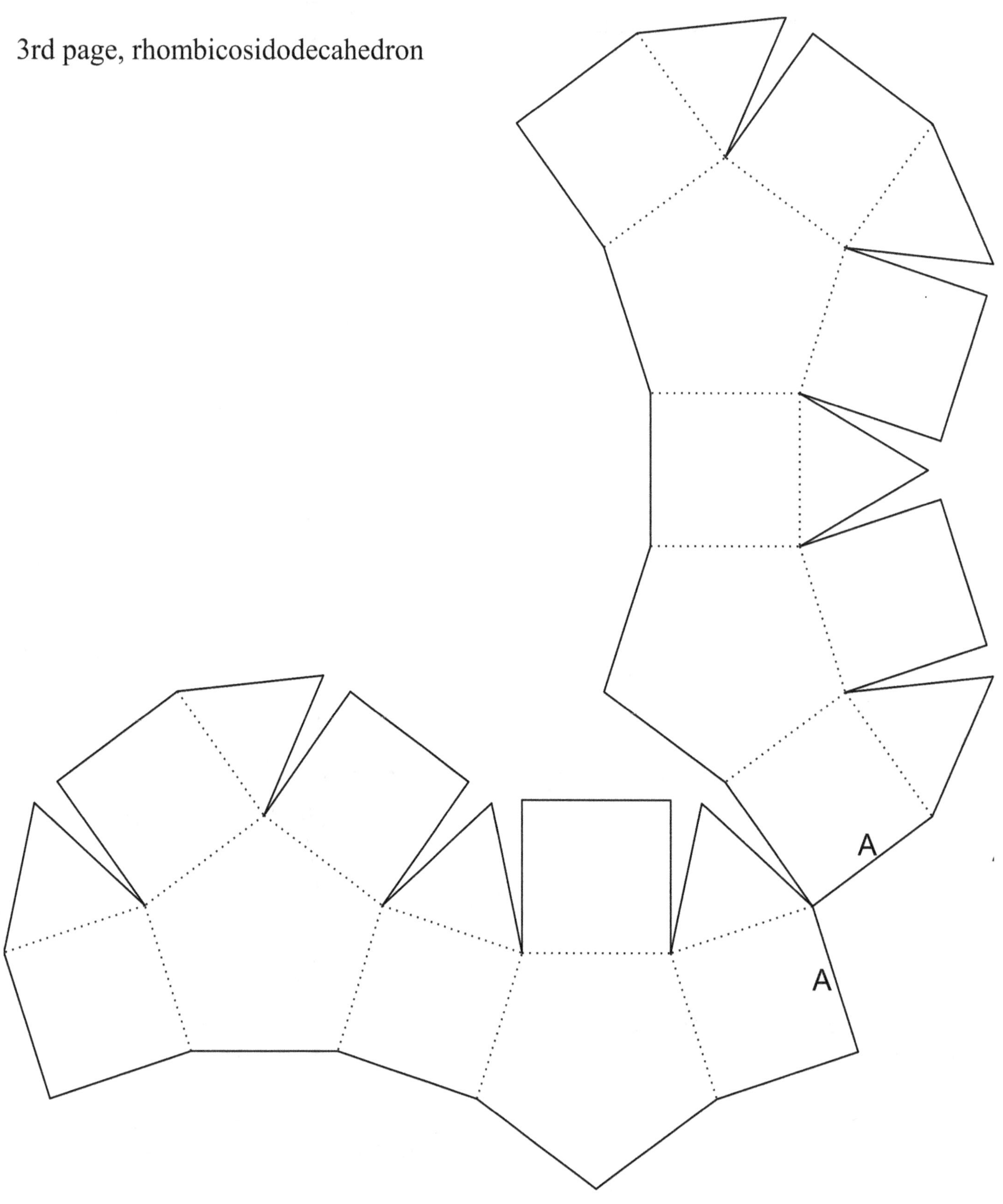

Geometric Nets Mega Project Book, Tabbed By David E. McAdams

Truncated icosidodecahedron

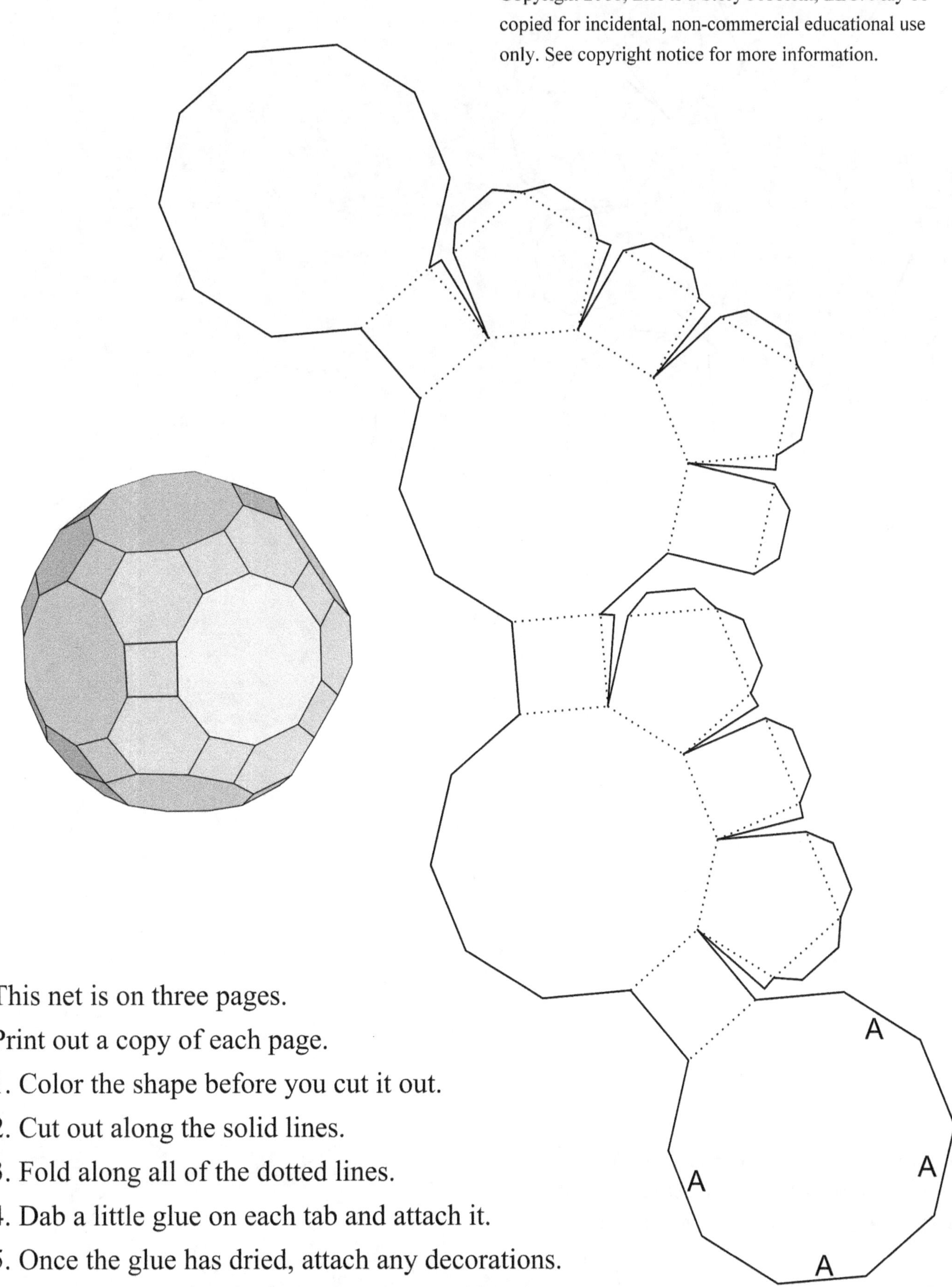

This net is on three pages.

Print out a copy of each page.

1. Color the shape before you cut it out.

2. Cut out along the solid lines.

3. Fold along all of the dotted lines.

4. Dab a little glue on each tab and attach it.

5. Once the glue has dried, attach any decorations.

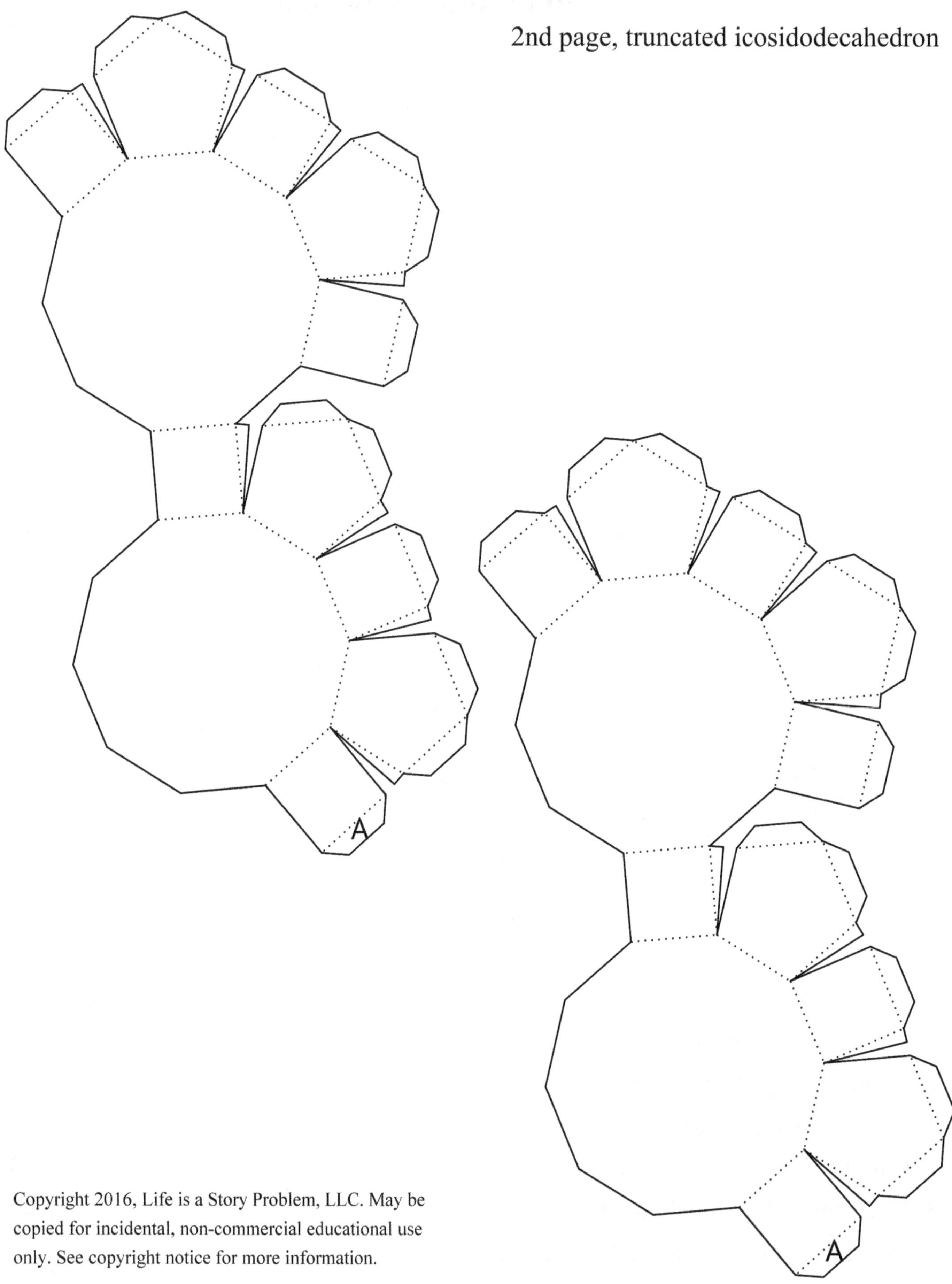

Geometric Nets Mega Project Book, Tabbed By David E. McAdams

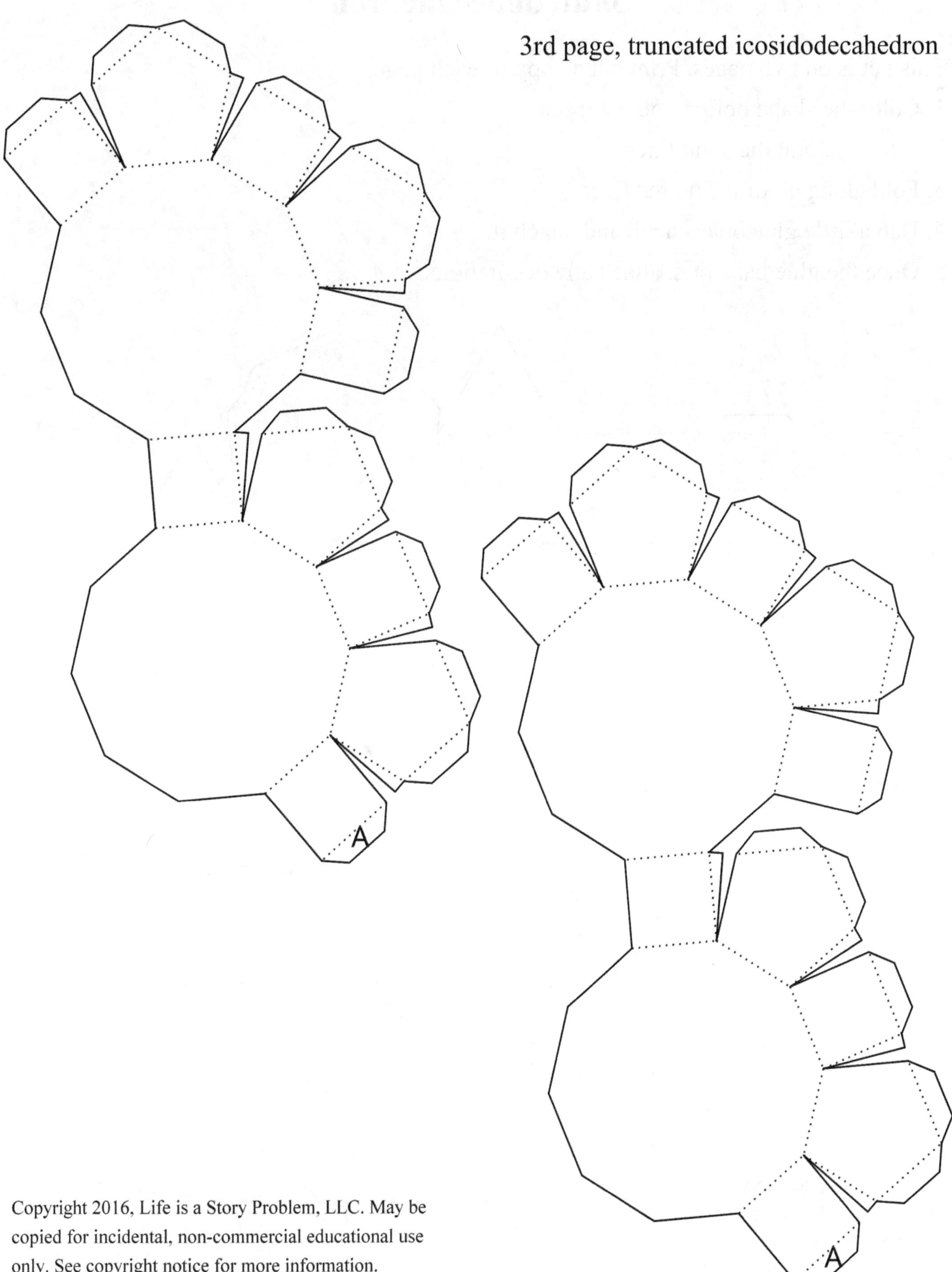
A
A

Snub dodecahedron

This net is on two pages. Print out a copy of each page.

1. Color the shape before you cut it out.

2. Cut out along the solid lines.

3. Fold along all of the dotted lines.

4. Dab a little glue on each tab and attach it.

5. Once the glue has dried, attach any decorations.

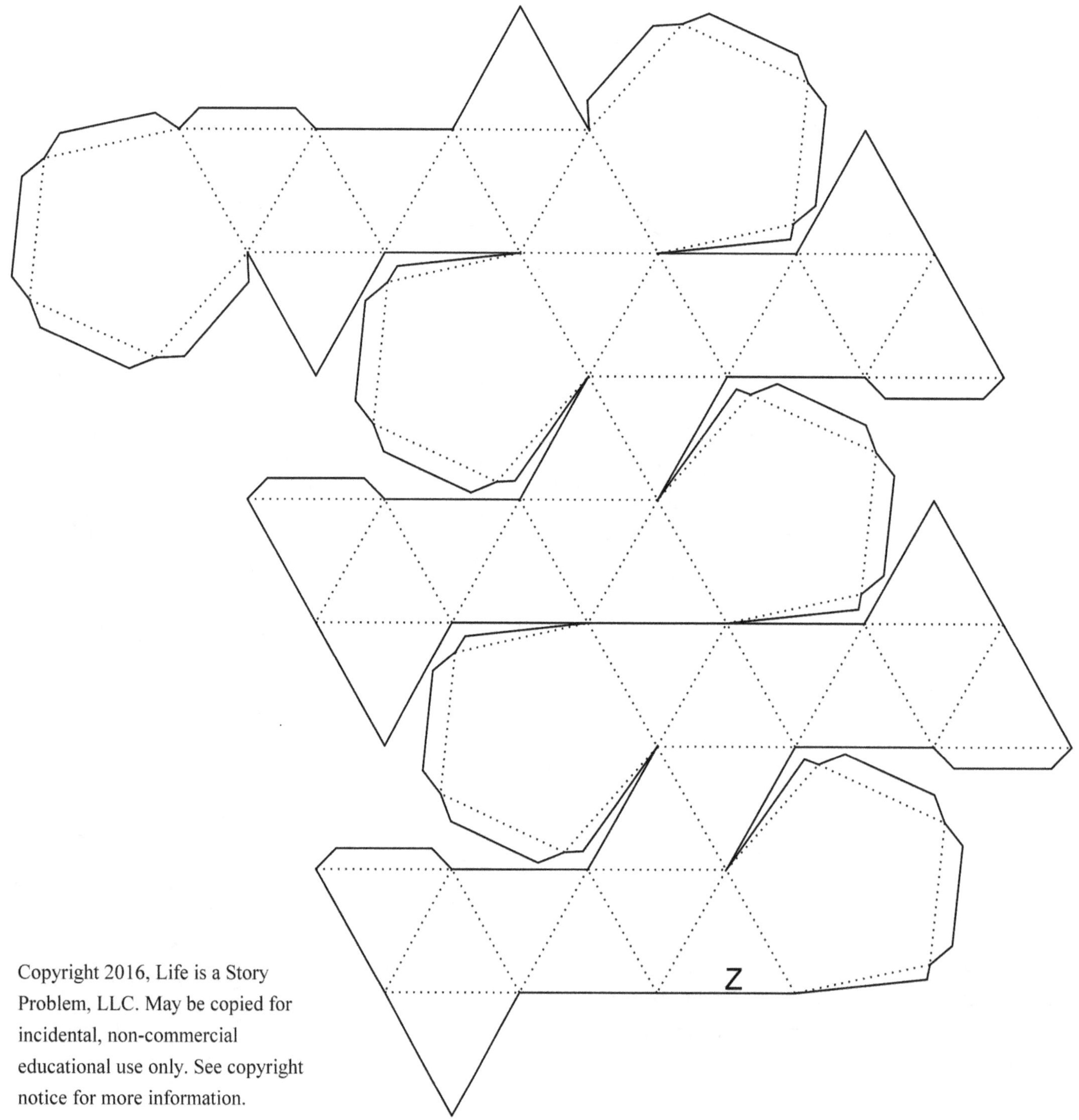

Geometric Nets Mega Project Book, Tabbed By David E. McAdams

2nd page, snub dodecahedron

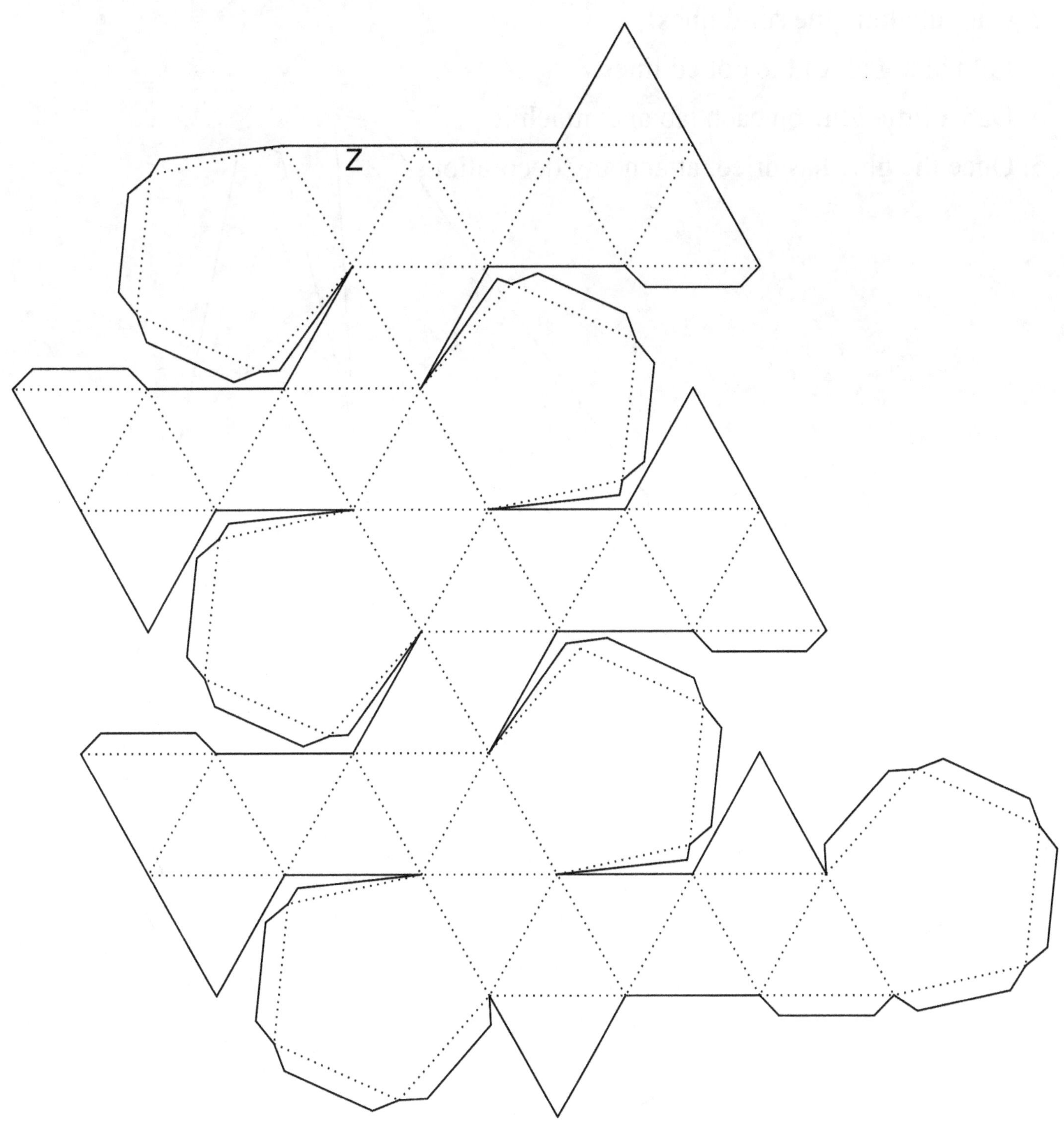

Triakis tetrahedron

1. Color the shape before you cut it out.
2. Cut out along the solid lines.
3. Fold along all of the dotted lines.
4. Dab a little glue on each tab and attach it.
5. Once the glue has dried, attach any decorations.

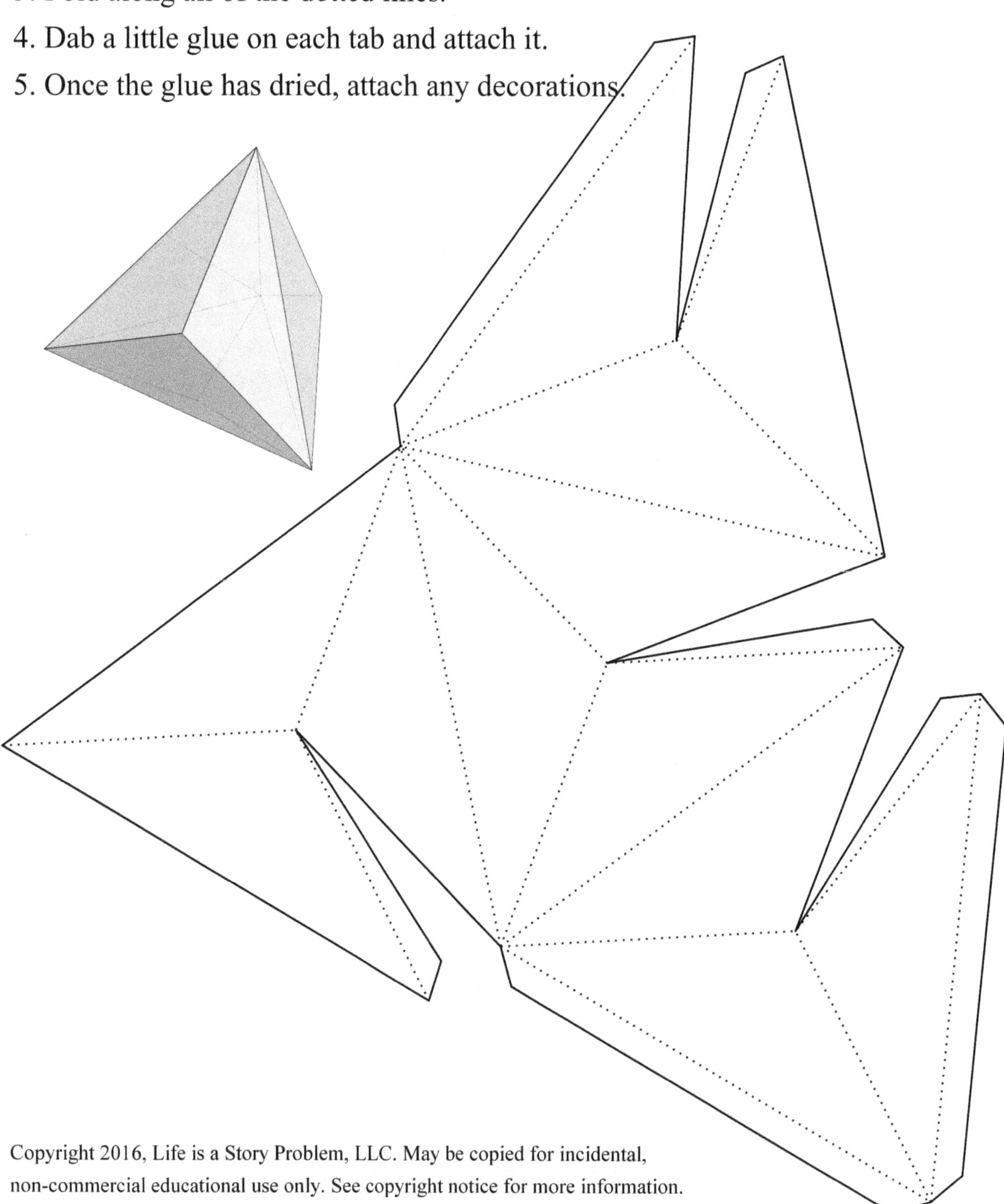

Triakis octahedron

1. Color the shape before you cut it out.

2. Cut out along the solid lines.

3. Fold along all of the dotted lines.

4. Dab a little glue on each tab and attach it.

5. Once the glue has dried, attach any decorations.

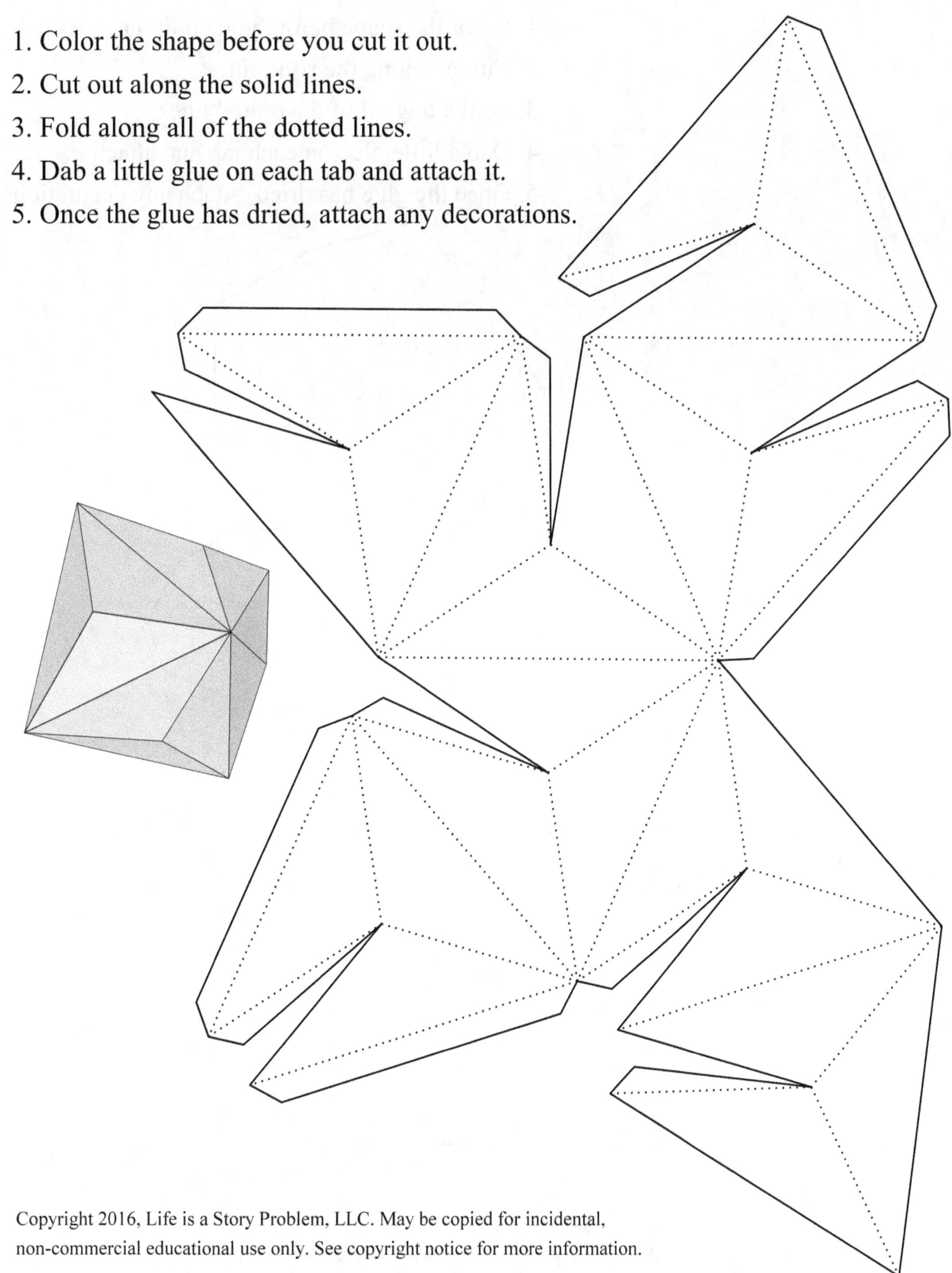

Disdyakis dodecahedron

1. Color the shape before you cut it out.

2. Cut out along the solid lines.

3. Fold along all of the dotted lines.

4. Dab a little glue on each tab and attach it.

5. Once the glue has dried, attach any decorations

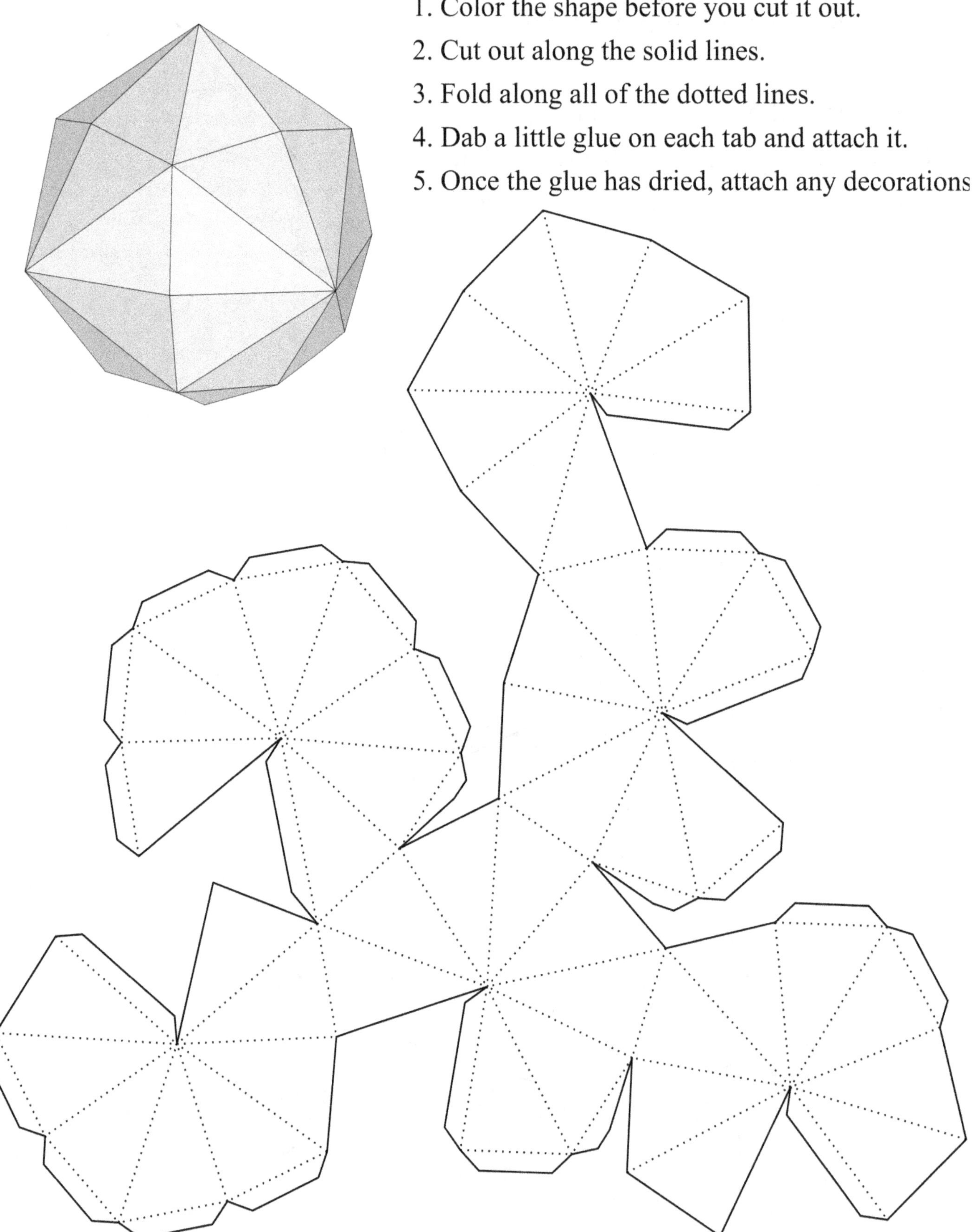

 Geometric Nets Mega Project Book, Tabbed By David E. McAdams

Tetrakis hexahedron

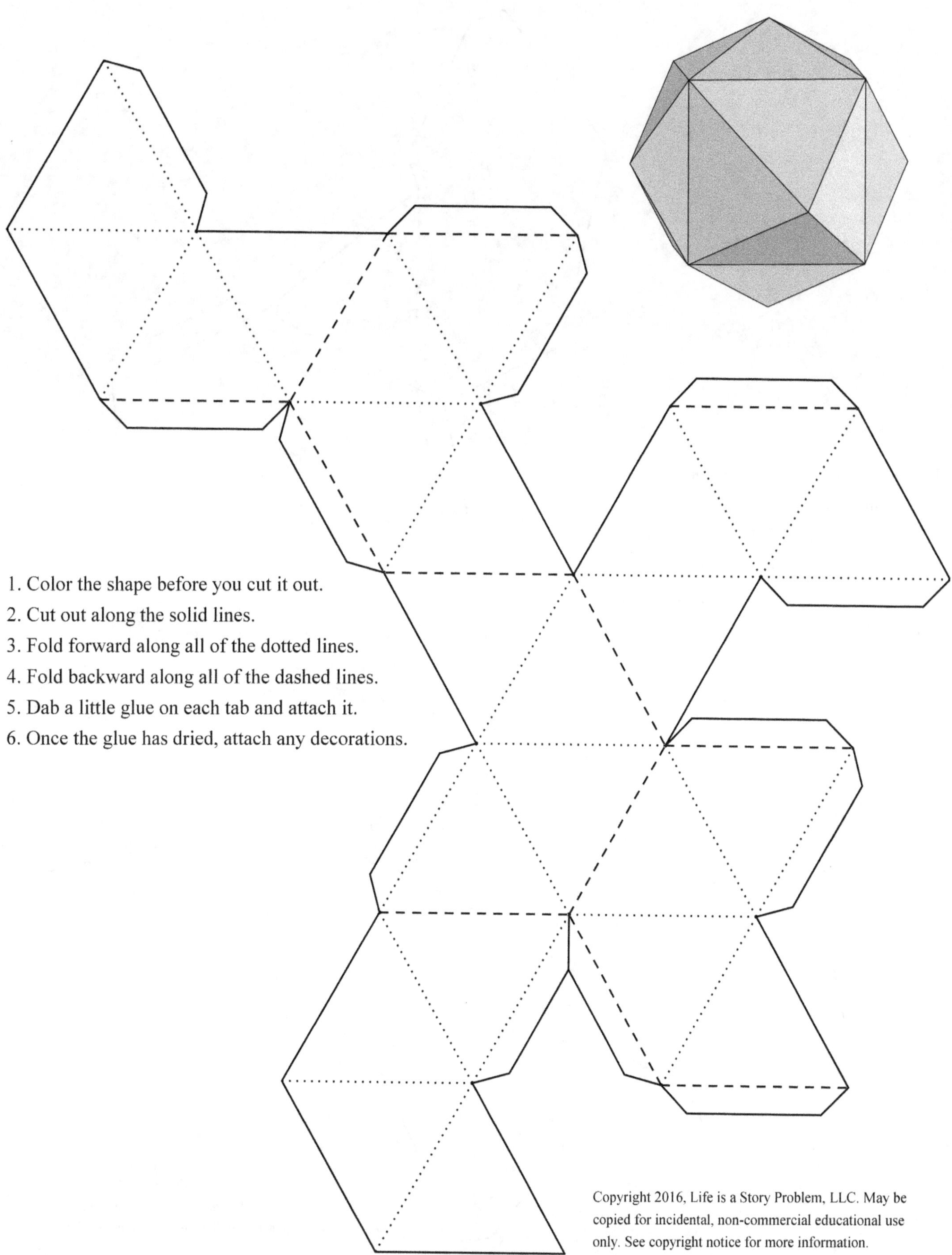

1. Color the shape before you cut it out.
2. Cut out along the solid lines.
3. Fold forward along all of the dotted lines.
4. Fold backward along all of the dashed lines.
5. Dab a little glue on each tab and attach it.
6. Once the glue has dried, attach any decorations.

Triakis icosahedron

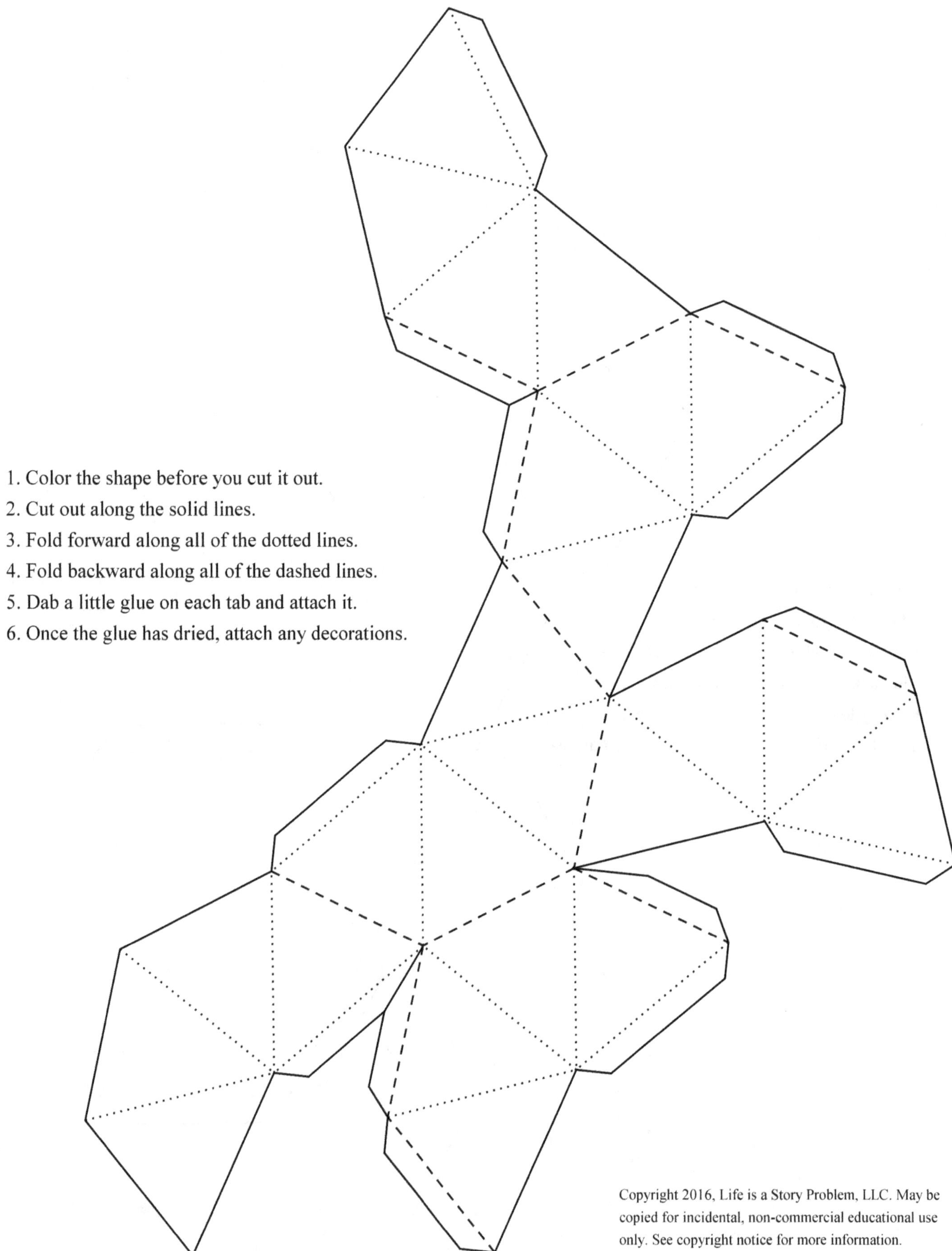

1. Color the shape before you cut it out.
2. Cut out along the solid lines.
3. Fold forward along all of the dotted lines.
4. Fold backward along all of the dashed lines.
5. Dab a little glue on each tab and attach it.
6. Once the glue has dried, attach any decorations.

    Geometric Nets Mega Project Book, Tabbed By David E. McAdams

Disdyakis triacontahedron

This net is on two pages. Print out a copy of each page.

1. Color the shape before you cut it out.
2. Cut out along the solid lines.
3. Fold along all of the dotted lines.
4. Dab a little glue on each tab and attach it.
5. Once the glue has dried, attach the decorations.

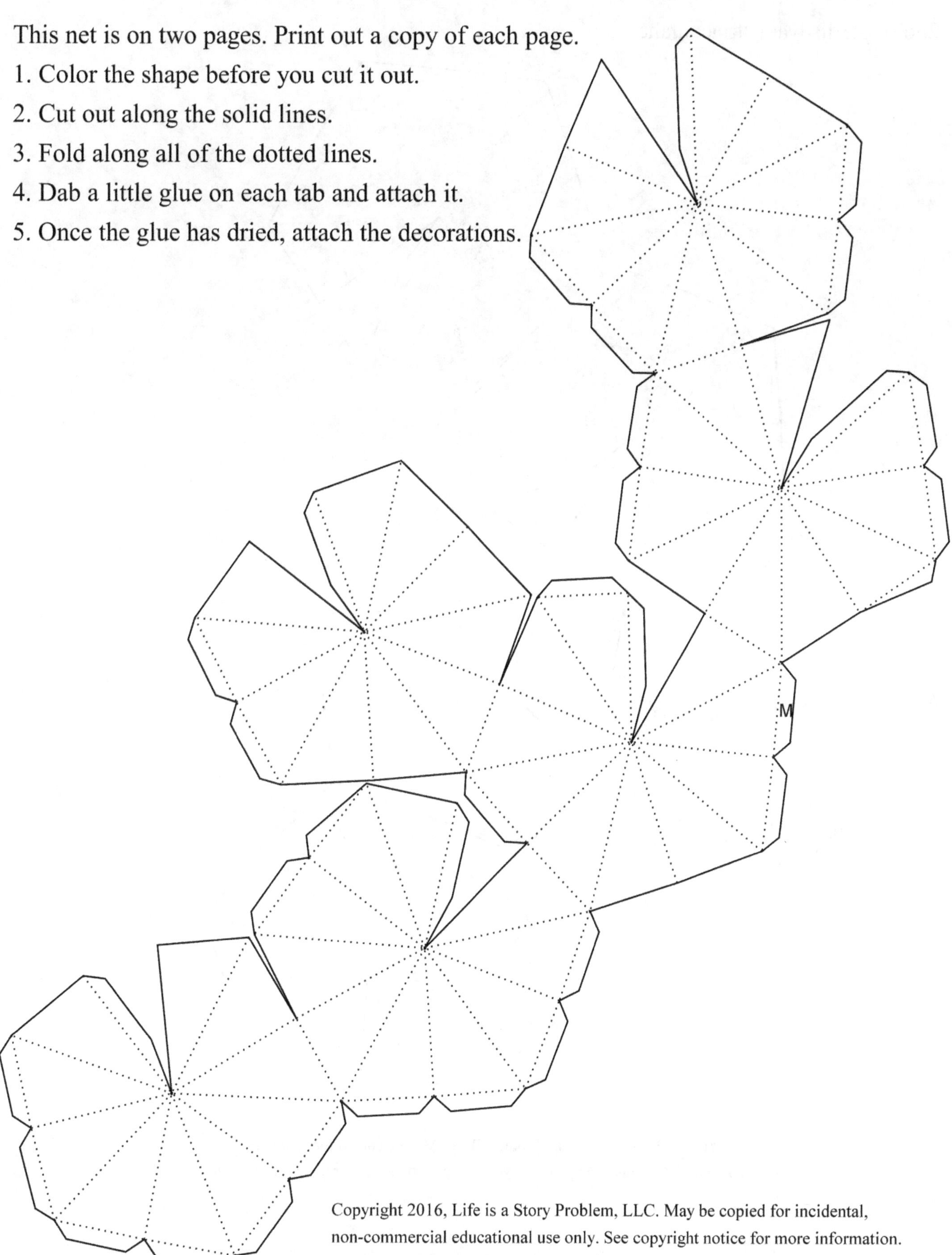

2nd page, disdyakis triacontahedron

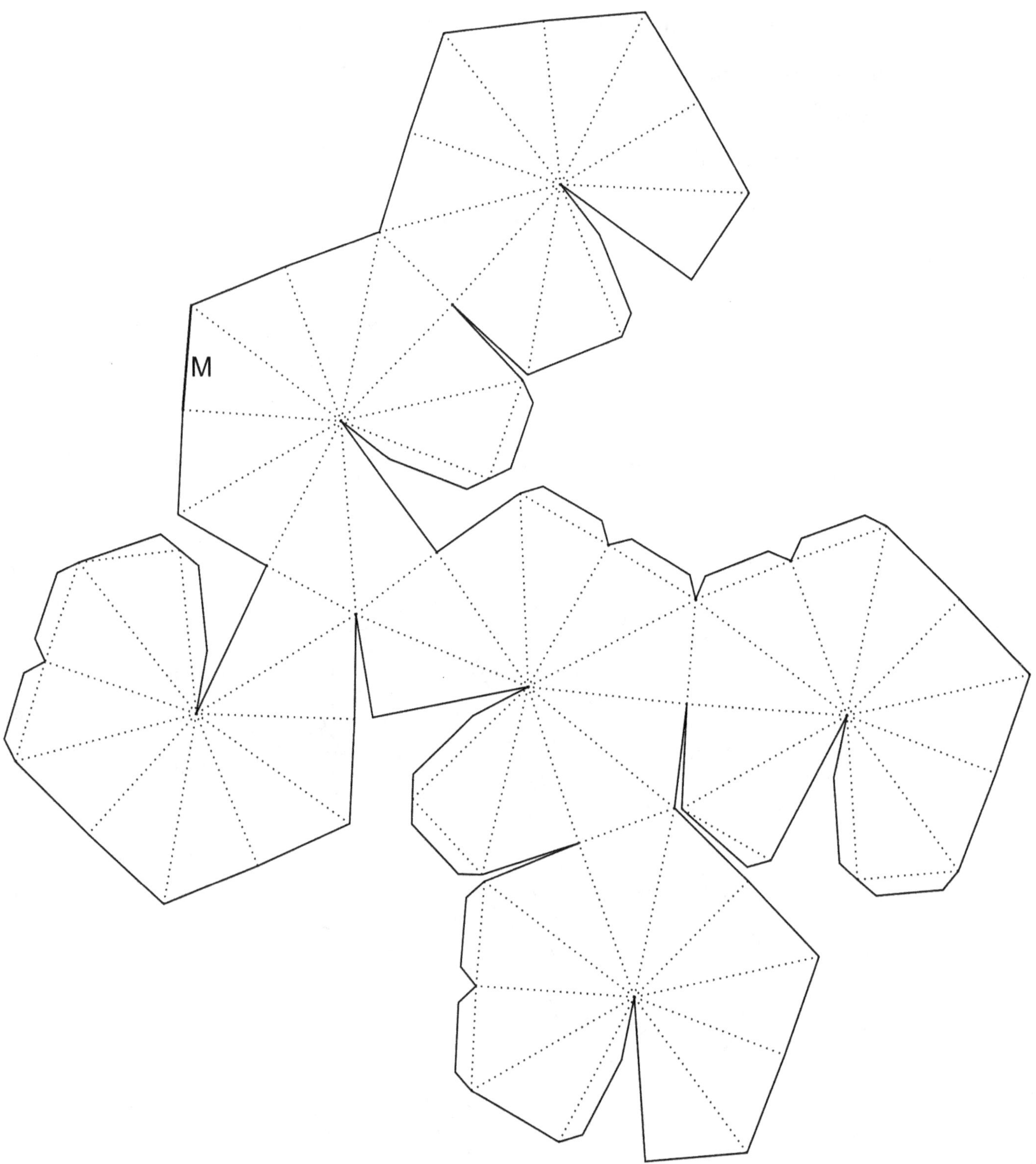

Geometric Nets Mega Project Book, Tabbed By David E. McAdams

Pentakis dodecahedron

This net is on three pages. Print out a copy of each pa

1. Color the shape before you cut it out.
2. Cut out along the solid lines.
3. Fold along all of the dotted lines.
4. Dab a little glue on each tab and attach it.
5. Once the glue has dried, attach the decorations.

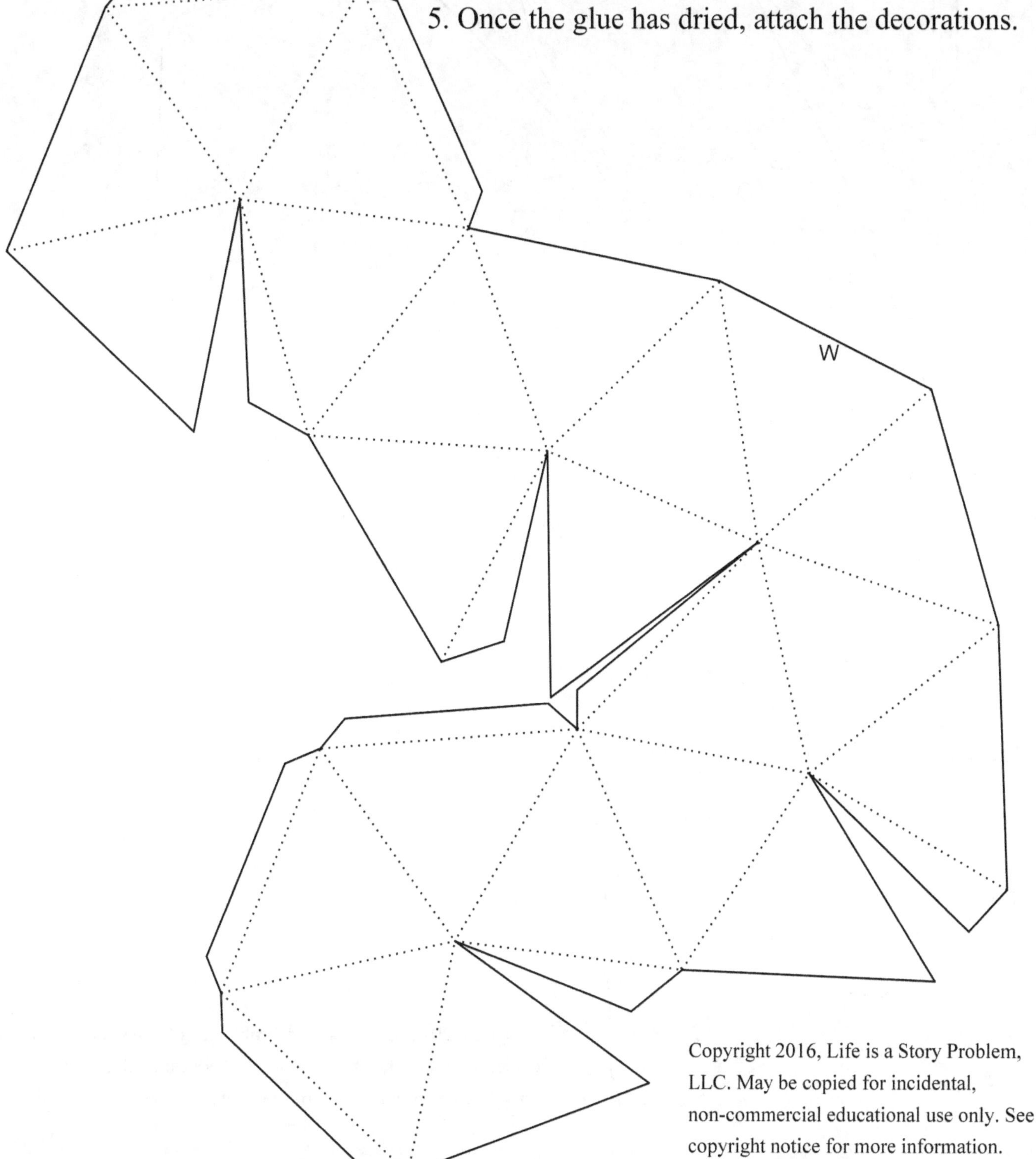

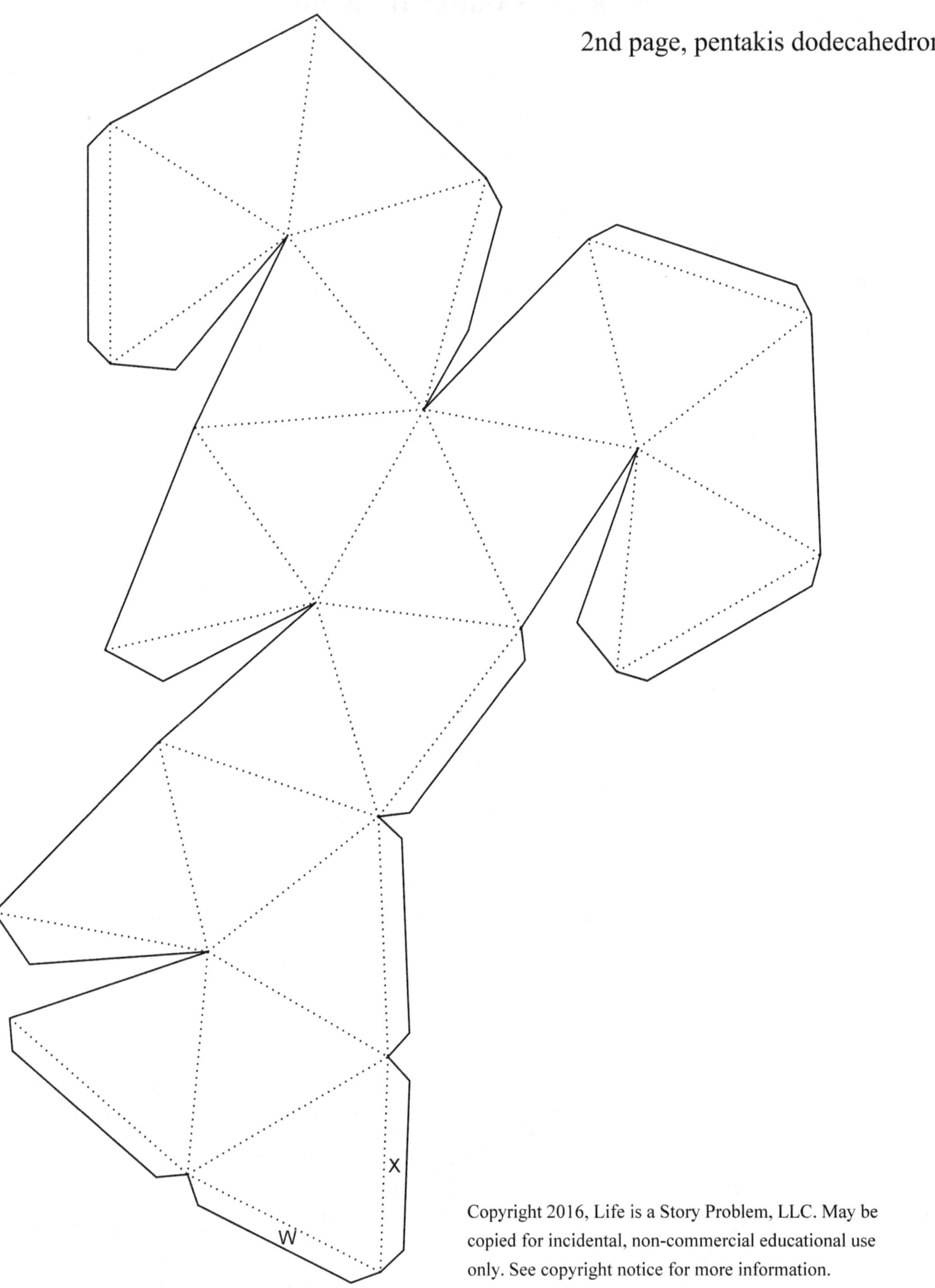

Geometric Nets Mega Project Book, Tabbed By David E. McAdams

3rd page, pentakis dodecahedron

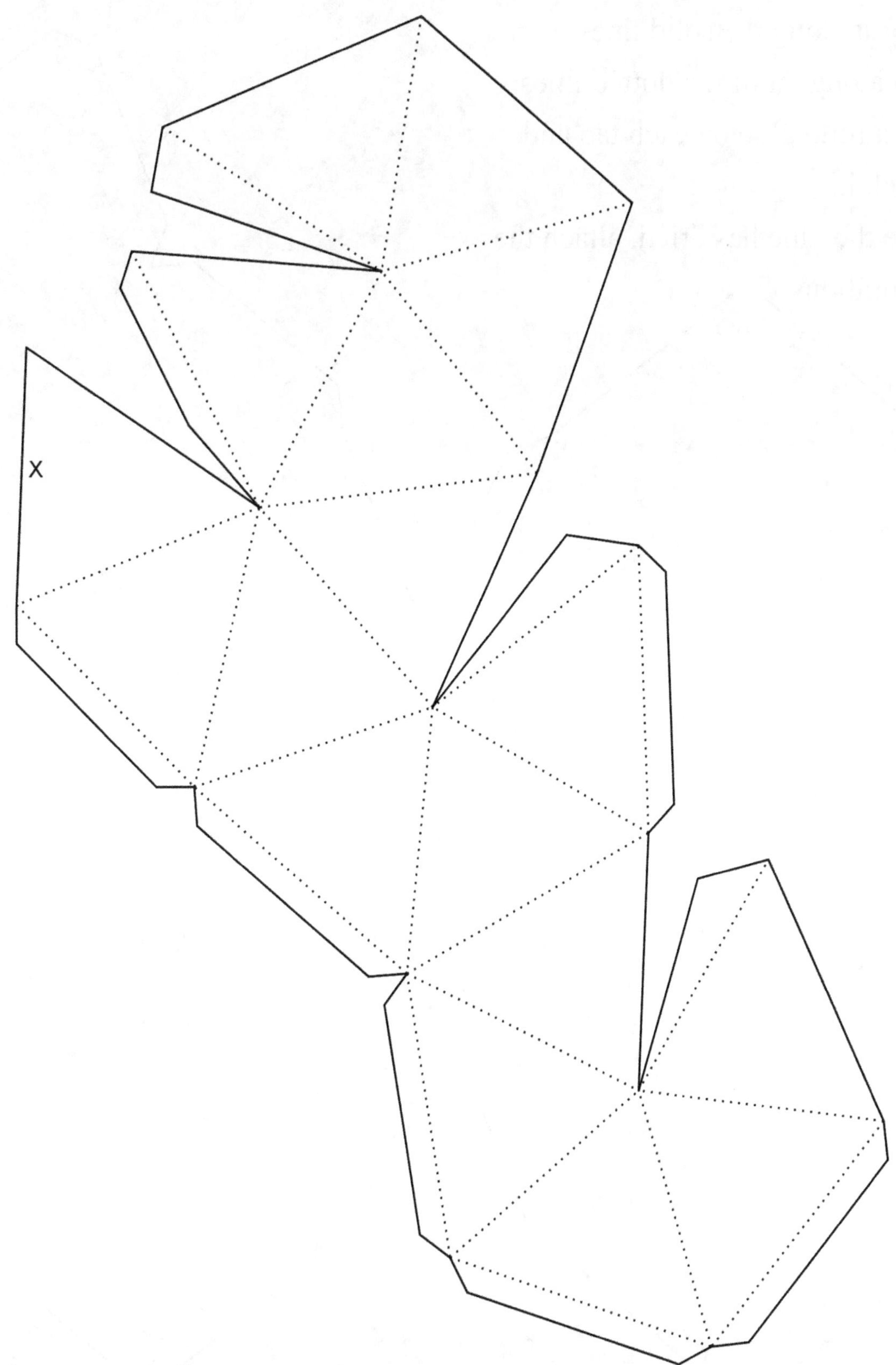

Rhombic dodecahedron

1. Color the shape before you cut it out.
2. Cut out along the solid lines.
3. Fold along all of the dotted lines.
4. Dab a little glue on each tab and attach it.
5. Once the glue has dried, attach the decorations.

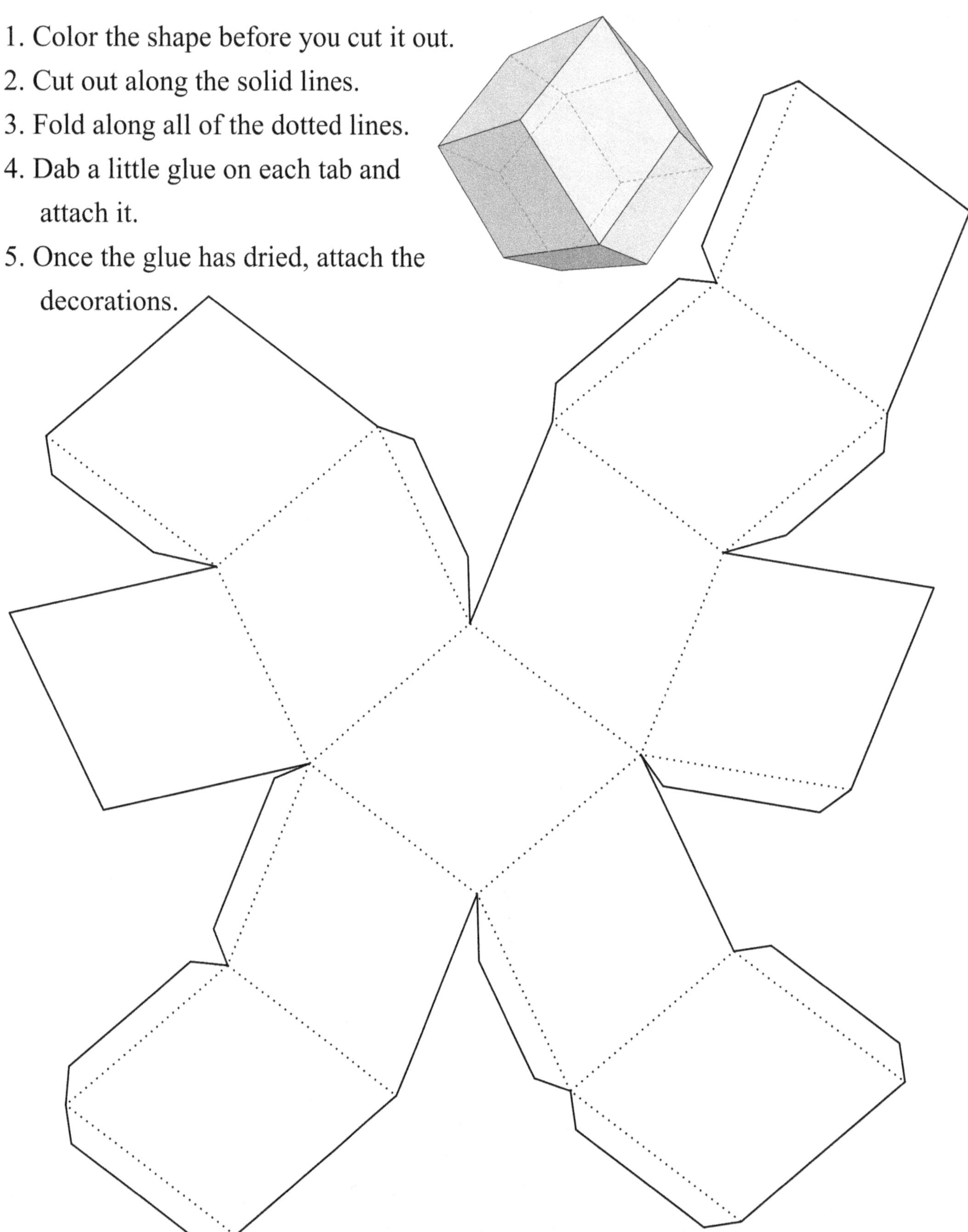

 Geometric Nets Mega Project Book, Tabbed By David E. McAdams

Rhombic triacontahedron

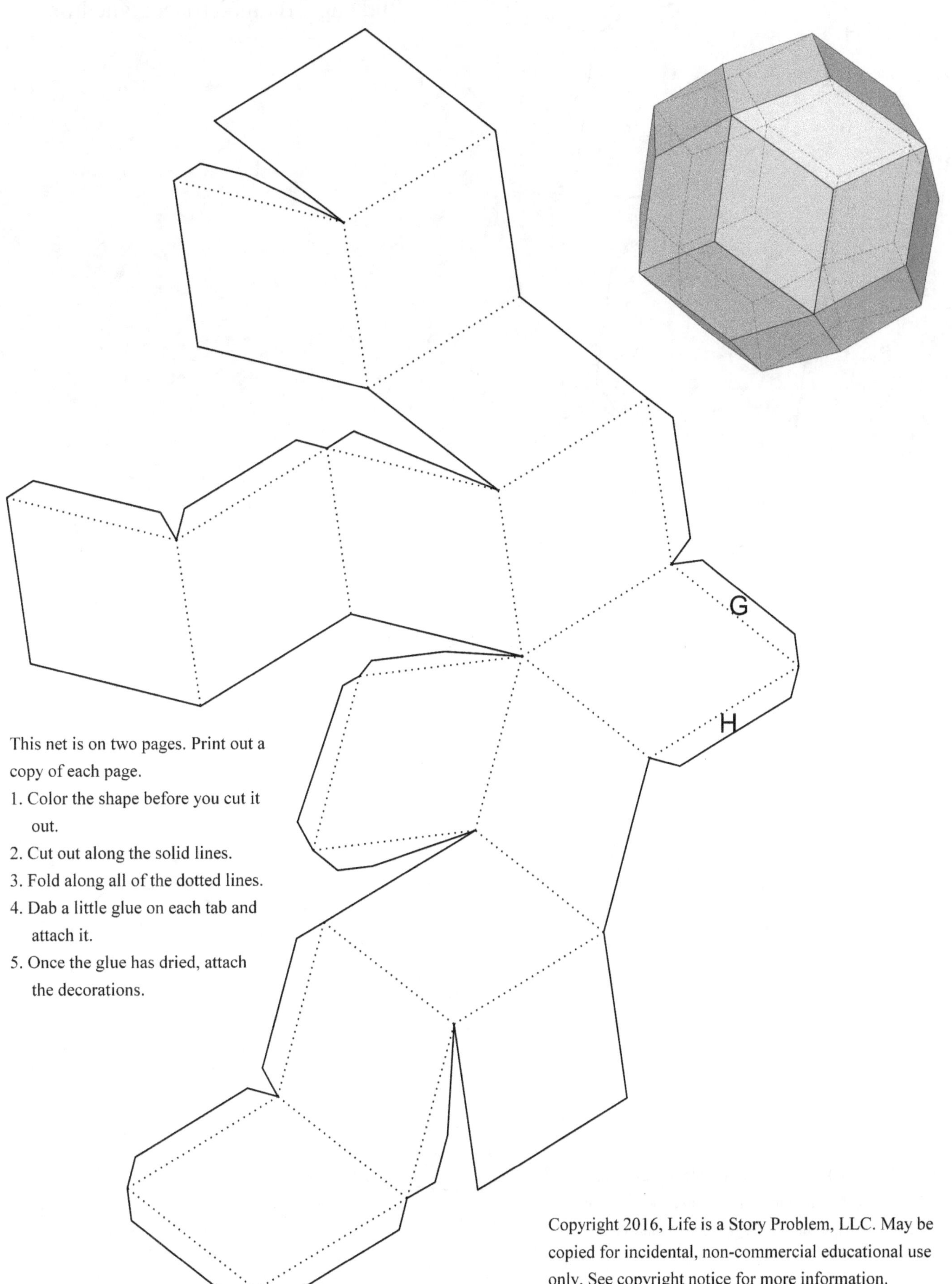

This net is on two pages. Print out a copy of each page.

1. Color the shape before you cut it out.
2. Cut out along the solid lines.
3. Fold along all of the dotted lines.
4. Dab a little glue on each tab and attach it.
5. Once the glue has dried, attach the decorations.

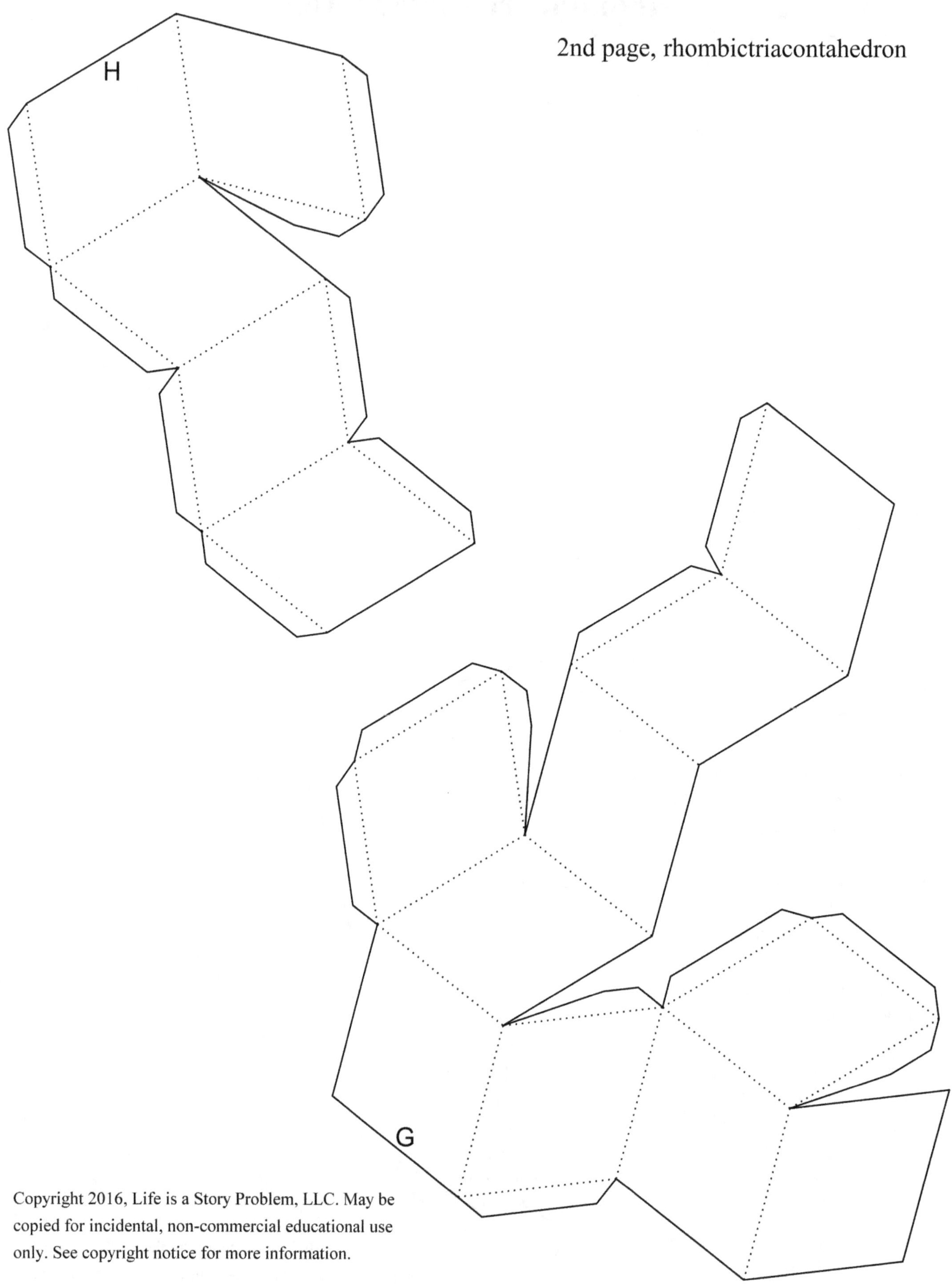

Geometric Nets Mega Project Book, Tabbed By David E. McAdams

Deltoidal icositetrahedron

1. Color the shape before you cut it out.
2. Cut out along the solid lines.
3. Fold along all of the dotted lines.
4. Dab a little glue on each tab and attach it.
5. Once the glue has dried, attach the decorations.

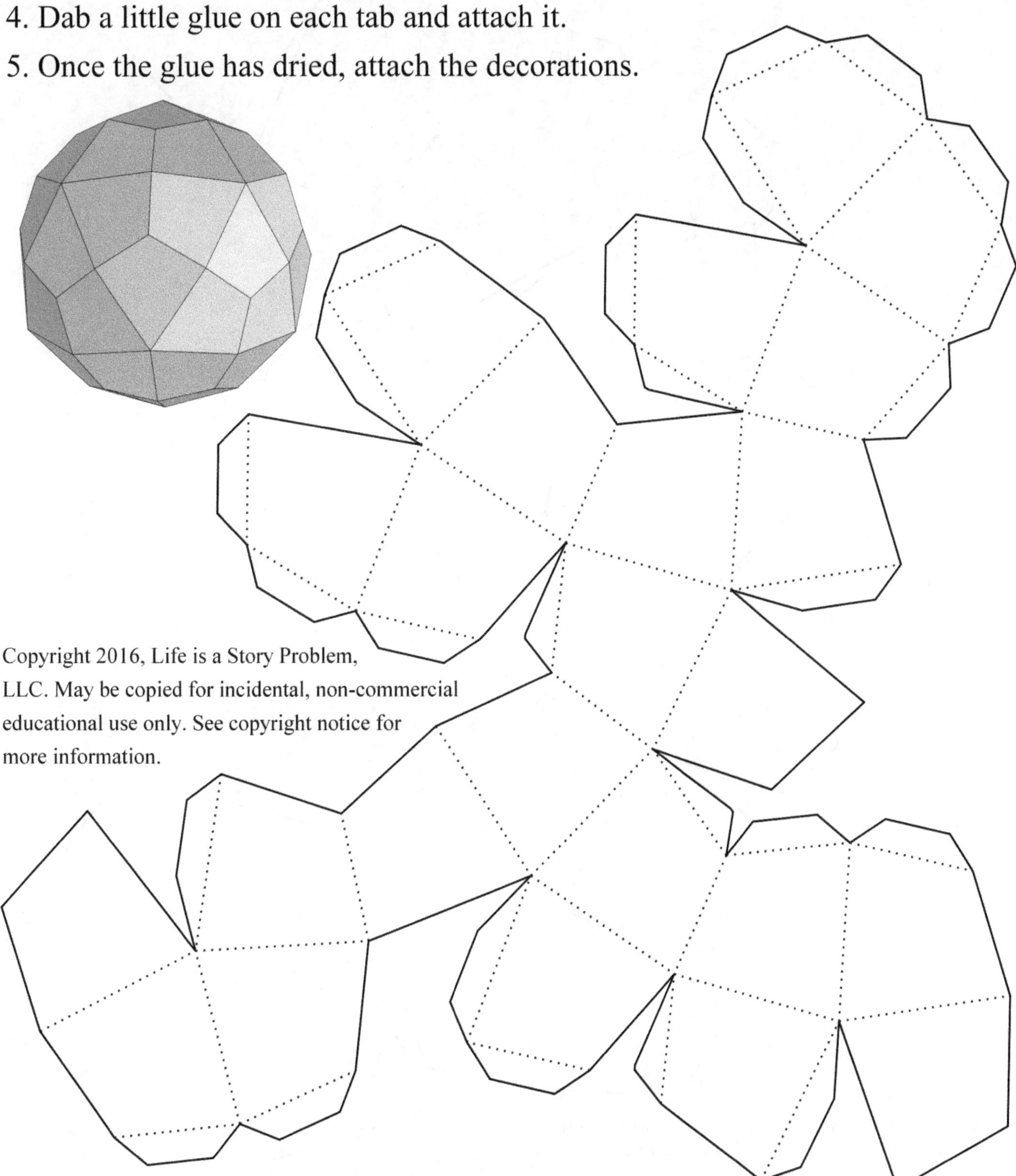

Deltoidal hexecontahedron

This net is on two pages. Print out a copy of each page.

1. Color the shape before you cut it out.
2. Cut out along the solid lines.
3. Fold along all of the dotted lines.
4. Dab a little glue on each tab and attach it.
5. Once the glue has dried, attach the decorations.

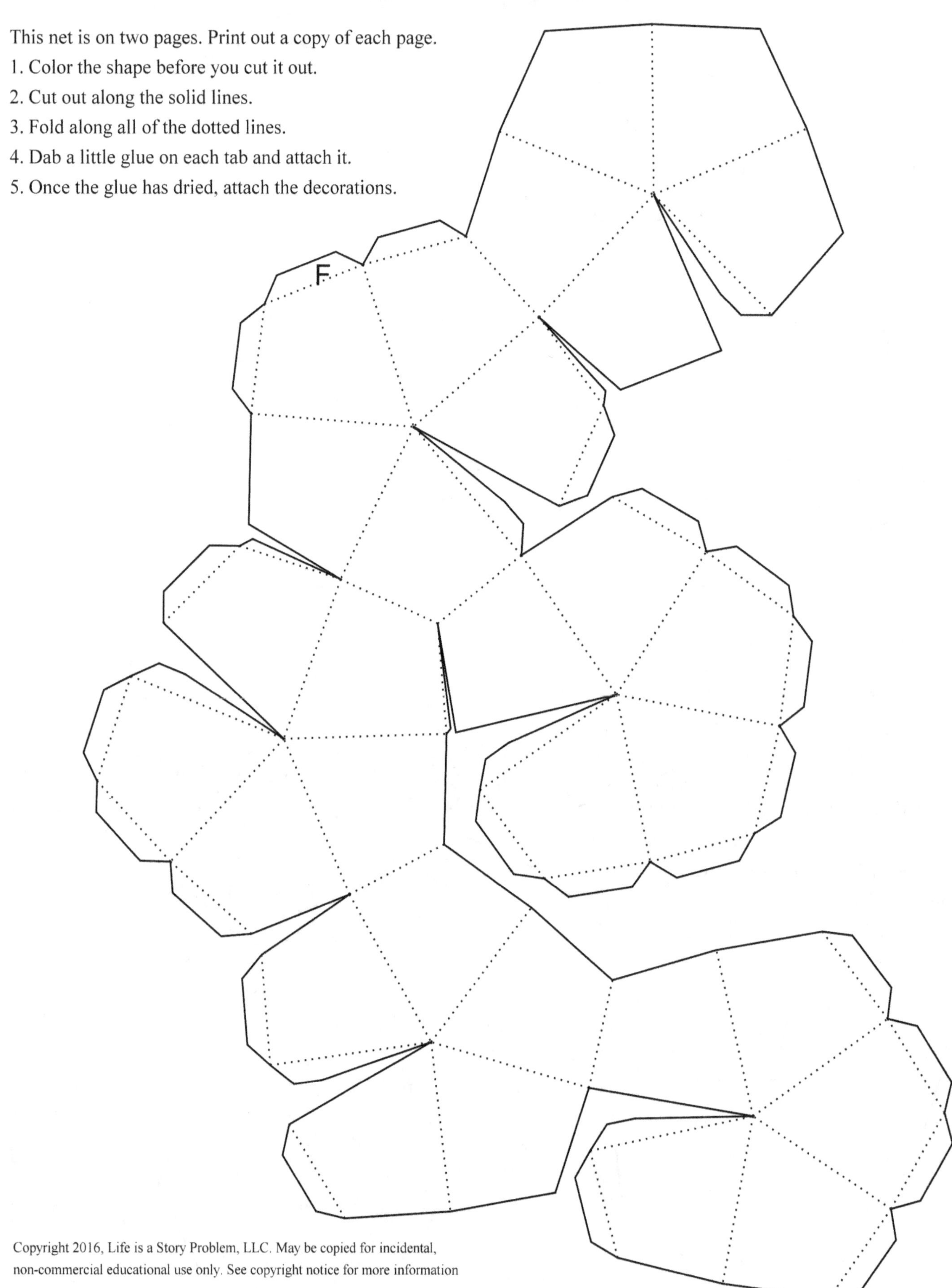

Geometric Nets Mega Project Book, Tabbed By David E. McAdams

Page 2, deltoidal hexecontahedron

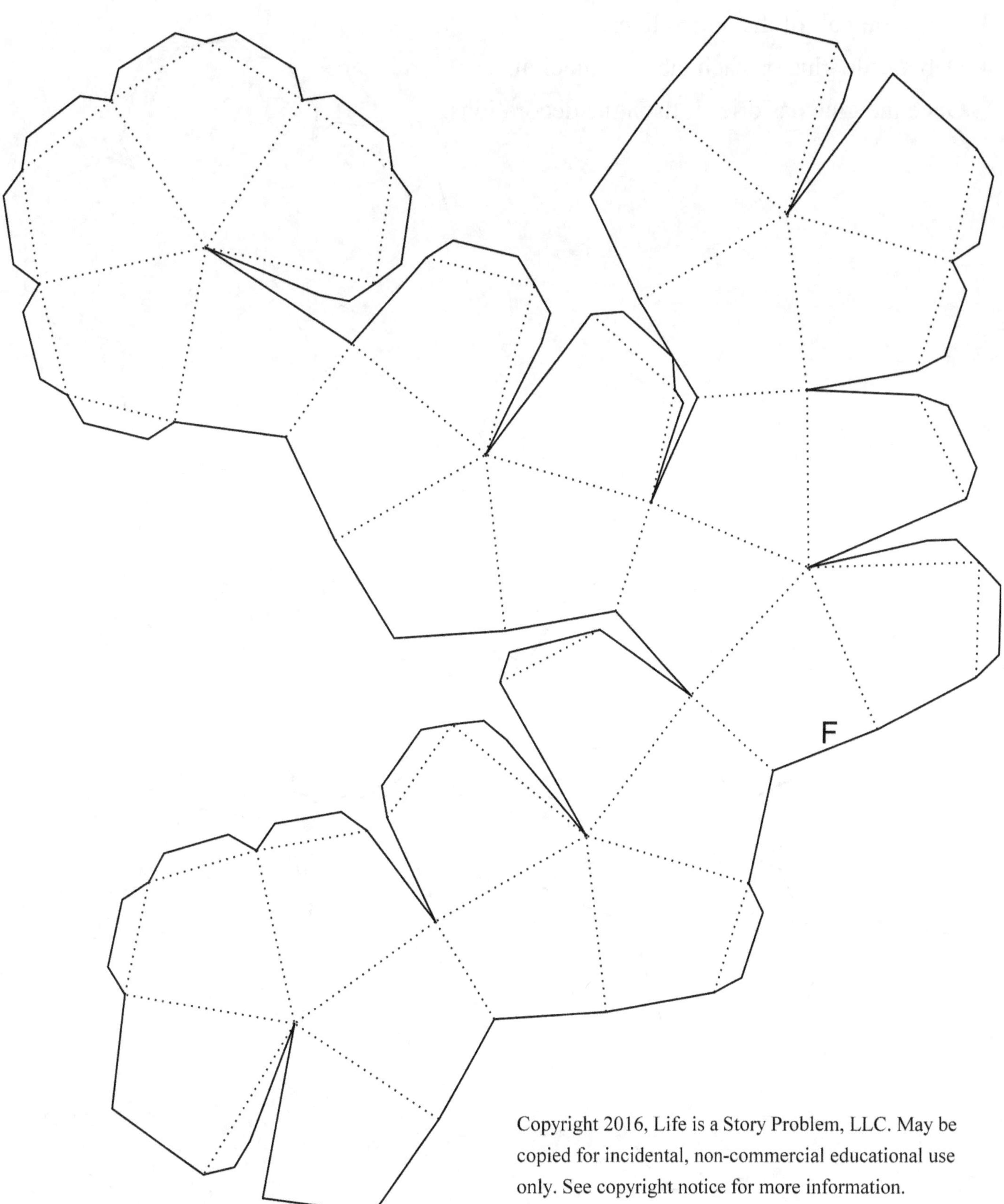

Pentagonal icositetrahedron

1. Color the shape before you cut it out.
2. Cut out along the solid lines.
3. Fold along all of the dotted lines.
4. Dab a little glue on each tab and attach it.
5. Once the glue has dried, attach the decorations.

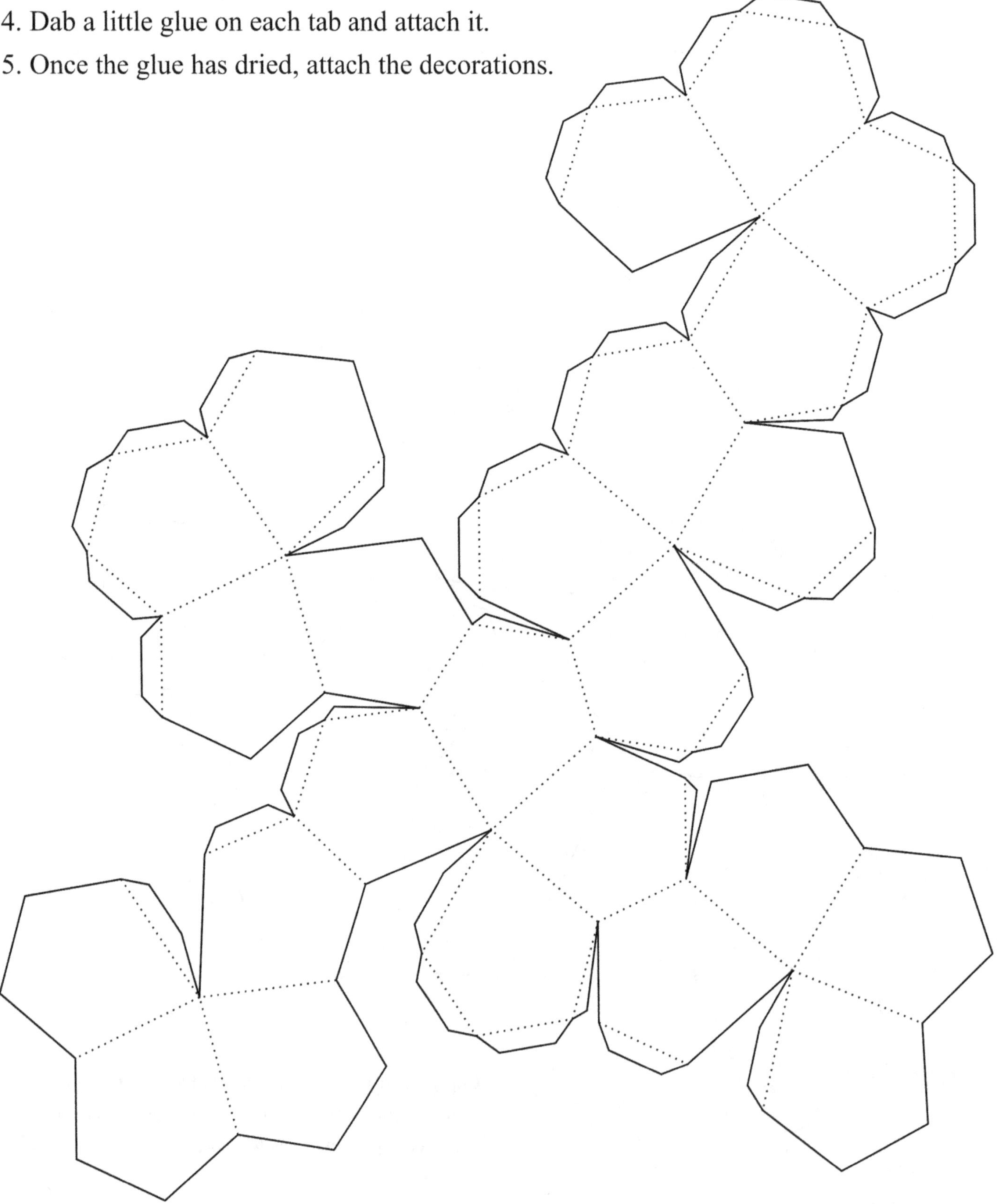

 Geometric Nets Mega Project Book, Tabbed By David E. McAdams

Pentagonal hexecontahedron

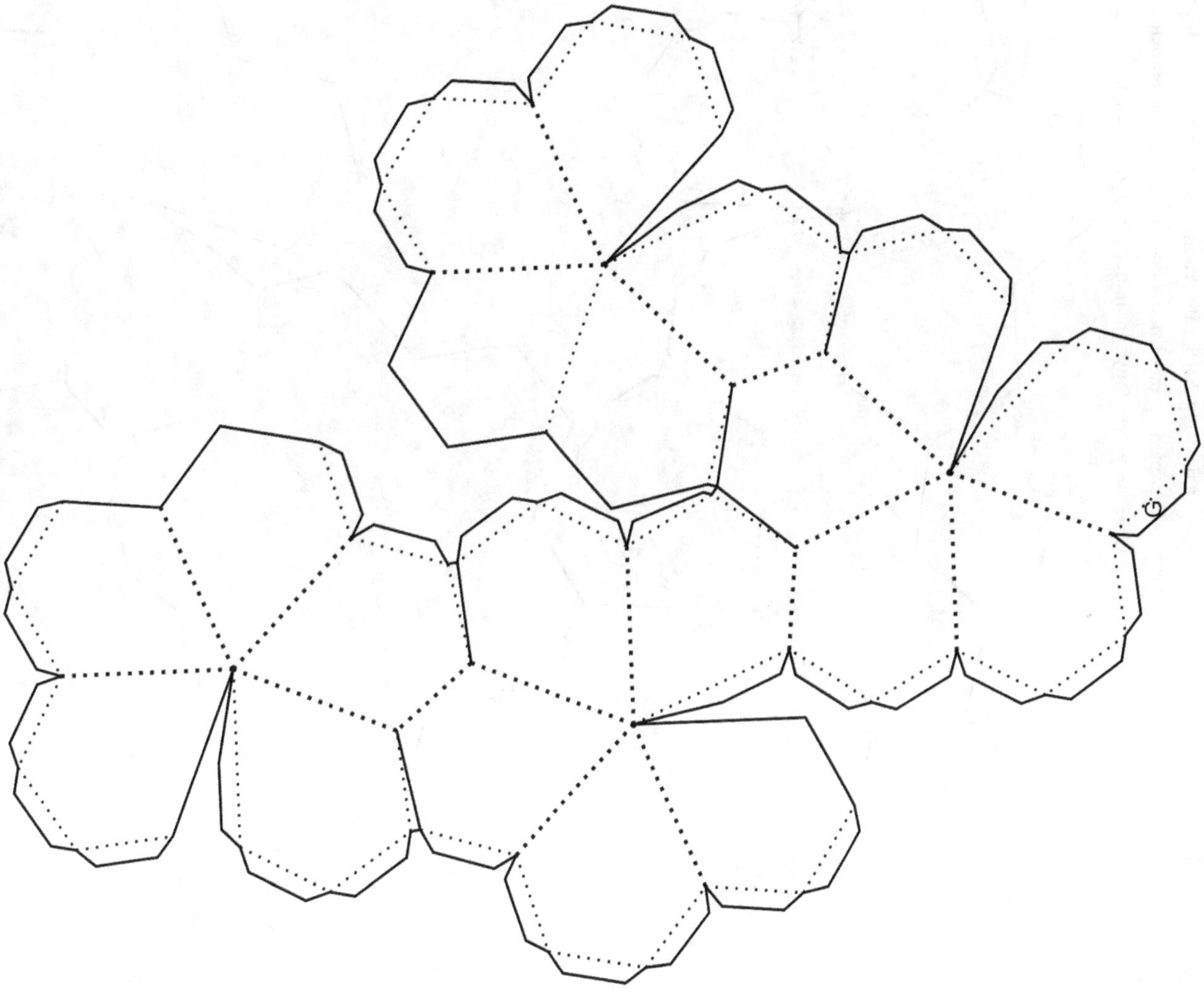

This net is on two pages. Print out a copy of each page.

1. Color the shape before you cut it out.

2. Cut out along the solid lines.

3. Fold along all of the dotted lines.

4. Dab a little glue on each tab and attach it.

5. Once the glue has dried, attach the decorations.

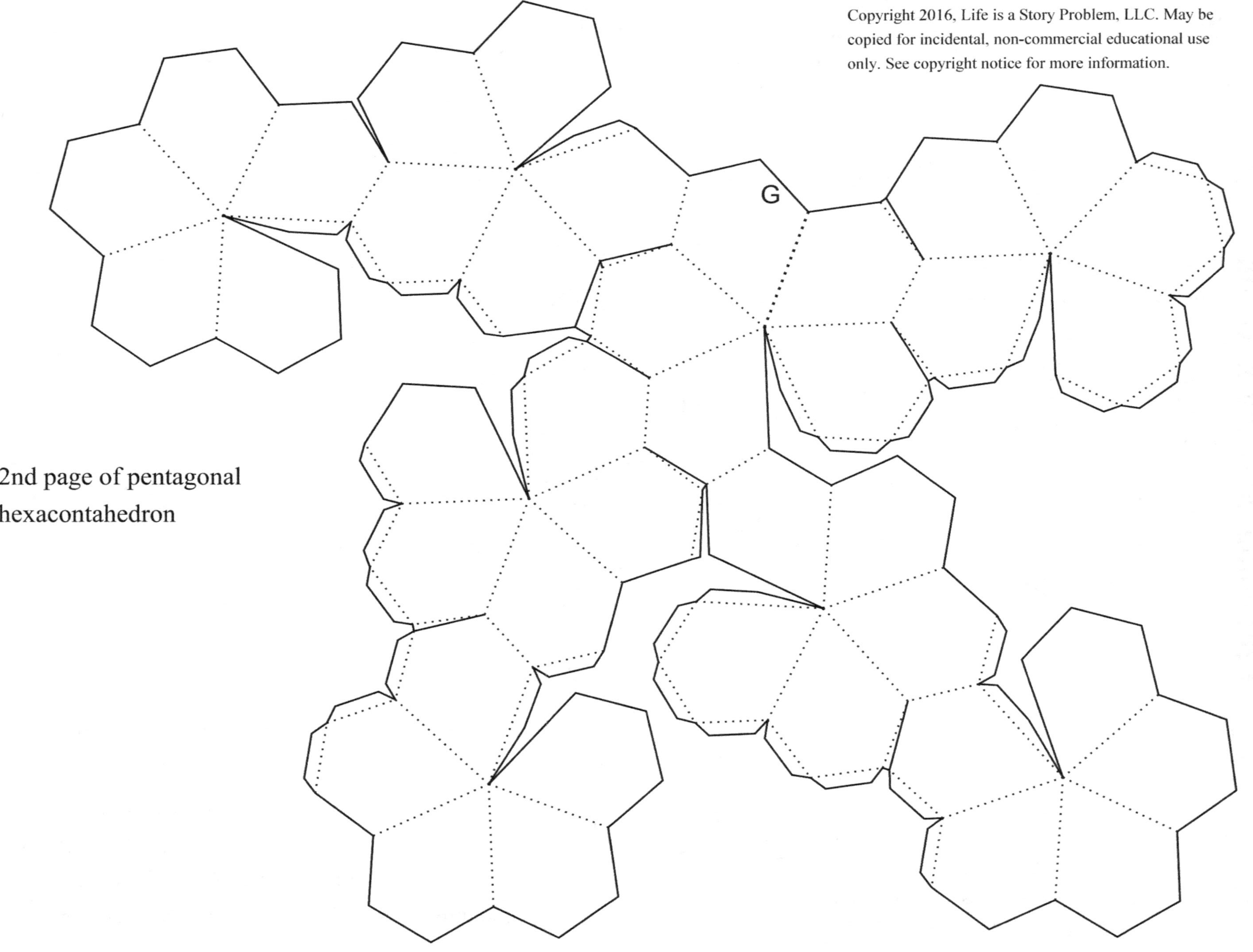

2nd page of pentagonal
hexacontahedron

Tetrahemihexahedron

1. Color the shape before you cut it out.
2. Cut out along the solid lines.
3. Fold forward along all of the dotted lines.
4. Fold backward along all of the dashed lines.
5. Dab a little glue on each tab and attach it.
6. Once the glue has dried, attach any
   decorations.

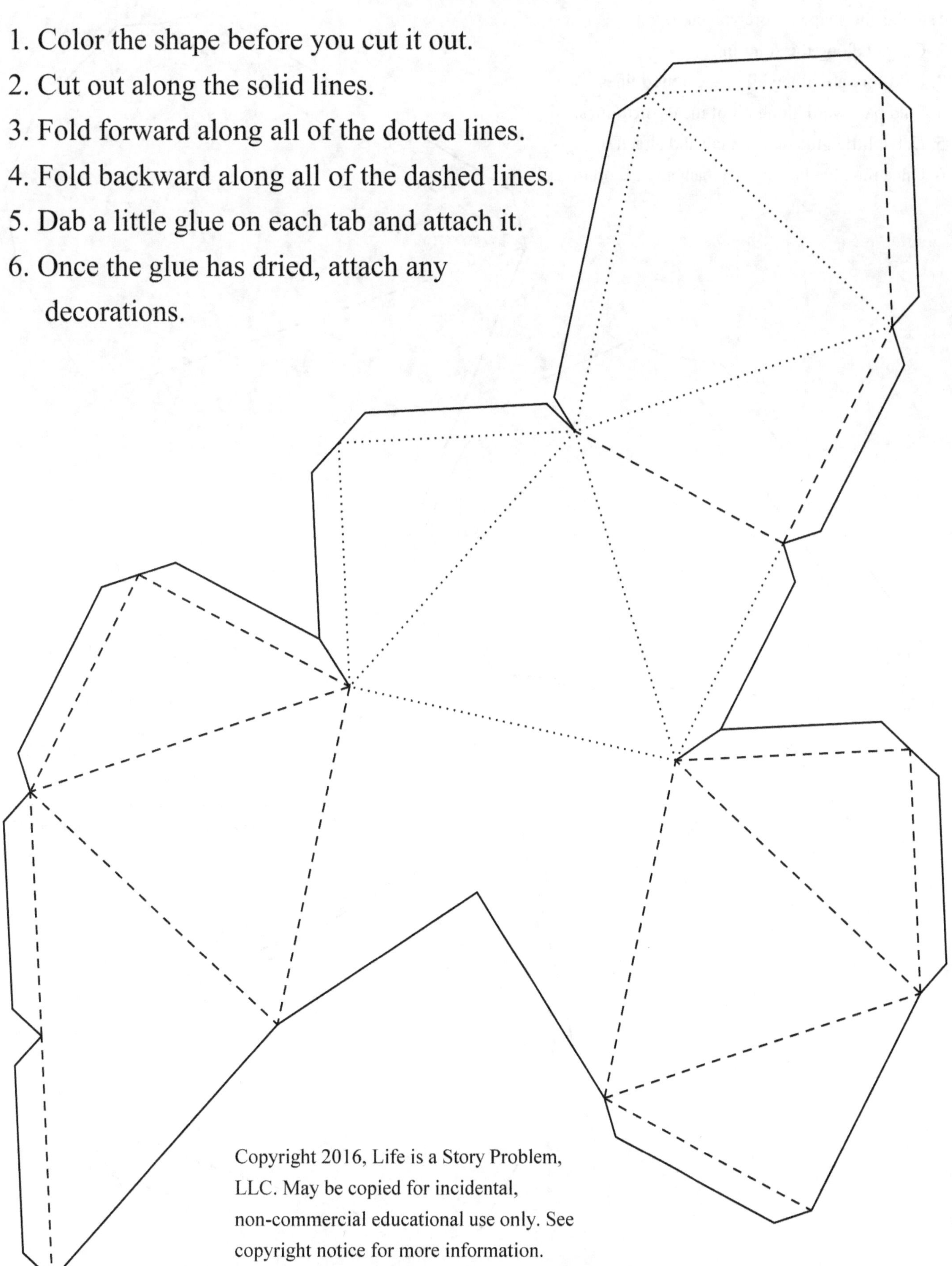

Pentagonal star pyramid

1. Color the shape before you cut it out.
2. Cut out along the solid lines.
3. Fold forward along all of the dotted lines.
4. Fold backward along all of the dashed lines.
5. Dab a little glue on each tab and attach it.
6. Once the glue has dried, attach any decorations.

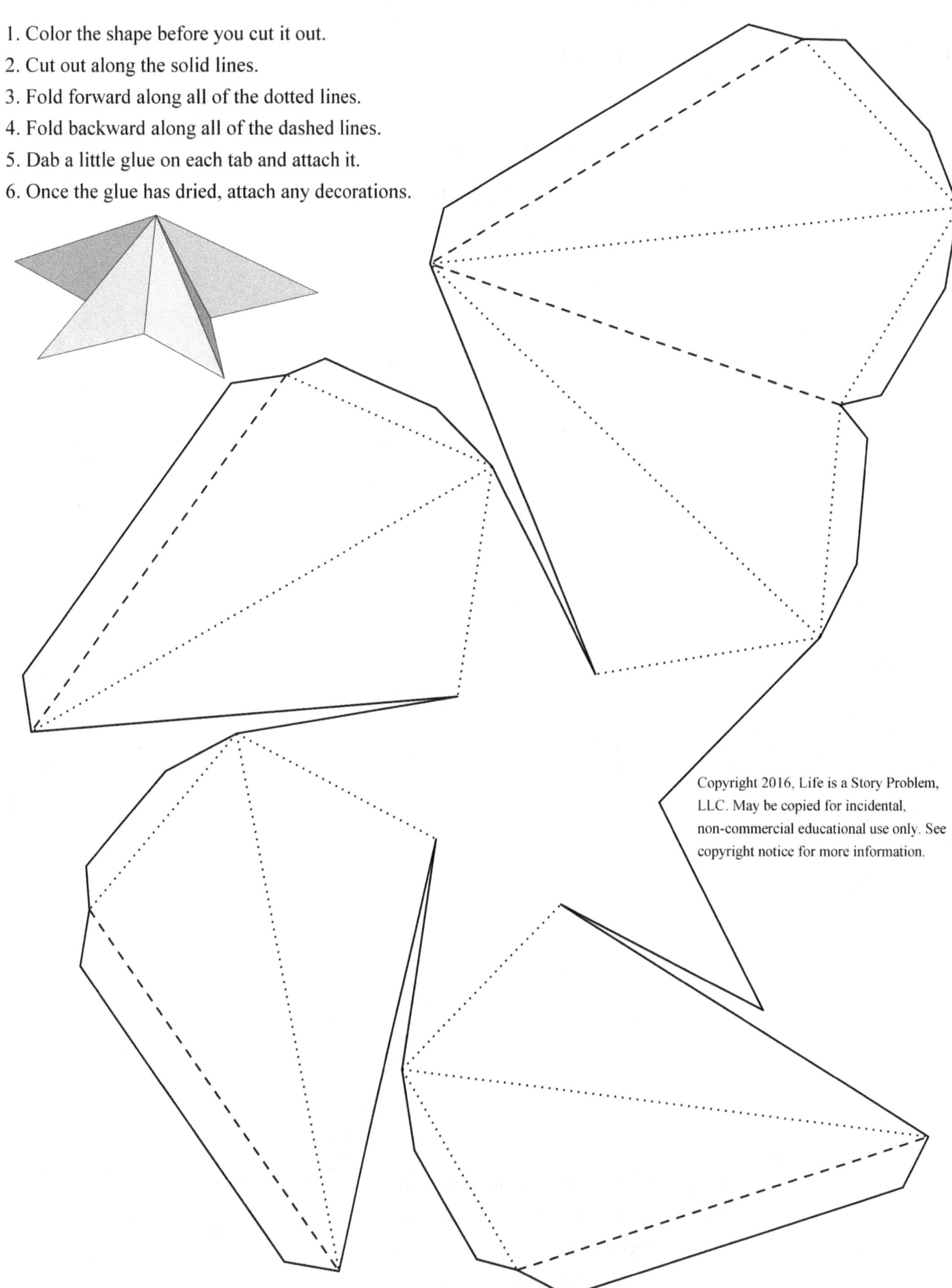

  Geometric Nets Mega Project Book, Tabbed By David E. McAdams

Pentagonal star bipyramid

1. Color the shape before you cut it out.
2. Cut out along the solid lines.
3. Fold forward along all of the dotted lines.
4. Fold backward along all of the dashed lines.
5. Dab a little glue on each tab and attach it.
6. Once the glue has dried, attach the decorations.

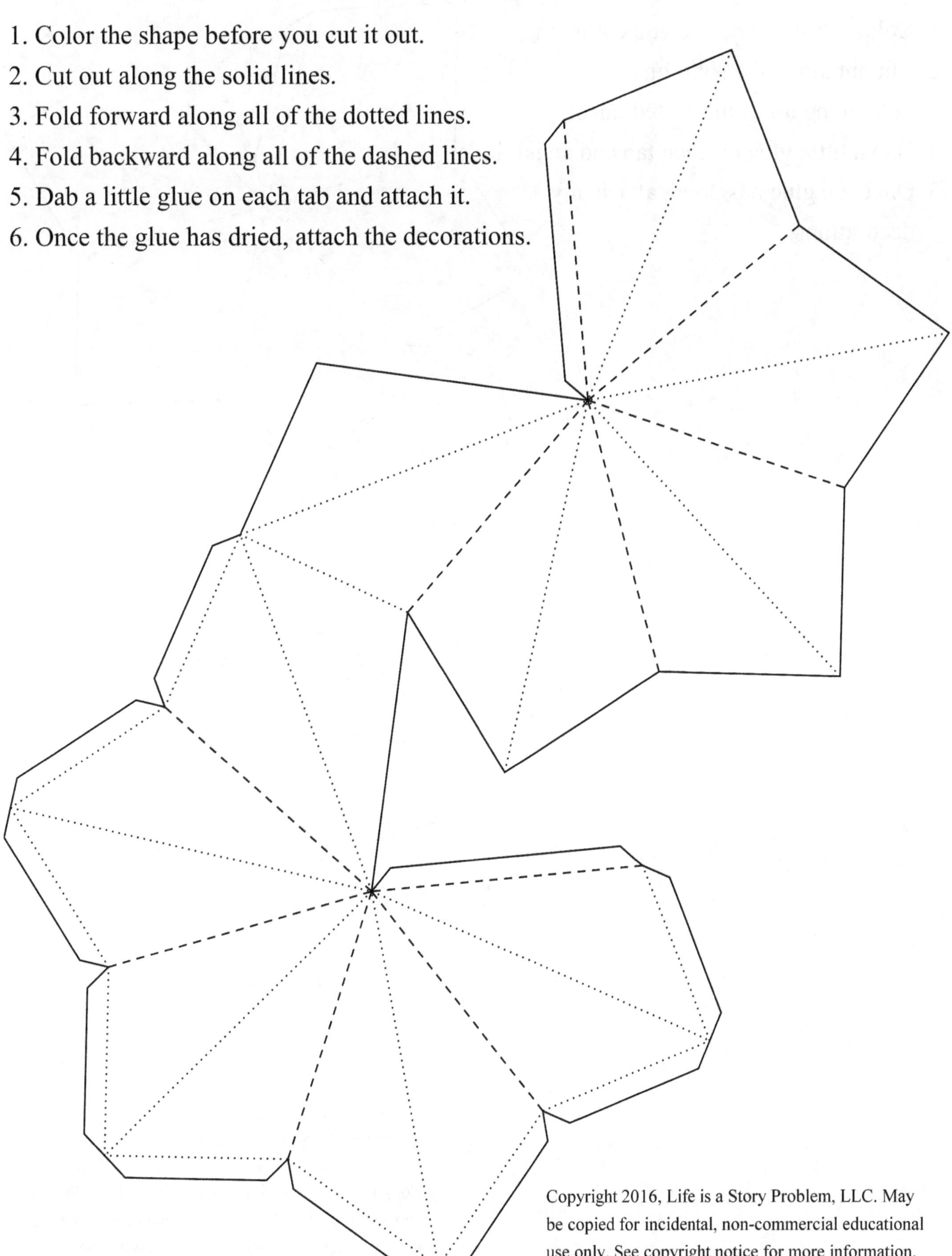

Elongated pentagonal star bipyramid

1. Color the shape before you cut it out.
2. Cut out along the solid lines.
3. Fold along all of the dotted lines.
4. Dab a little glue on each tab and attach it.
5. Once the glue has dried, attach any decorations.

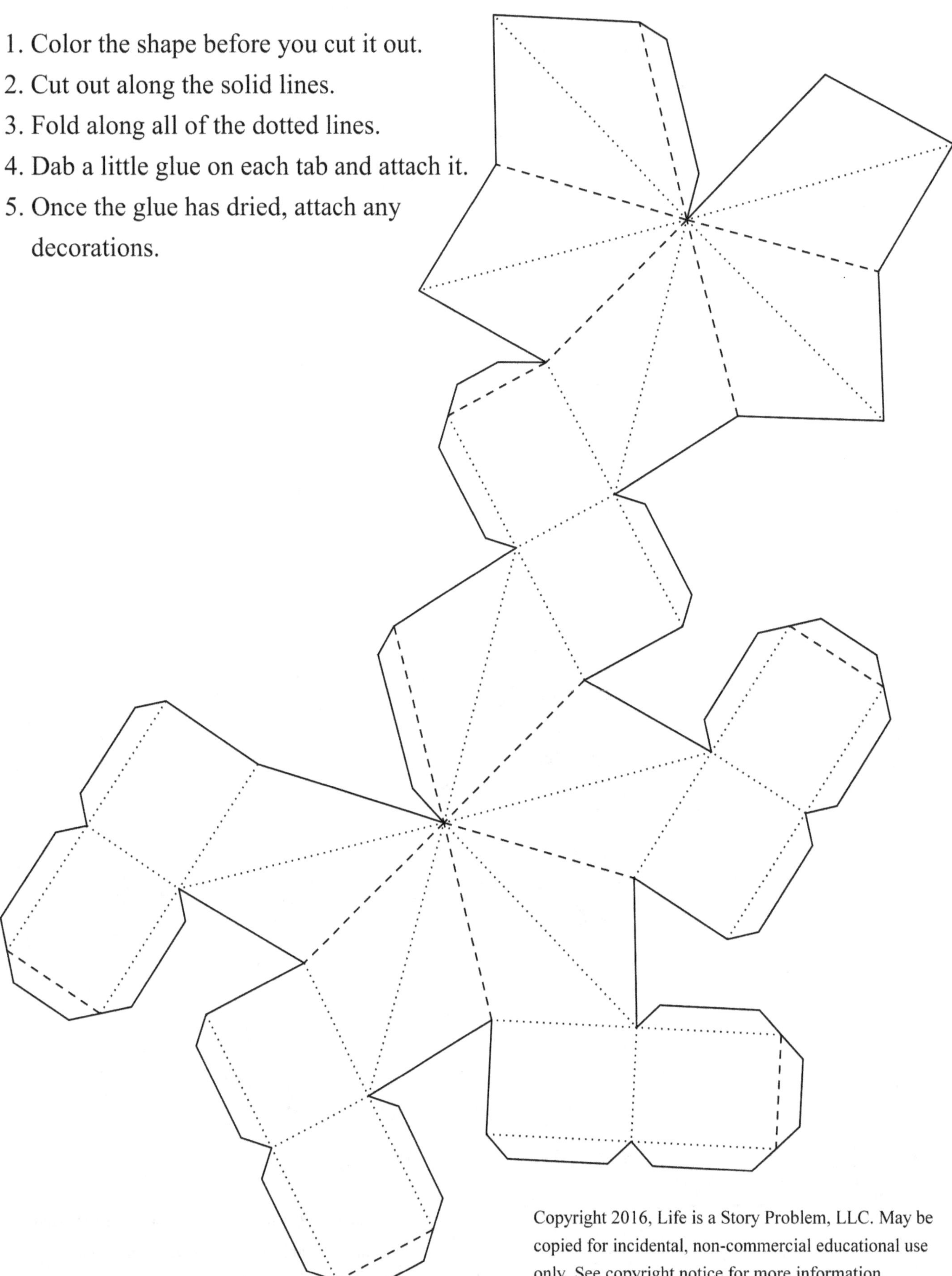

 Geometric Nets Mega Project Book, Tabbed By David E. McAdams

Truncated pentagonal star pyramid

1. Color the shape before you cut it out.

2. Cut out along the solid lines.

3. Fold along all of the dotted lines.

4. Dab a little glue on each tab and attach it.

5. Once the glue has dried, attach any decorations.

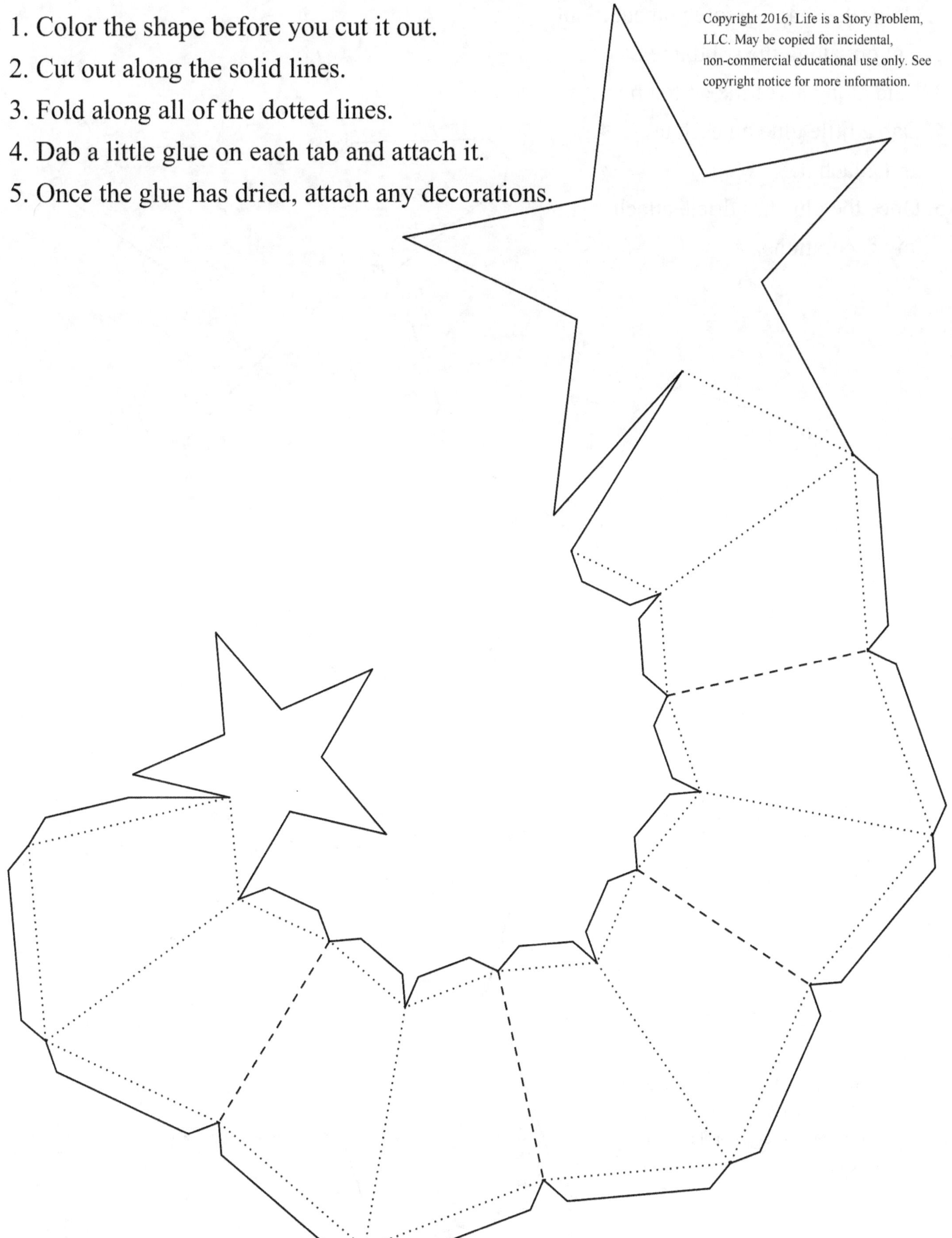

Pentagonal star prism

1. Color the shape before you cut it out.
2. Cut out along the solid lines.
3. Fold along all of the dotted lines.
4. Dab a little glue on each tab
   and attach it.
5. Once the glue has dried, attach
   any decorations.

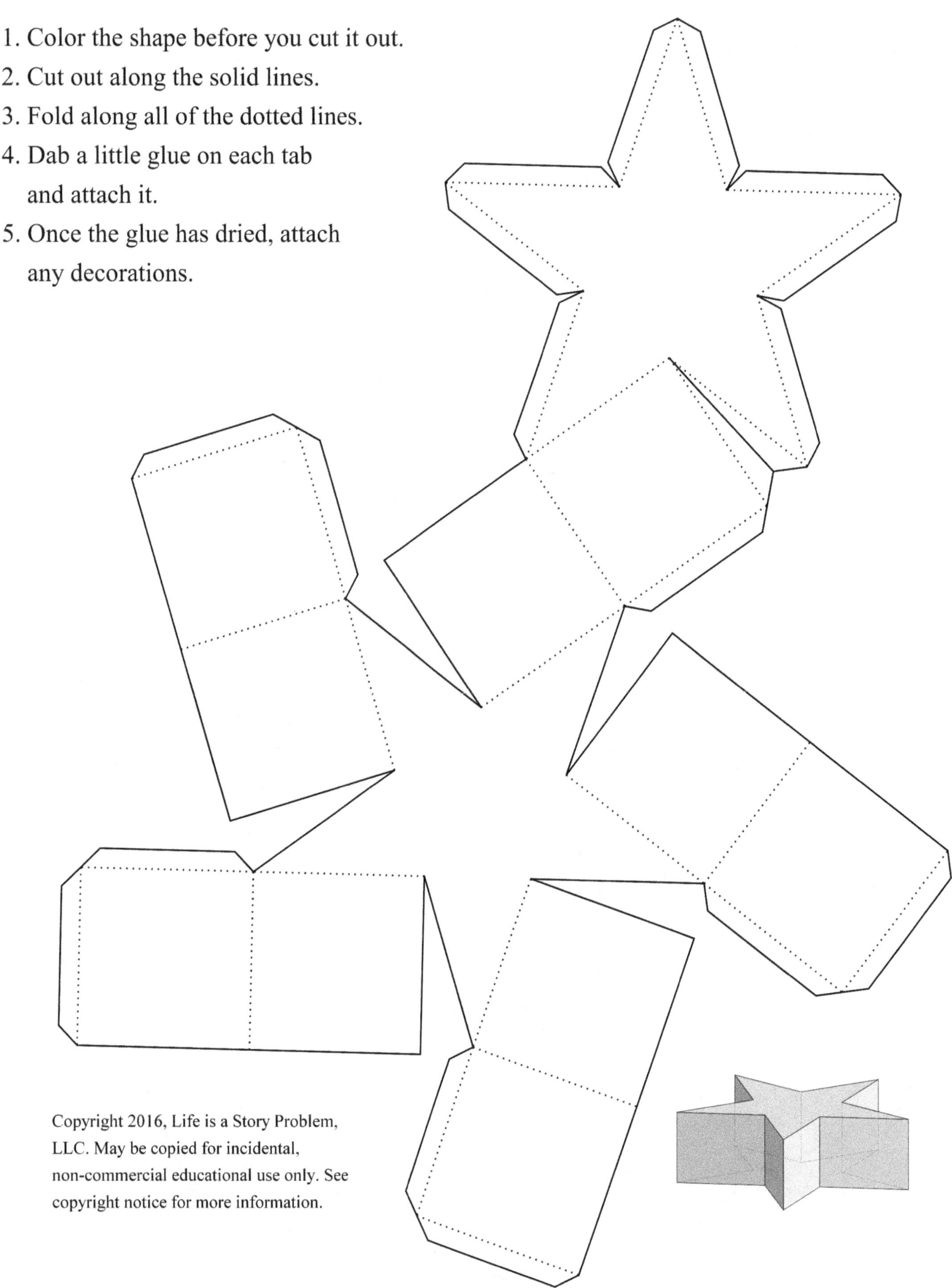

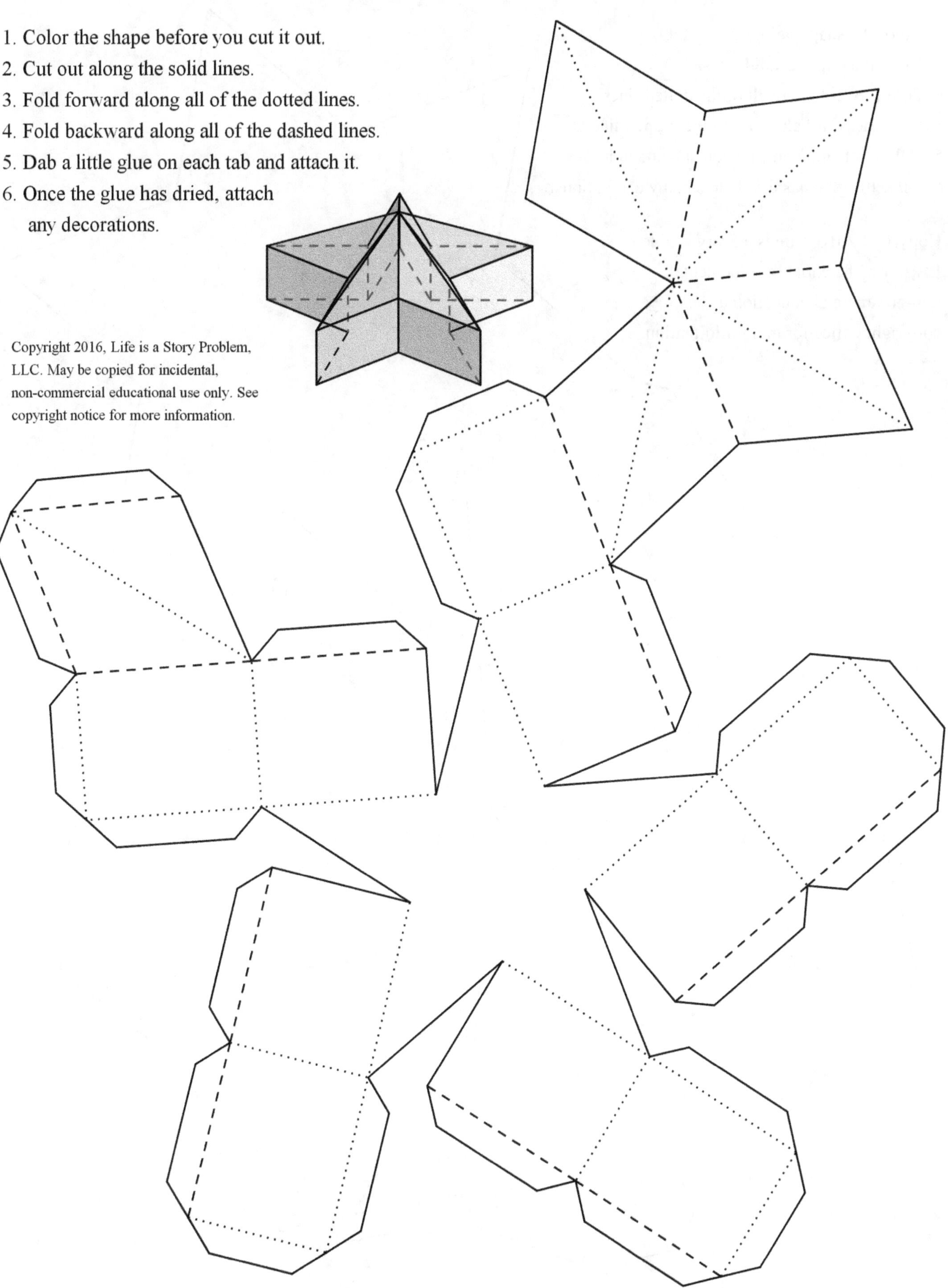

Elongated pentagonal star pyramid

1. Color the shape before you cut it out.
2. Cut out along the solid lines.
3. Fold forward along all of the dotted lines.
4. Fold backward along all of the dashed lines.
5. Dab a little glue on each tab and attach it.
6. Once the glue has dried, attach any decorations.

Hexagonal star pyramid

1. Color the shape before you cut it out.
2. Cut out along the solid lines.
3. Fold forward along all of the dotted lines.
4. Fold backward along all of the dashed lines.
5. Dab a little glue on each tab and attach it.
6. Once the glue has dried, attach any decorations.

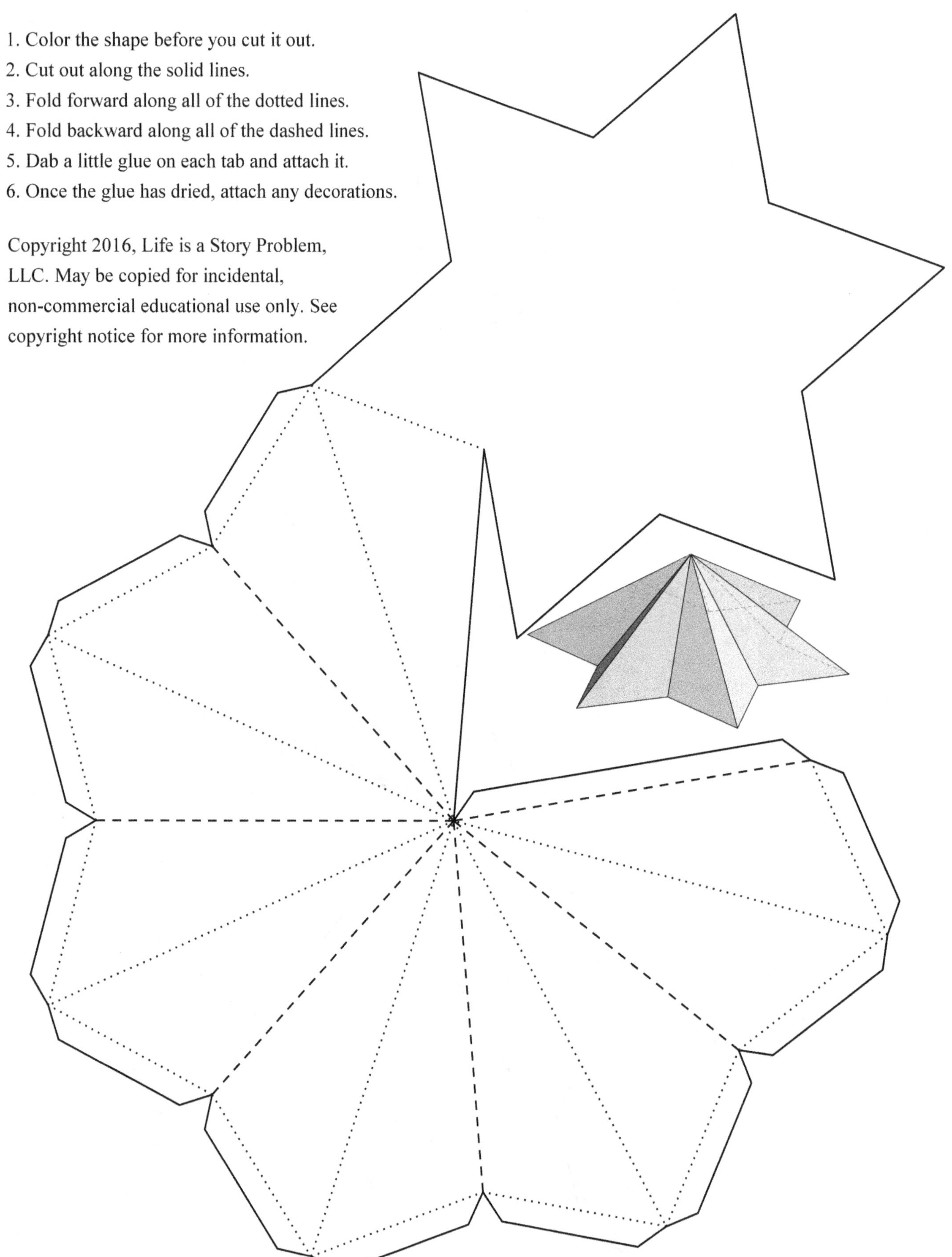

 Geometric Nets Mega Project Book, Tabbed By David E. McAdams

Hexagonal star bipyramid

1. Color the shape before you cut it out.
2. Cut out along the solid lines.
3. Fold forward along all of the dotted lines.
4. Fold backward along all of the dashed lines.
5. Dab a little glue on each tab
   and attach it.
6. Once the glue has dried,
   attach any decorations.

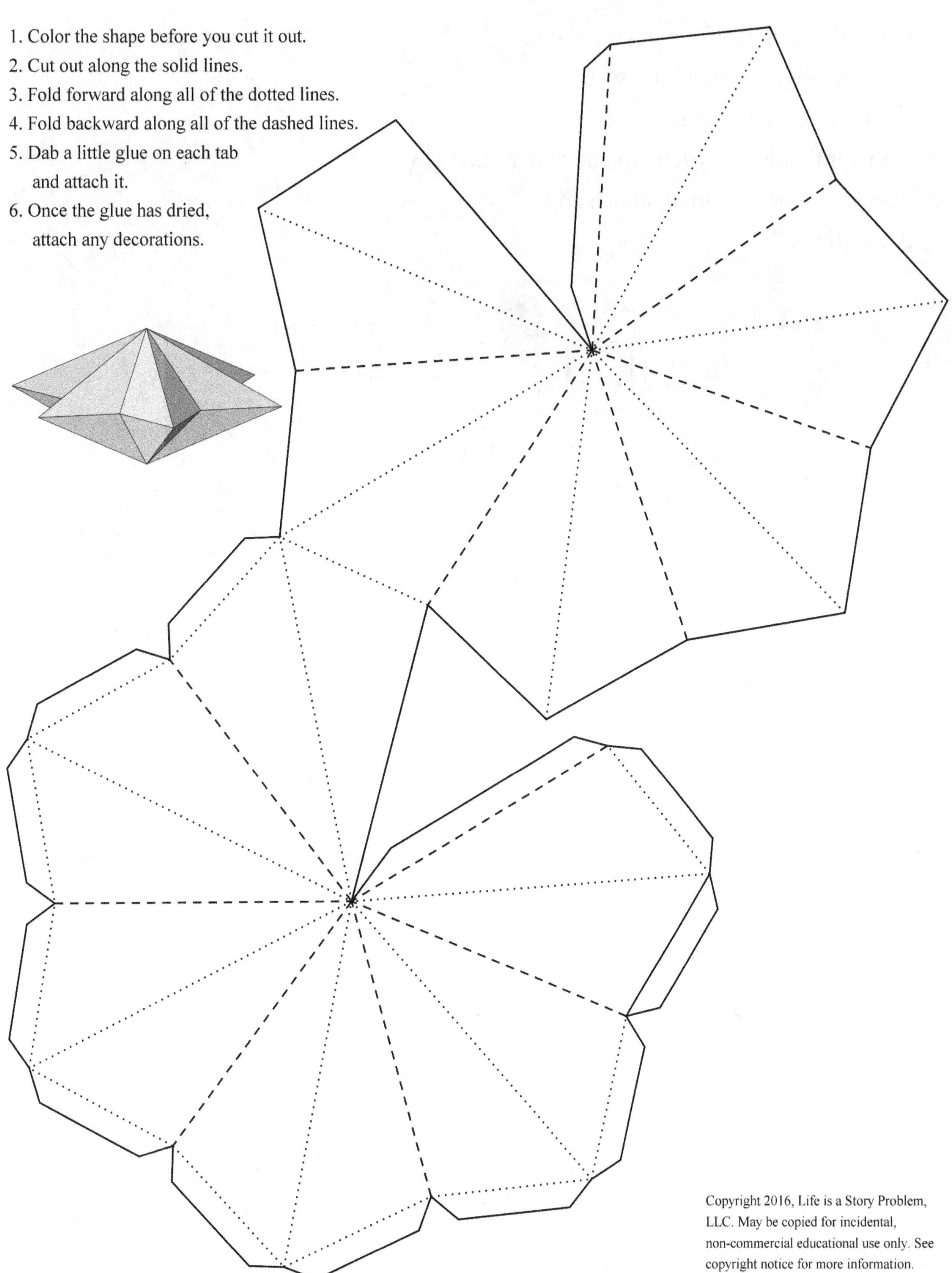

Hexagonal star prism

1. Color the shape before you cut it out.
2. Cut out along the solid lines.
3. Fold along all of the dotted lines.
4. Dab a little glue on each tab and attach it.
5. Once the glue has dried, attach any decorations

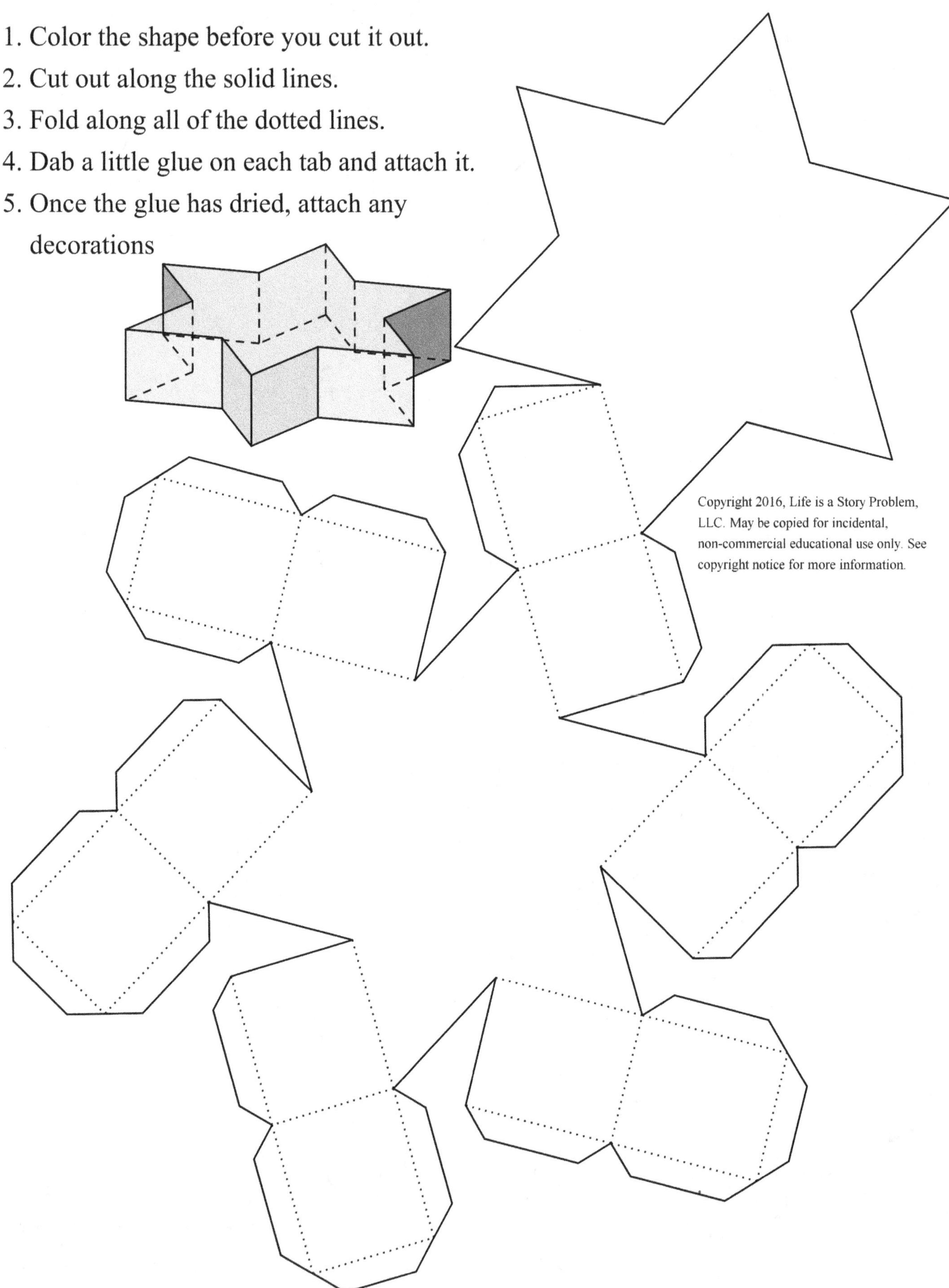

 Geometric Nets Mega Project Book, Tabbed By David E. McAdams

Hexagonal star antiprism

1. Color the shape before you cut it out.
2. Cut out along the solid lines.
3. Discard the shaded areas.
5. Fold forward along all of the dotted lines.
6. Fold backward along all of the dashed lines.
7. Dab a little glue on each tab and attach it.
8. Once the glue has dried, attach any decorations.

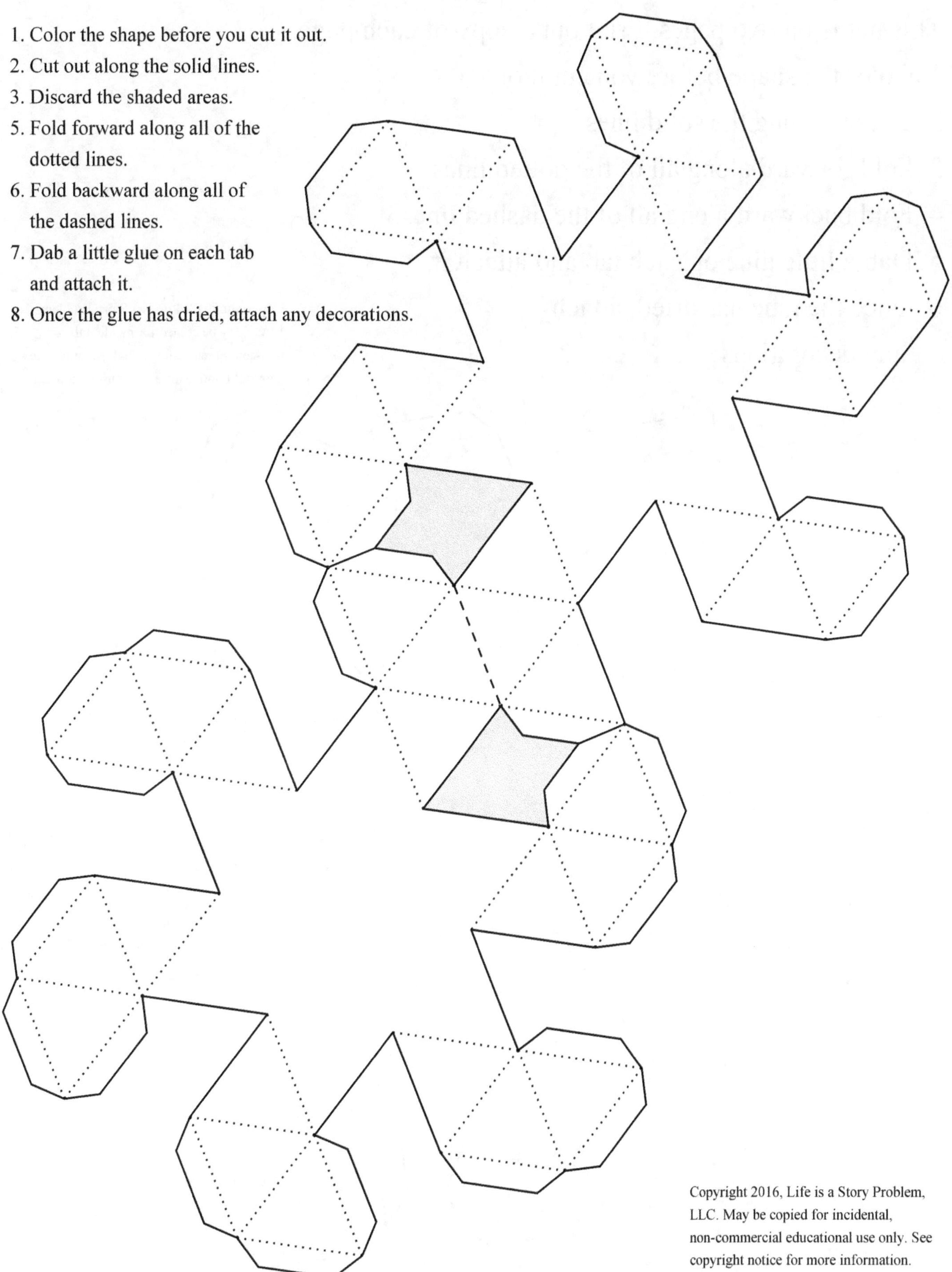

Augmented hexagonal star elongated antiprism

This net is on two pages. Print out a copy of each page.

1. Color the shape before you cut it out.

2. Cut out along the solid lines.

3. Fold forward along all of the dotted lines.

4. Fold backward along all of the dashed lines.

5. Dab a little glue on each tab and attach it.

6. Once the glue has dried, attach any decorations.

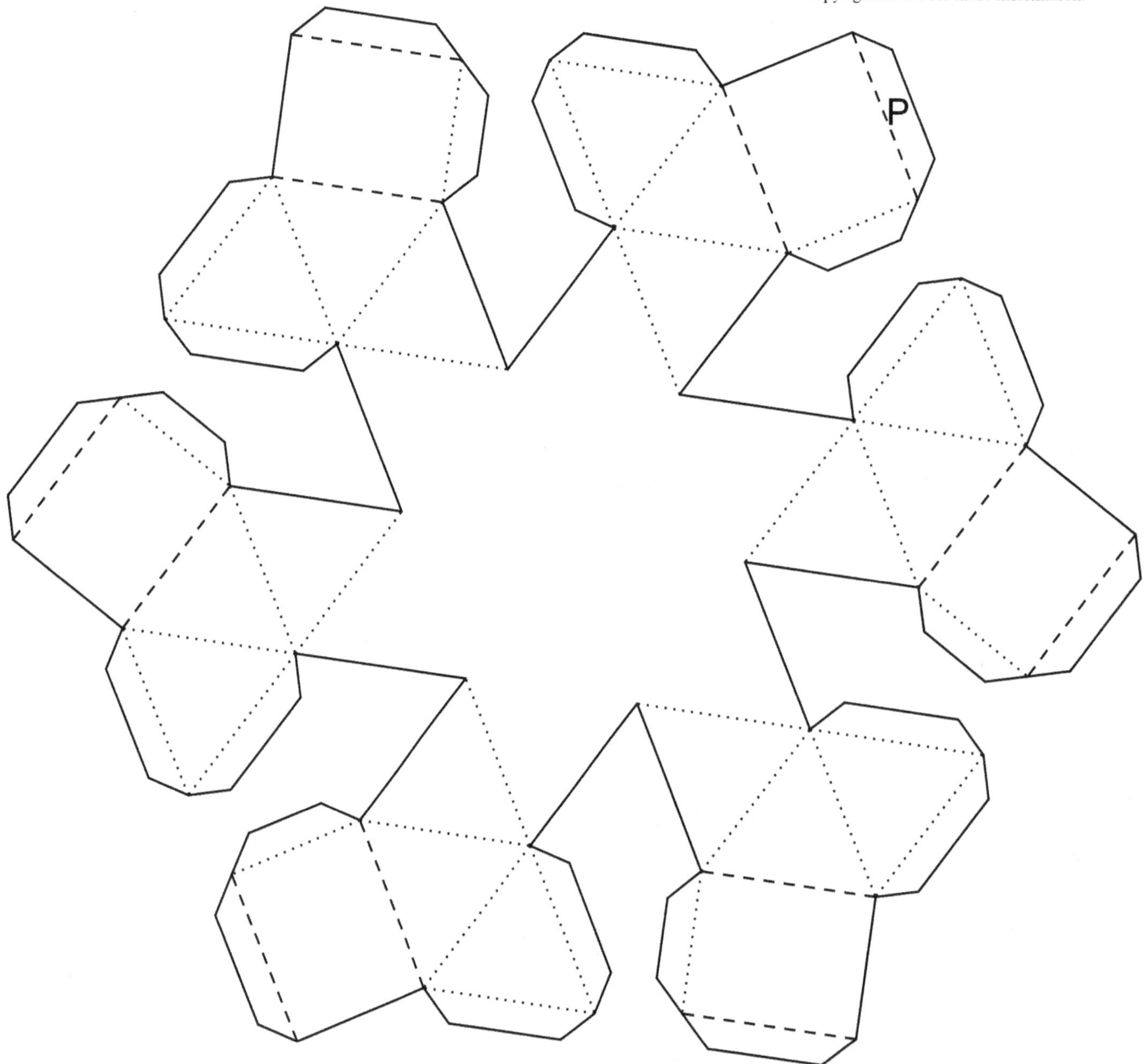

 Geometric Nets Mega Project Book, Tabbed By David E. McAdams

2nd page of Augmented hexagonal star elongated antiprism

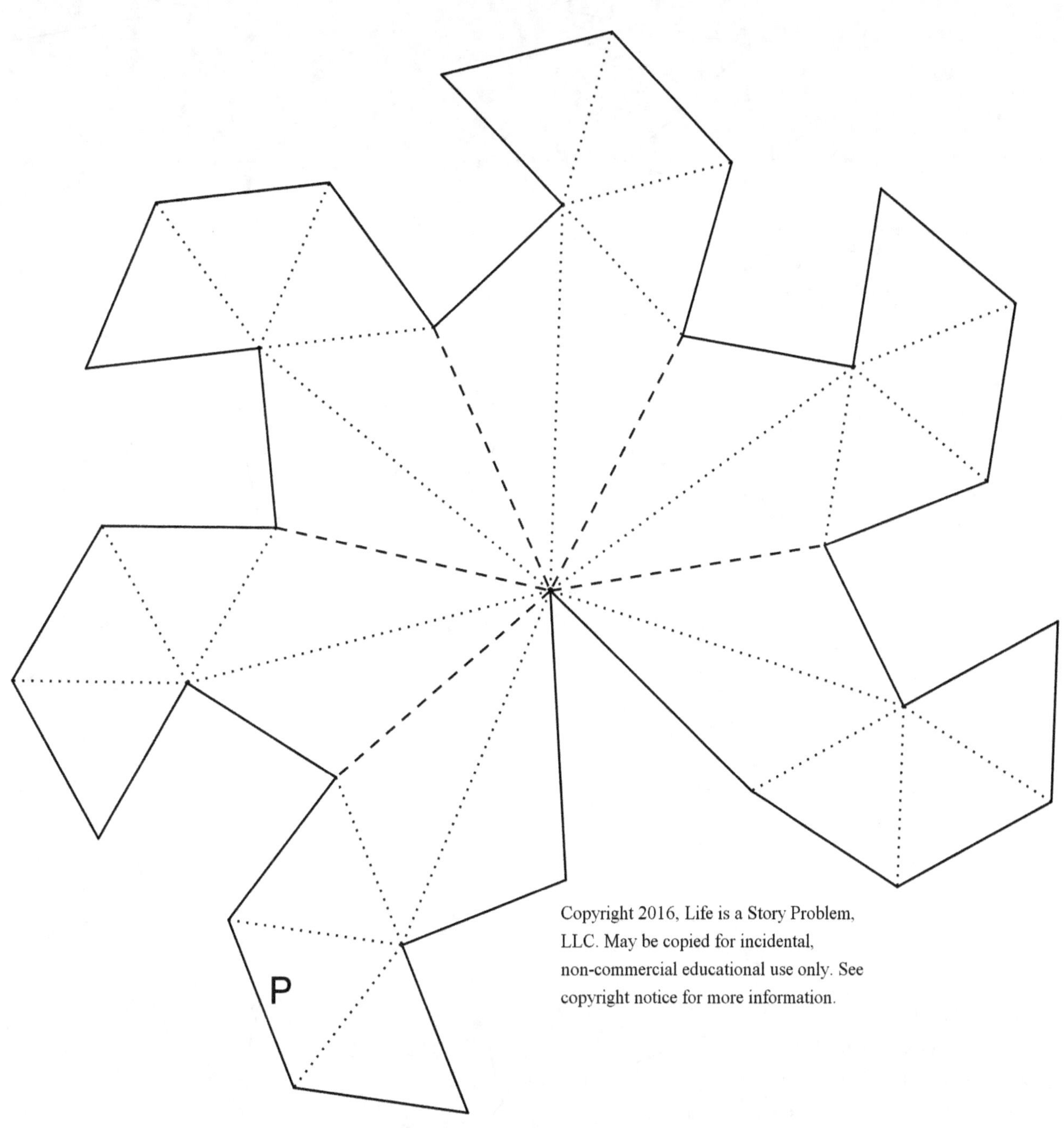

Cubed hexagonal star

This net is on three pages. Print out a copy of each page.

1. Color the shape before you cut it out.
2. Cut out along the solid lines.
3. Fold forward along all of the dotted lines.
4. Fold backward along all of the dashed lines.
5. Dab a little glue on each tab and attach it.
6. Once the glue has dried, attach any decorations.

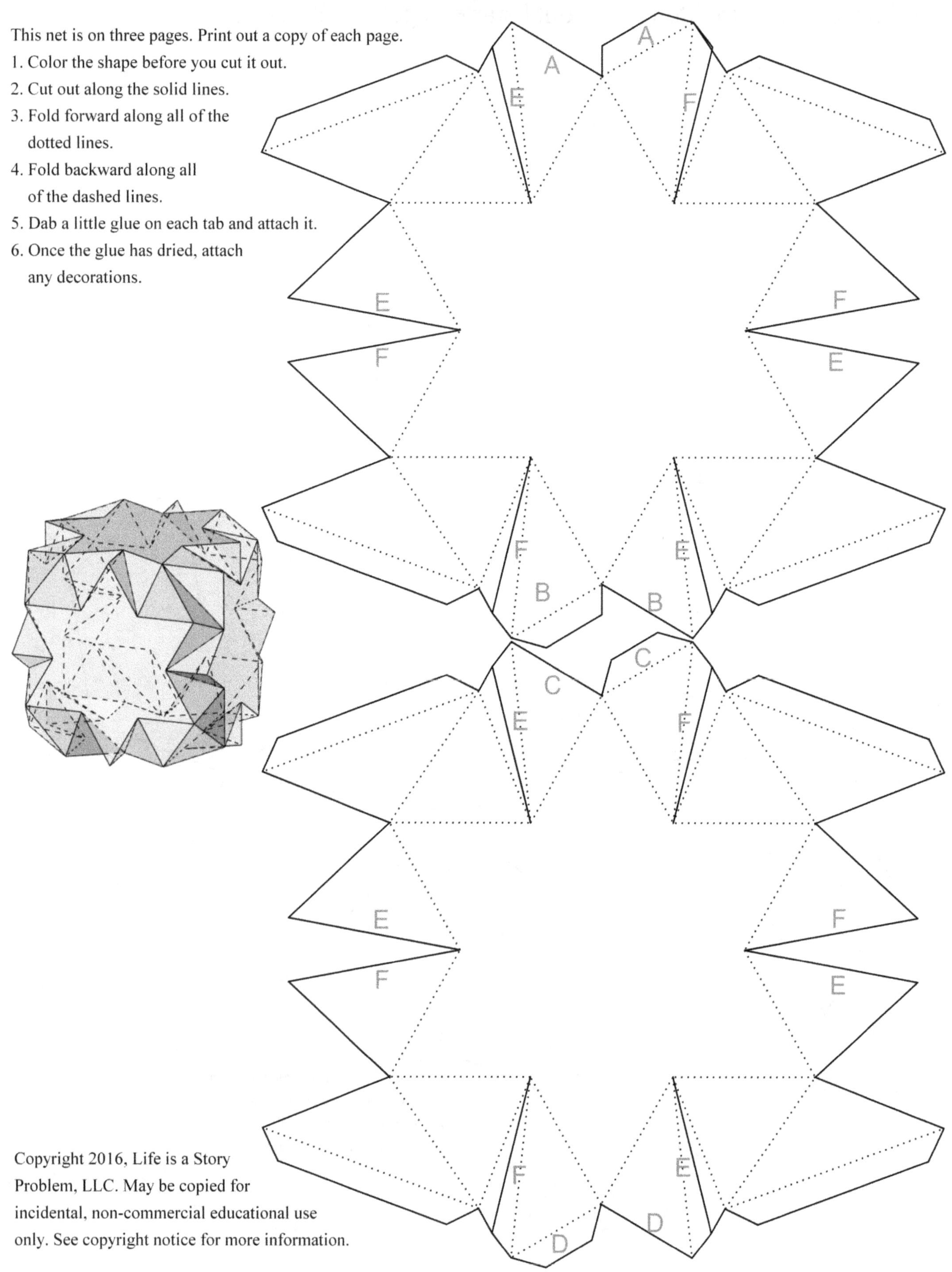

 Geometric Nets Mega Project Book, Tabbed By David E. McAdams

2nd or 3rd page, Cubed hexagonal star

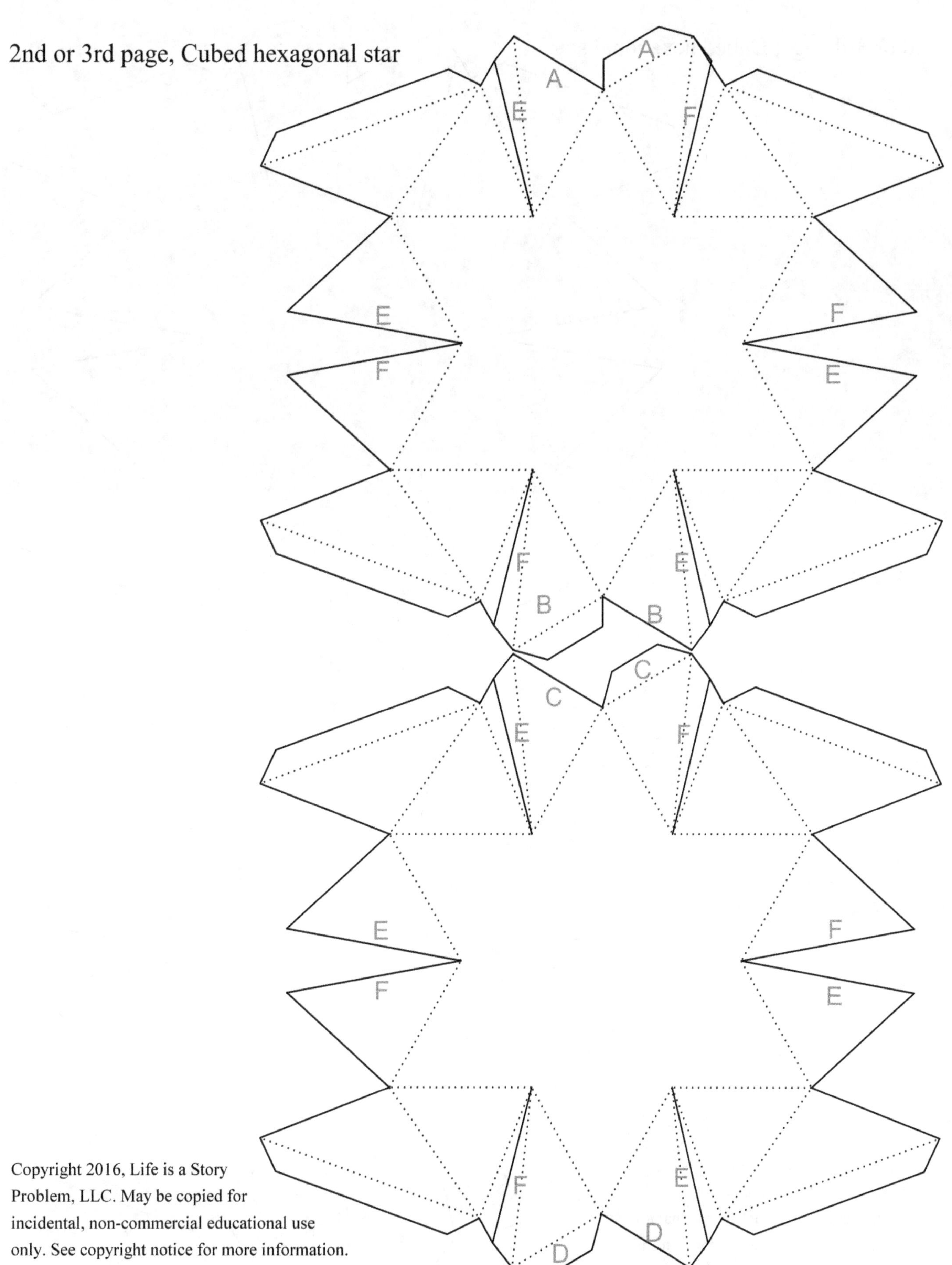

2nd or 3rd page, Cubed hexagonal star

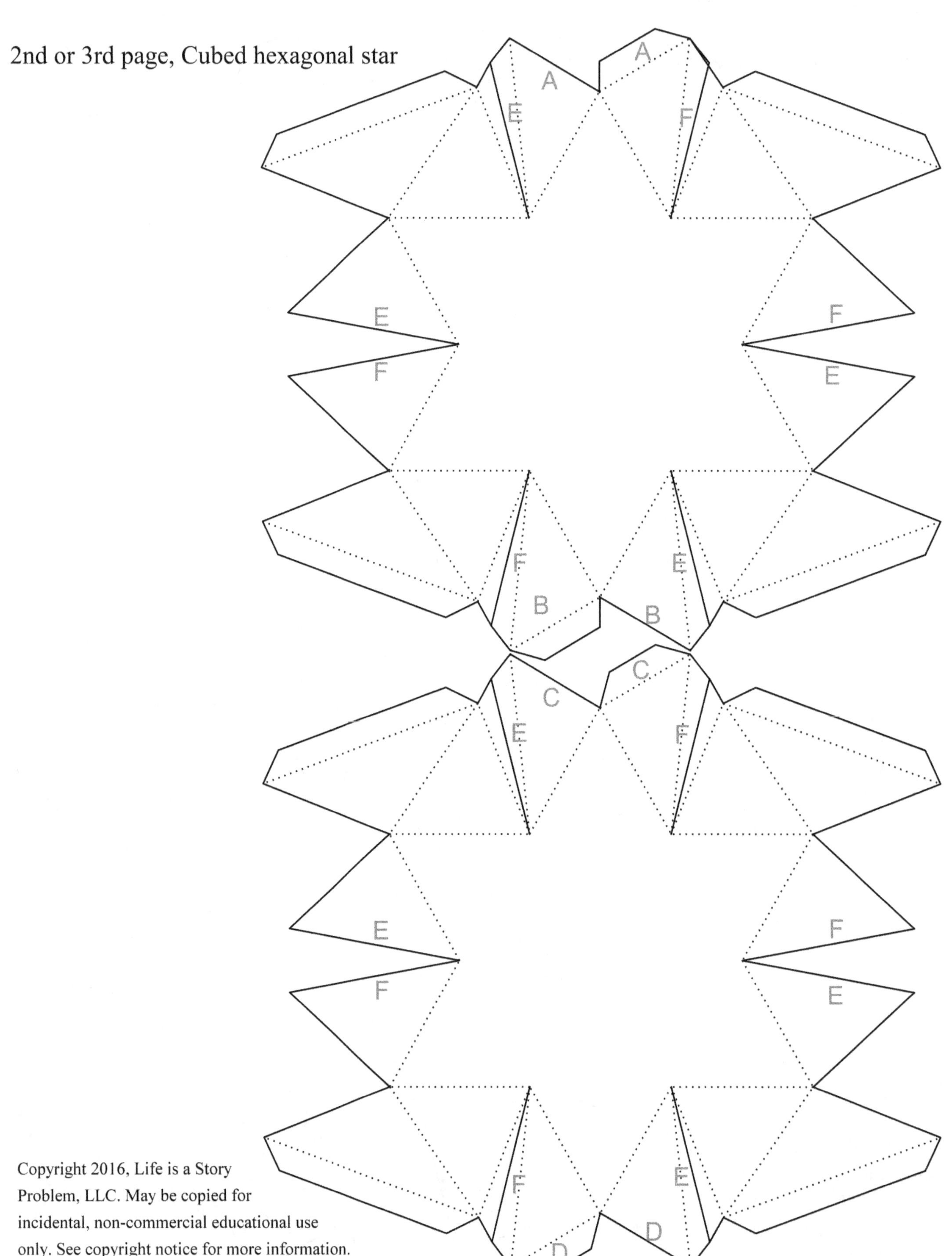

Geometric Nets Mega Project Book, Tabbed By David E. McAdams

Cubed octagonal star

2nd page, Cubed octagonal star

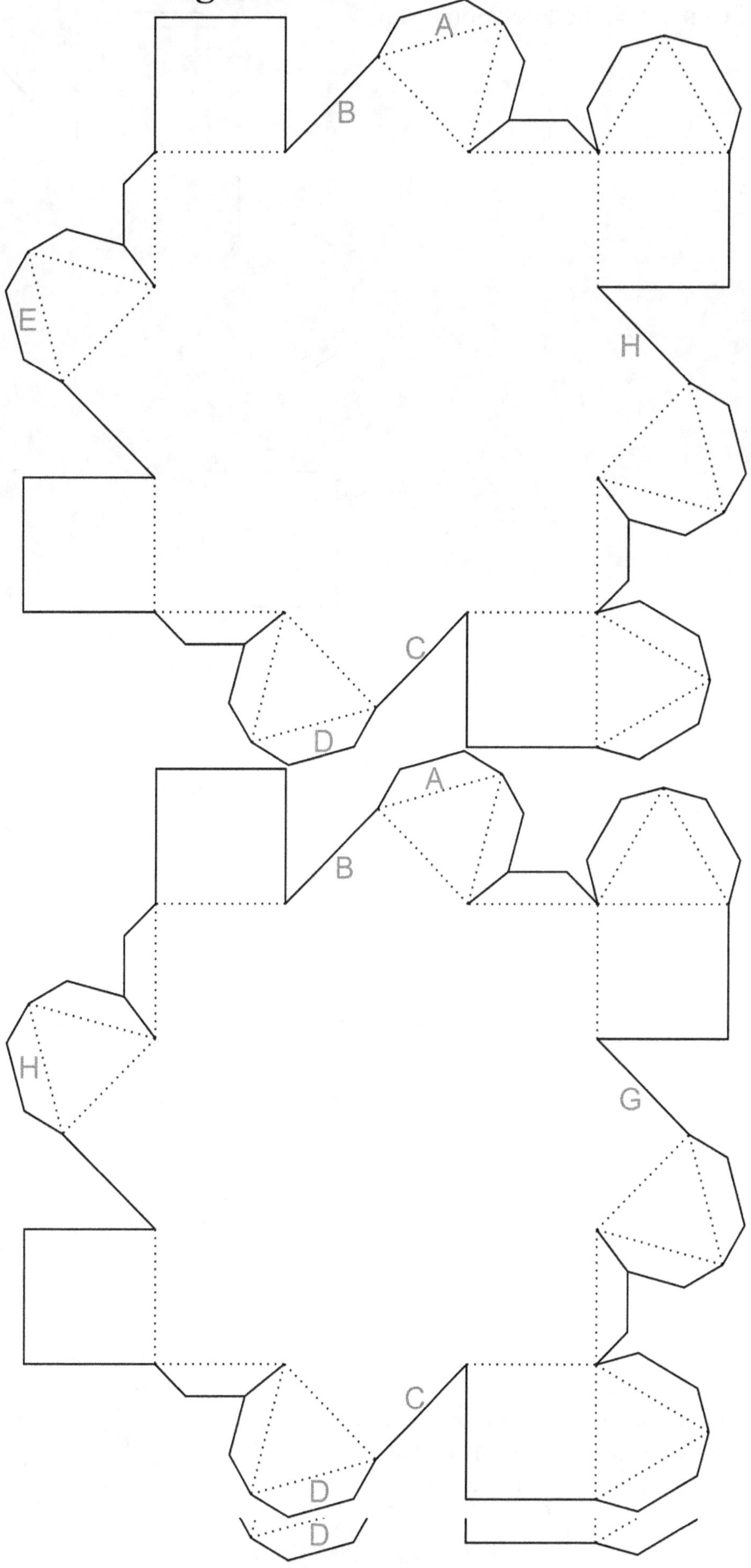

3rd page, Cubed octagonal star

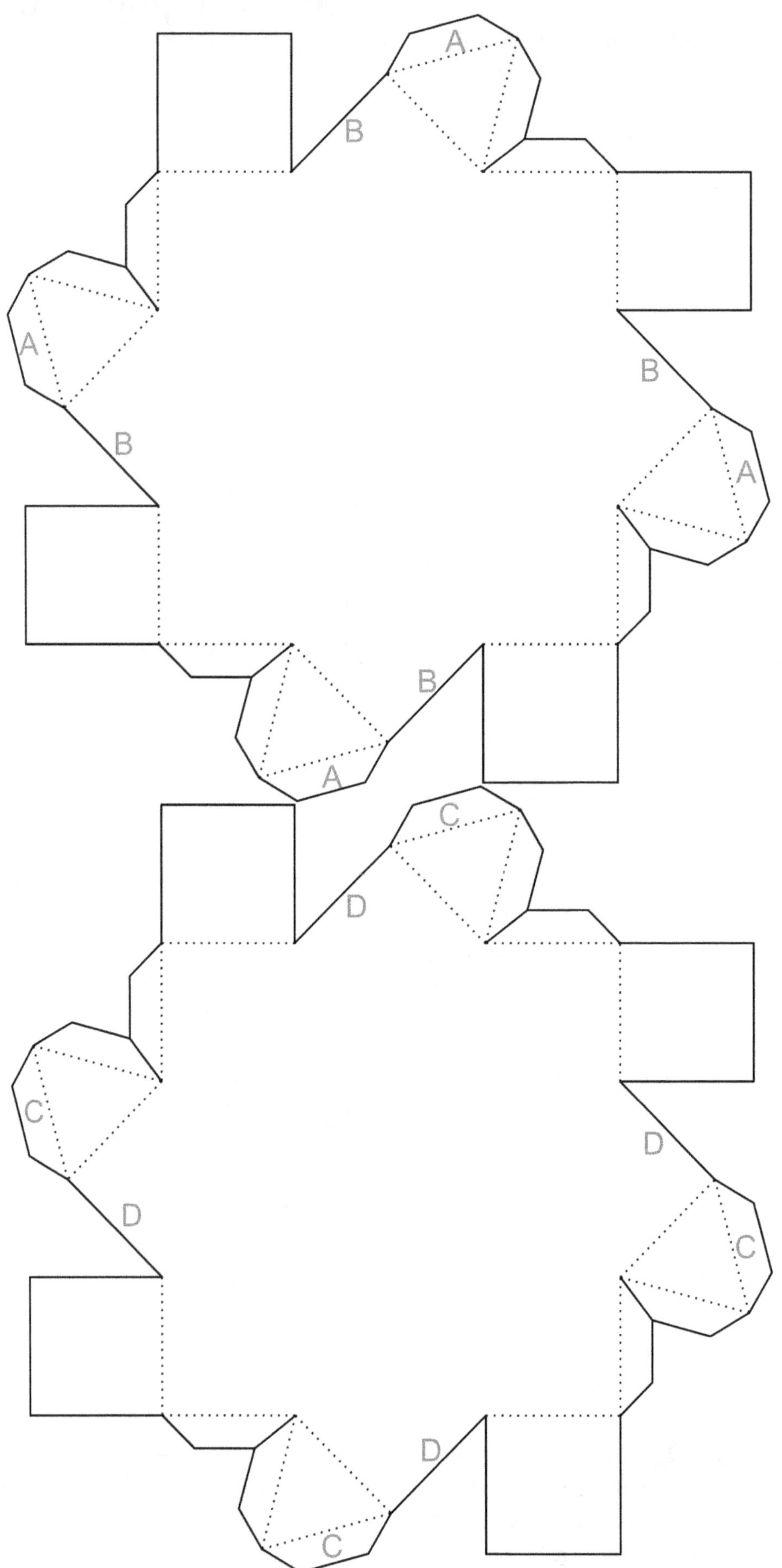

 Geometric Nets Mega Project Book, Tabbed By David E. McAdams

Rhombohedron

1. Color the shape before you cut it out.

2. Cut out along the solid lines.

3. Fold along all of the dotted lines.

4. Dab a little glue on each tab and attach it.

5. Once the glue has dried, attach any decorations.

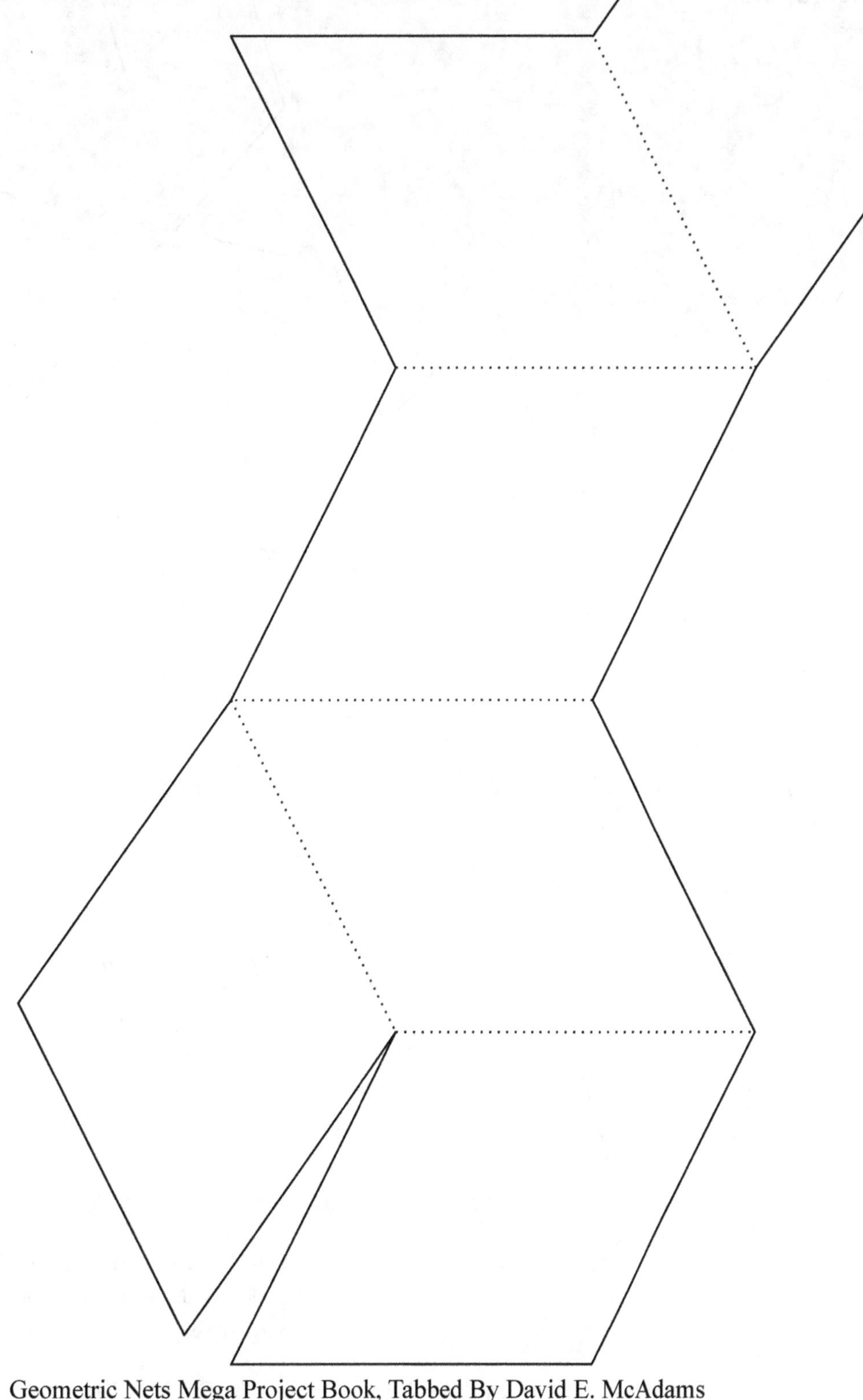

Pyramid augmented dodecahedron

1. Color the shape before you cut it out.
2. Cut out along the solid lines.
3. Fold along all of the dotted lines.
4. Dab a little glue on each tab and attach it.
5. Once the glue has dried, attach any decorations.

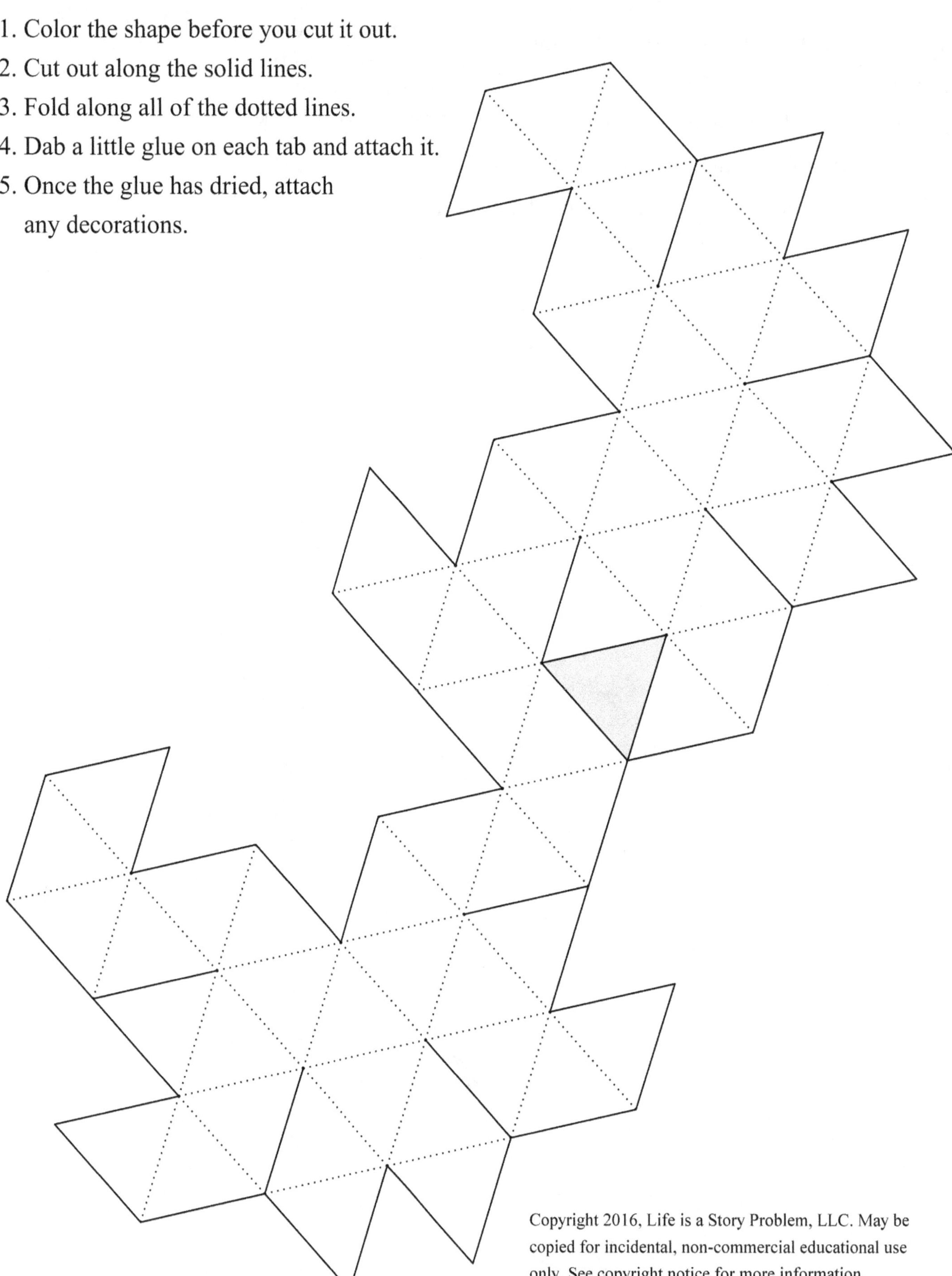

 Geometric Nets Mega Project Book, Tabbed By David E. McAdams

Octahemioctahedron

1. Color the shape before you cut it out.
2. Cut out along the solid lines.
3. Fold forward along all of the dotted lines.
4. Fold backward along all of the dashed lines.
5. Dab a little glue on each tab and attach it.
6. Once the glue has   dried, attach any decorations.

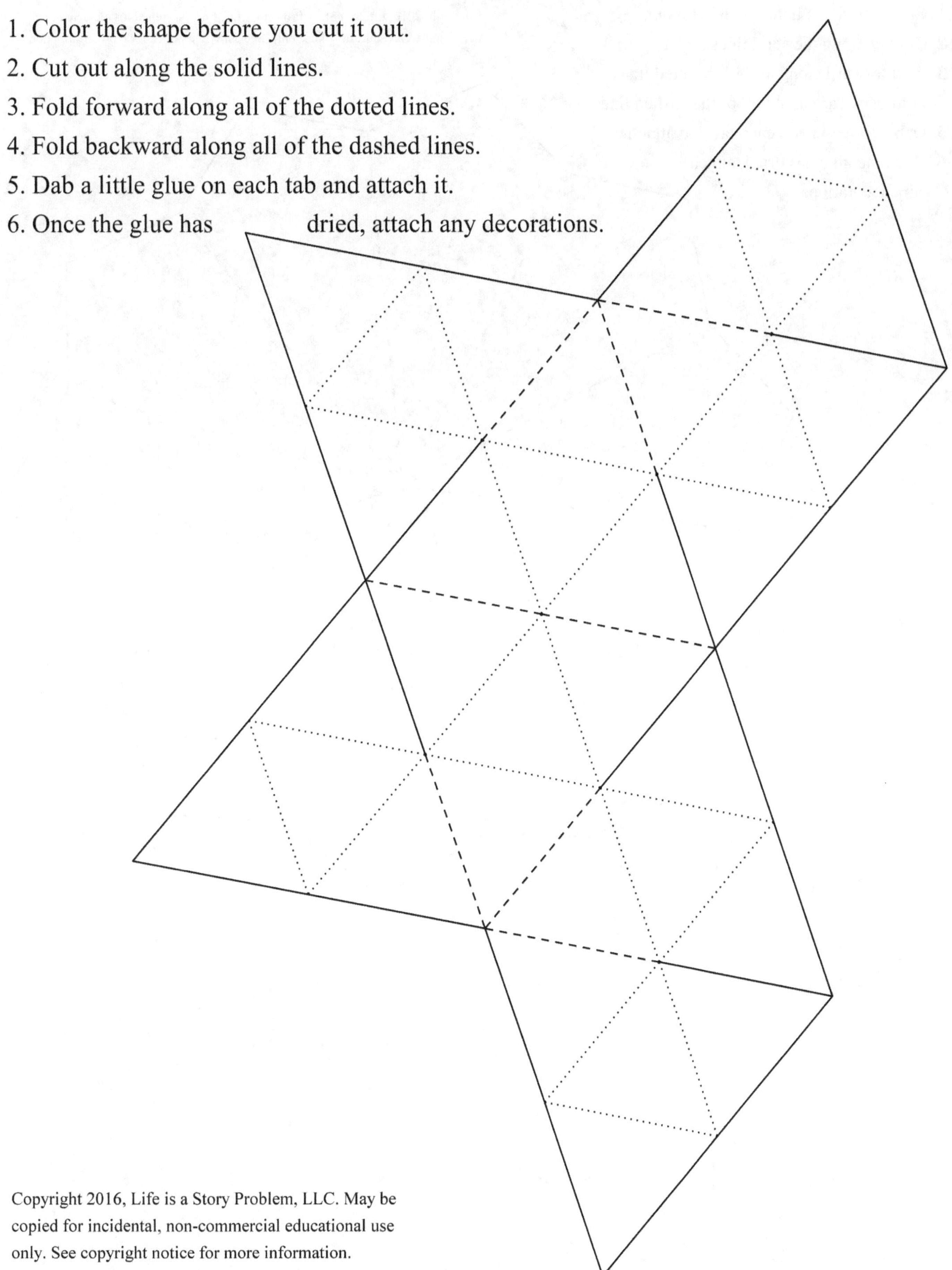

Decagonal star pyramid

1. Color the shape before you cut it out.
2. Cut out along the solid lines.
3. Fold forward along all of the dotted lines.
4. Fold backward along all of the dashed lines.
5. Dab a little glue on each tab and attach it.
6. Once the glue has dried, attach any decorations.

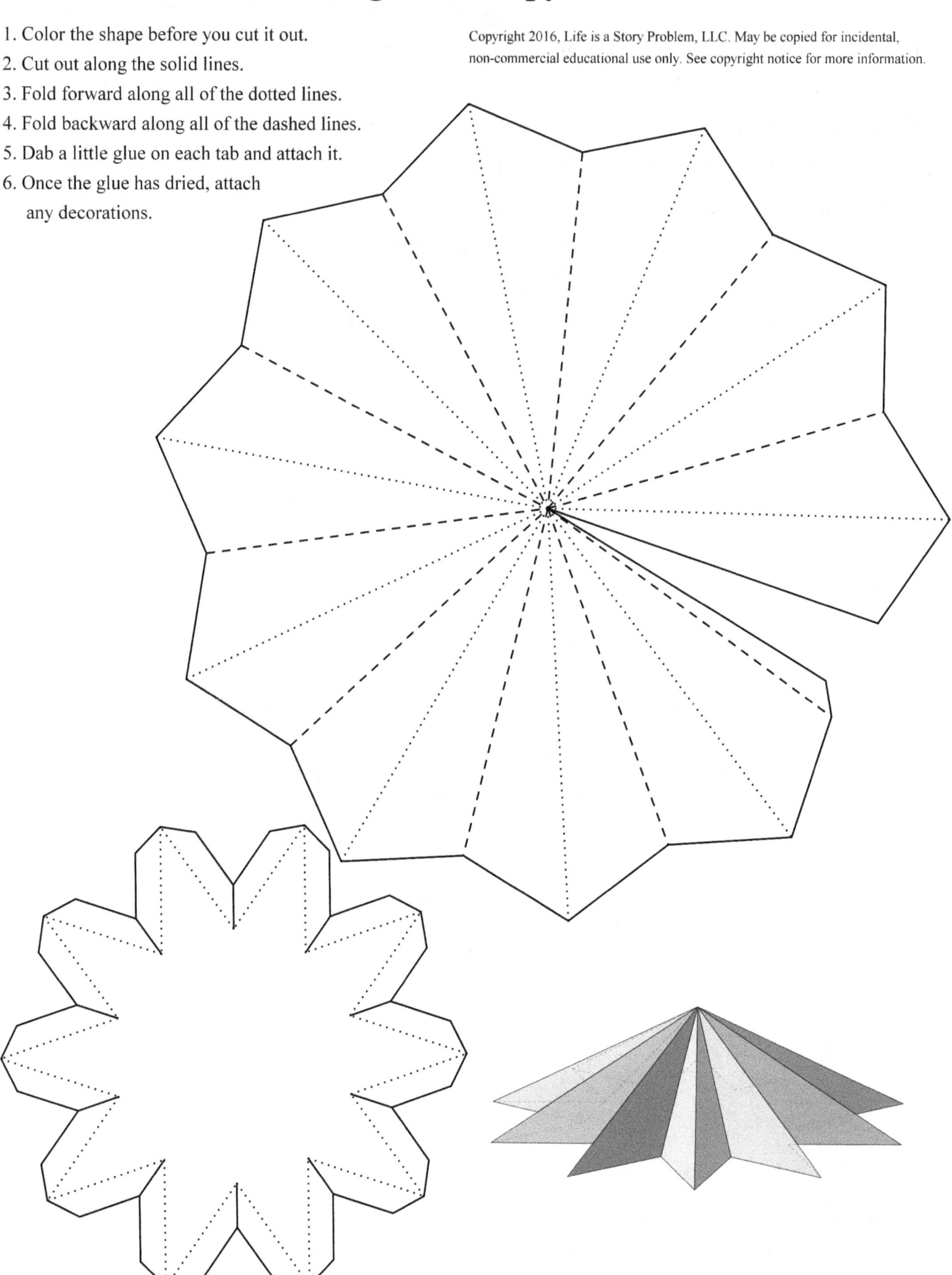

 Geometric Nets Mega Project Book, Tabbed By David E. McAdams

Great dodecahedron

1. Color the shape before you cut it out.
2. Cut out along the solid lines.
3. Fold forward along all of the dotted lines.
4. Fold backward along all of the dashed lines.
5. Dab a little glue on each tab and attach it.
6. Once the glue has dried, attach any decorations.

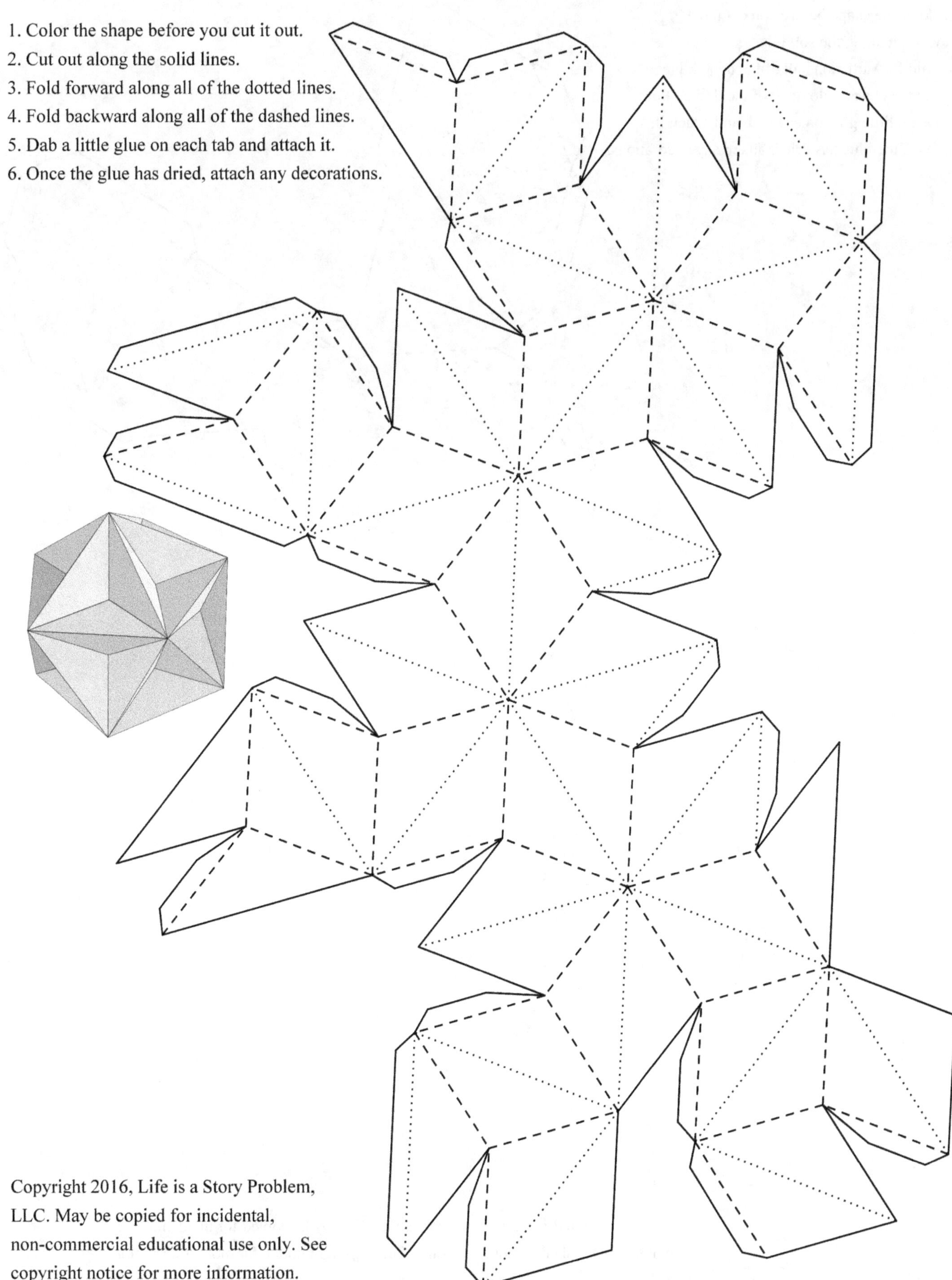

Small stellated dodecahedron

1. Color the shape before you cut it out.
2. Cut out along the solid lines.
3. Fold forward along all of the dotted lines.
4. Fold backward along all of the dashed lines.
5. Dab a little glue on each tab and attach it.
6. Once the glue has dried, attach any decorations.

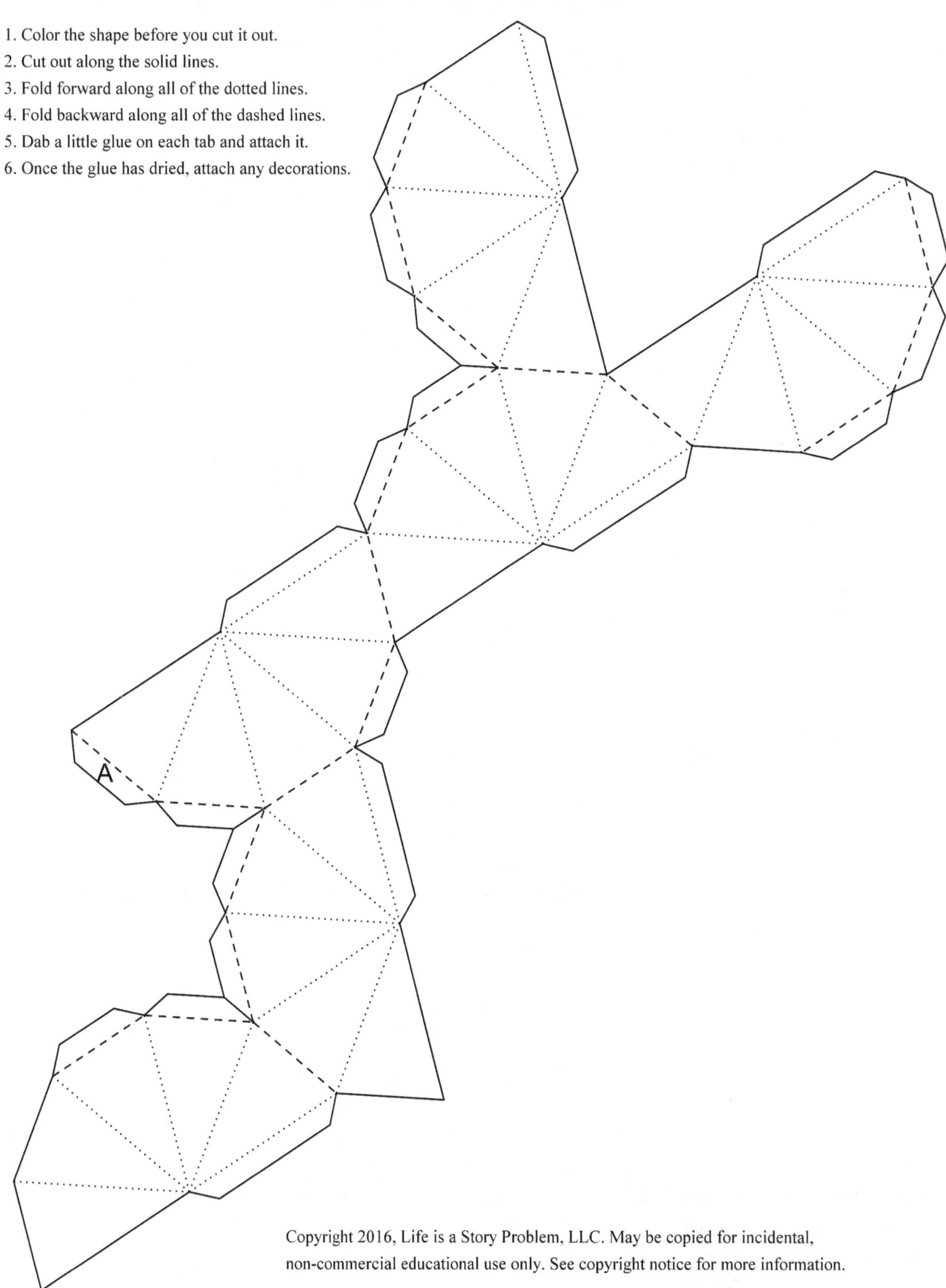

Geometric Nets Mega Project Book, Tabbed By David E. McAdams

Page 2 of small stellated dodecahedron.

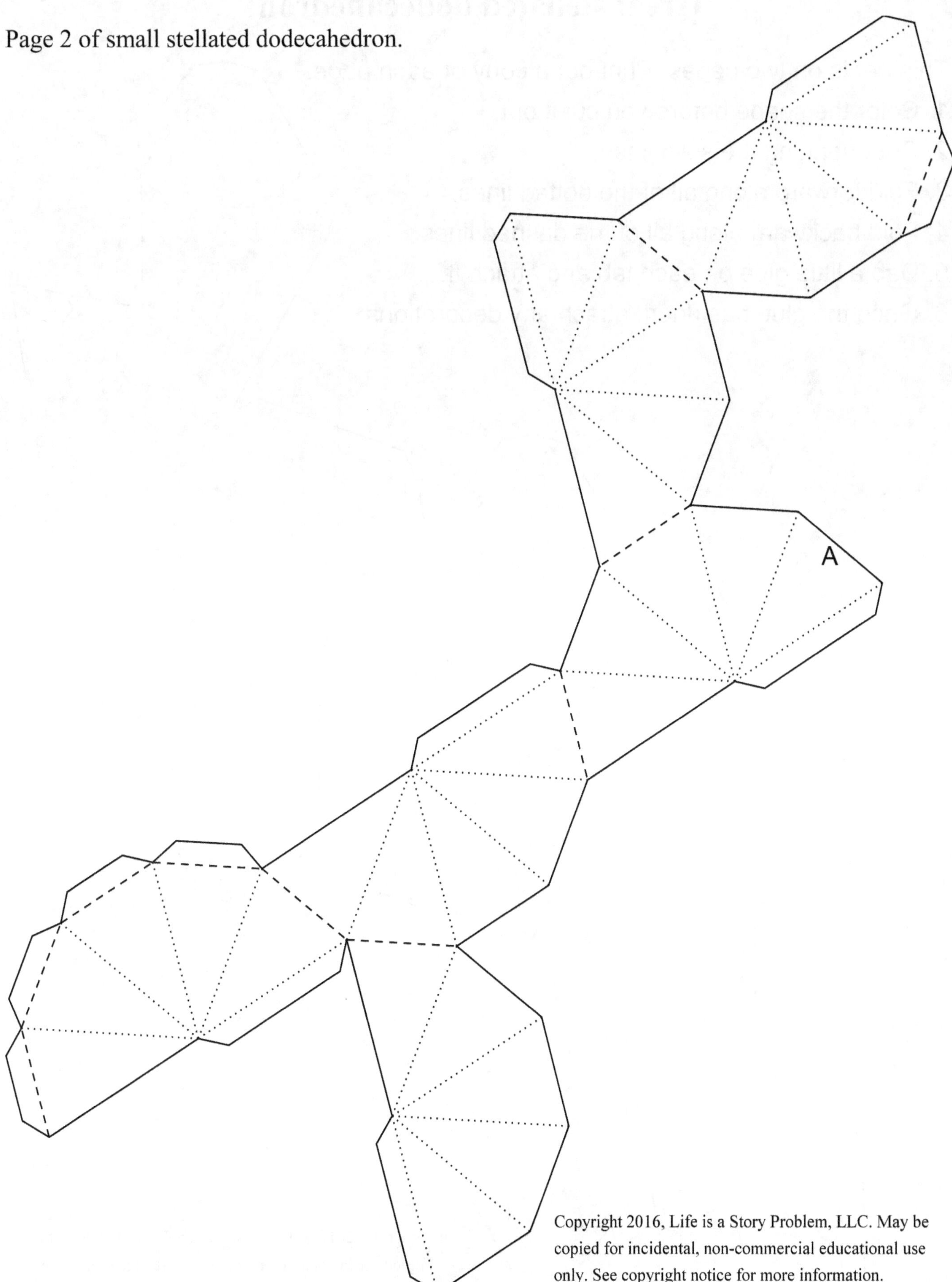

Great stellated dodecahedron

This net is on two pages. Print out a copy of each page.

1. Color the shape before you cut it out.
2. Cut out along the solid lines.
3. Fold forward along all of the dotted lines.
4. Fold backward along all of the dashed lines.
5. Dab a little glue on each tab and attach it.
6. Once the glue has dried, attach any decorations.

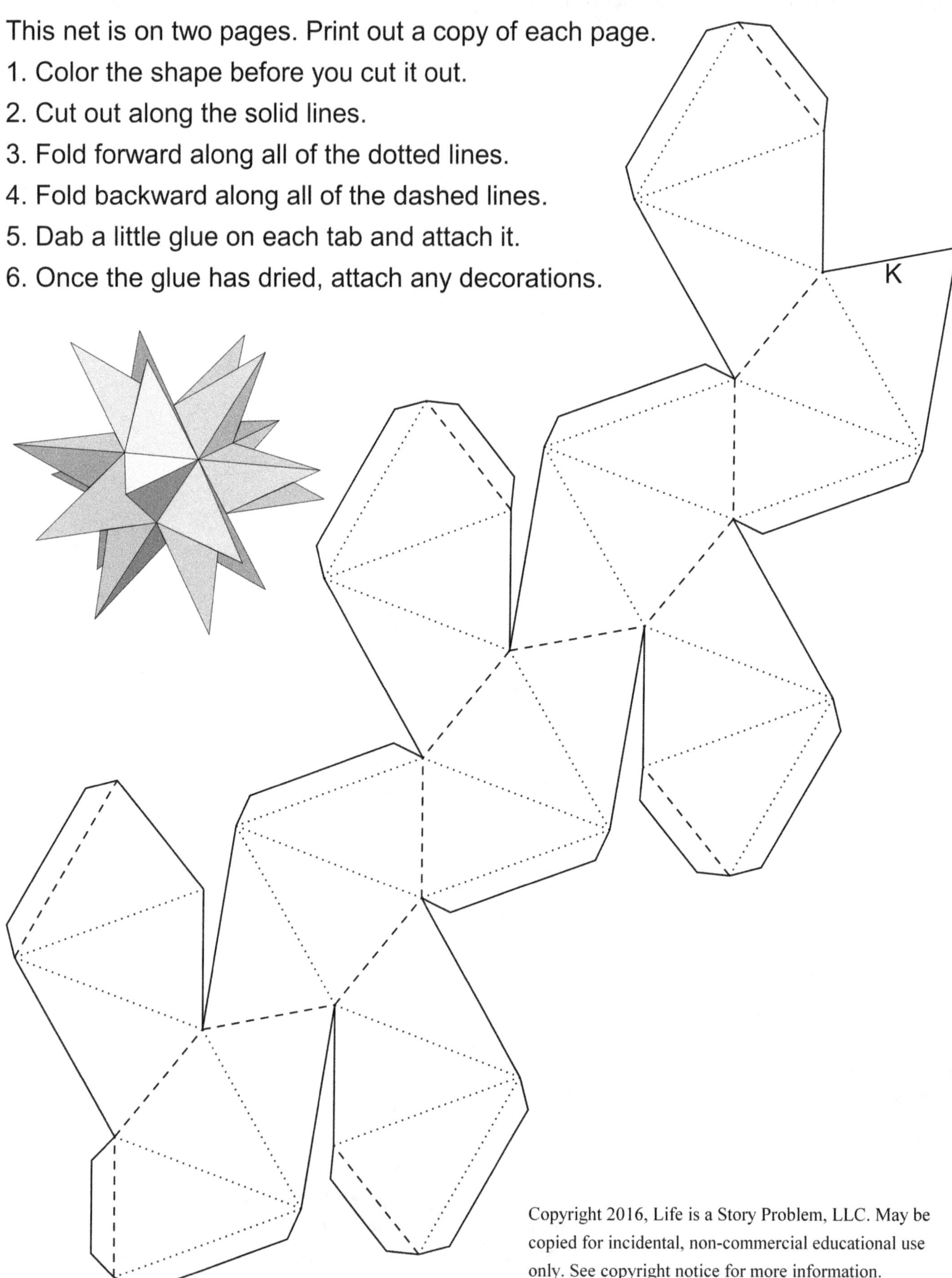

Geometric Nets Mega Project Book, Tabbed By David E. McAdams

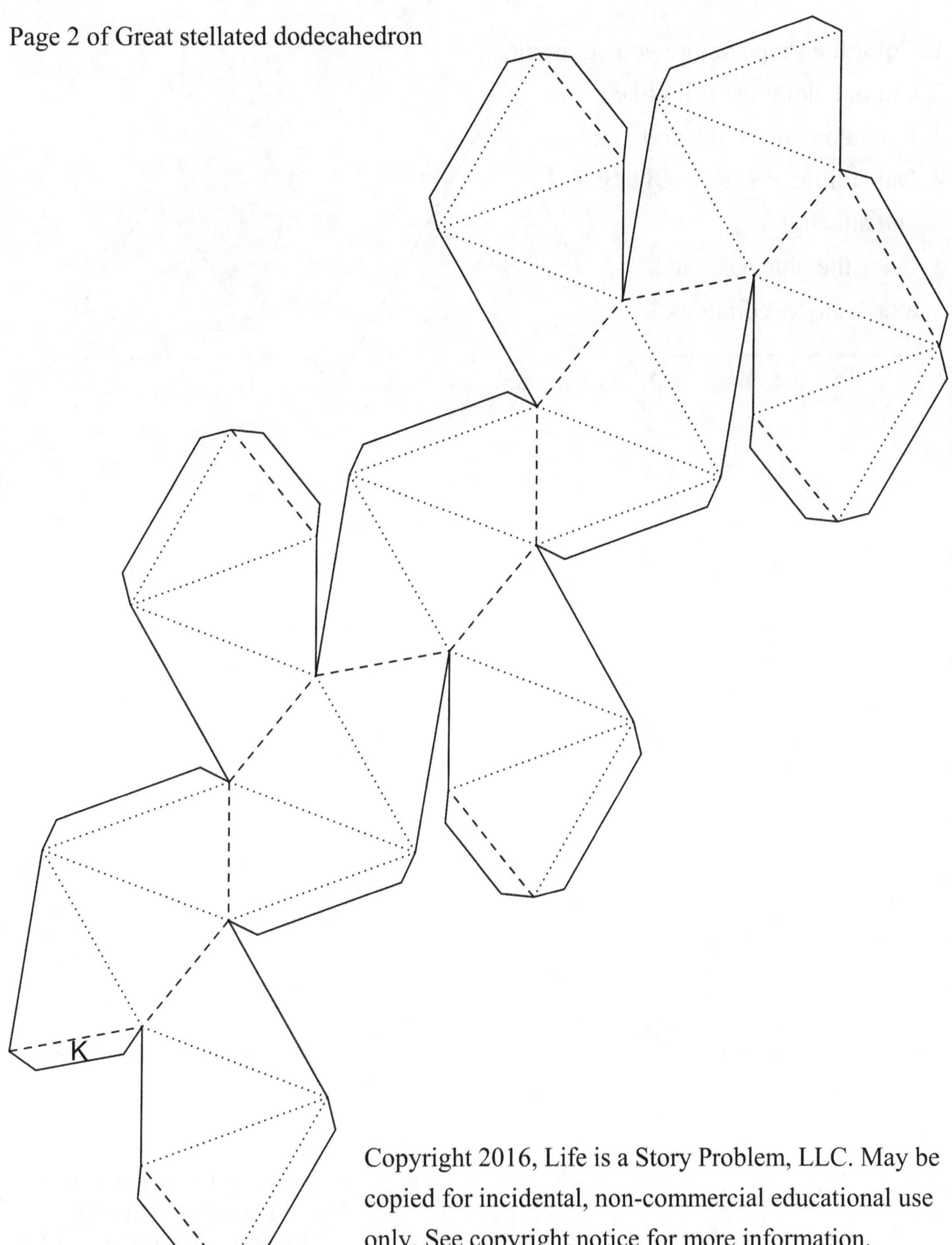
K

Triangular prism

1. Color the shape before you cut it out.

2. Cut out along the solid lines.

3. Fold along all of the dotted lines.

4. Dab a little glue on each tab
   and attach it.

5. Once the glue has dried,
   attach any decorations.

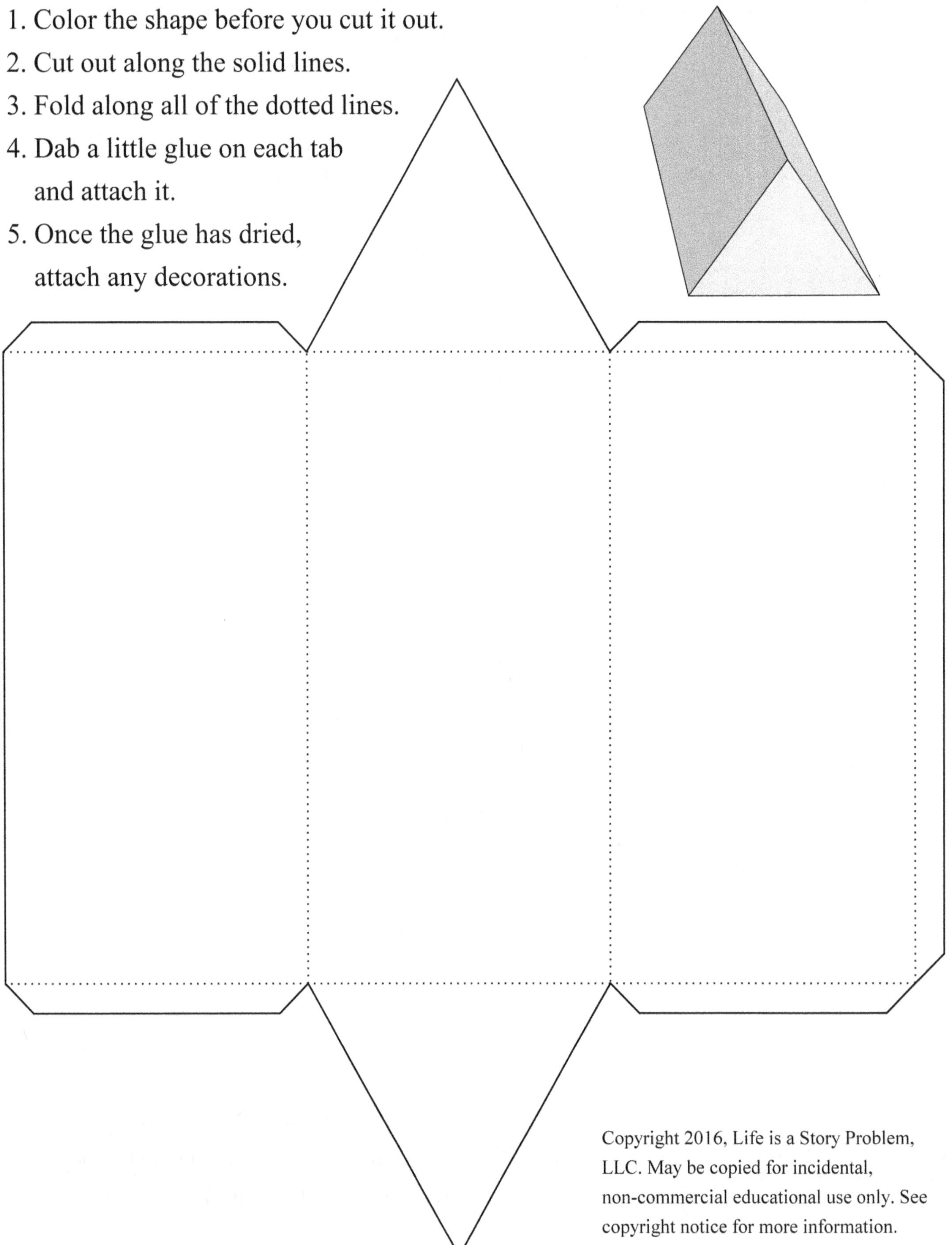

 Geometric Nets Mega Project Book, Tabbed By David E. McAdams

Elongated triangular antiprism

1. Color the shape before you cut it out.
2. Cut out along the solid lines.
3. Fold forward along all of the dotted lines.
4. Fold backward along all of the dashed lines.
5. Dab a little glue on each tab and attach it.
6. Once the glue has dried, attach any decorations.

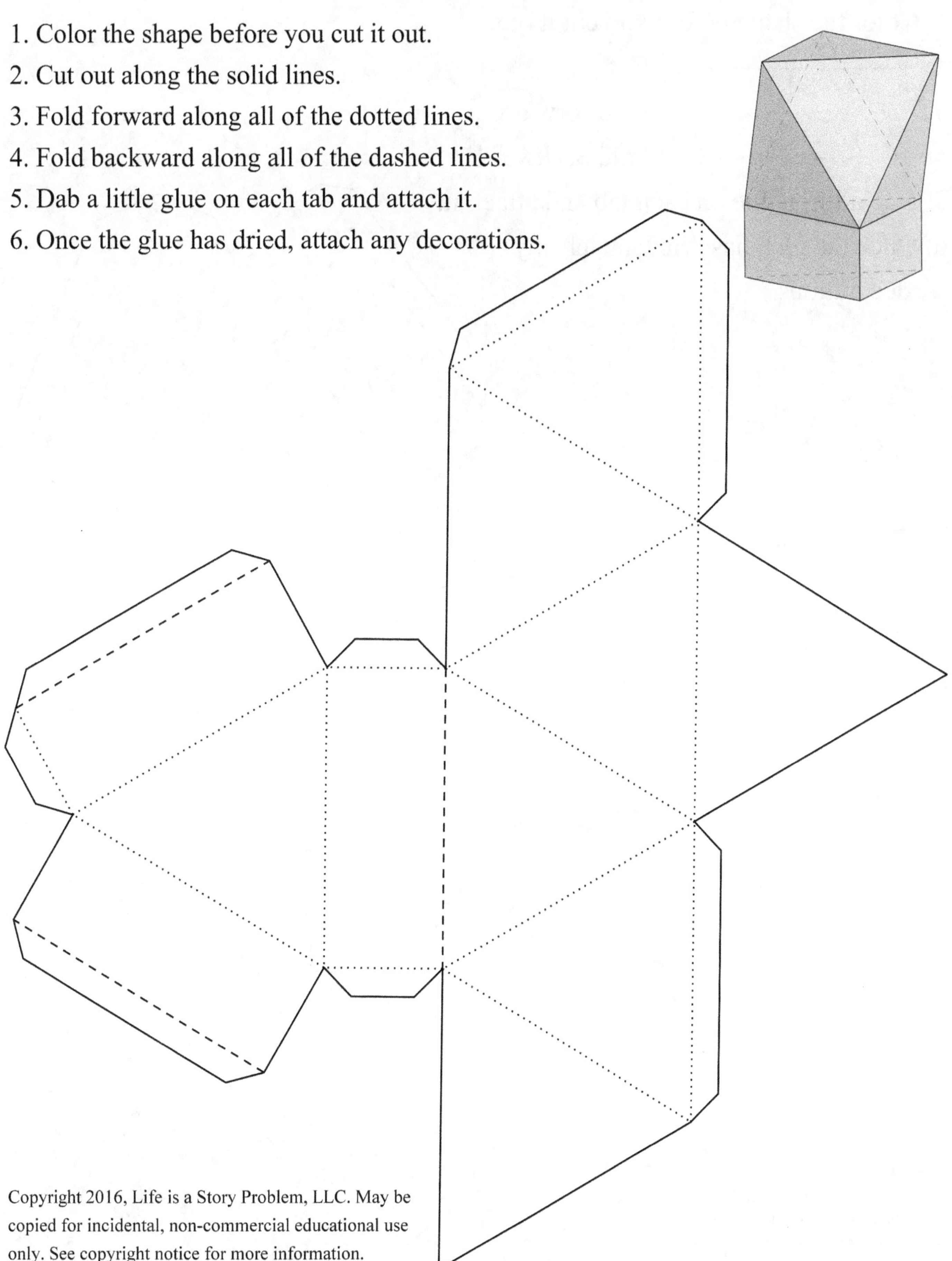

Bielongated triangular antiprism

1. Color the shape before you cut it out.
2. ——————————— Cut.
3. .... Fold forward.
4. - - - - - - - - - - - Fold backward.
5. Dab a little glue on each tab and attach it.
6. Once the glue has dried, attach any decorations.

Geometric Nets Mega Project Book, Tabbed By David E. McAdams

Twisted triangular prism

1. Color the shape before you cut it out.
2. Cut out along the solid lines.
3. Fold forward along all of the dotted lines.
4. Fold backward along all of the dashed lines.
5. Dab a little glue on each tab and attach it.
6. Once the glue has dried, attach decorations.

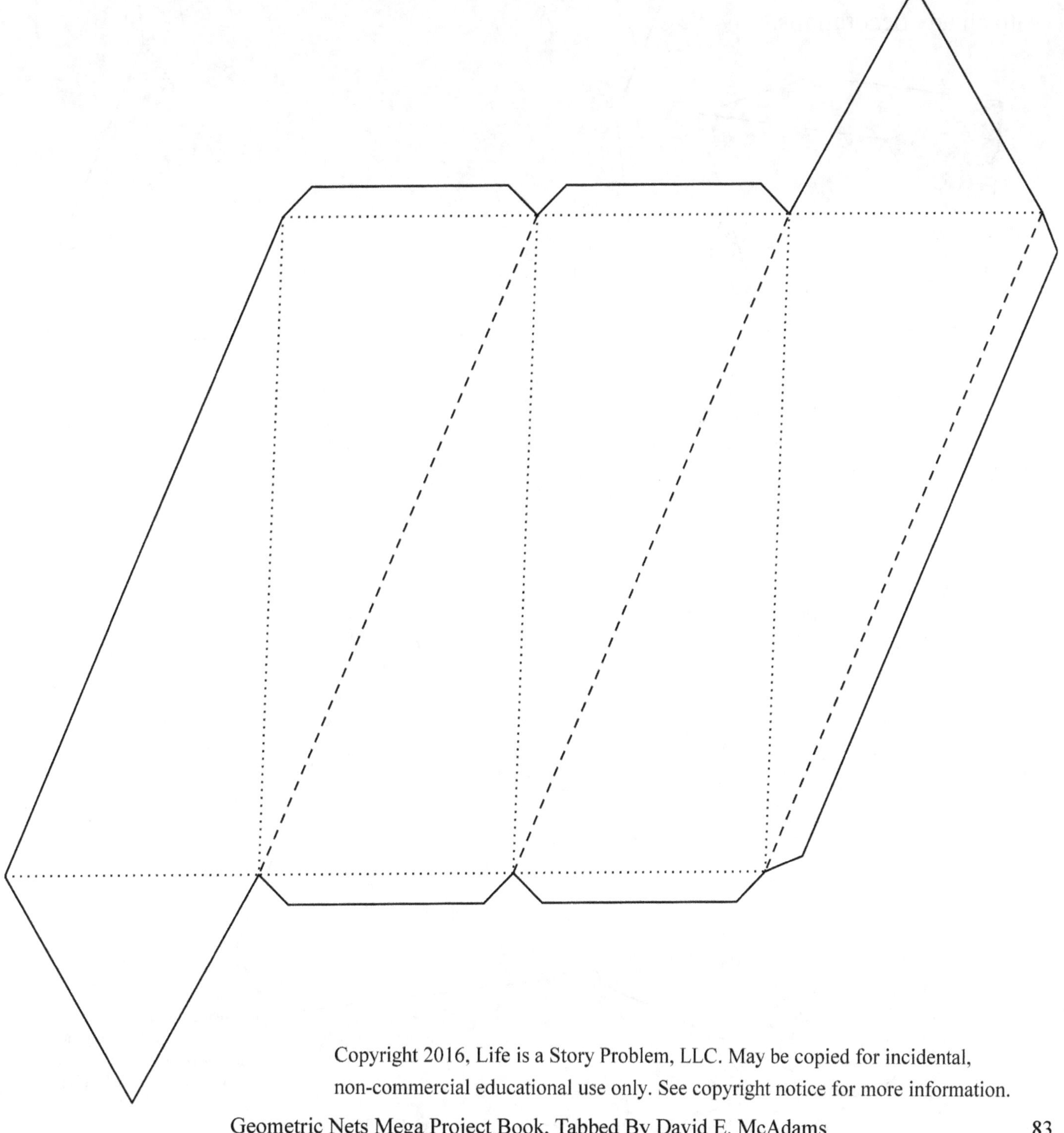

Square antiprism

1. Color the shape before you cut it out.
2. Cut out along the solid lines.
3. Fold along all of the dotted lines.
4. Dab a little glue on each tab and attach it.
5. Once the glue has dried, attach any decorations.

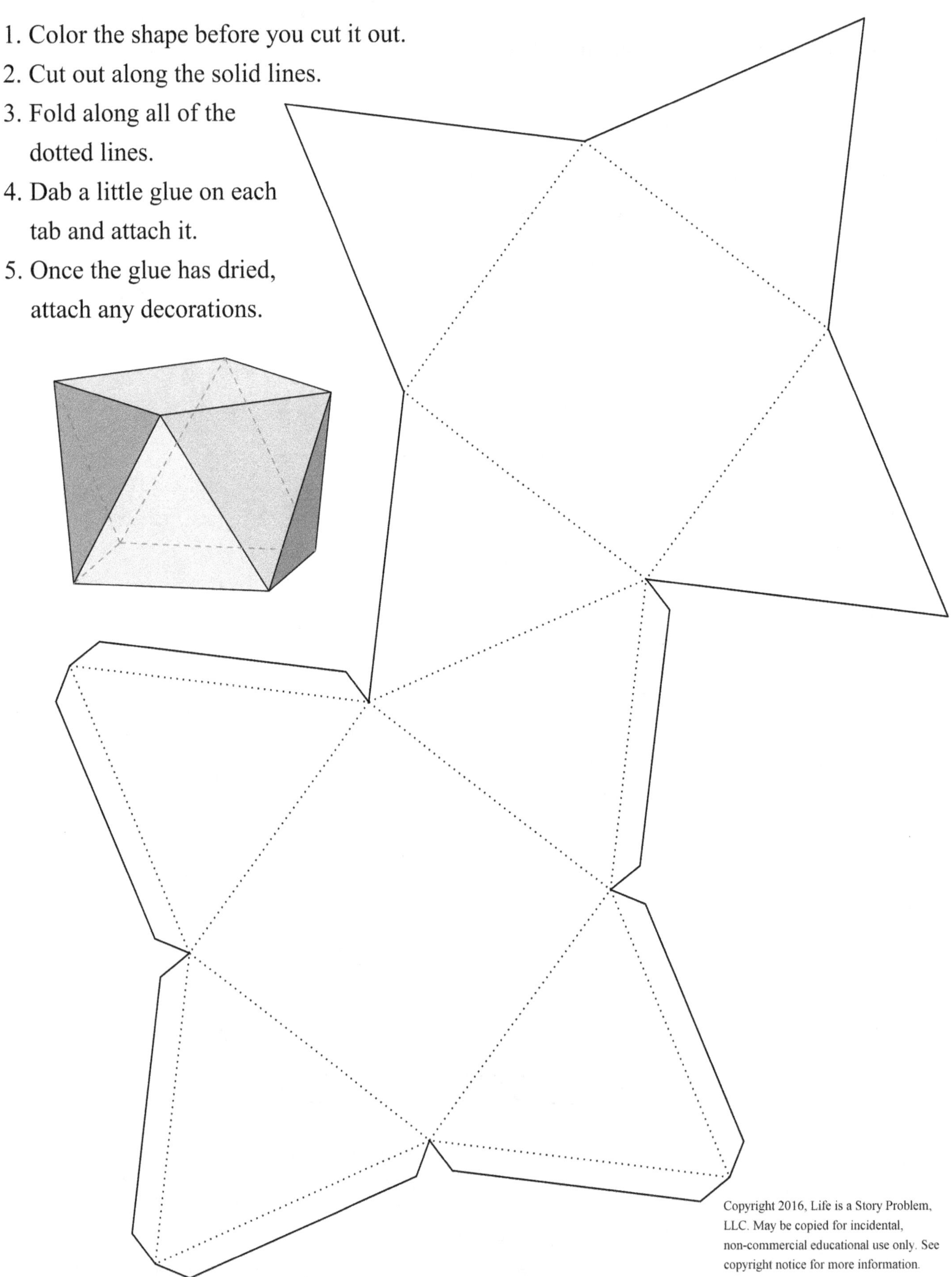

 Geometric Nets Mega Project Book, Tabbed By David E. McAdams

Square double antiprism

1. Color the shape before you cut it out.
2. Cut out along the solid lines.
3. Fold forward along all of the dotted lines.
4. Fold backward along all of the dashed lines.
5. Dab a little glue on each tab and attach it.
6. Once the glue has dried, attach any
   decorations.

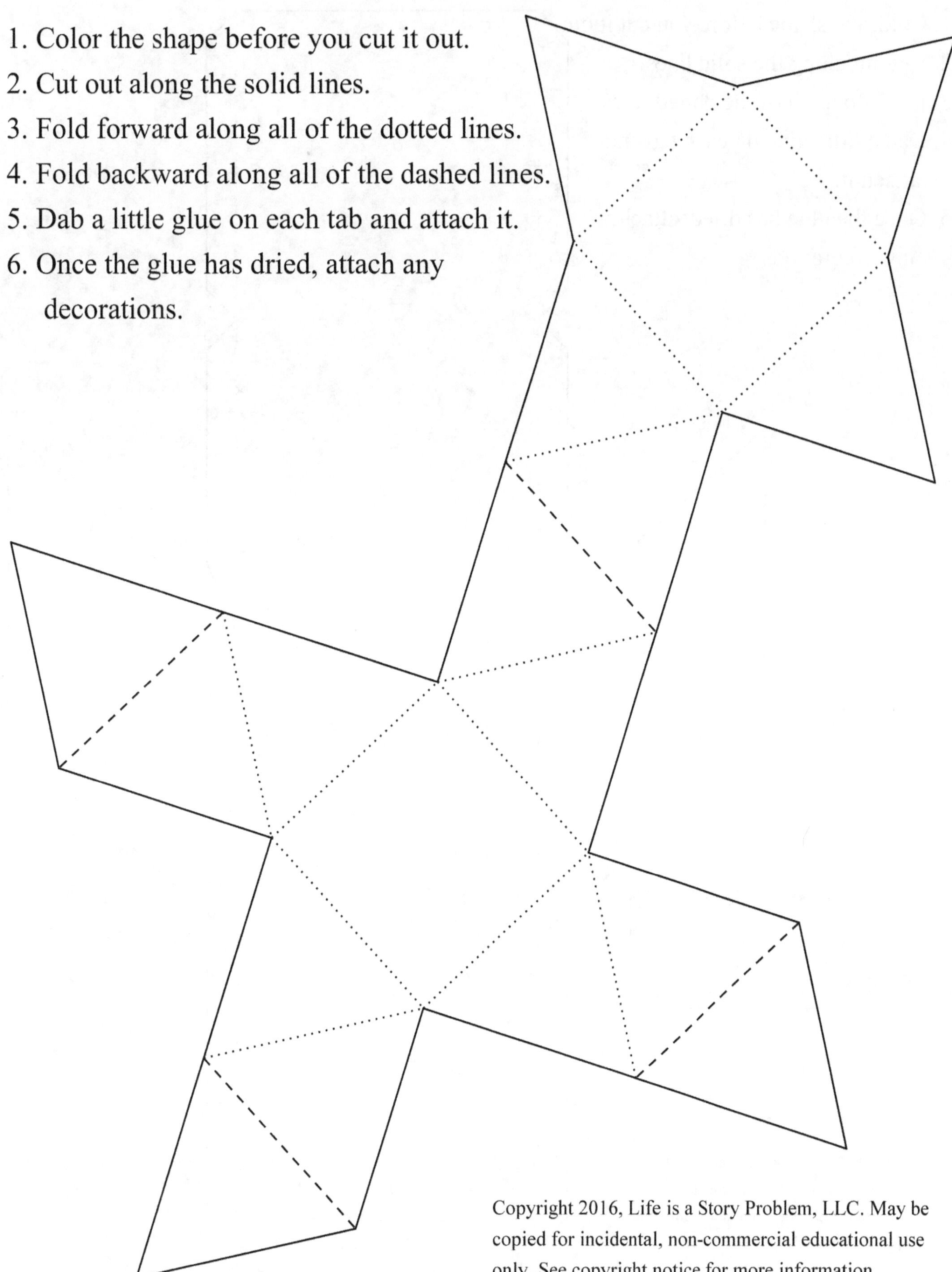

Rhombic prism

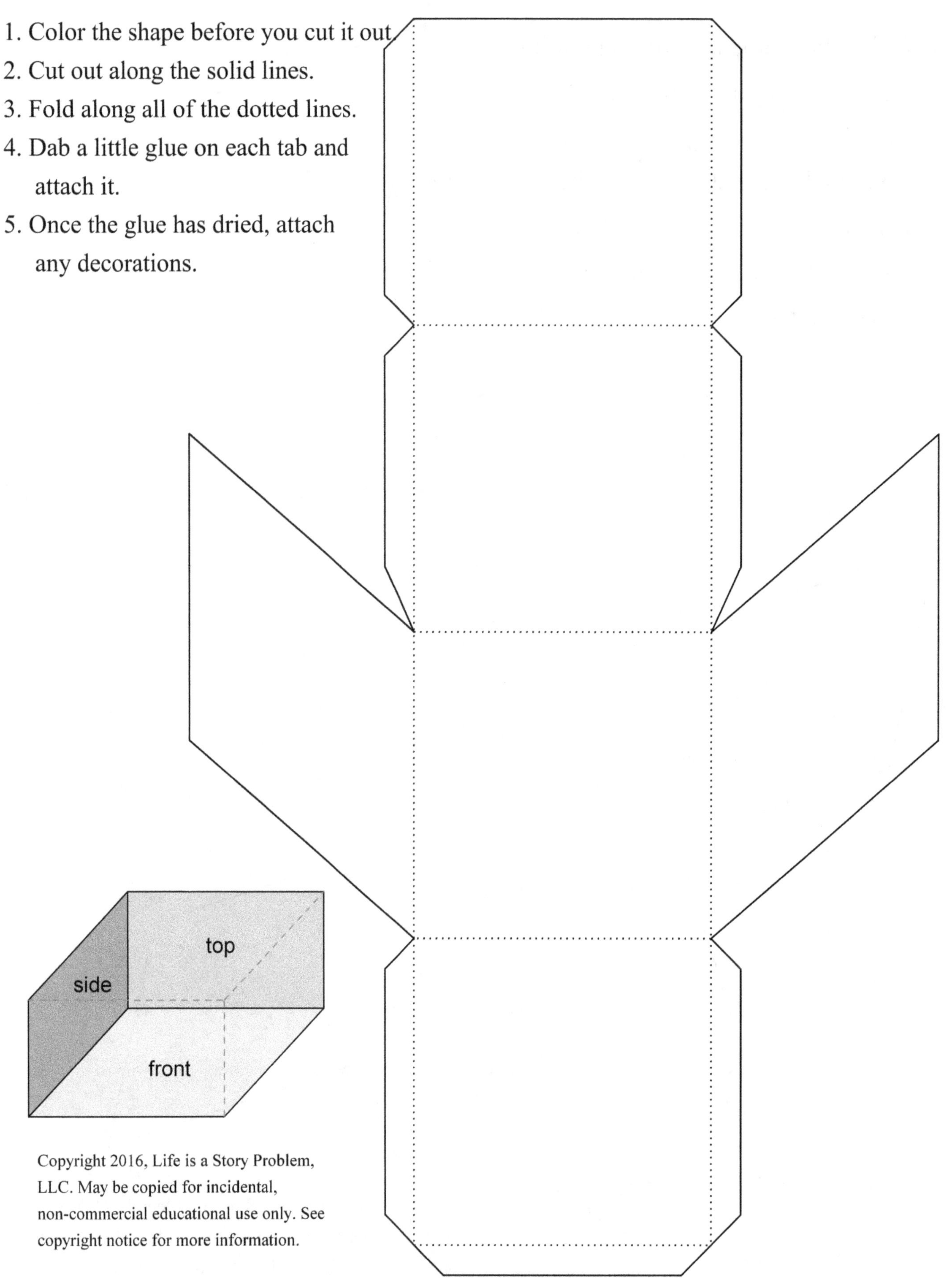

1. Color the shape before you cut it out.
2. Cut out along the solid lines.
3. Fold along all of the dotted lines.
4. Dab a little glue on each tab and attach it.
5. Once the glue has dried, attach any decorations.

Geometric Nets Mega Project Book, Tabbed By David E. McAdams

Pentagonal prism

1. Color the shape before you cut it out.
2. Cut out along the solid lines.
3. Fold along all of the dotted lines.
4. Dab a little glue on each tab and attach it.
5. Once the glue has dried, attach
   any decorations.

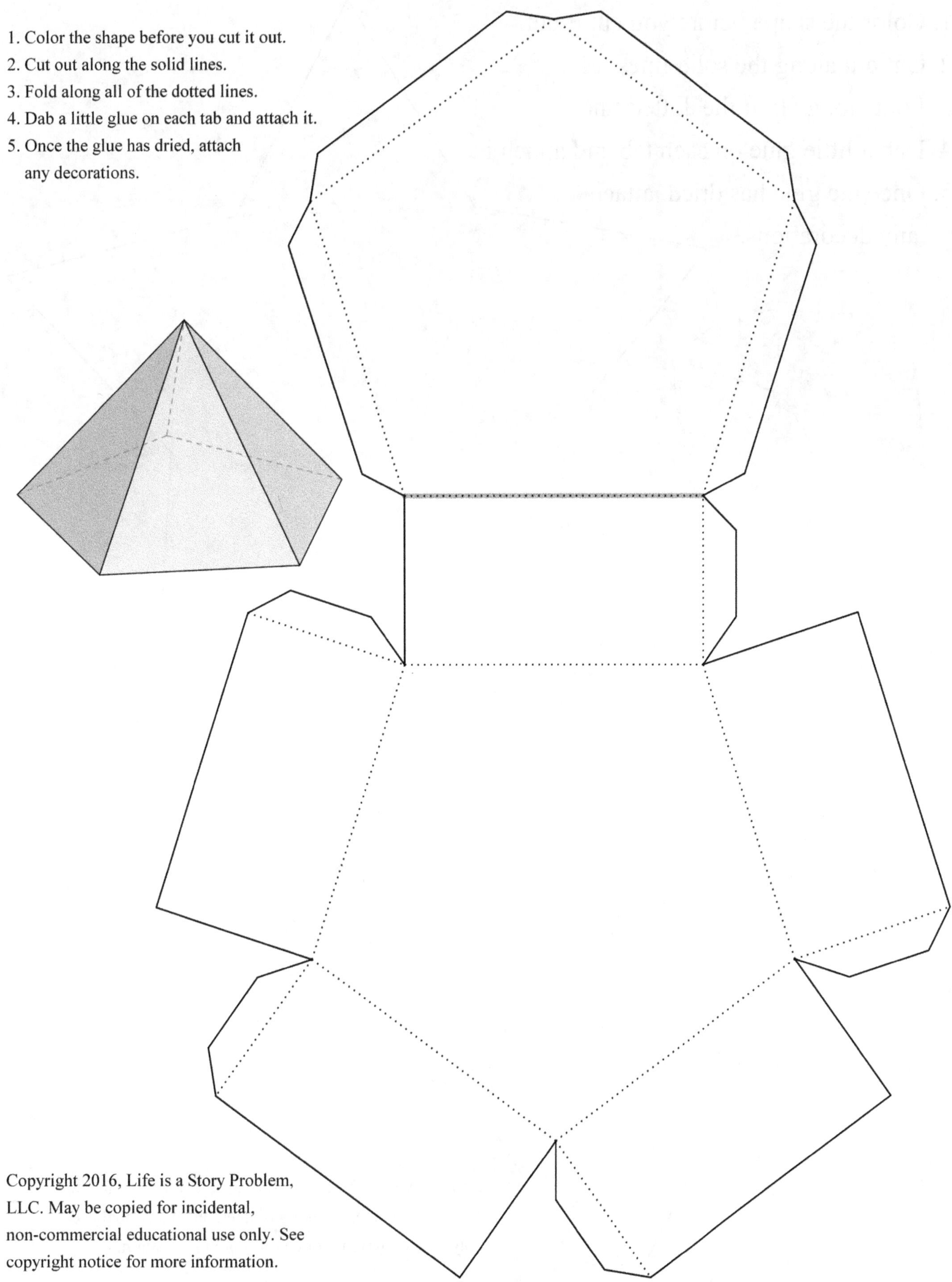

Pentagonal antiprism

1. Color the shape before you cut it out.
2. Cut out along the solid lines.
3. Fold along all of the dotted lines.
4. Dab a little glue on each tab and attach it.
5. Once the glue has dried, attach
   any decorations.

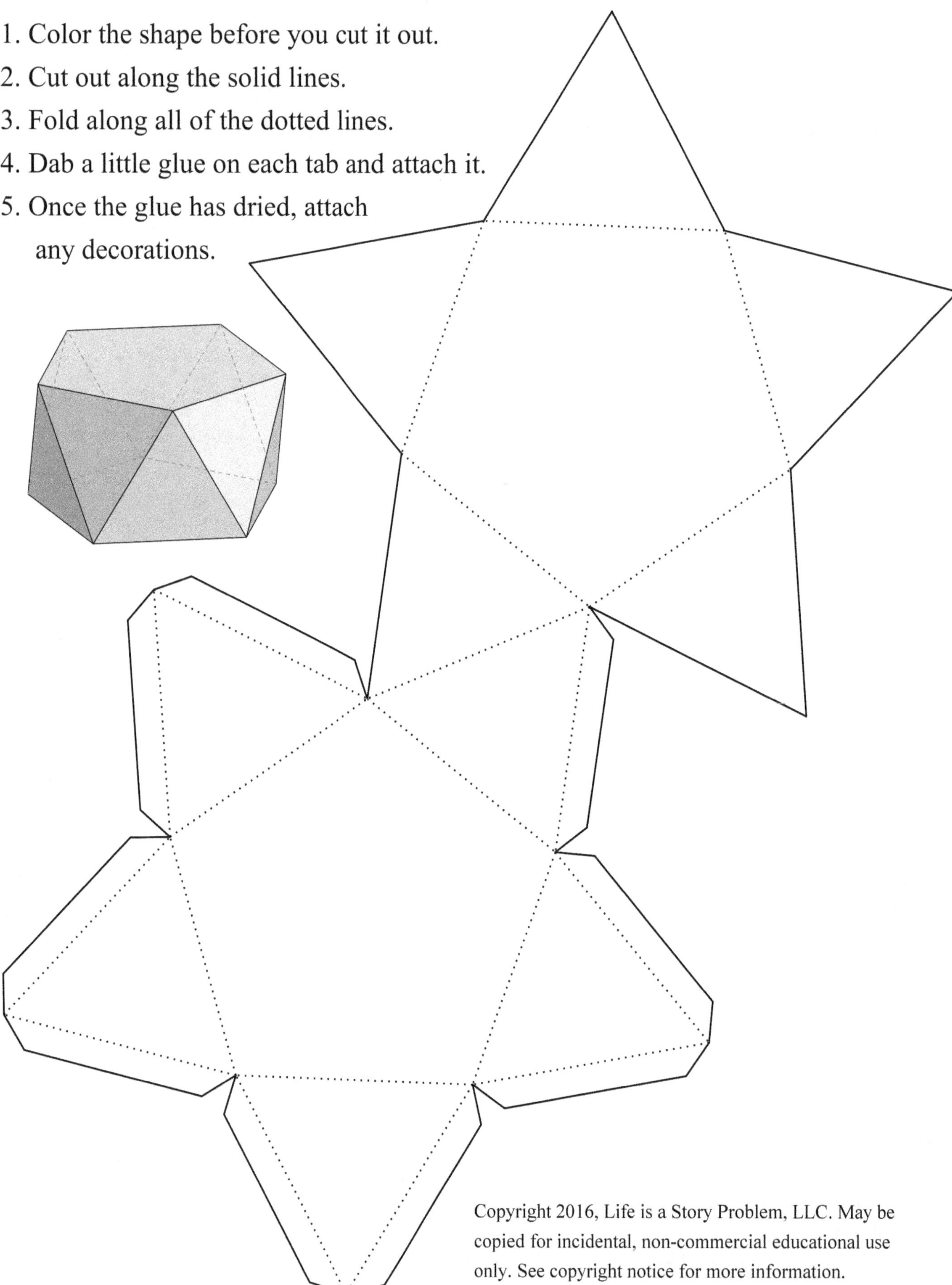

 Geometric Nets Mega Project Book, Tabbed By David E. McAdams

Pentagonal star antiprism

1. Color the shape before you cut it out.
2. Cut out along the solid lines
3. Fold forward along all of the dotted lines.
4. Fold backward along all of the dashed lines.
5. Dab a little glue on each tab and attach it.
6. Once the glue has dried, attach any decorations

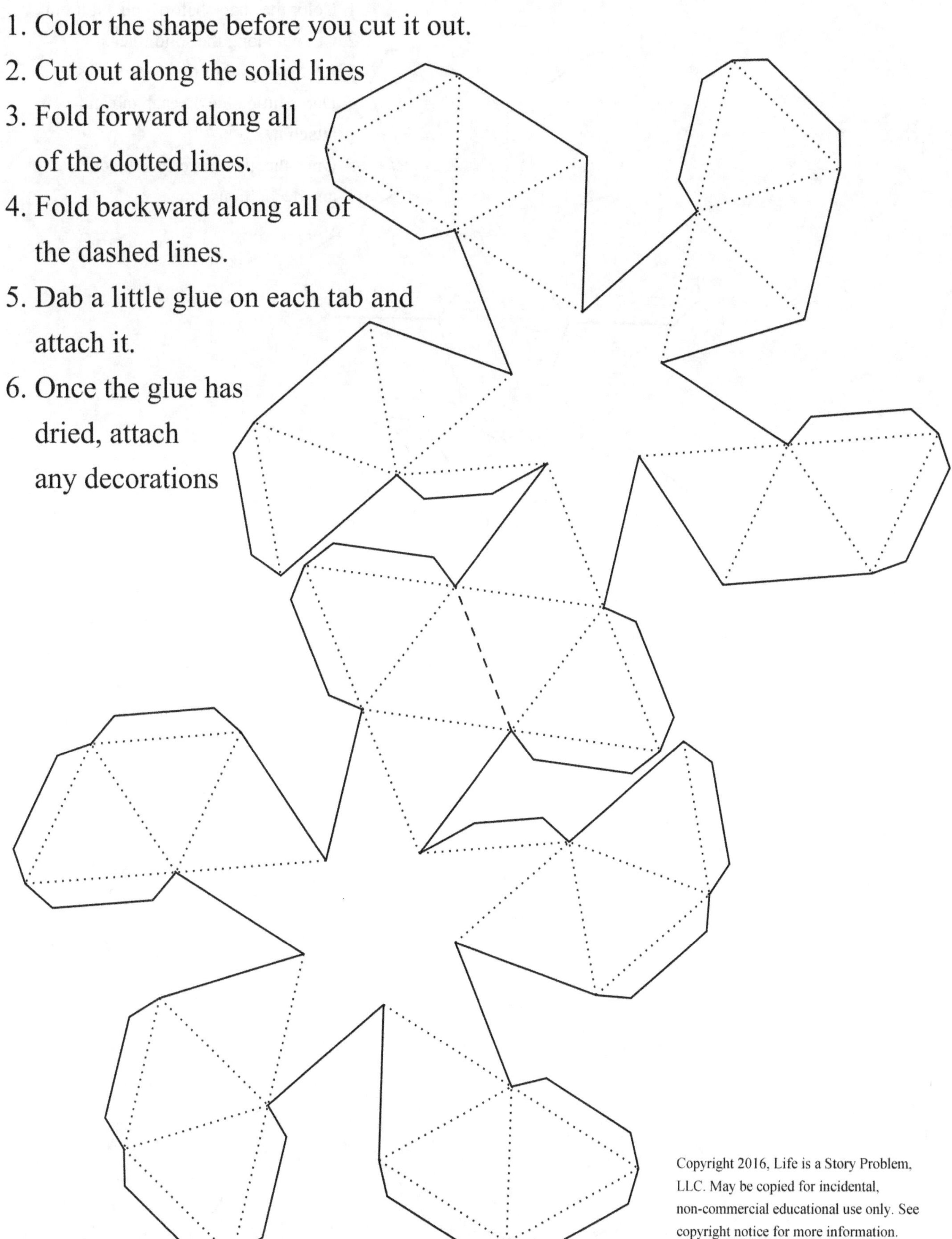

Pentagon to pentagonal star pseudo-prism

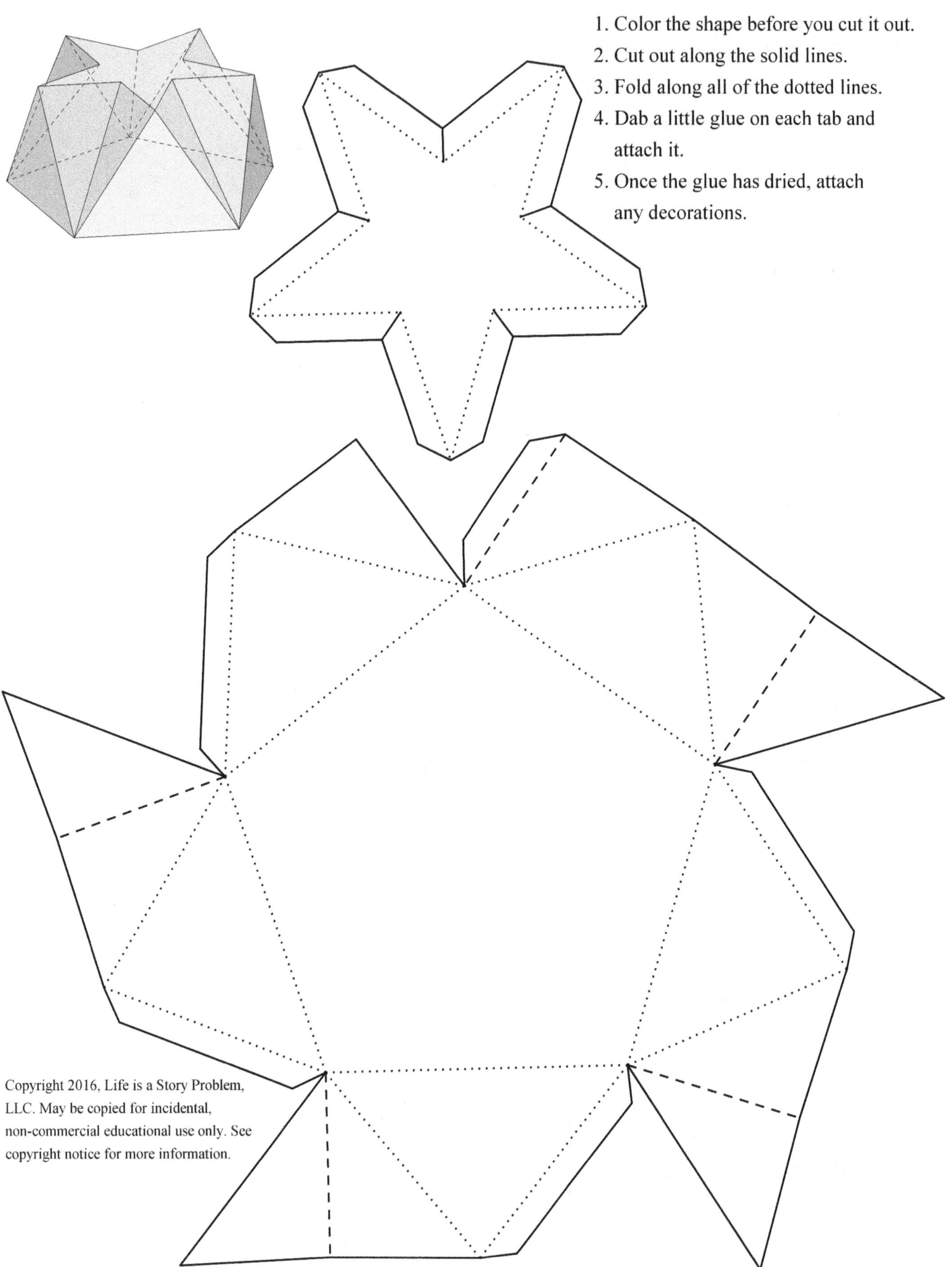

1. Color the shape before you cut it out.
2. Cut out along the solid lines.
3. Fold along all of the dotted lines.
4. Dab a little glue on each tab and attach it.
5. Once the glue has dried, attach any decorations.

Geometric Nets Mega Project Book, Tabbed By David E. McAdams

Hexagonal prism

1. Color the shape before you cut it out.
2. Cut out along the solid lines.
3. Fold along all of the dotted lines.
4. Dab a little glue on each tab and attach it.
5. Once the glue has dried, attach any decorations.

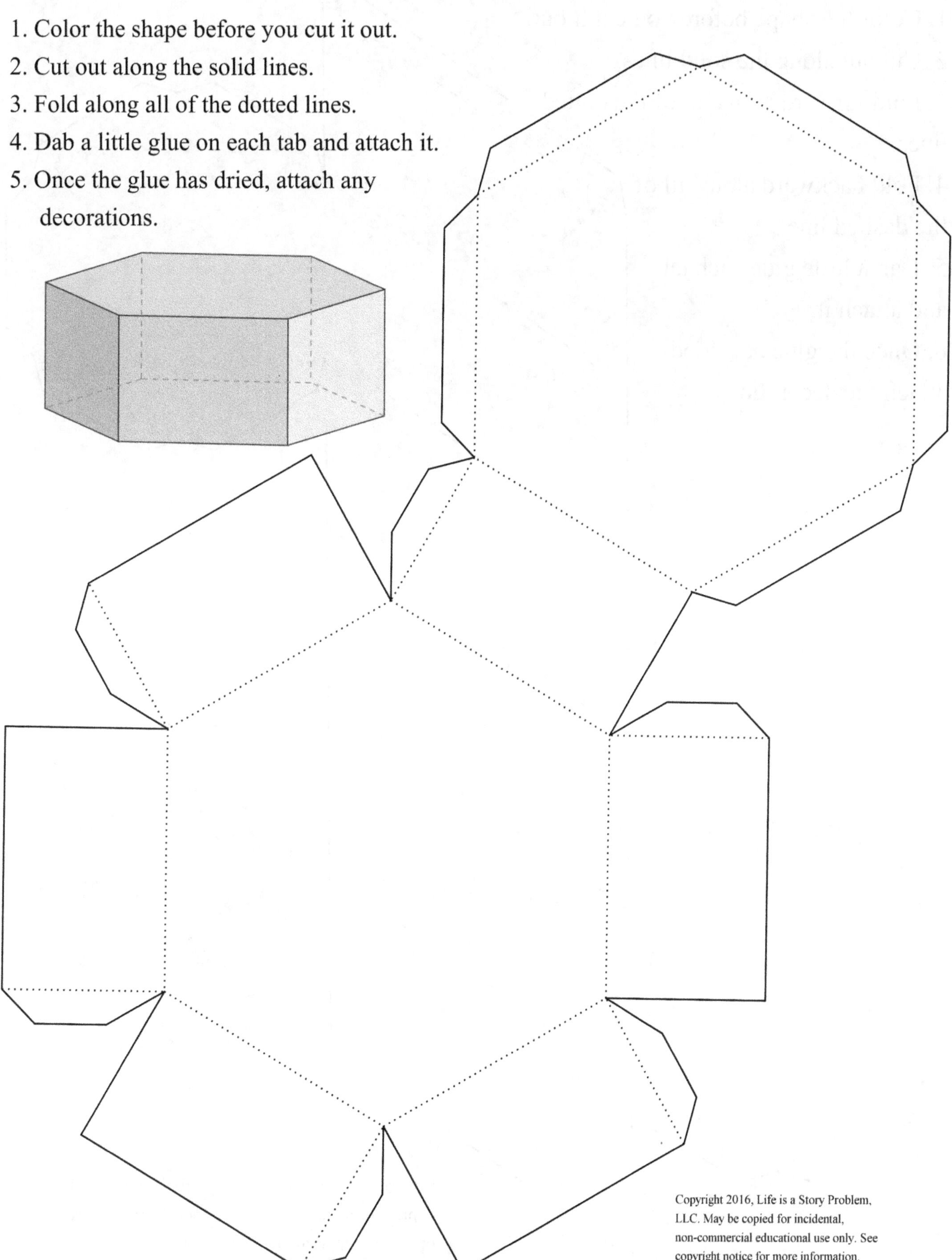

Twisted hexagonal prism

1. Color the shape before you cut it out.

2. Cut out along the solid lines.

3. Fold forward along all of the dotted lines.

4. Fold backward along all of the dashed lines.

5. Dab a little glue each tab and attach it.

6. Once the glue has dried, attach the decorations.

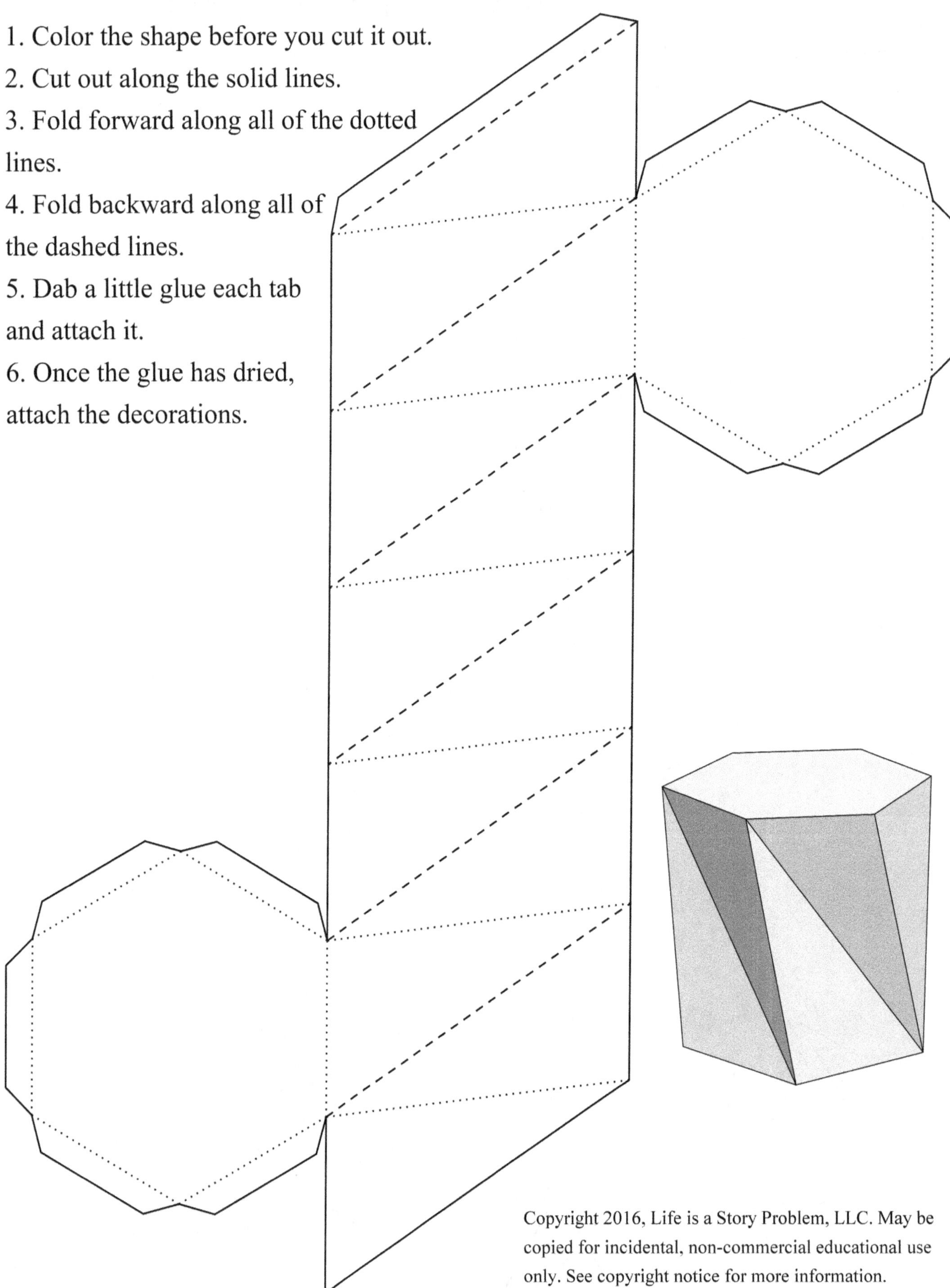

 Geometric Nets Mega Project Book, Tabbed By David E. McAdams

XTruncated hexagonal antiprism

1. Color the shape before you cut it out.

2. Cut out along the solid lines.

3. Fold along all of the dotted lines.

4. Dab a little glue on each tab and attach it.

5. Once the glue has dried, attach any decorations.

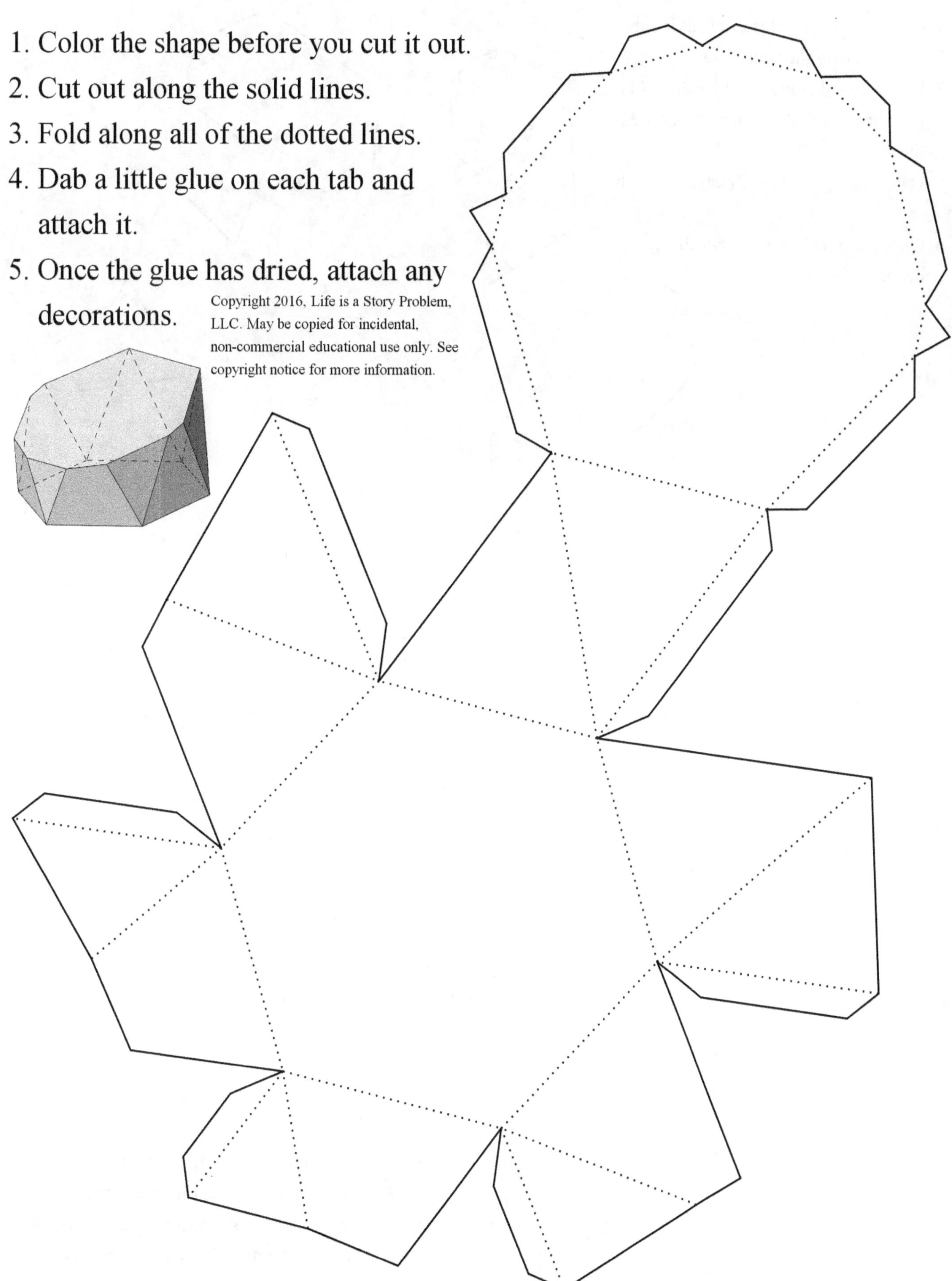

Heptagonal star antiprism

1. Color the shape before you cut it out.
2. Cut out along the solid lines.
3. Fold forward along all of the dotted lines.
4. Fold backward along all of the dashed lines.
5. Dab a little glue on each tab and attach it.
6. Once the glue has dried, attach any decorations.

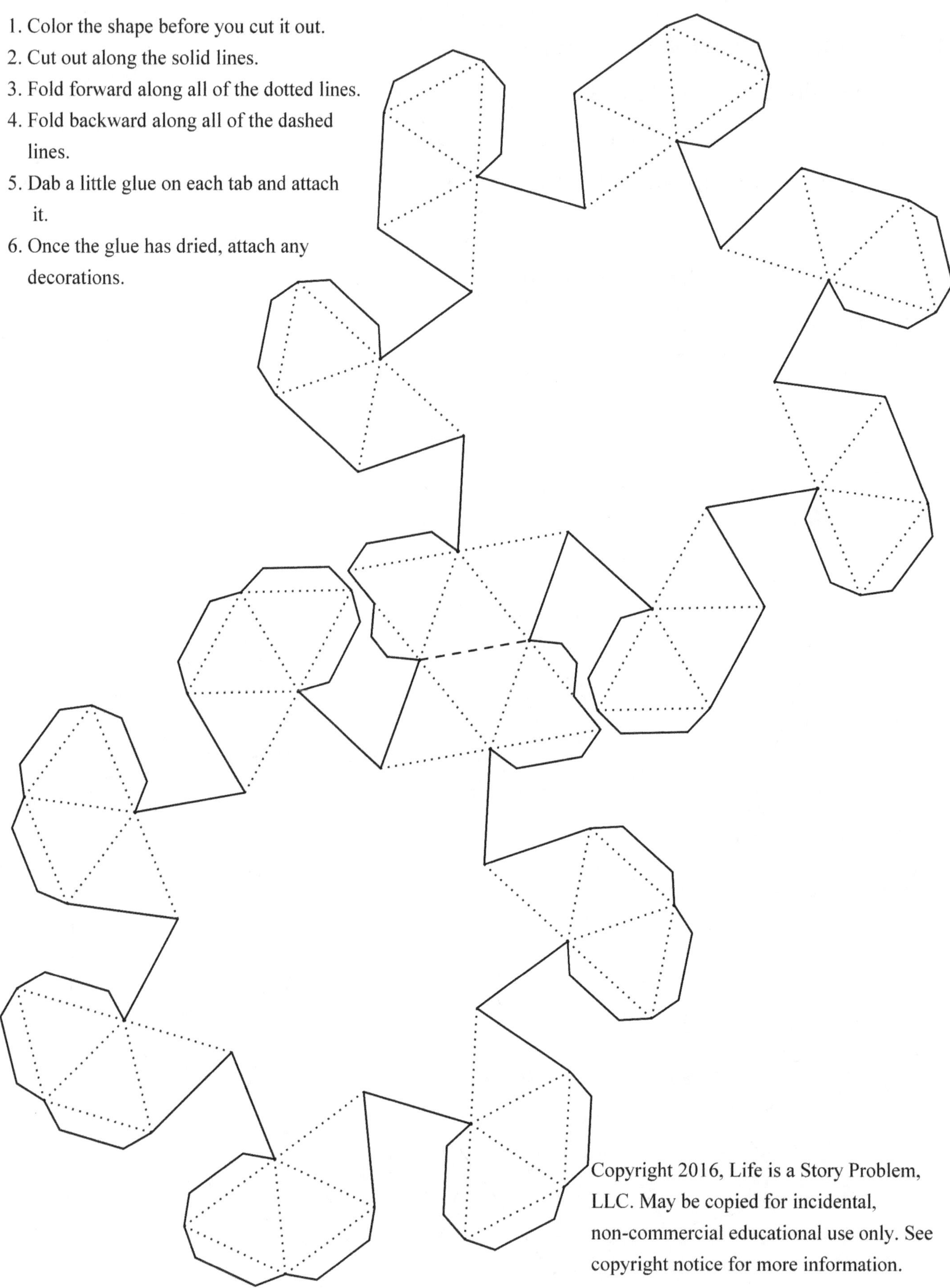

 Geometric Nets Mega Project Book, Tabbed By David E. McAdams

Oblique heptagonal antiprism

1. Color the shape before you cut it out.
2. Cut out along the solid lines.
3. Fold along all of the dotted lines.
4. Dab a little glue on each tab and attach it.
5. Once the glue has dried, attach any decorations.

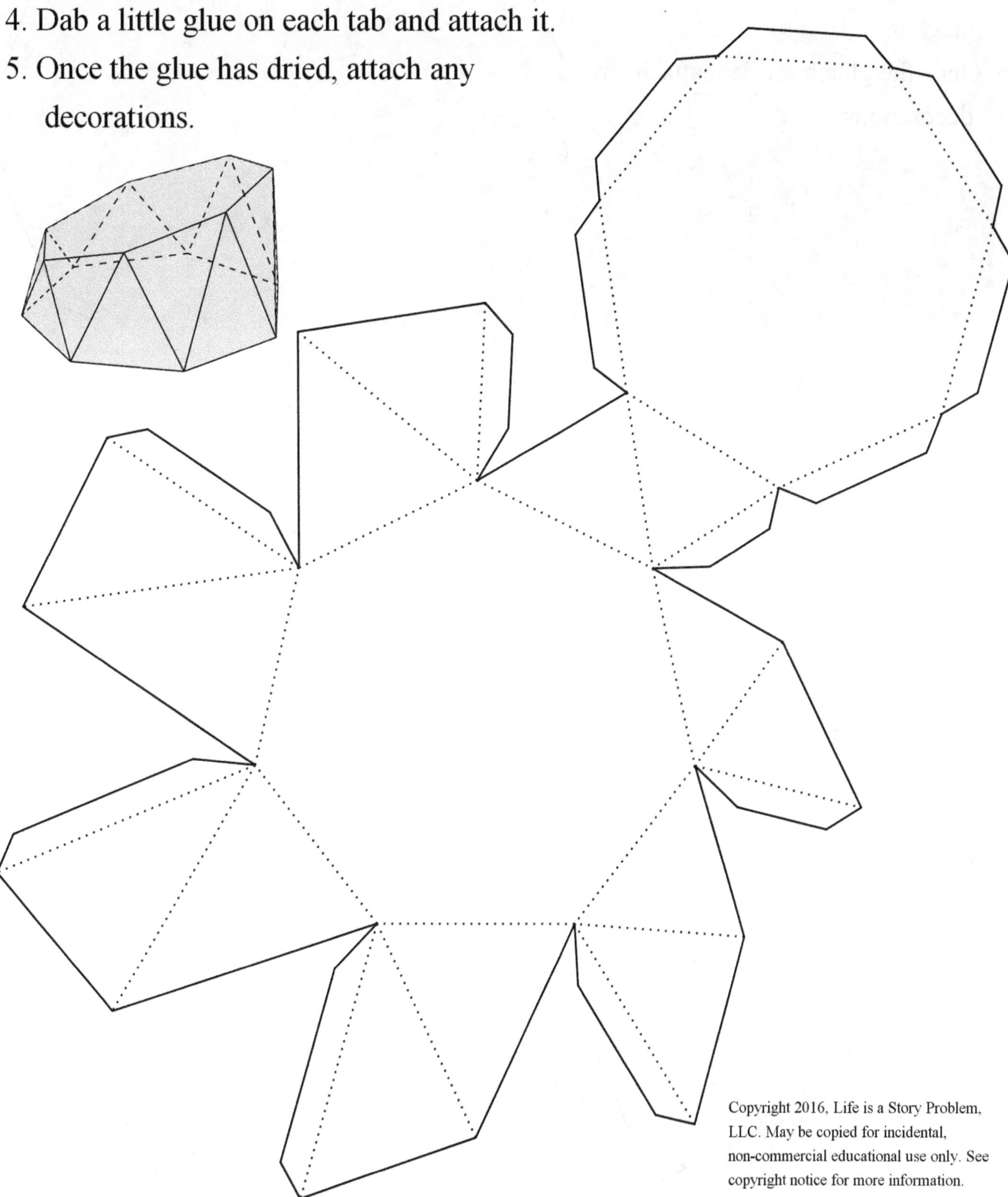

Octagonal prism

1. Color the shape before you cut it out.
2. Cut out along the solid lines.
3. Fold along all of the dotted lines.
4. Dab a little glue on each tab and attach it.
5. Once the glue has dried, attach any decorations.

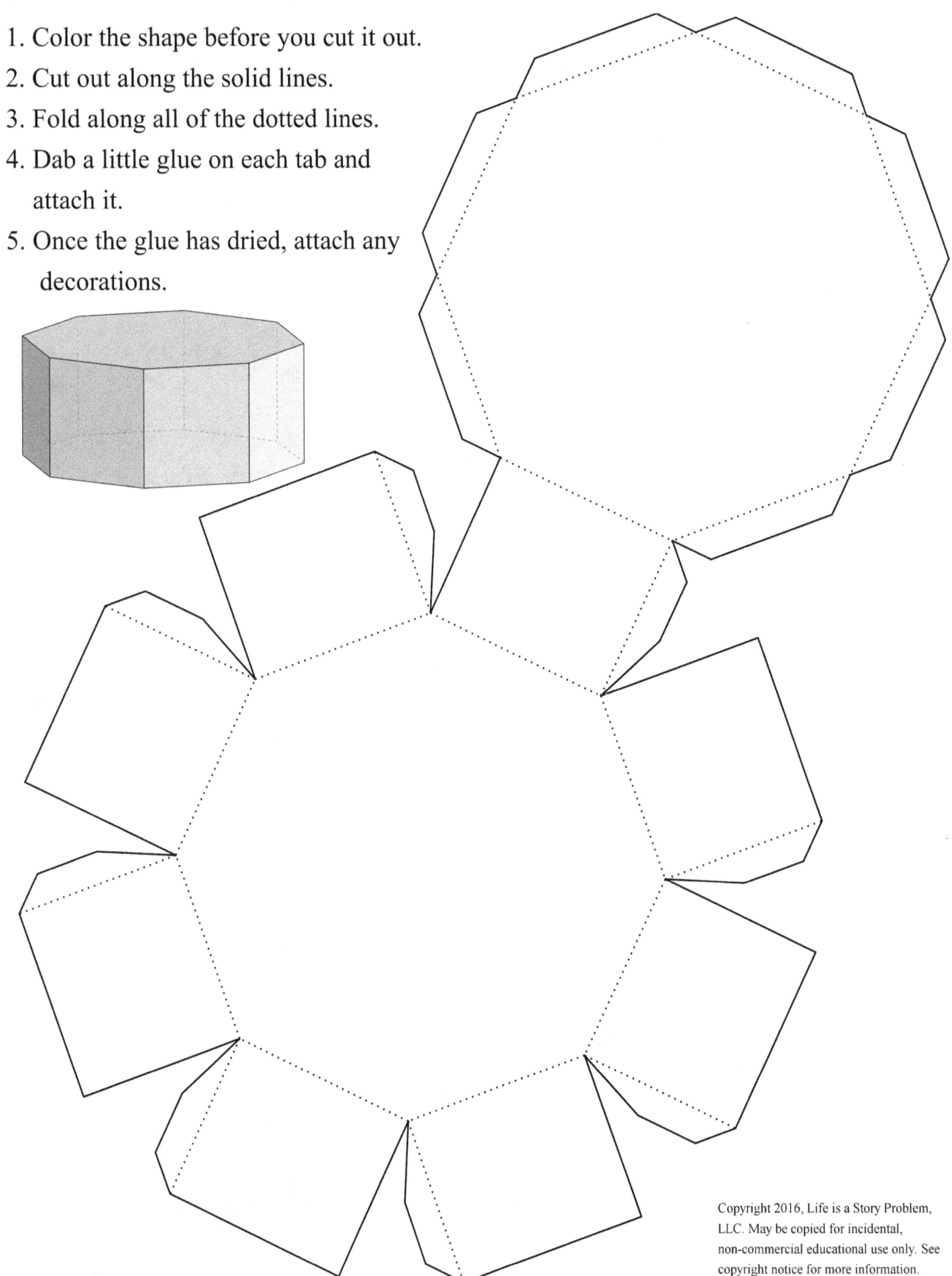

 Geometric Nets Mega Project Book, Tabbed By David E. McAdams

Octagonal antiprism

1. Color the shape before you cut it out.
2. Cut out along the solid lines.
3. Fold along all of the dotted lines.
4. Dab a little glue on each tab and attach it.
5. Once the glue has dried, attach any decorations.

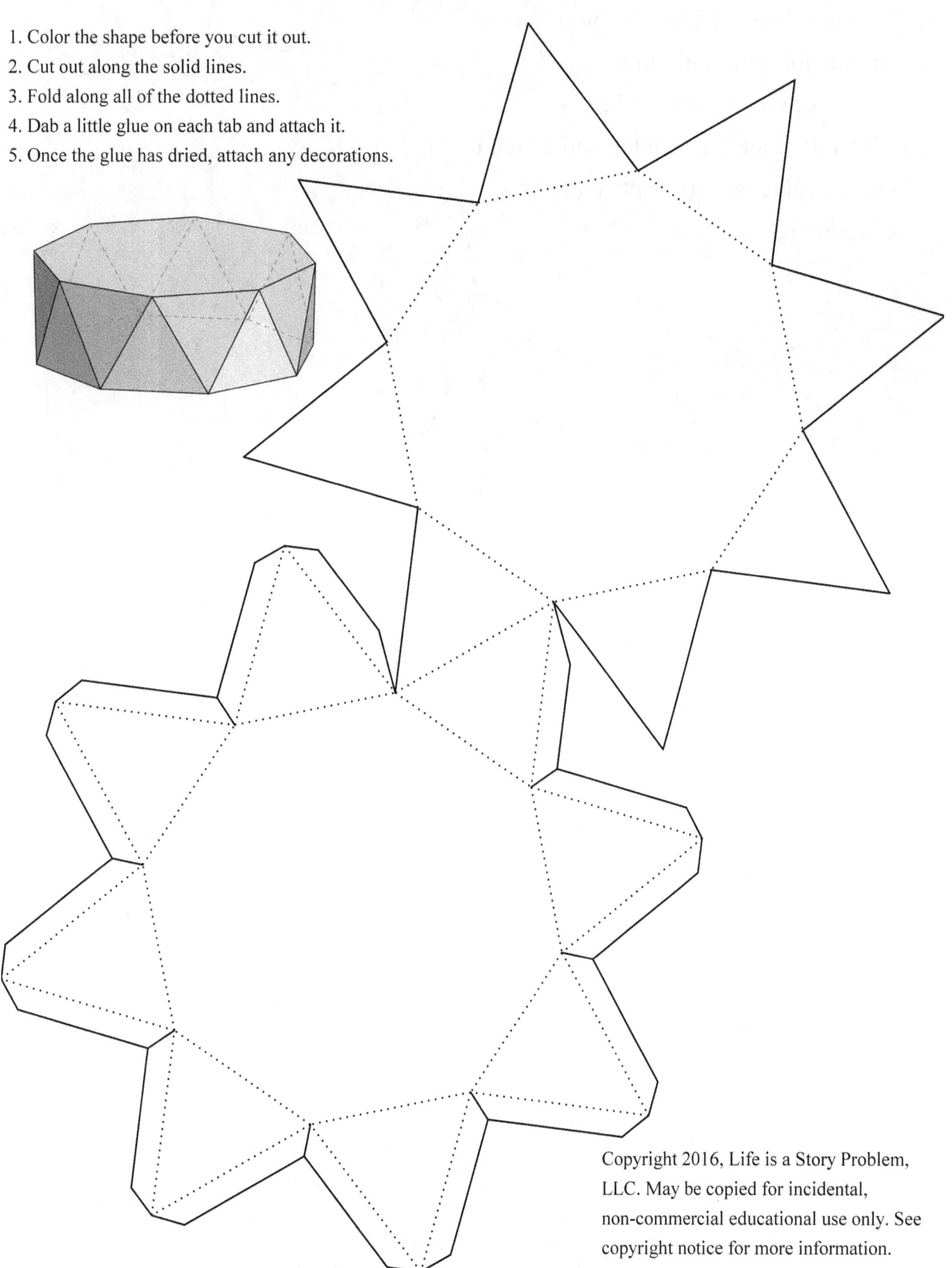

Decagonal prism

1. Color the shape before you cut it out.

2. Cut out along the solid lines.

3. Fold along all of the dotted lines.

4. Dab a little glue on each tab and attach it.

5. Once the glue has dried, attach any decorations.

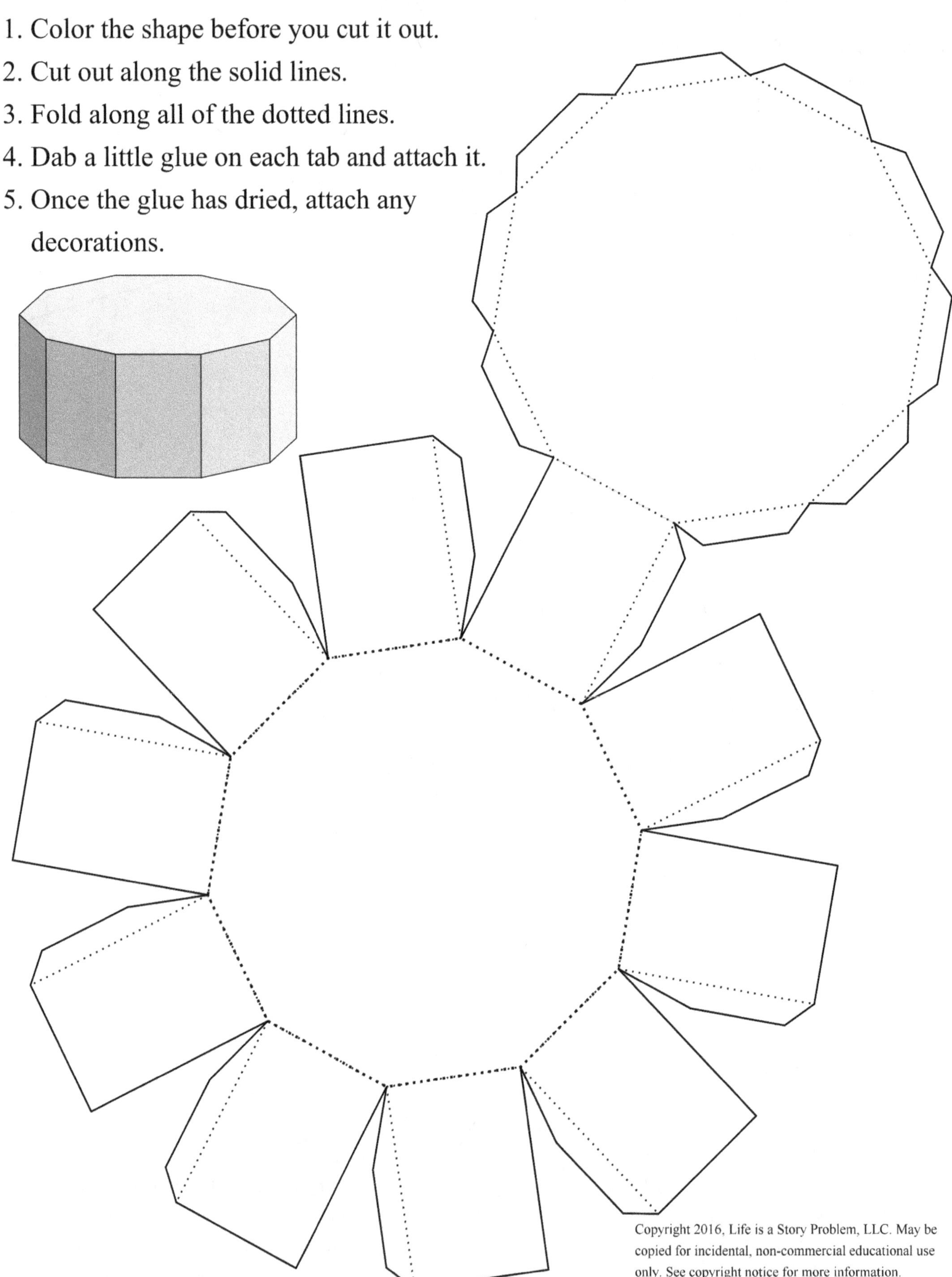

 Geometric Nets Mega Project Book, Tabbed By David E. McAdams

Decagonal antiprism

1. Color the shape before you cut it out.
2. Cut out along the solid lines.
3. Fold along all of the dotted lines.
4. Dab a little glue on each tab and
   attach it.
5. Once the glue has dried, attach any
   decorations.

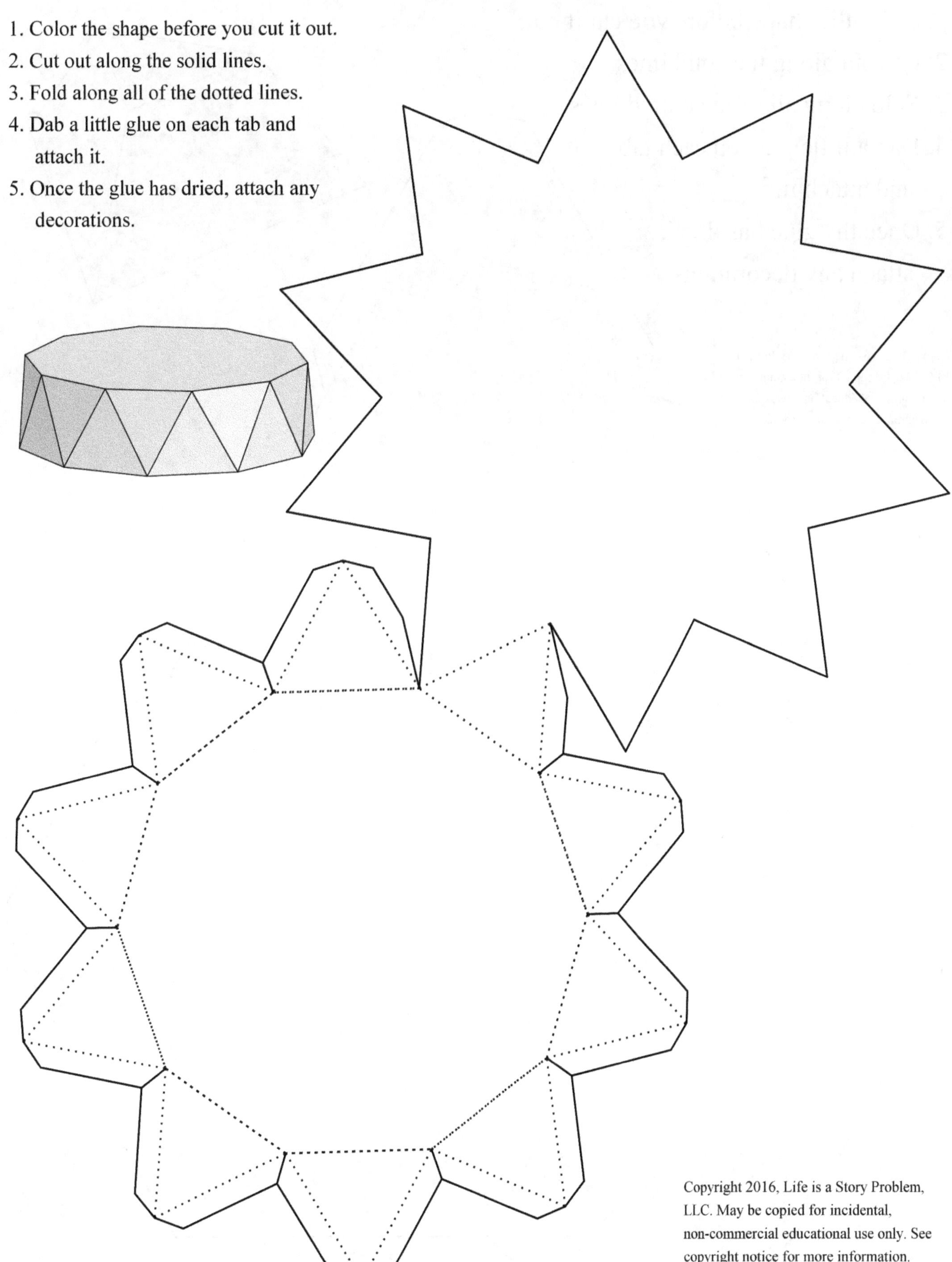

Frustum of a triangular pyramid

1. Color the shape before you cut it out.

2. Cut out along the solid lines.

3. Fold along all of the dotted lines.

4. Dab a little glue on each tab
   and attach it.

5. Once the glue has dried,
   attach any decorations.

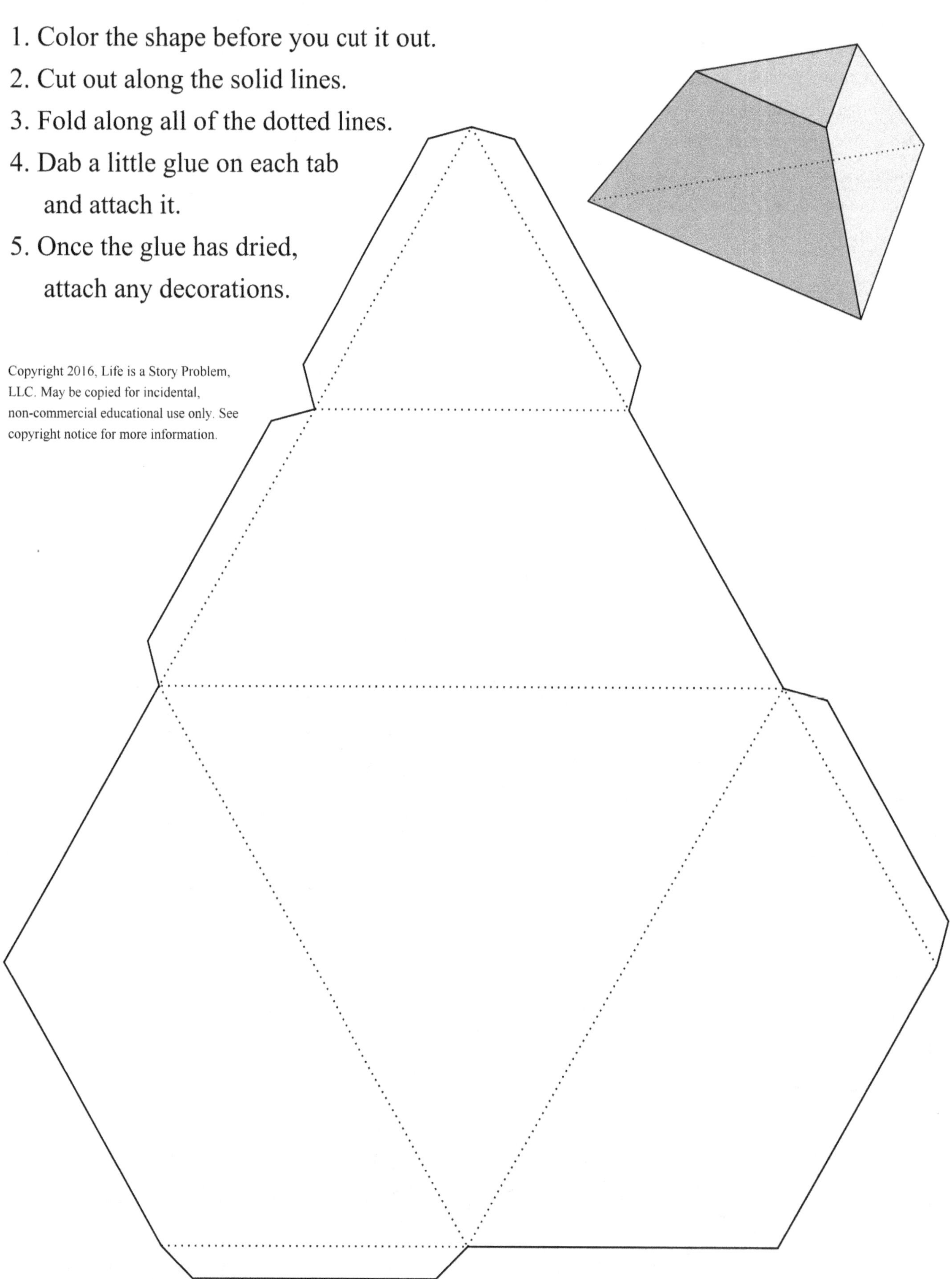

 Geometric Nets Mega Project Book, Tabbed By David E. McAdams

Frustum of a square pyramid

1. Color the shape before you cut it out.
2. Cut out along the solid lines.
3. Fold along all of the dotted lines.
4. Dab a little glue on each tab and attach it.
5. Once the glue has dried, attach any decorations.

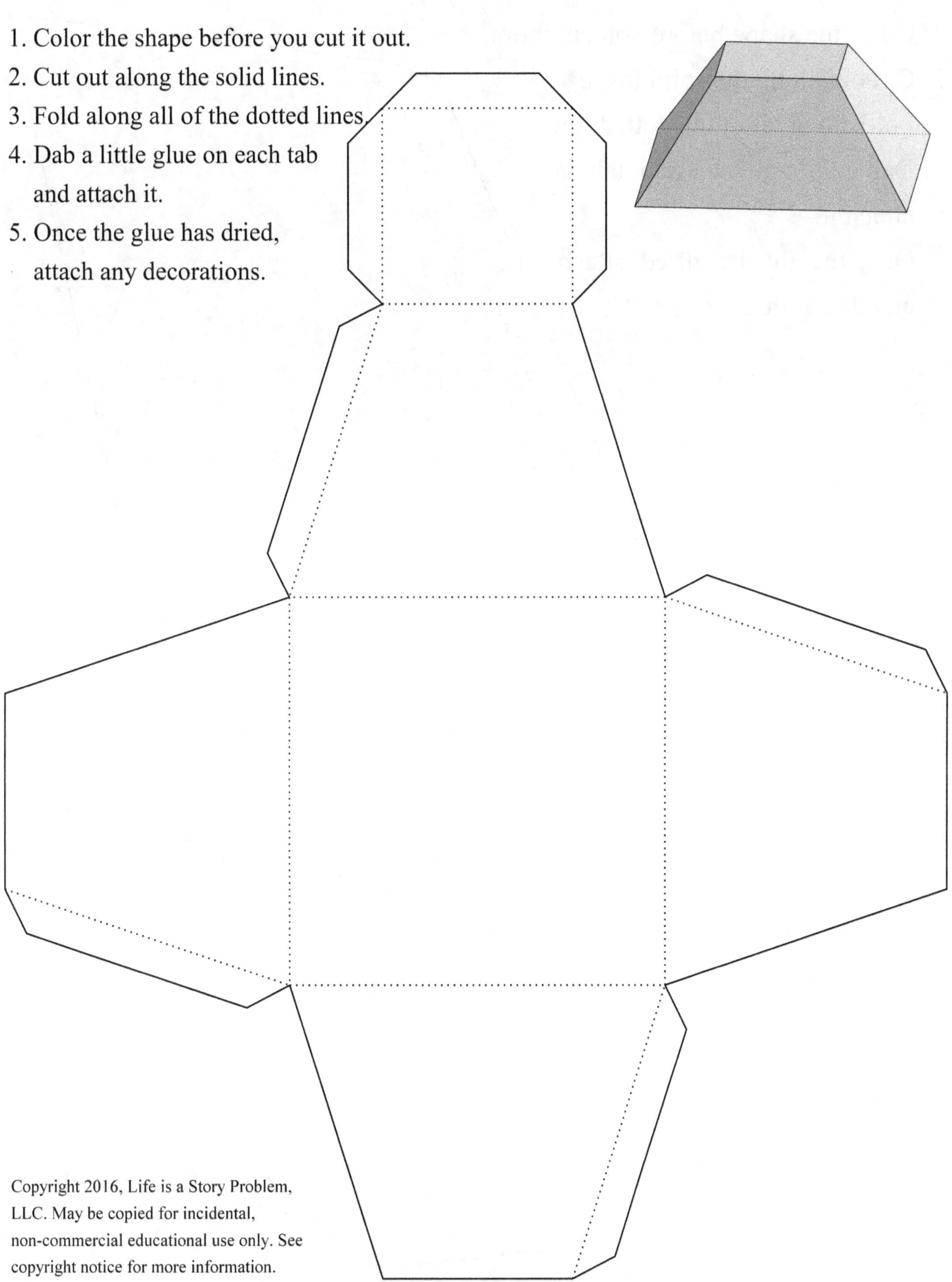

Pyramid of an concave quadrilateral

1. Color the shape before you cut it out.
2. Cut out along the solid lines.
3. Fold along all of the dotted lines.
4. Dab a little glue on each tab and attach it.
5. Once the glue has dried, attach any decorations.

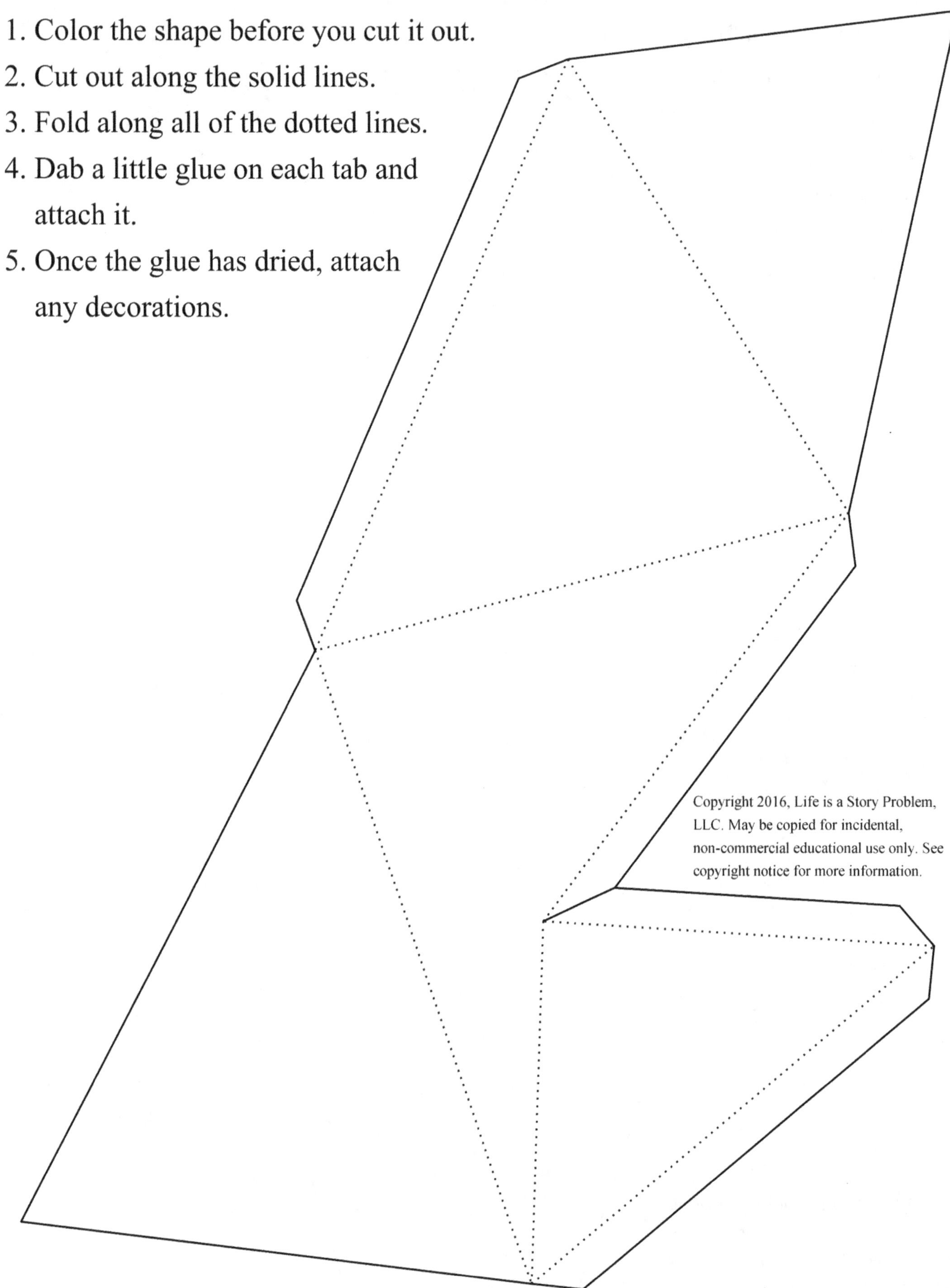

 Geometric Nets Mega Project Book, Tabbed By David E. McAdams

Hexagonal pyramid

1. Color the shape before you cut it out.
2. Cut out along the solid lines.
3. Fold along all of the dotted lines.
4. Dab a little glue on each tab and attach it.
5. Once the glue has dried, attach any decorations.

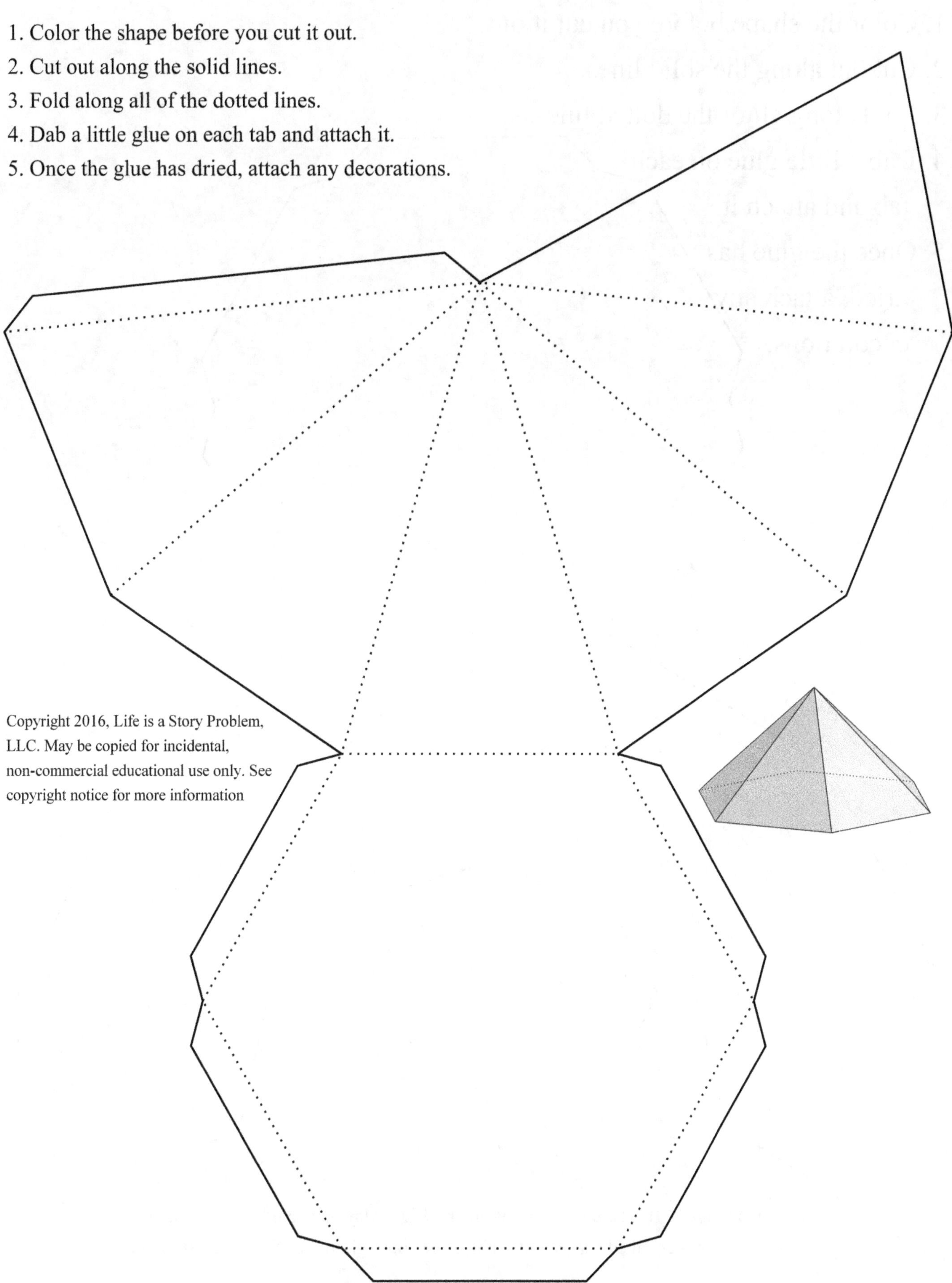

Hexagonal antipyramid

1. Color the shape before you cut it out.
2. Cut out along the solid lines.
3. Fold along all of the dotted lines.
4. Dab a little glue on each tab and attach it.
5. Once the glue has dried, attach any decorations.

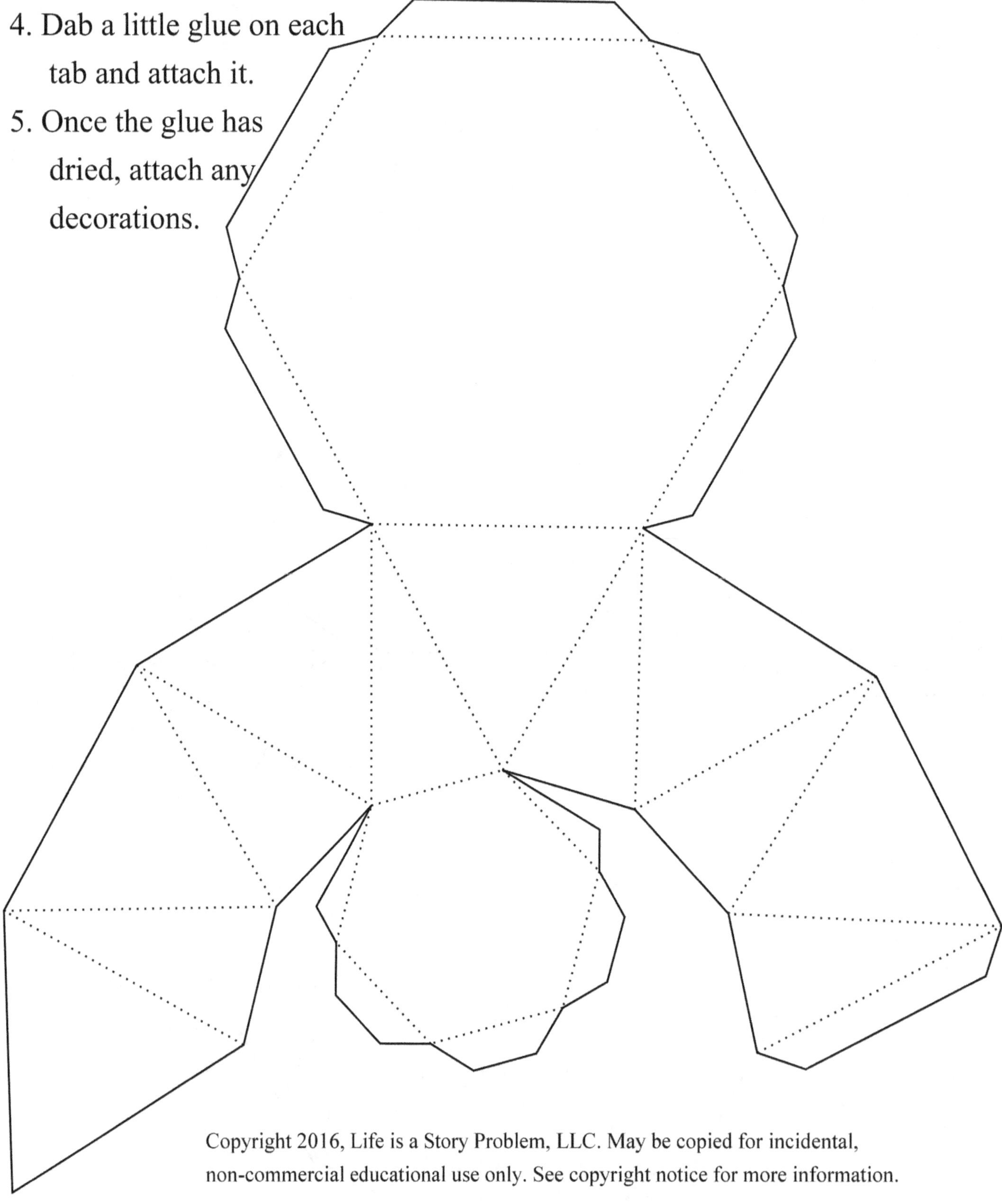

 Geometric Nets Mega Project Book, Tabbed By David E. McAdams

Heptagonal pyramid

1. Color the shape before you cut it out.
2. Cut out along the solid lines.
3. Fold along all of the dotted lines.
4. Dab a little glue on each tab
   and attach it.
5. Once the glue has dried,
   attach any decorations.

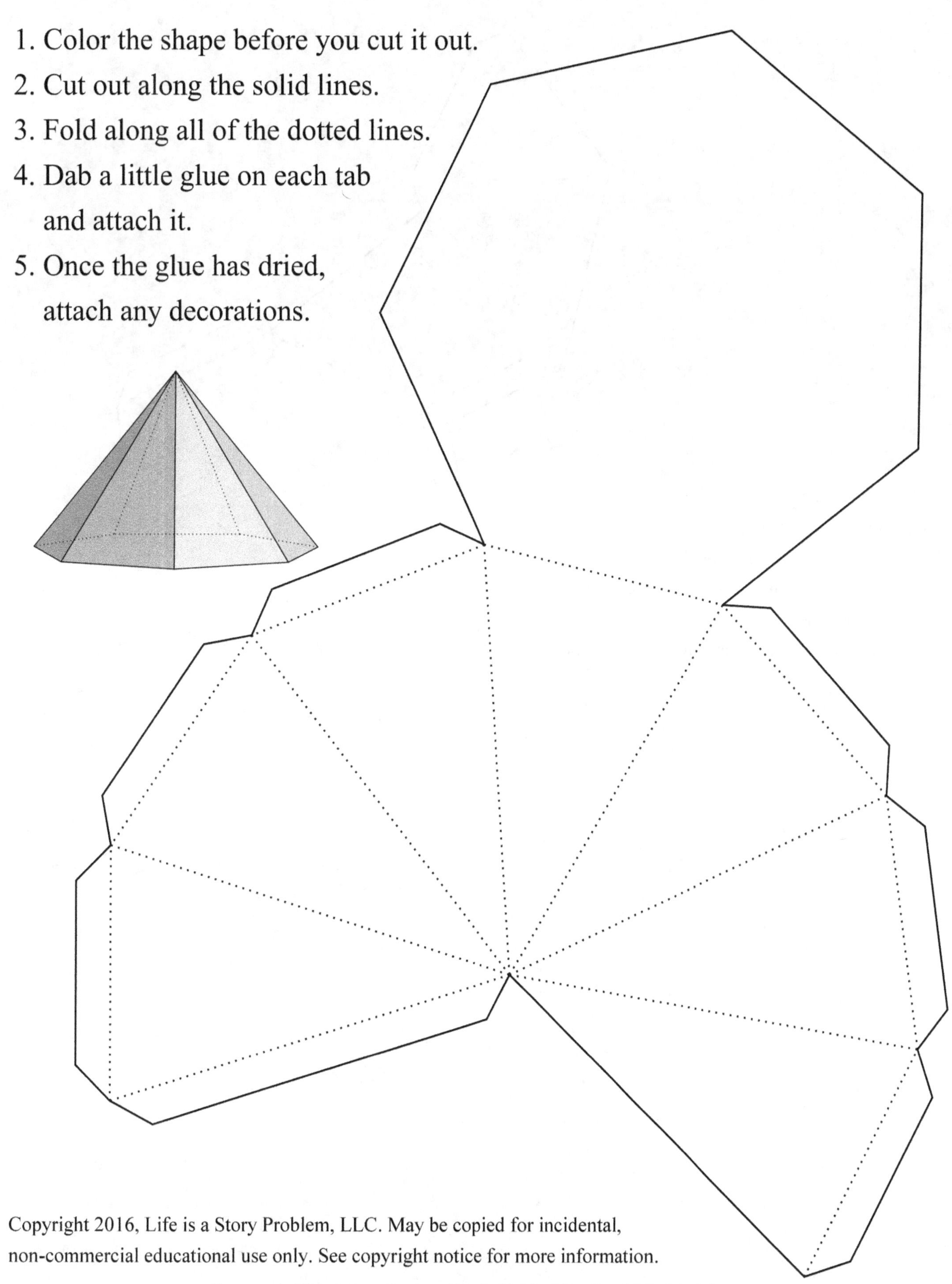

Stellated cube 2

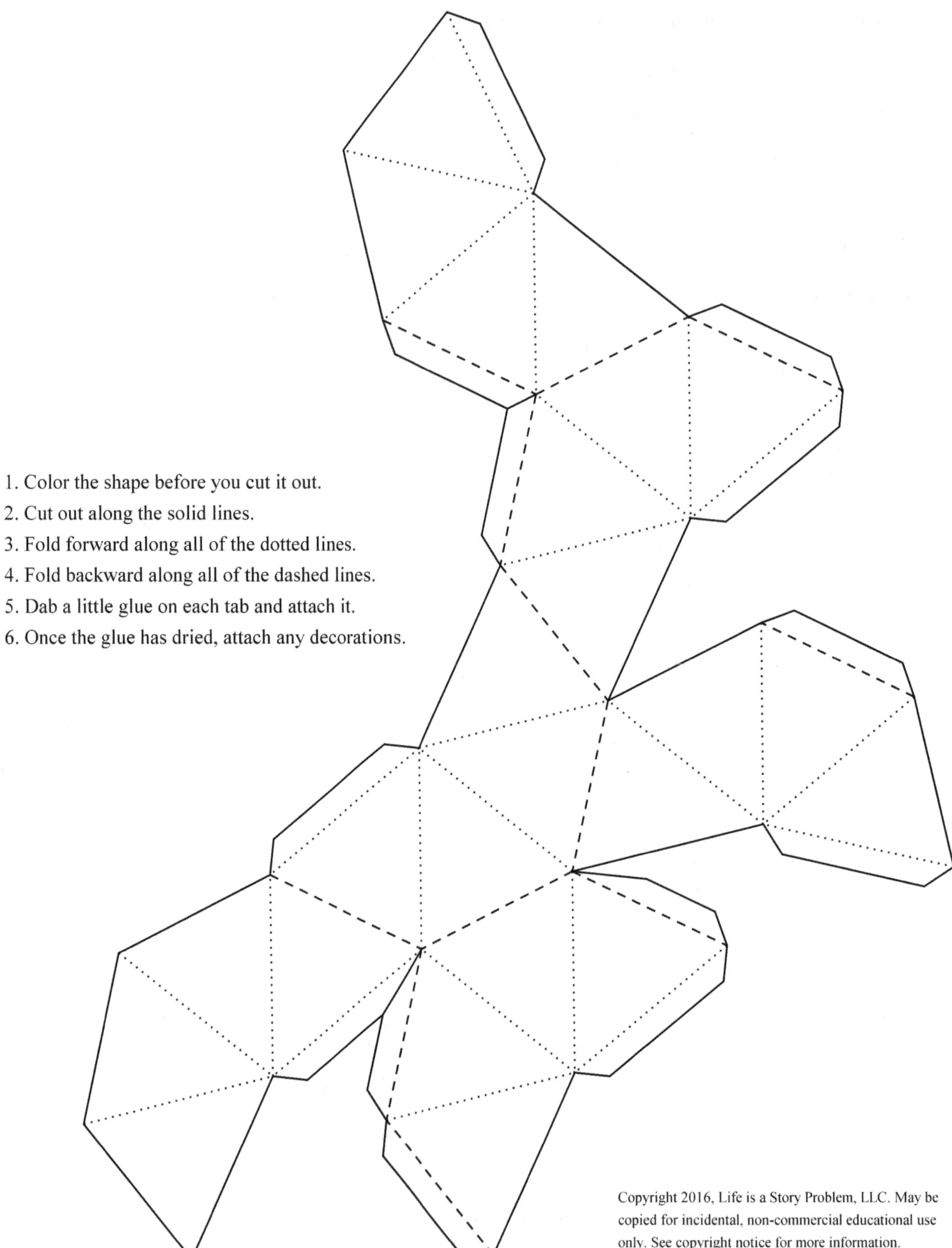

1. Color the shape before you cut it out.
2. Cut out along the solid lines.
3. Fold forward along all of the dotted lines.
4. Fold backward along all of the dashed lines.
5. Dab a little glue on each tab and attach it.
6. Once the glue has dried, attach any decorations.

Stellated cube 3

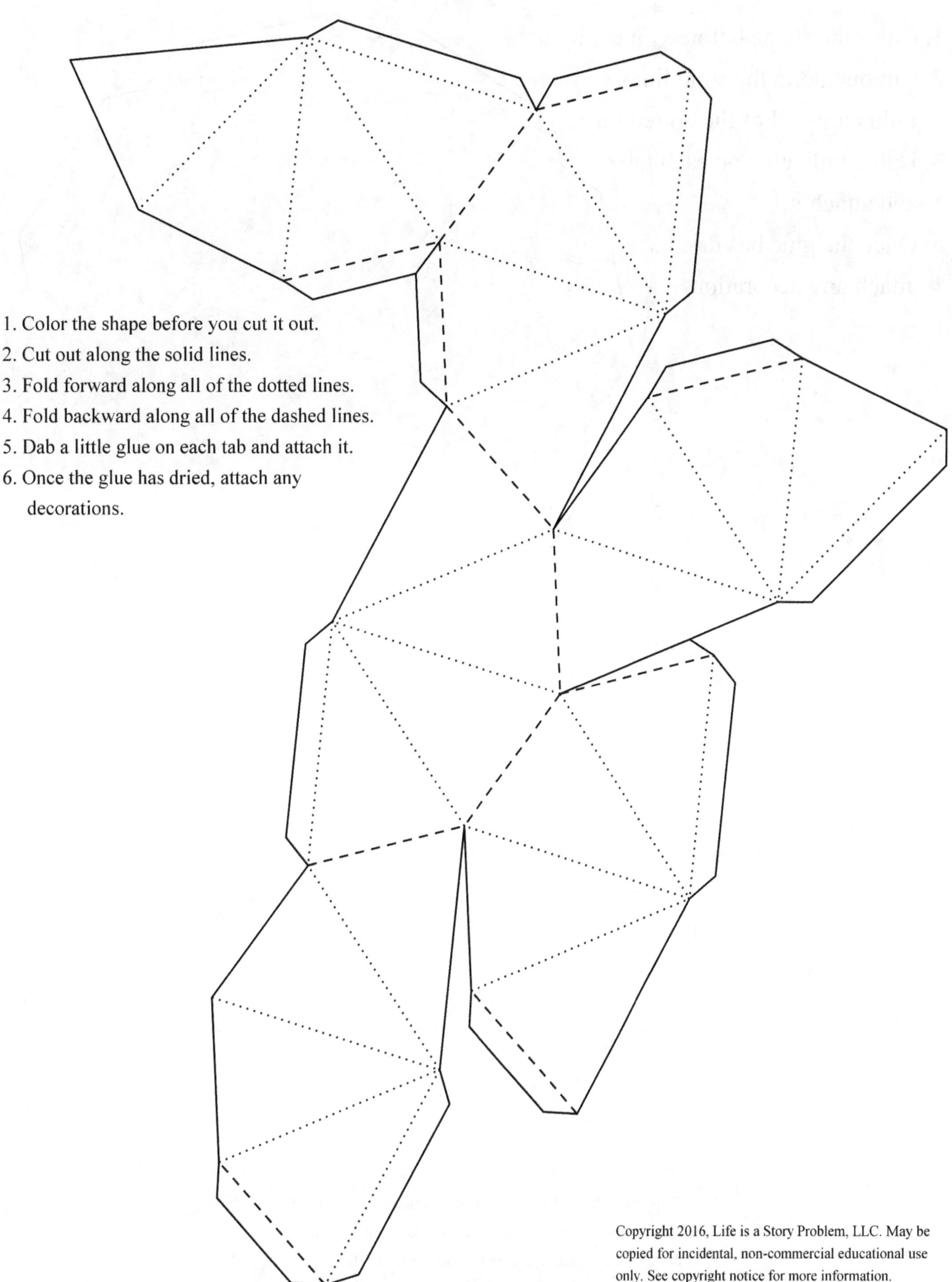

1. Color the shape before you cut it out.
2. Cut out along the solid lines.
3. Fold forward along all of the dotted lines.
4. Fold backward along all of the dashed lines.
5. Dab a little glue on each tab and attach it.
6. Once the glue has dried, attach any
   decorations.

Octagonal pyramid

1. Color the shape before you cut it out.
2. Cut out along the solid lines.
3. Fold along all of the dotted lines.
4. Dab a little glue on each tab and attach it.
5. Once the glue has dried, attach any decorations.

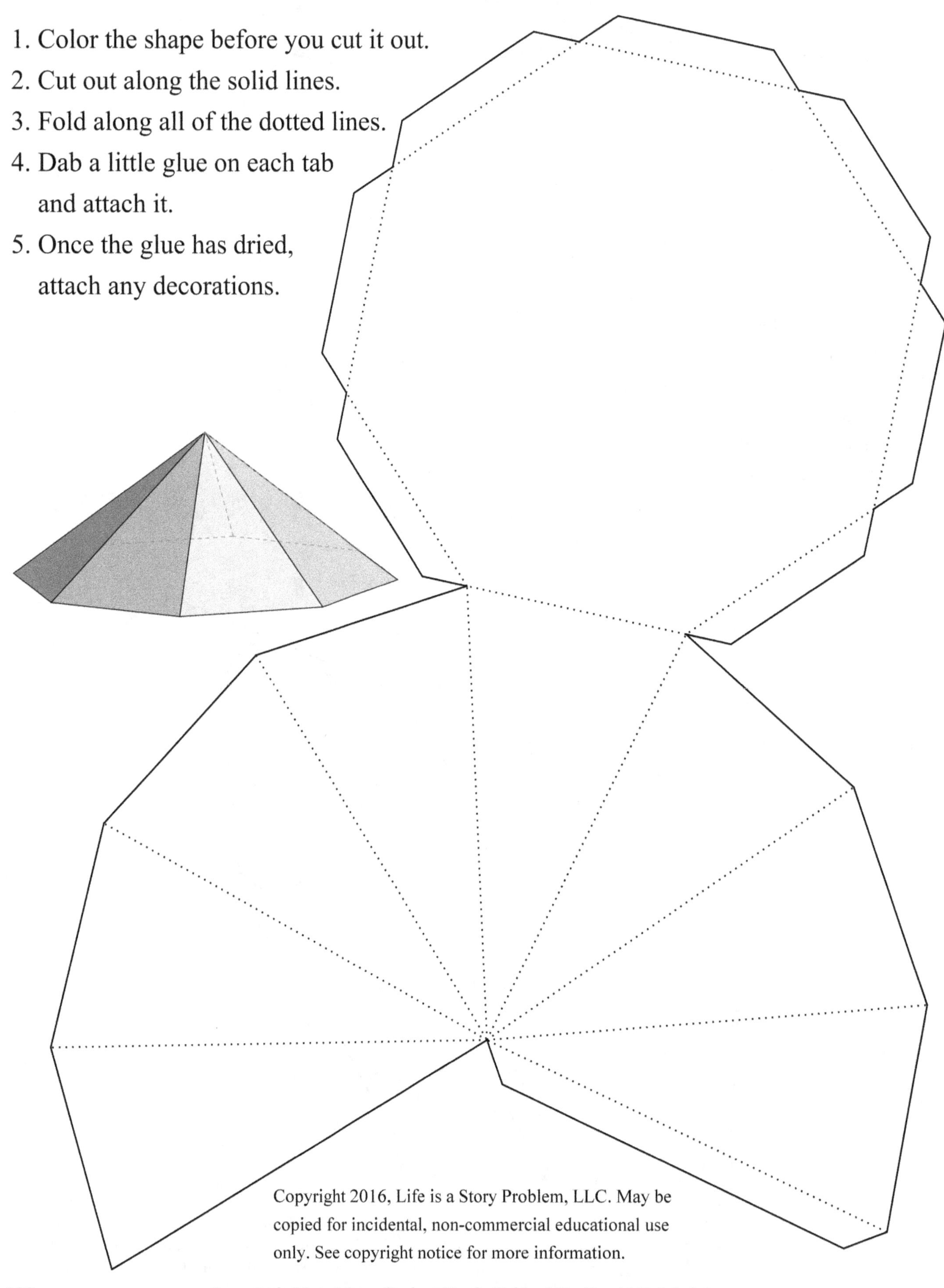

 Geometric Nets Mega Project Book, Tabbed By David E. McAdams

Gyroelongated octagonal pyramid

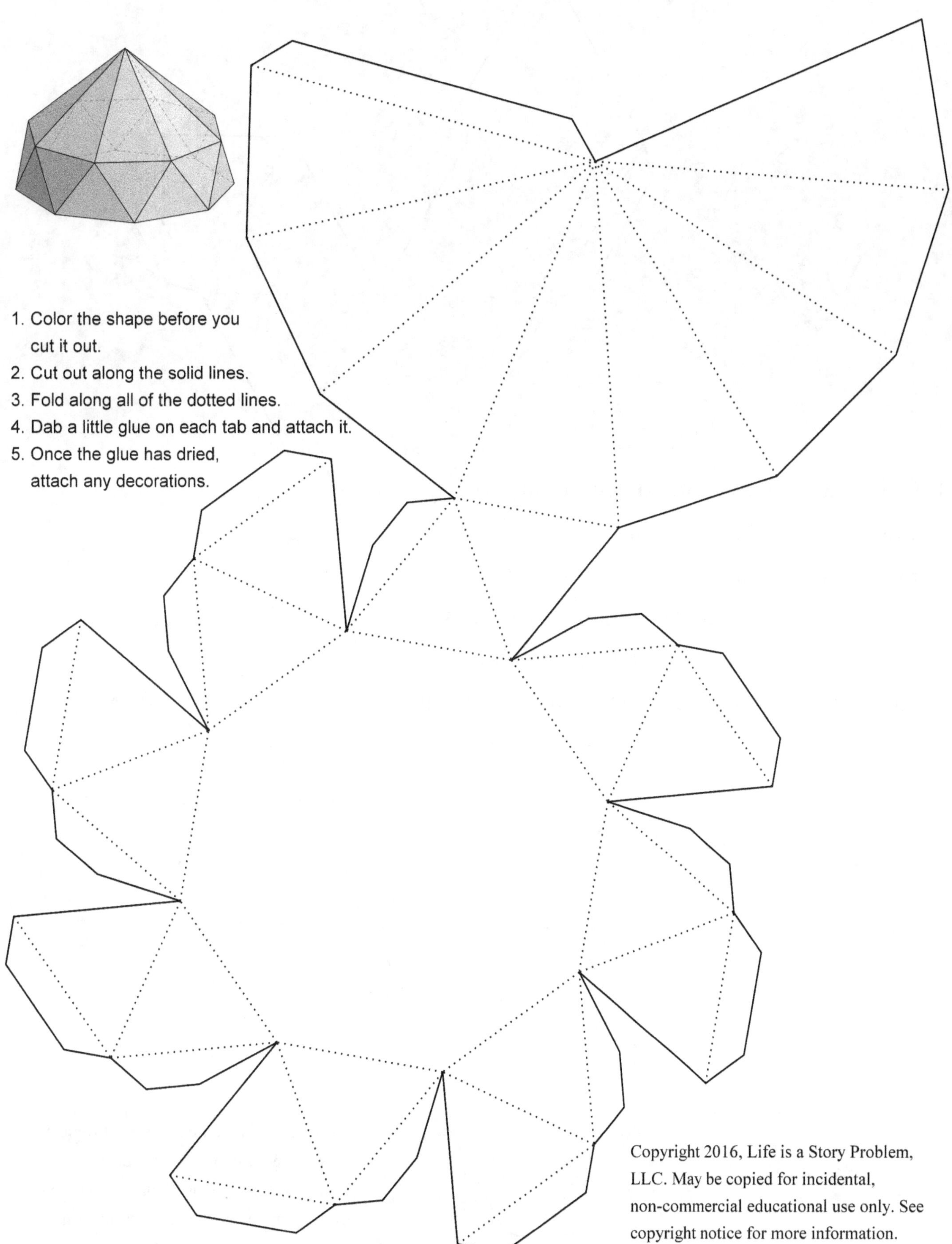

1. Color the shape before you
   cut it out.
2. Cut out along the solid lines.
3. Fold along all of the dotted lines.
4. Dab a little glue on each tab and attach it.
5. Once the glue has dried,
   attach any decorations.

Stellated octahedron

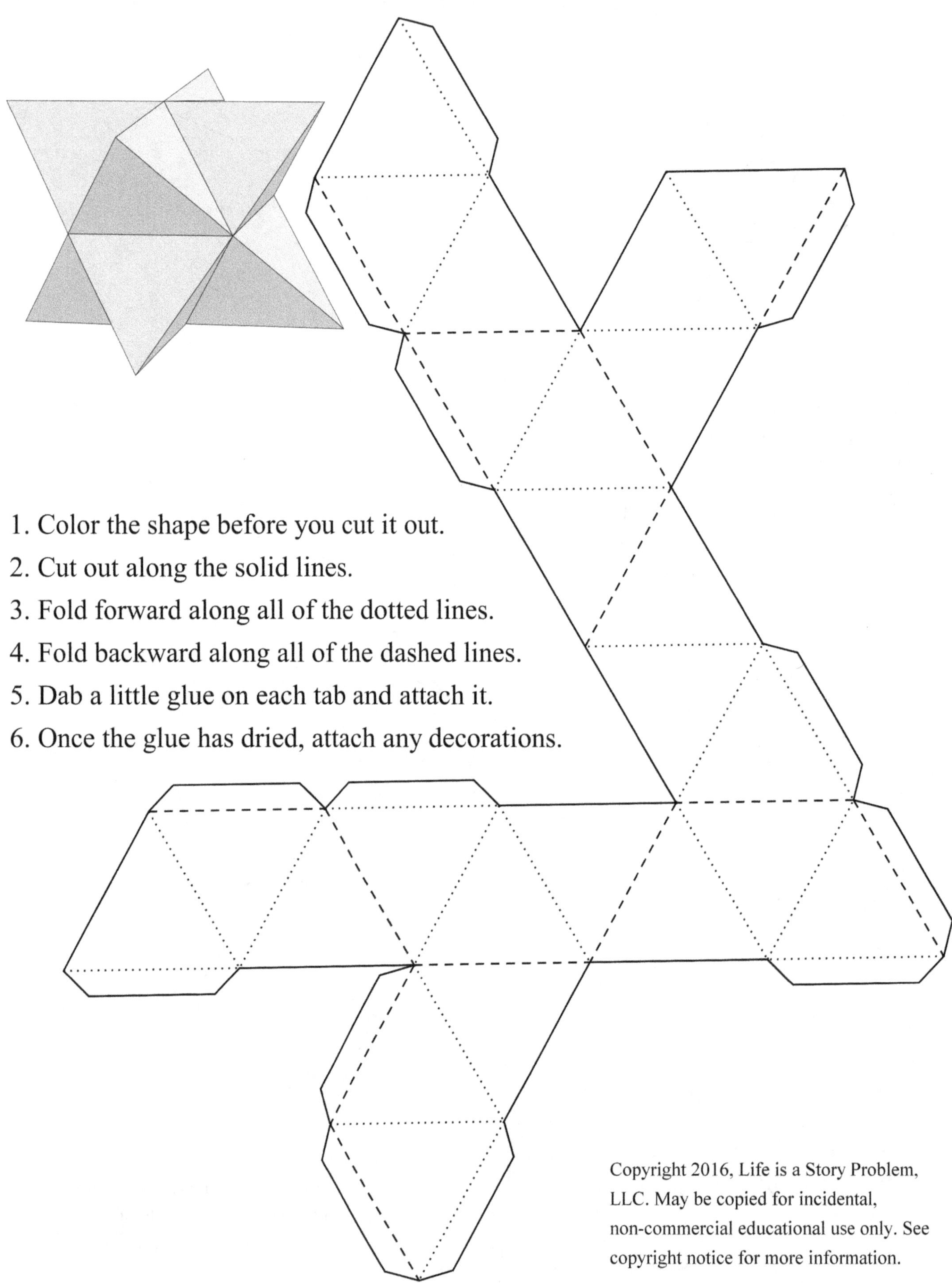

1. Color the shape before you cut it out.
2. Cut out along the solid lines.
3. Fold forward along all of the dotted lines.
4. Fold backward along all of the dashed lines.
5. Dab a little glue on each tab and attach it.
6. Once the glue has dried, attach any decorations.

Frustum of a decagonal pyramid

1. Color the shape before you cut it out.
2. Cut out along the solid lines.
3. Fold along all of the dotted lines.
4. Dab a little glue on each tab and attach it.
5. Once the glue has dried, attach any decorations.

Decapentagonal bipyramid

1. Color the shape before you cut it out.
2. Cut out along the solid lines.
3. Fold along all of the dotted lines.
4. Dab a little glue on each tab and attach it.
5. Once the glue has dried, attach any decorations.

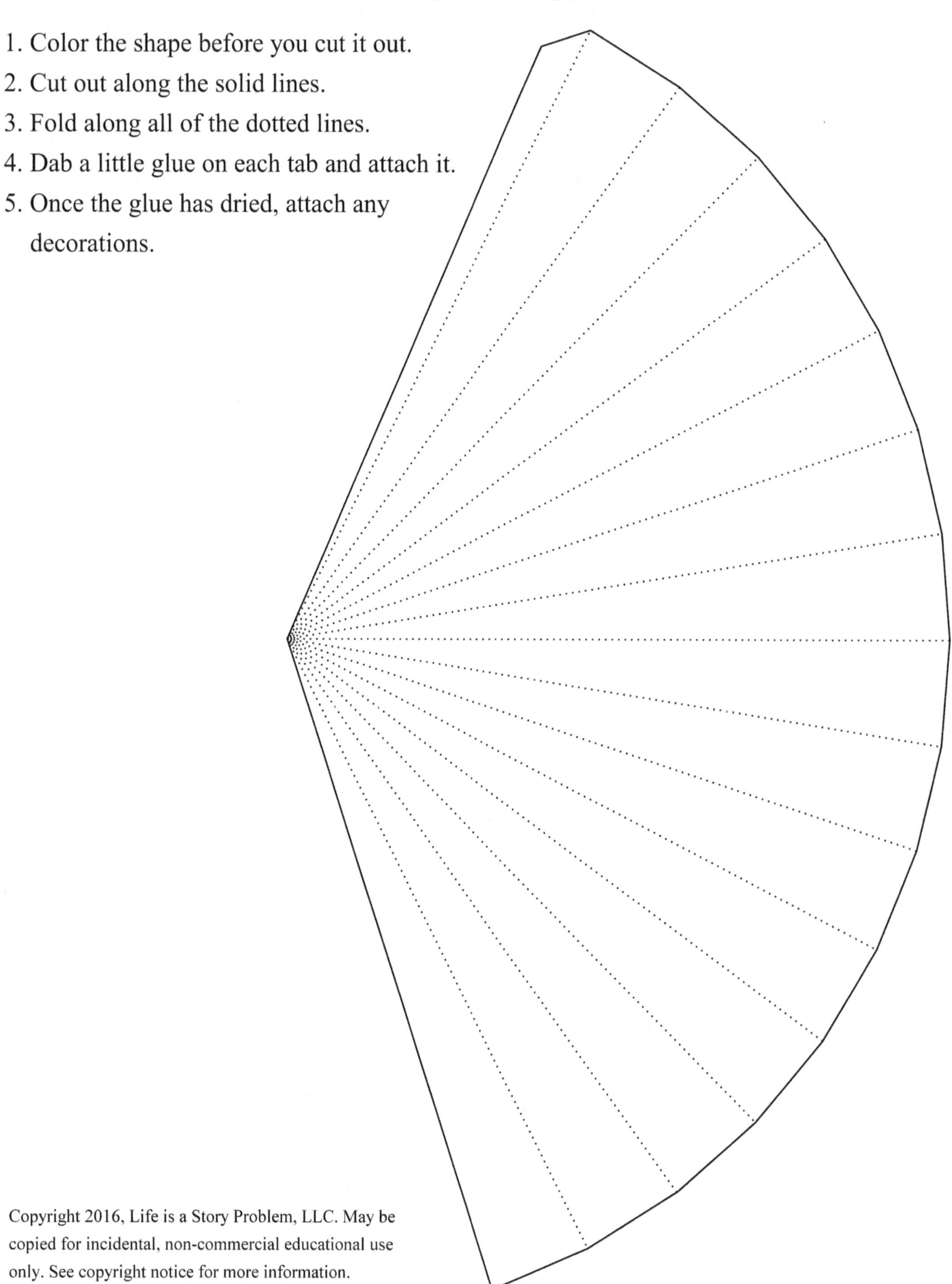

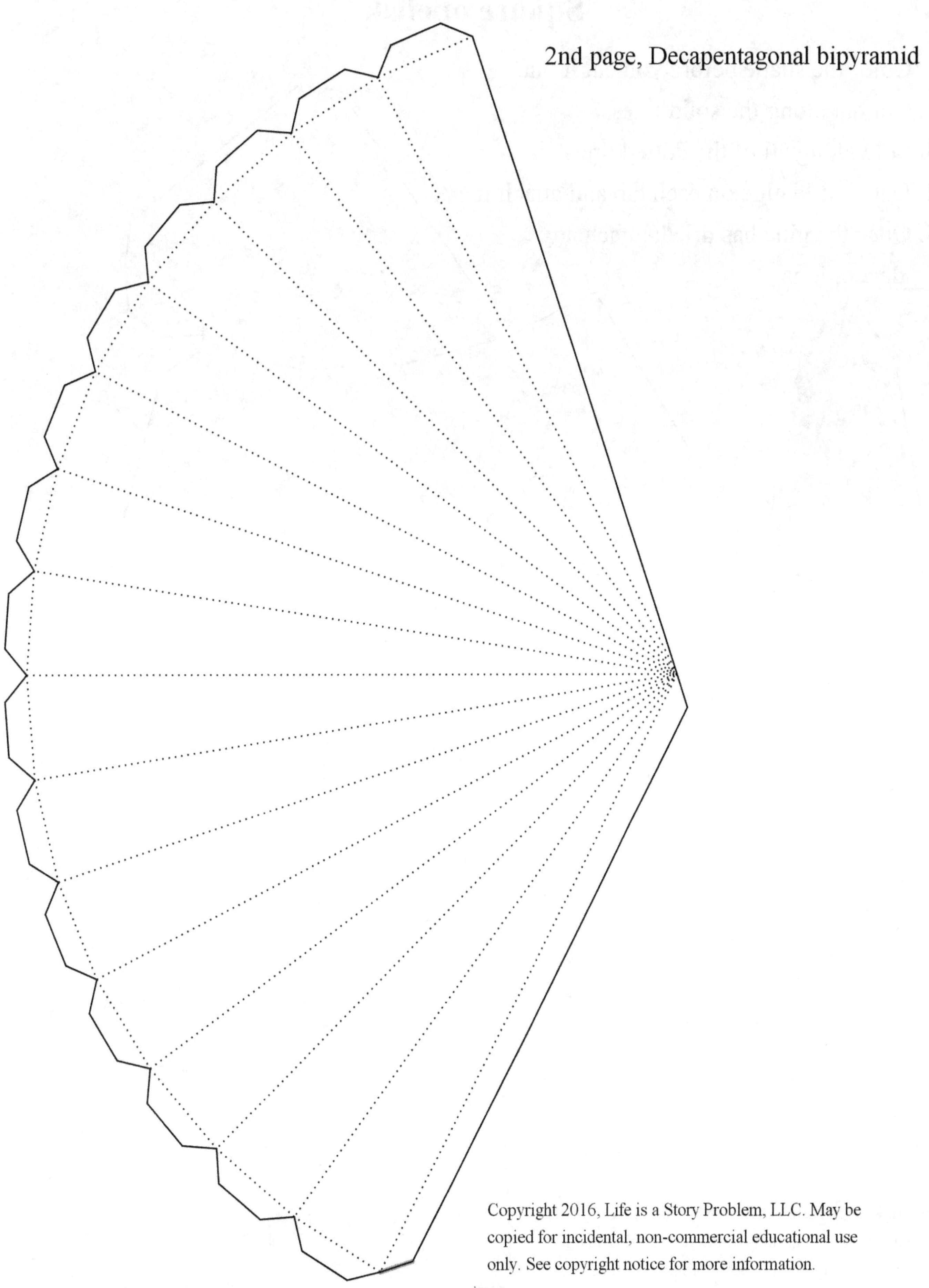

2nd page, Decapentagonal bipyramid

Square obelisk

1. Color the shape before you cut it out.

2. Cut out along the solid lines.

3. Fold along all of the dotted lines.

4. Dab a little glue on each tab and attach it.

5. Once the glue has dried, attach any
   decorations.

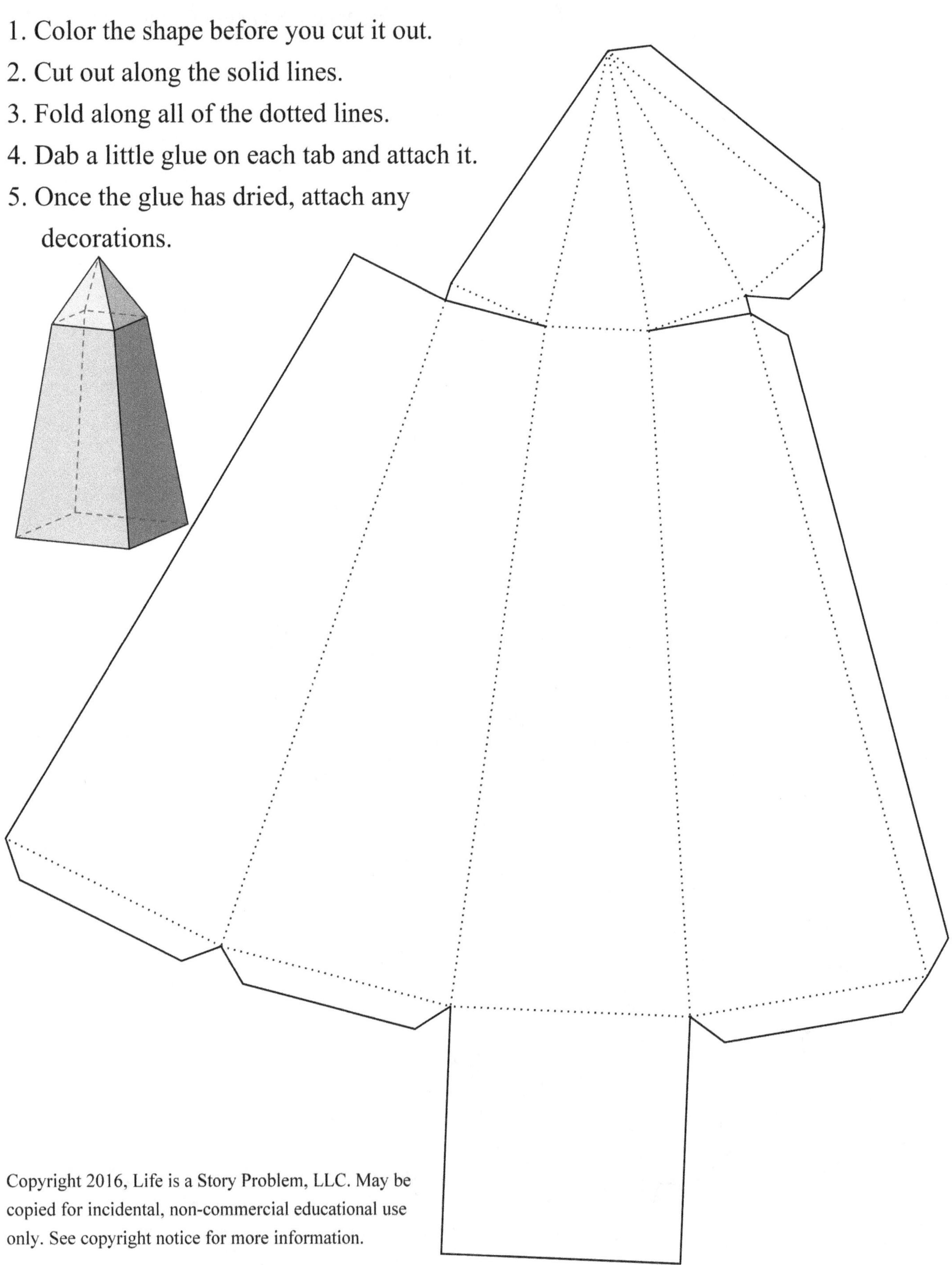

Octagonal obelisk

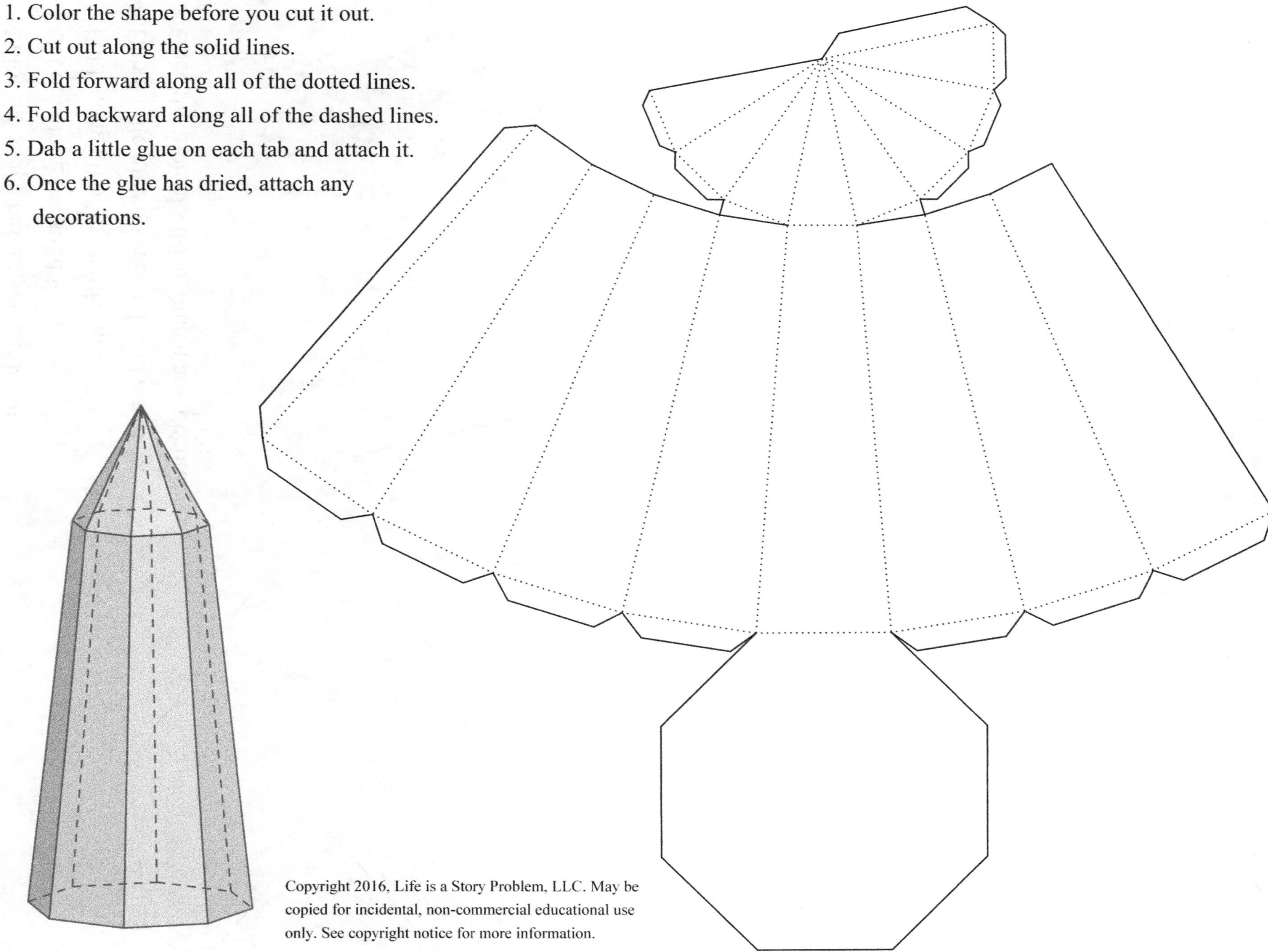

1. Color the shape before you cut it out.
2. Cut out along the solid lines.
3. Fold forward along all of the dotted lines.
4. Fold backward along all of the dashed lines.
5. Dab a little glue on each tab and attach it.
6. Once the glue has dried, attach any decorations.

Square trapezohedron

1. Color the shape before you cut it out.
2. Cut out along the solid lines.
3. Fold along all of the dotted lines.
4. Dab a little glue on each tab and attach it.
5. Once the glue has dried, attach any decorations.

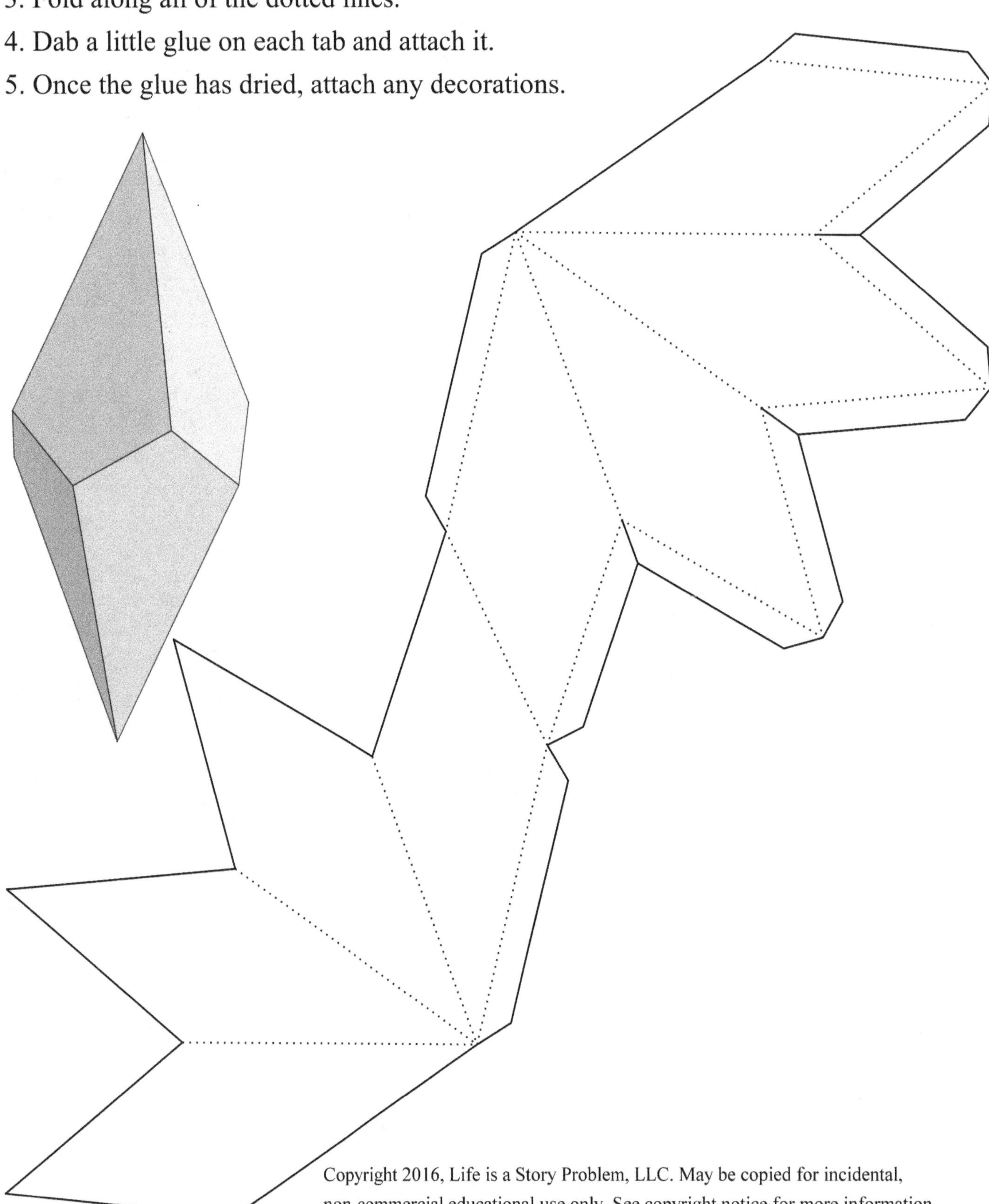

 Geometric Nets Mega Project Book, Tabbed By David E. McAdams

Trigonal trapezohedron

1. Color the shape before you cut it out.

2. Cut out along the solid lines.

3. Fold along all of the dotted lines.

4. Dab a little glue on each tab and attach it.

5. Once the glue has dried, attach any decorations.

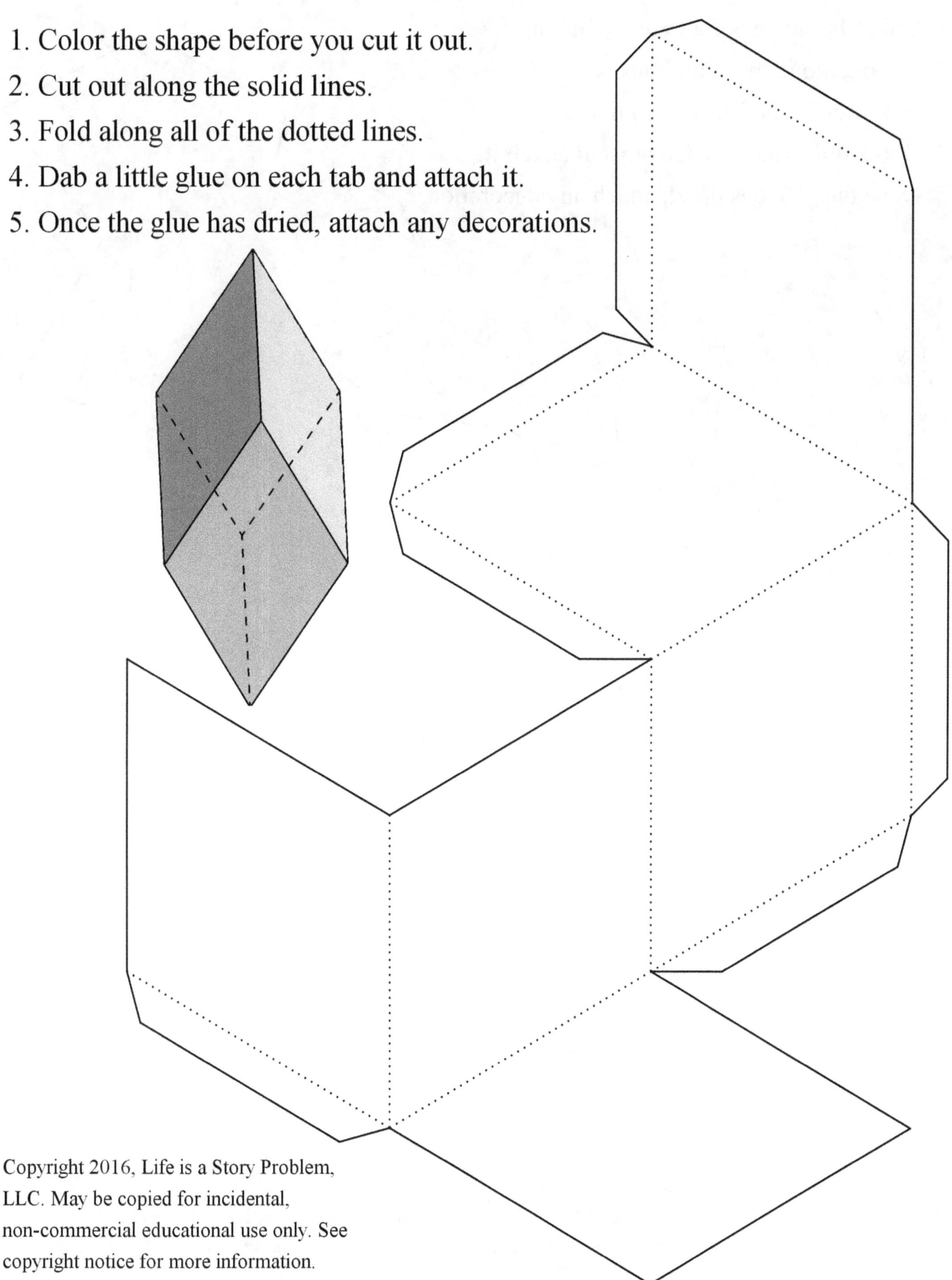

Truncated square trapezohedron

1. Color the shape before you cut it out.
2. Cut out along the solid lines.
3. Fold along all of the dotted lines.
4. Dab a little glue on each tab and attach it.
5. Once the glue has dried, attach any decorations.

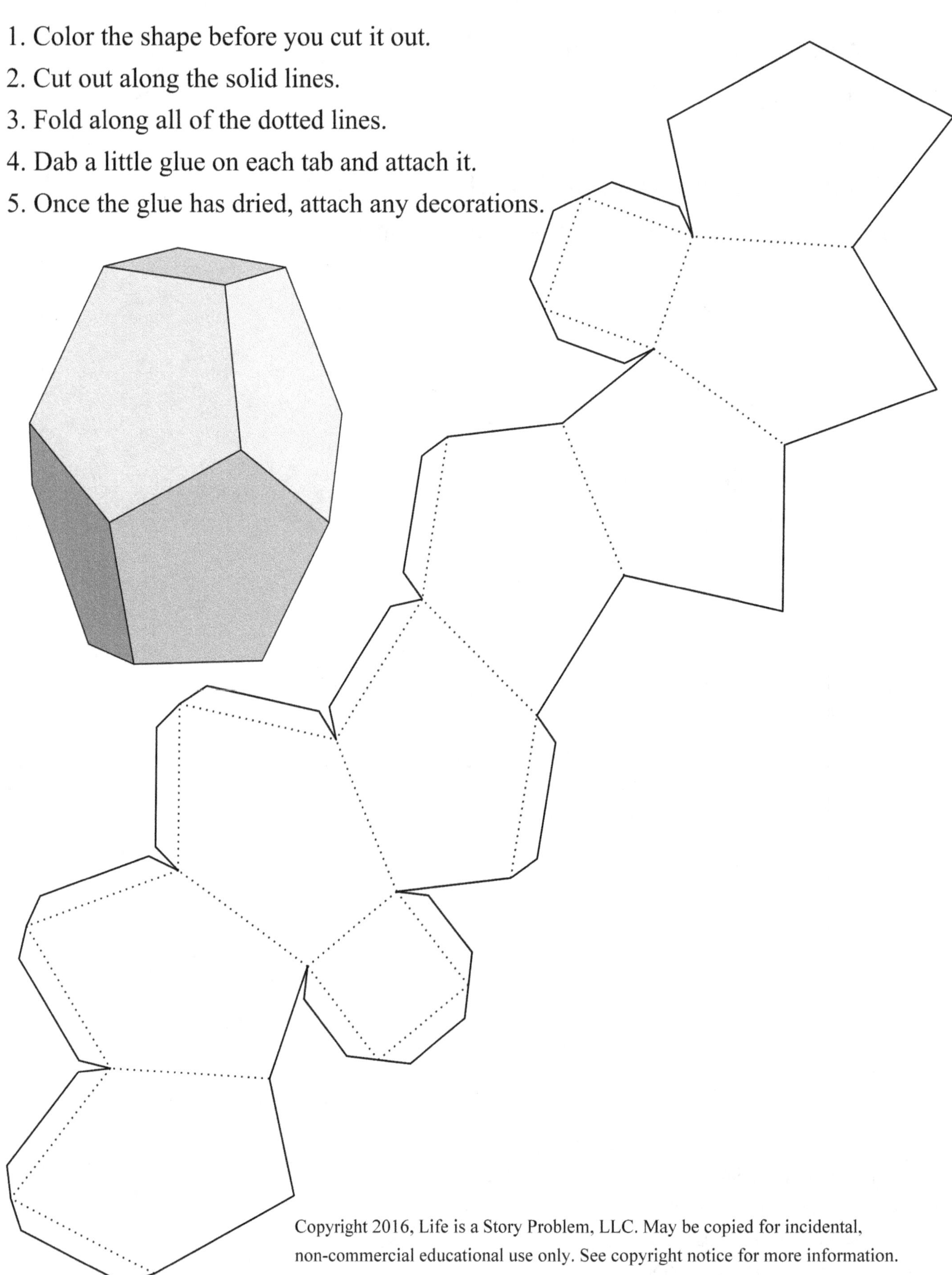

    Geometric Nets Mega Project Book, Tabbed By David E. McAdams

Chamfered tetrahedron

1. Color the shape before you cut it out.
2. Cut out along the solid lines.
3. Fold along all of the dotted lines.
4. Dab a little glue on each tab and attach it.
5. Once the glue has dried, attach any decorations.

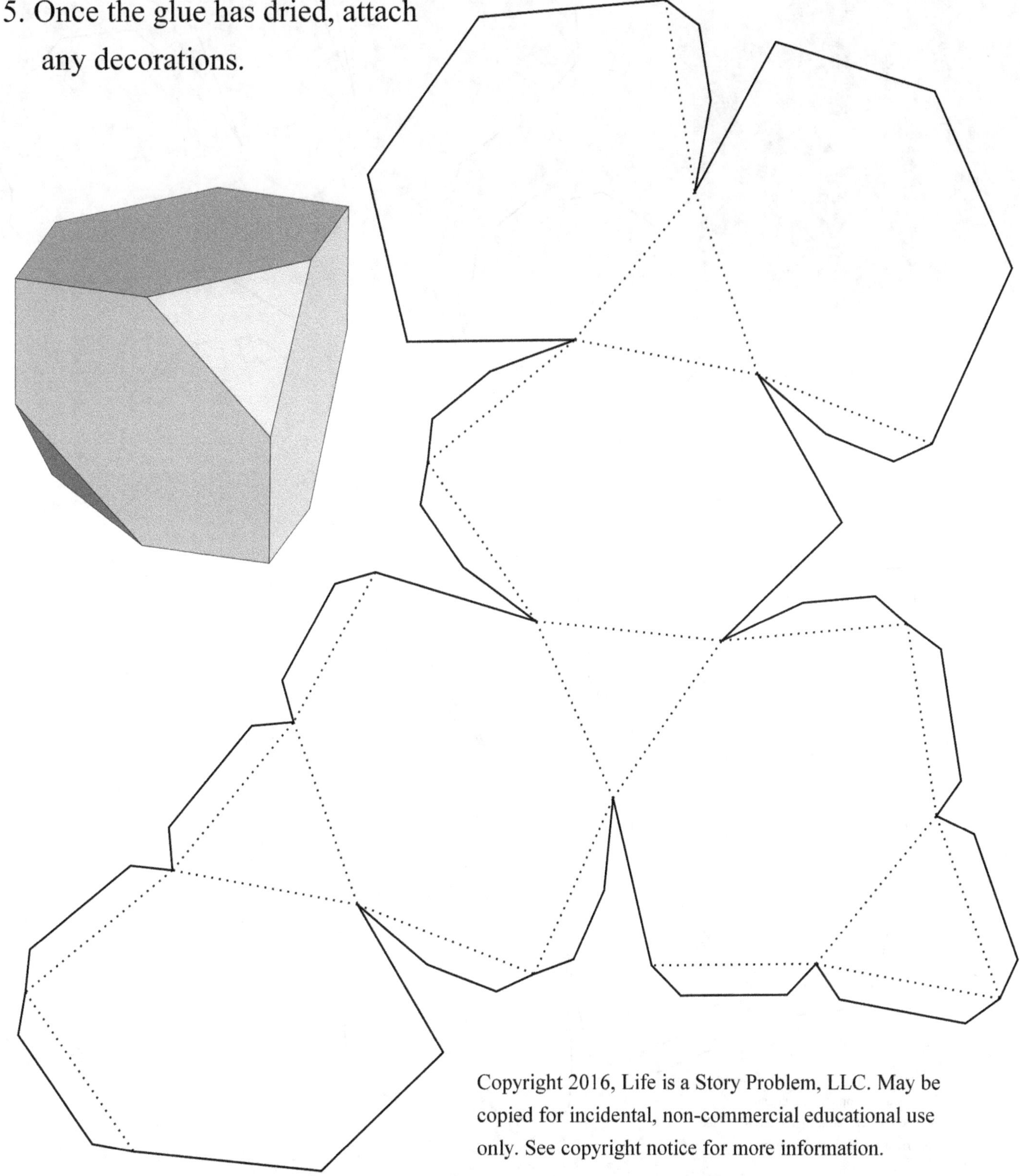

Chamfered cube

1. Color the shape before you cut it out.
2. Cut out along the solid lines.
3. Fold along all of the dotted lines.
4. Dab a little glue on each tab and attach it.
5. Once the glue has dried, attach any decorations.

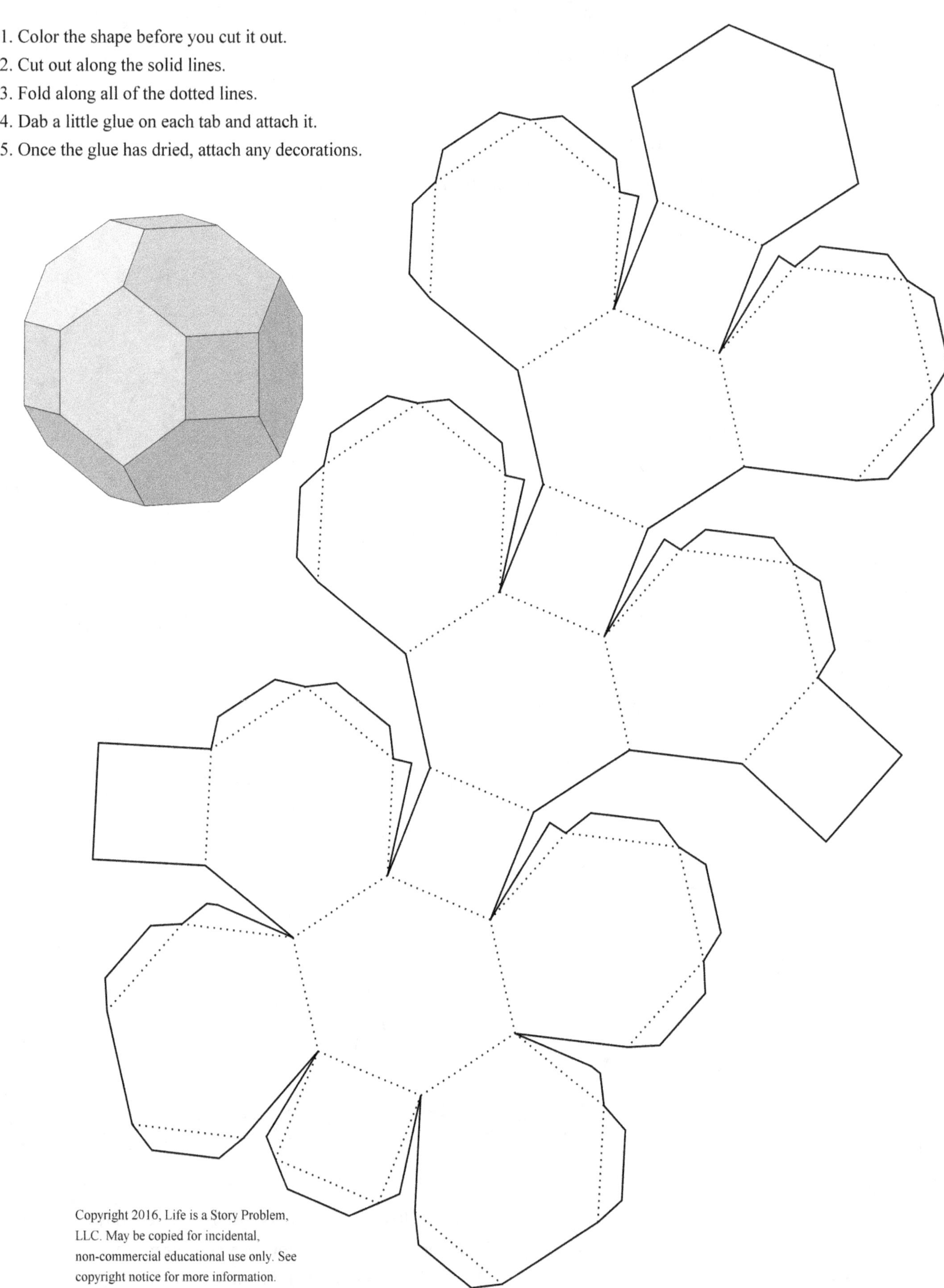

     Geometric Nets Mega Project Book, Tabbed By David E. McAdams

Chamfered dodecahedron

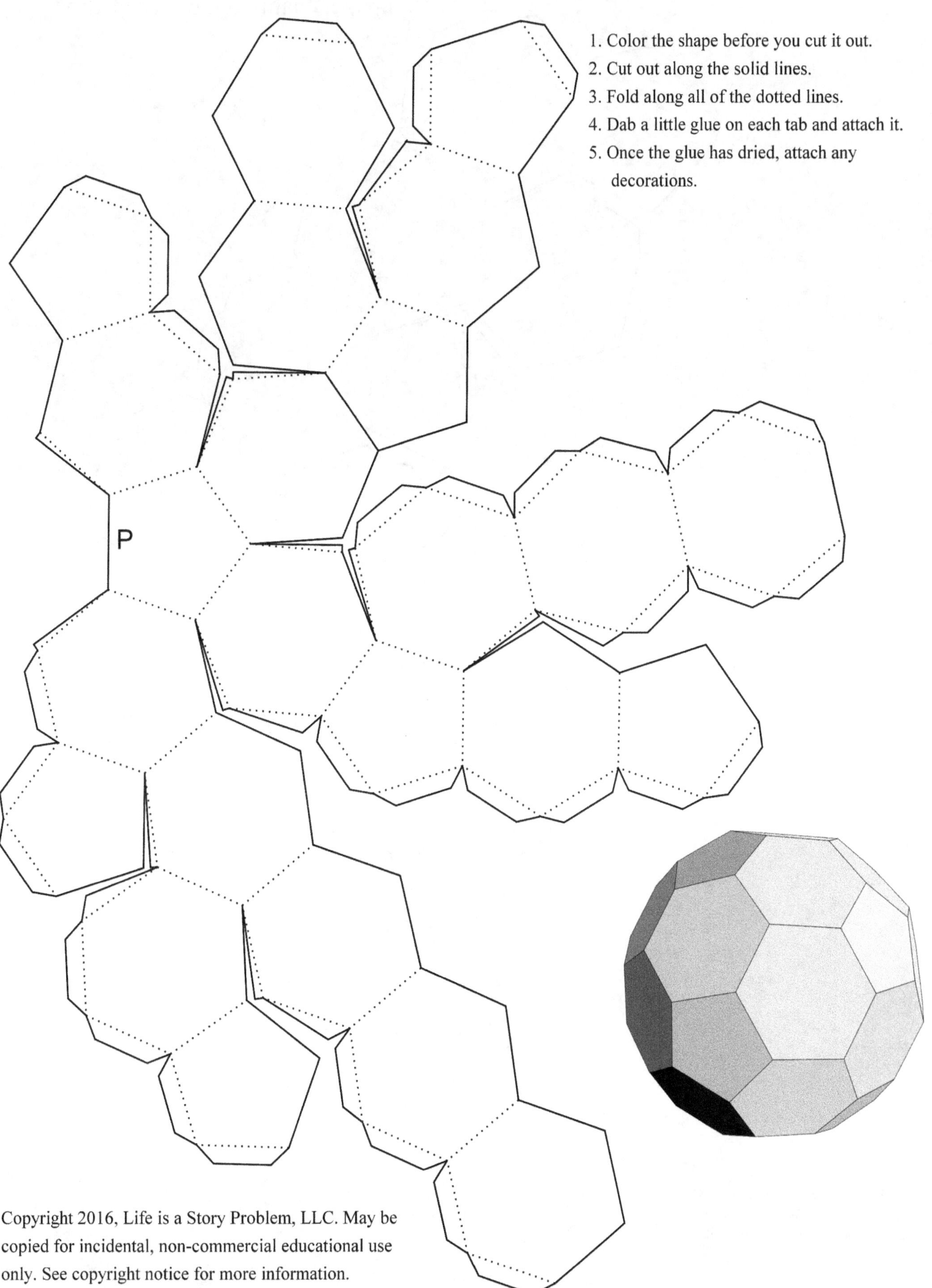

1. Color the shape before you cut it out.
2. Cut out along the solid lines.
3. Fold along all of the dotted lines.
4. Dab a little glue on each tab and attach it.
5. Once the glue has dried, attach any decorations.

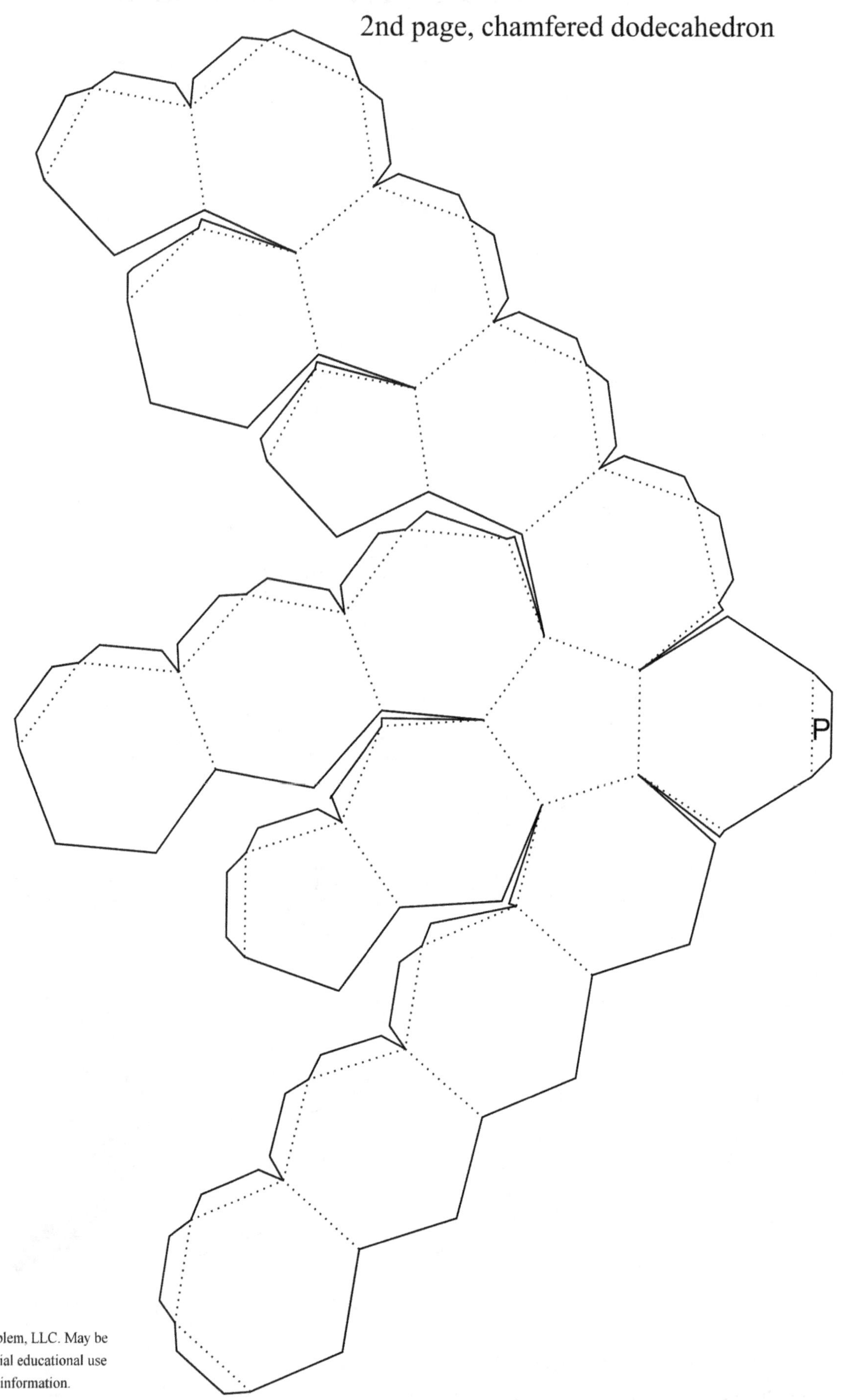

Geometric Nets Mega Project Book, Tabbed By David E. McAdams

Pentagonal deltohedron

1. Color the shape before you cut it out.
2. Cut out along the solid lines.
3. Fold along all of the dotted lines.
4. Dab a little glue on each tab and attach it.
5. Once the glue has dried, attach
   any decorations.

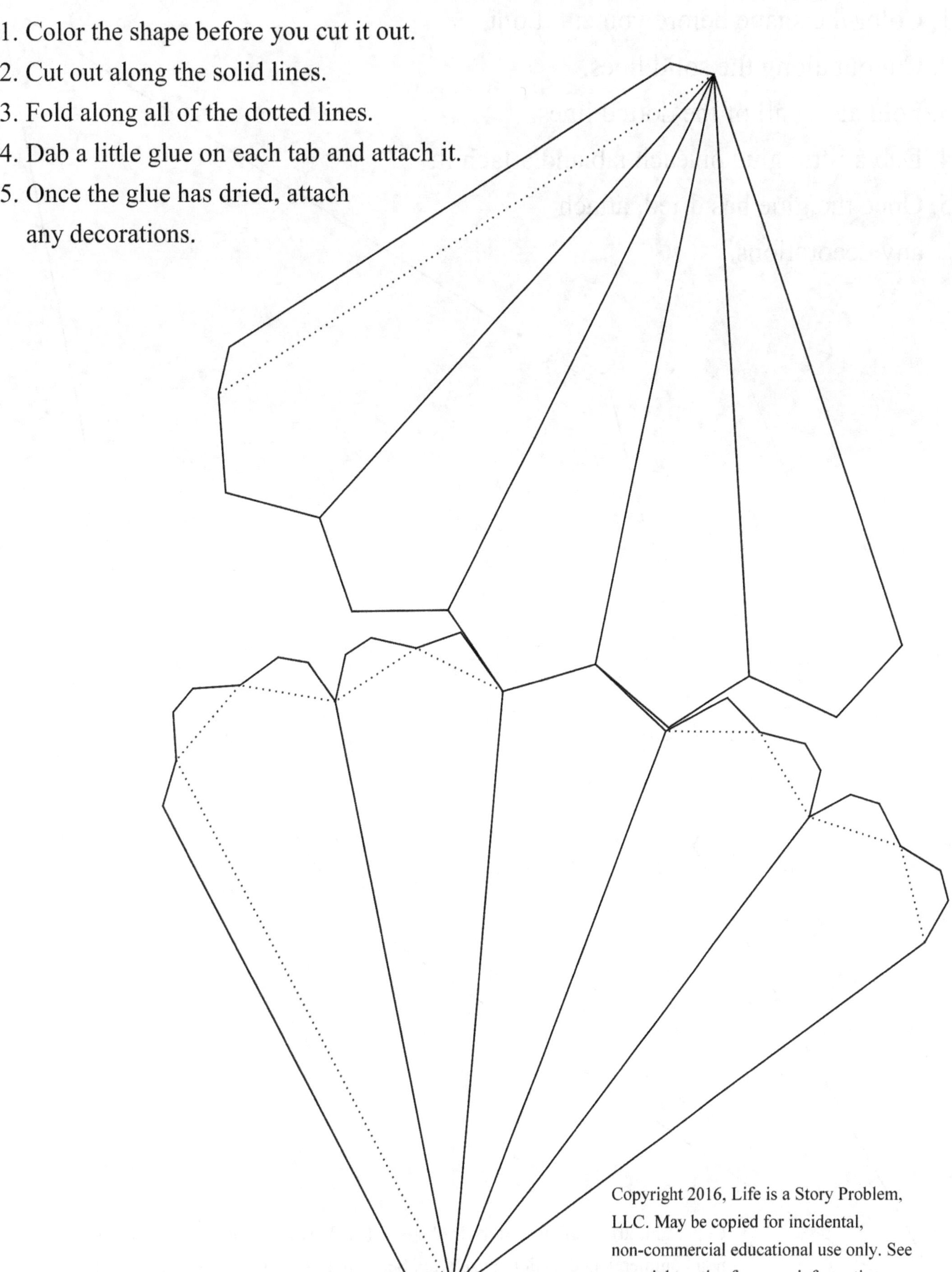

Oblique triangular pyramid

1. Color the shape before you cut it out.

2. Cut out along the solid lines.

3. Fold along all of the dotted lines.

4. Dab a little glue on each tab and attach it.

5. Once the glue has dried, attach
   any decorations.

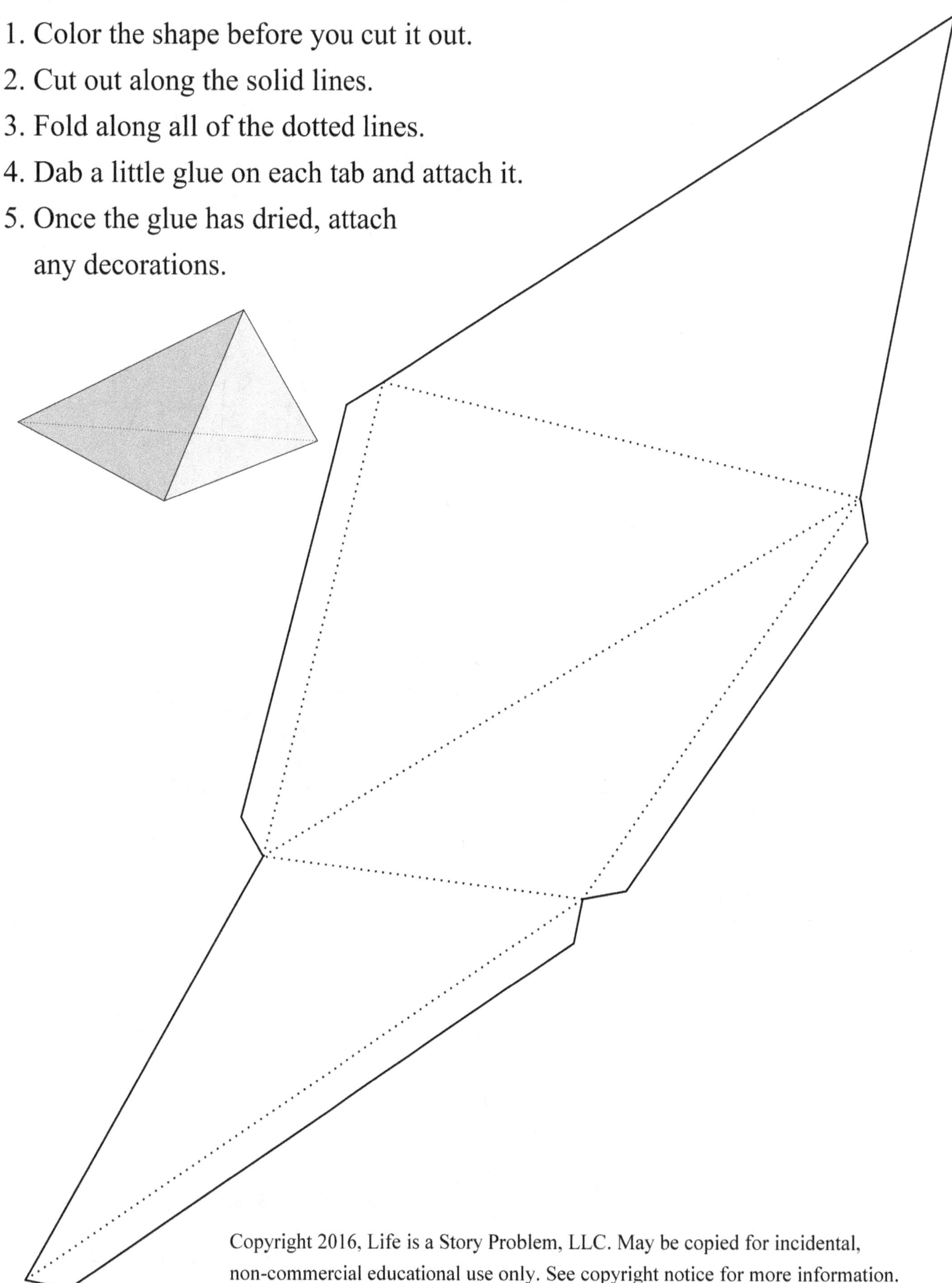

Geometric Nets Mega Project Book, Tabbed By David E. McAdams

Oblique square pyramid

1. Color the shape before you cut it out.
2. Cut out along the solid lines.
3. Fold along all of the dotted lines.
4. Dab a little glue on each tab and attach it.
5. Once the glue has dried, attach any decorations.

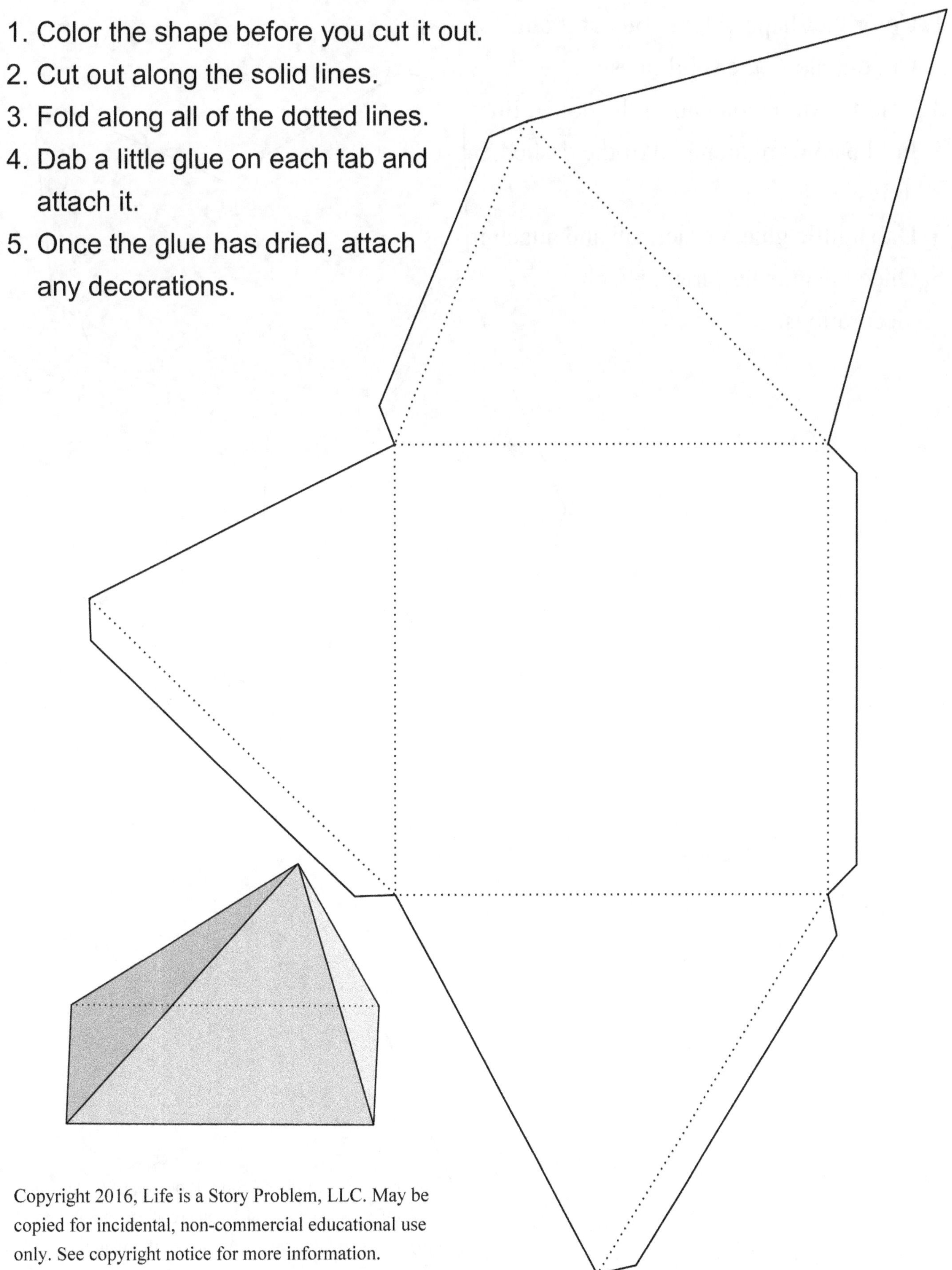

Twisted square pyramid

1. Color the shape before you cut it out.
2. Cut out along the solid lines.
3. Fold forward along all of the dotted lines.
4. Fold backward along all of the dashed lines.
5. Dab a little glue on each tab and attach it.
6. Once the glue has dried, attach decorations.

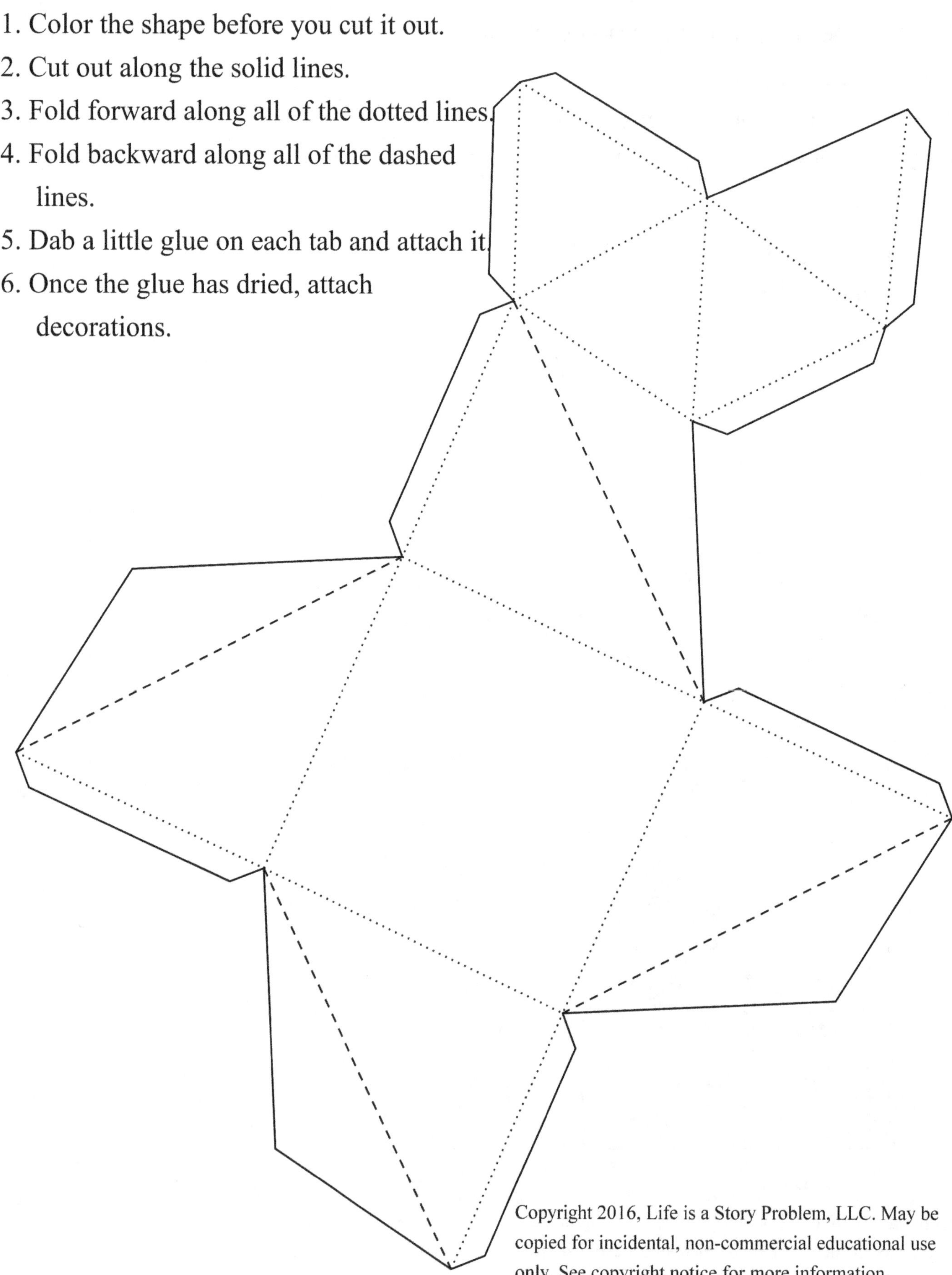

 Geometric Nets Mega Project Book, Tabbed By David E. McAdams

Rectangular pyramid

1. Color the shape before you cut it out.
2. Cut out along the solid lines.
3. Fold along all of the dotted lines.
4. Dab a little glue on each tab and attach it.
5. Once the glue has dried, attach any decorations.

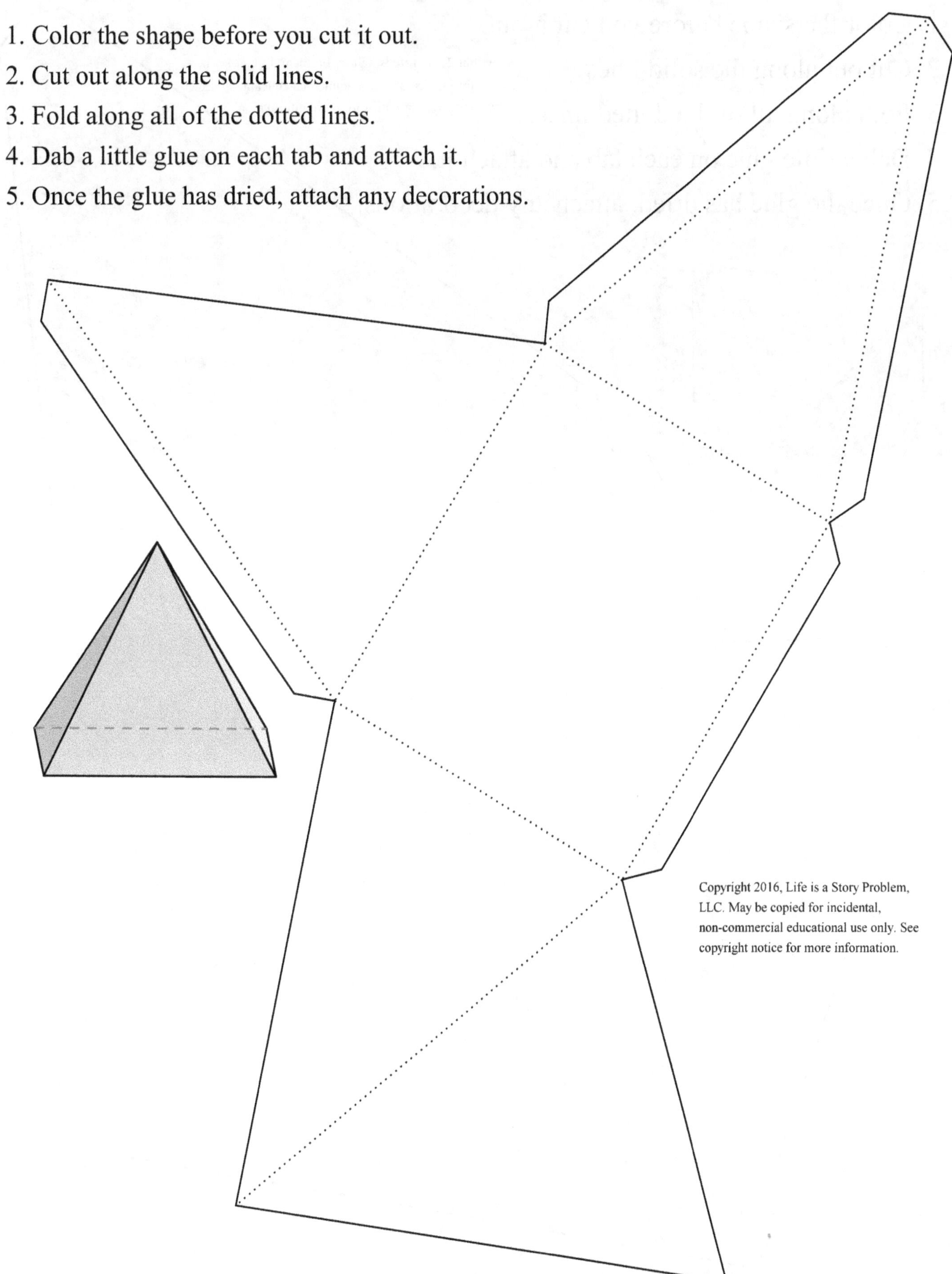

Triangular pentahedron

1. Color the shape before you cut it out.

2. Cut out along the solid lines.

3. Fold along all of the dotted lines.

4. Dab a little glue on each tab and attach it.

5. Once the glue has dried, attach any decorations.

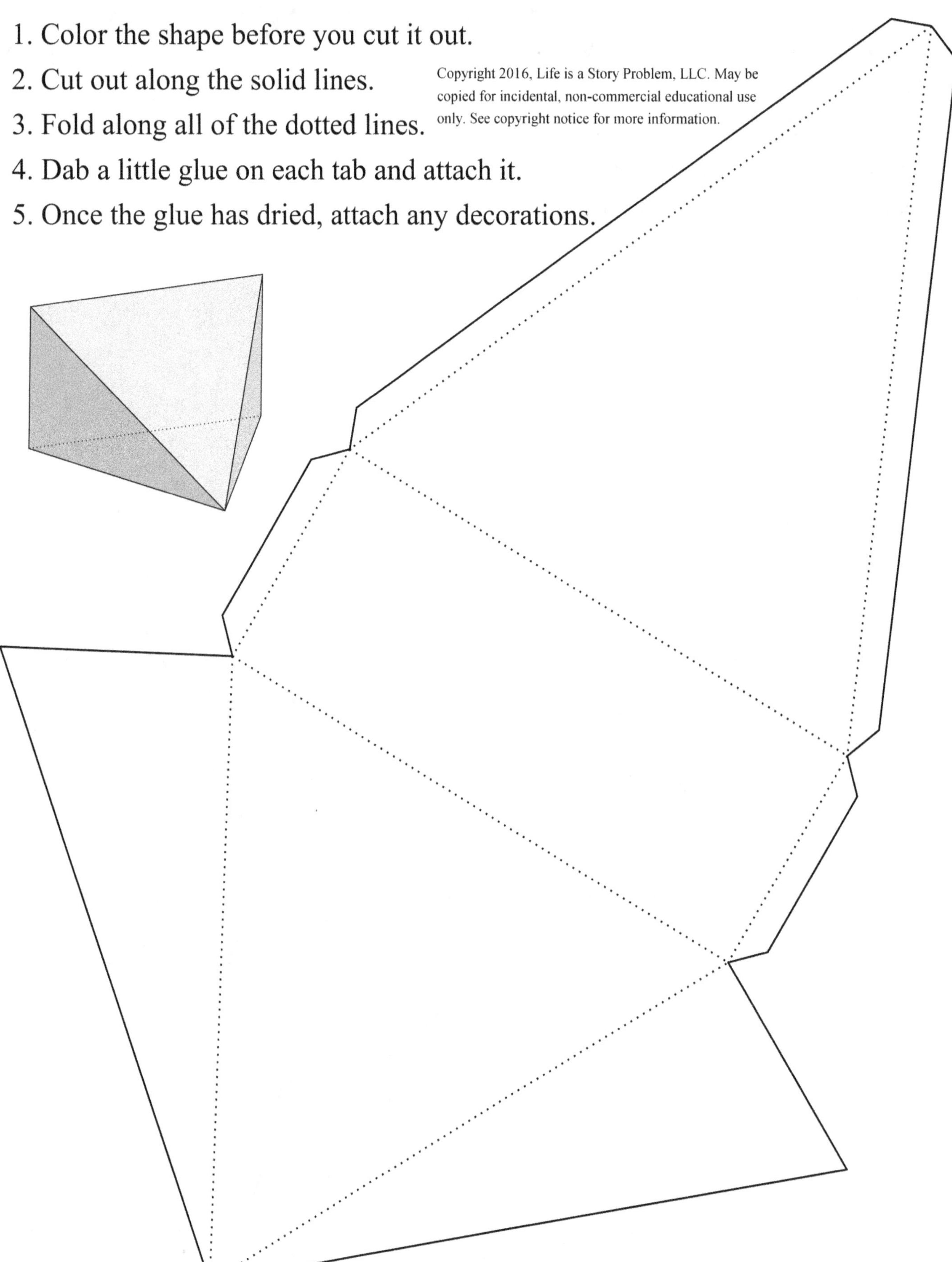

Geometric Nets Mega Project Book, Tabbed By David E. McAdams

Edge contracted icosahedron

1. Color the shape before you cut it out.
2. Cut out along the solid lines.
3. Fold along all of the dotted lines.
4. Dab a little glue on each tab and attach it.
5. Once the glue has dried, attach any decorations.

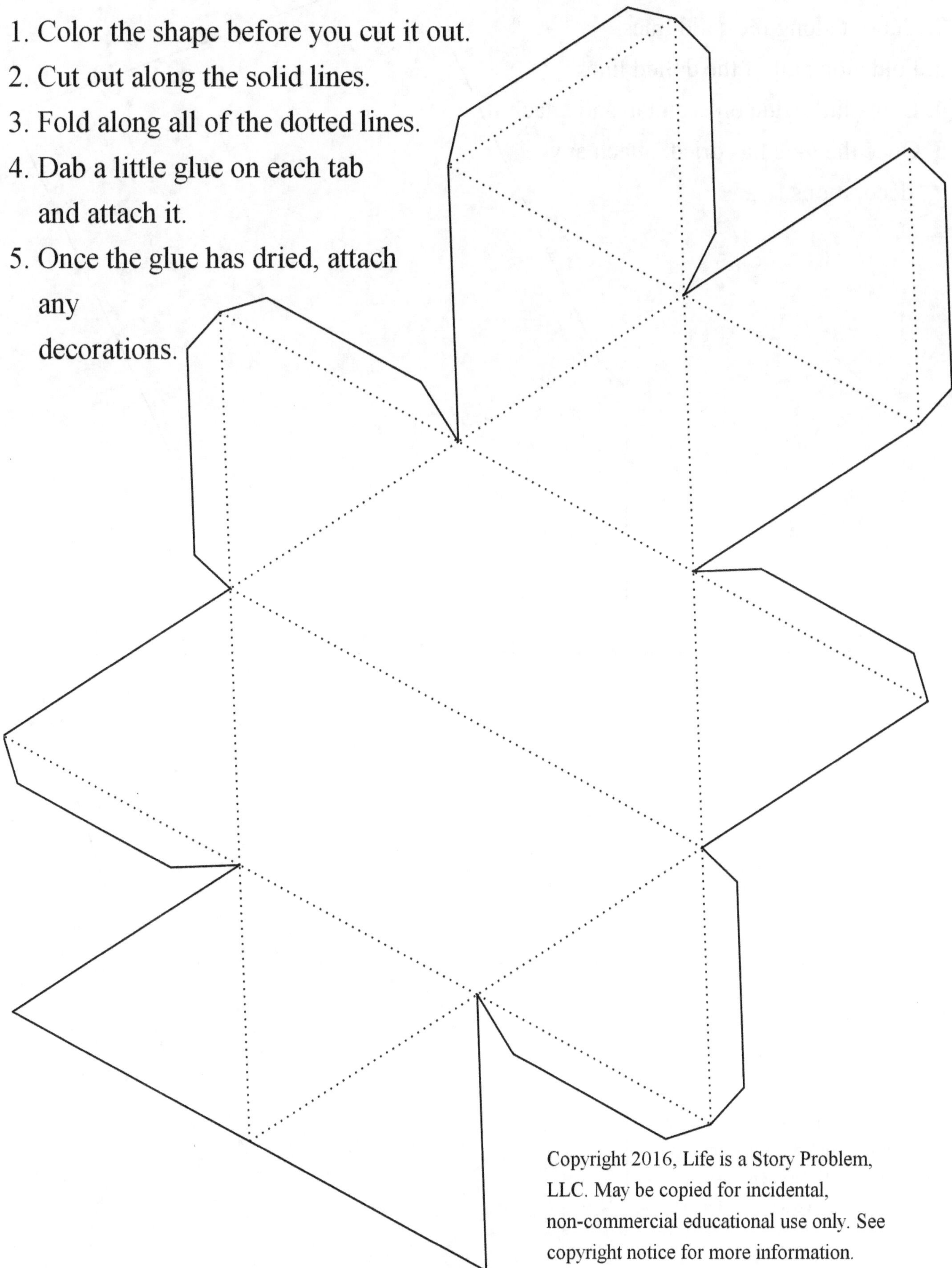

Hexahedron 4,4,4,4,3,3

1. Color the shape before you cut it out.

2. Cut out along the solid lines.

3. Fold along all of the dotted lines.

4. Dab a little glue on each tab and attach it.

5. Once the glue has dried, attach any decorations.

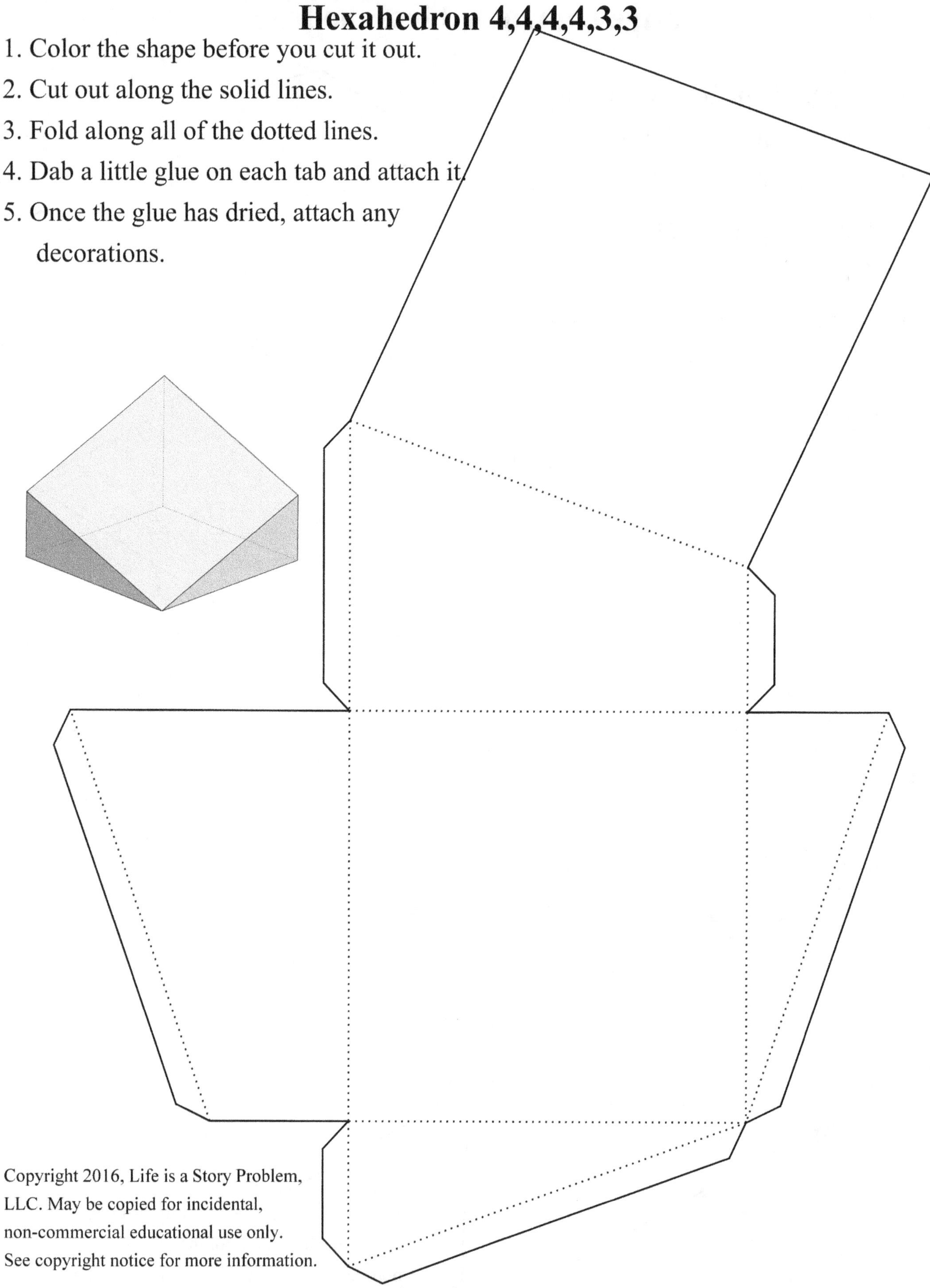

Hexahedron 5,4,4,3,3,3

1. Color the shape before you cut it out.
2. Cut out along the solid lines.
3. Fold along all of the dotted lines.
4. Dab a little glue on each tab and attach it.
5. Once the glue has dried, attach any decorations.

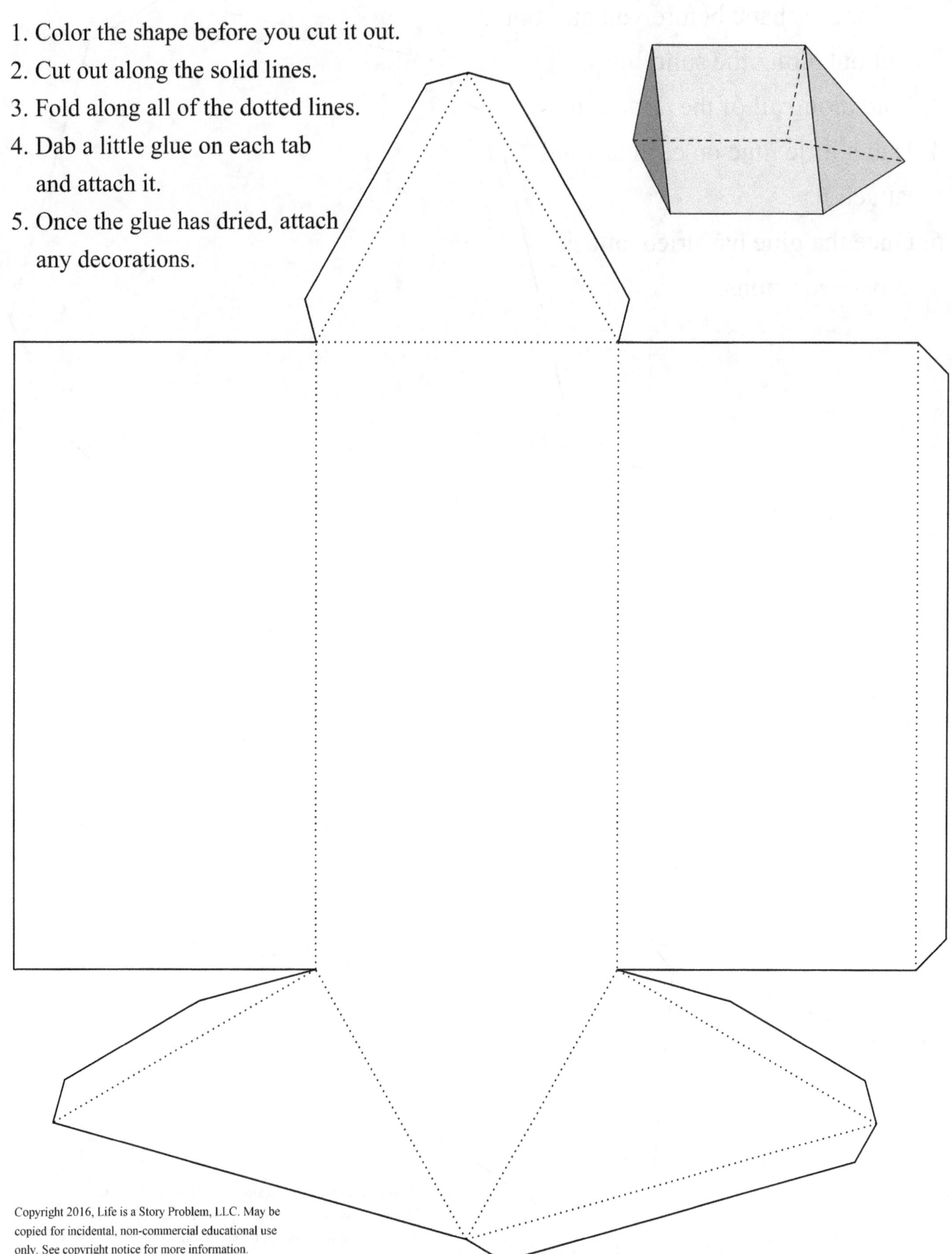

Hexahedron 5,5,4,4,3,3

1. Color the shape before you cut it out.
2. Cut out along the solid lines.
3. Fold along all of the dotted lines.
4. Dab a little glue on each tab and attach it.
5. Once the glue has dried, attach any decorations.

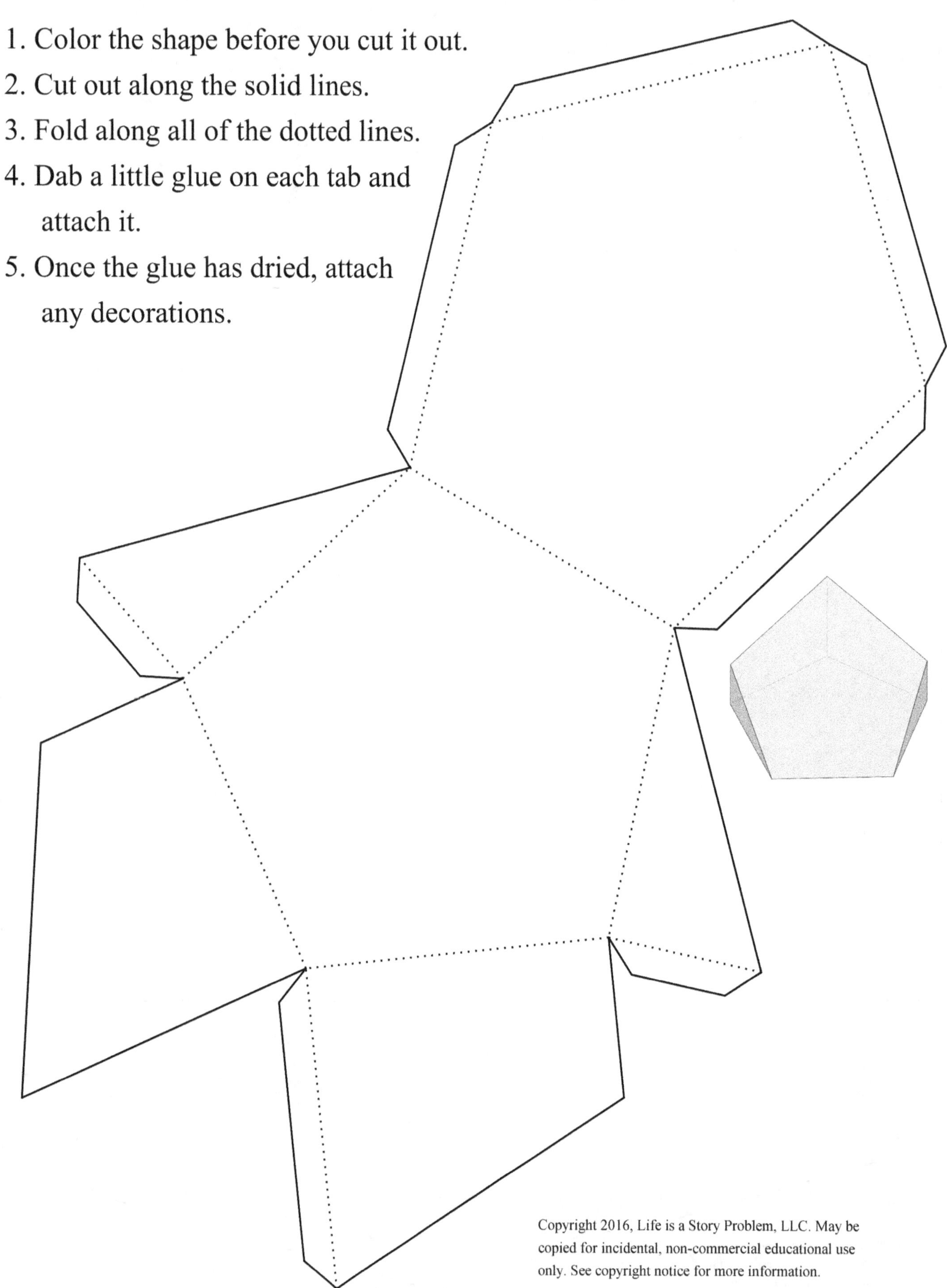

Geometric Nets Mega Project Book, Tabbed By David E. McAdams

Heptahedron 4,4,4,3,3,3,3

1. Color the shape before you cut it out.
2. Cut out along the solid lines.
3. Fold along all of the dotted lines.
4. Dab a little glue on each tab and attach it.
5. Once the glue has dried, attach any decorations.

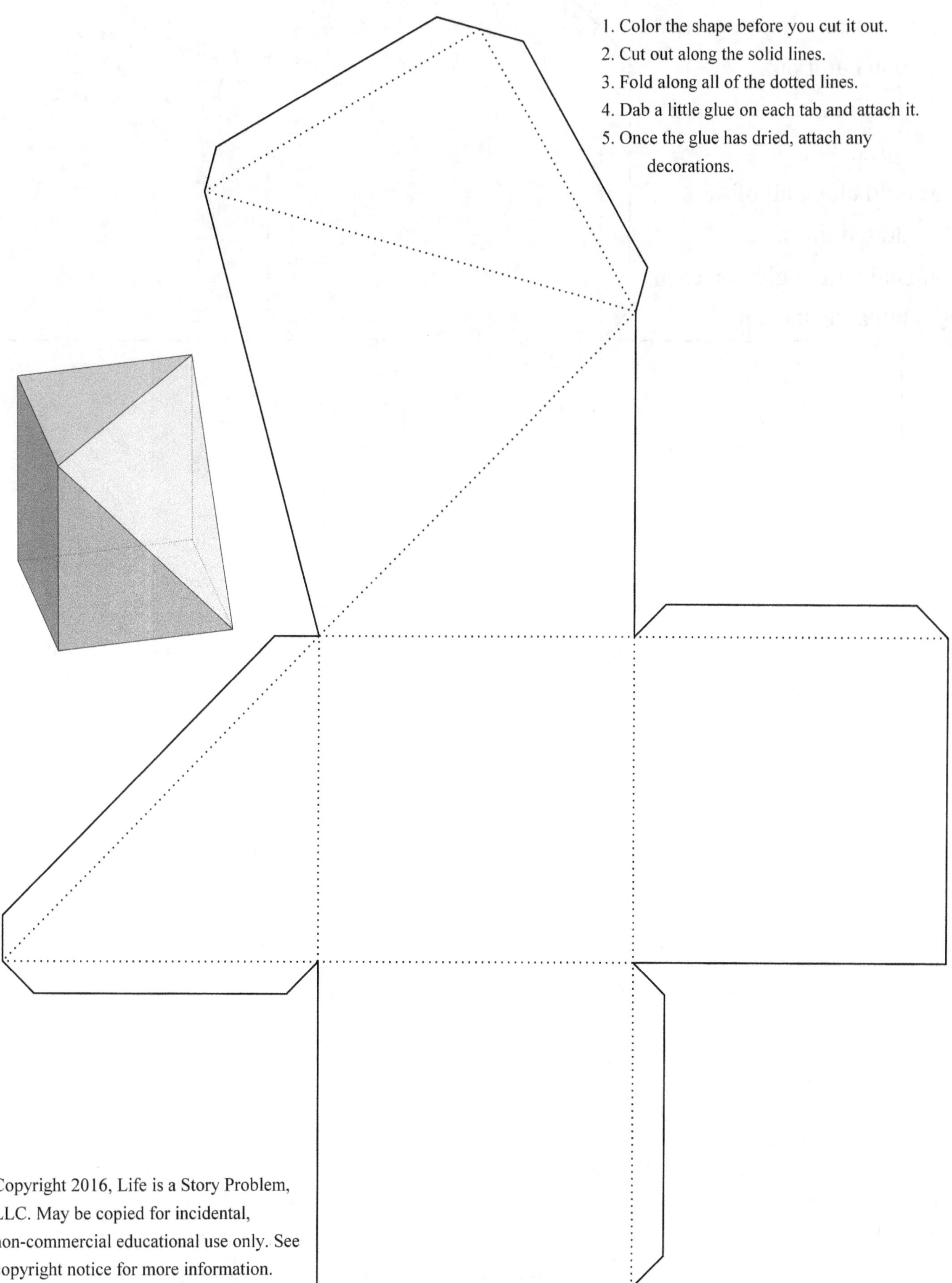

Heptahedron 5,5,5,4,4,4,3

1. Color the shape before you cut it out.

2. Cut out along the solid lines.

3. Fold along all of the dotted lines.

4. Dab a little glue on each tab and attach it.

5. Once the glue has dried, attach any decorations.

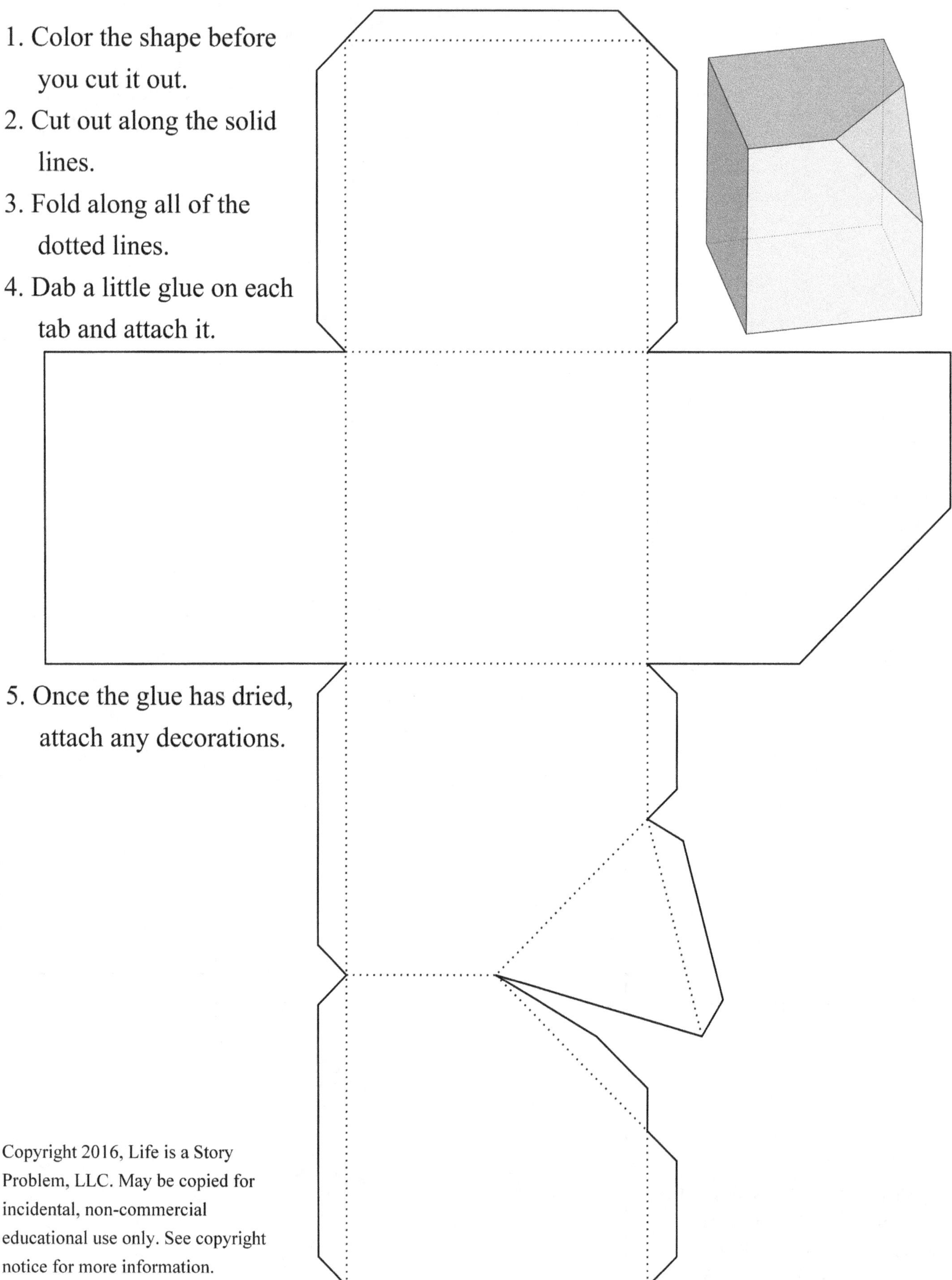

Geometric Nets Mega Project Book, Tabbed By David E. McAdams

Heptahedron 6,6,4,4,4,3,3

1. Color the shape before you cut it out.
2. Cut out along the solid lines.
3. Fold along all of the dotted lines.
4. Dab a little glue on each tab and attach it.
5. Once the glue has dried, attach any
   decorations.

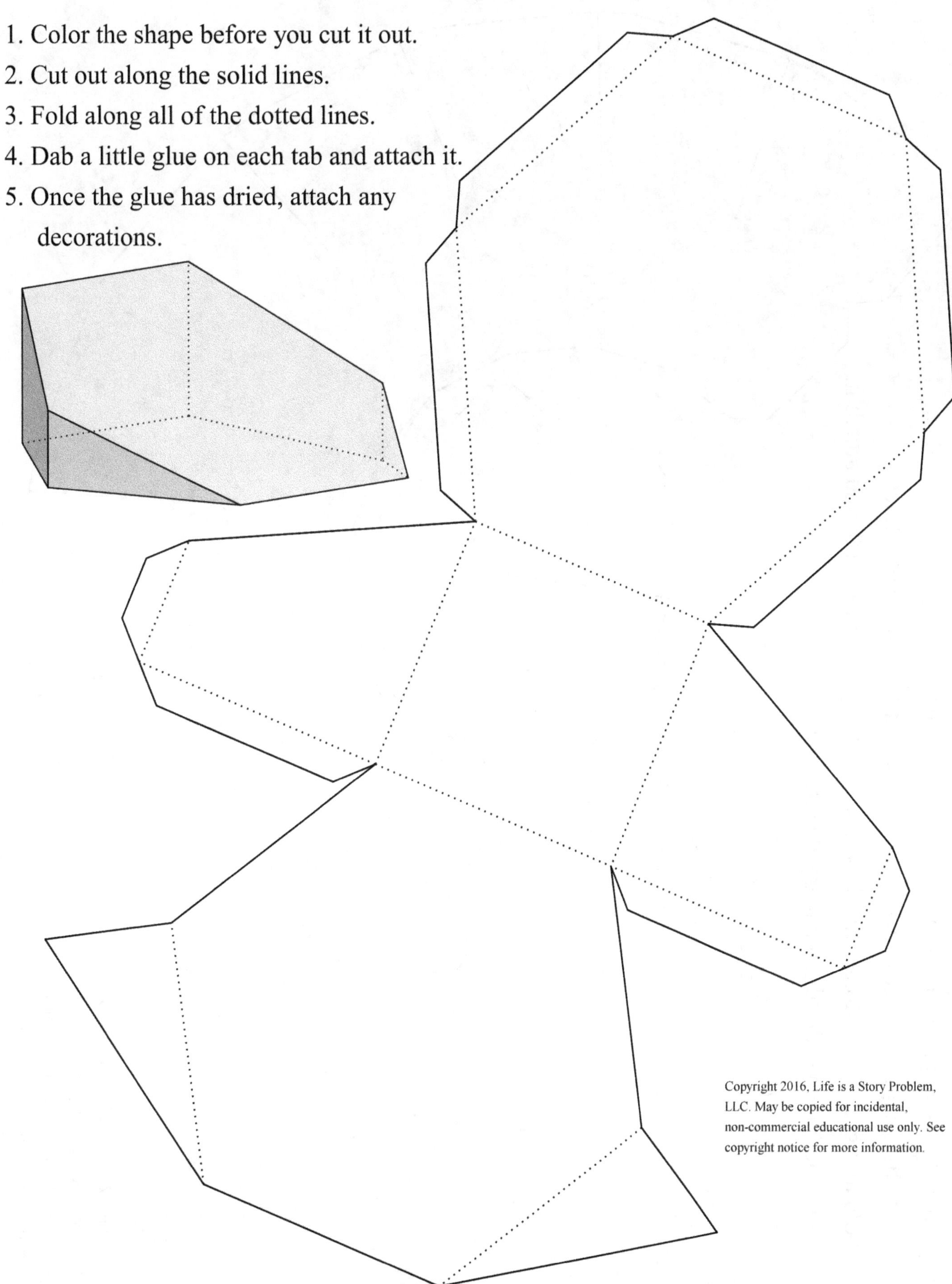

This net is on two pages. Print out a copy of each page.

1. Color the shape before you cut it out. 2. Cut out along the solid lines.

3. Fold along all of the dotted lines. 4. Dab a little glue on each tab and attach it.

5. Once the glue has dried, attach any decorations.

Geometric Nets Mega Project Book, Tabbed By David E. McAdams

Hebdomicontadissaedron

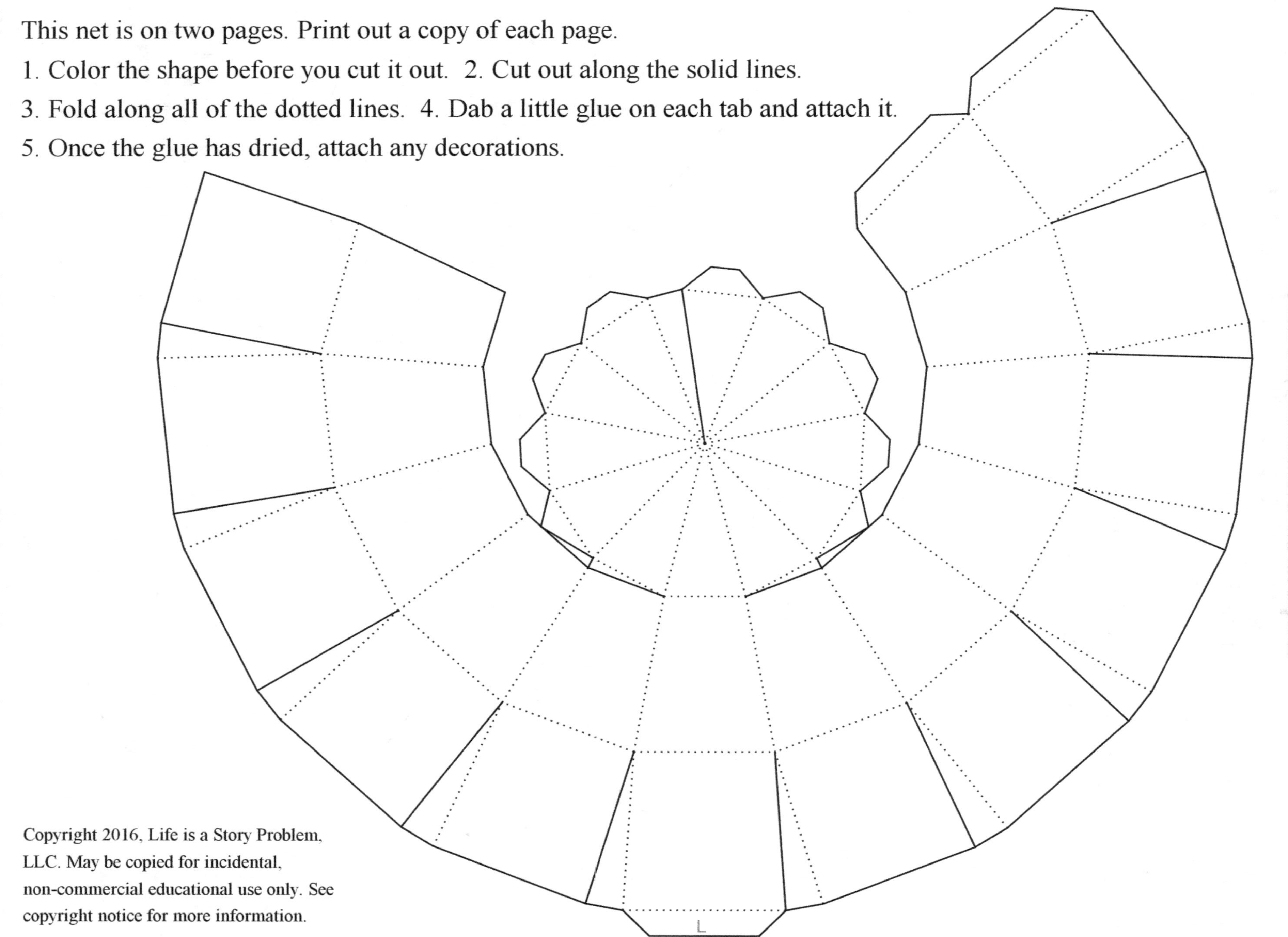

2nd page, Hebdomicontadissaedron

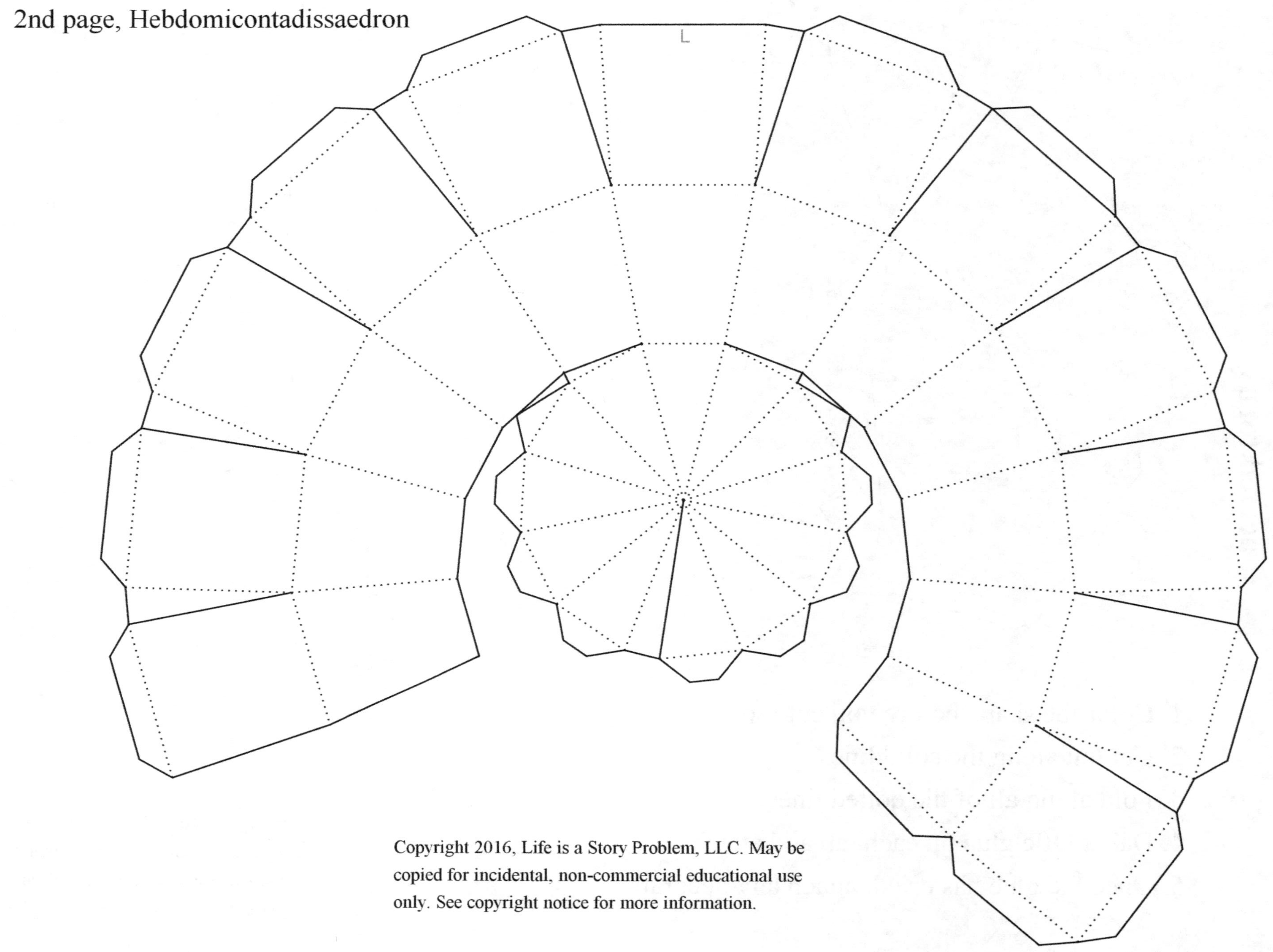

Faceted torus

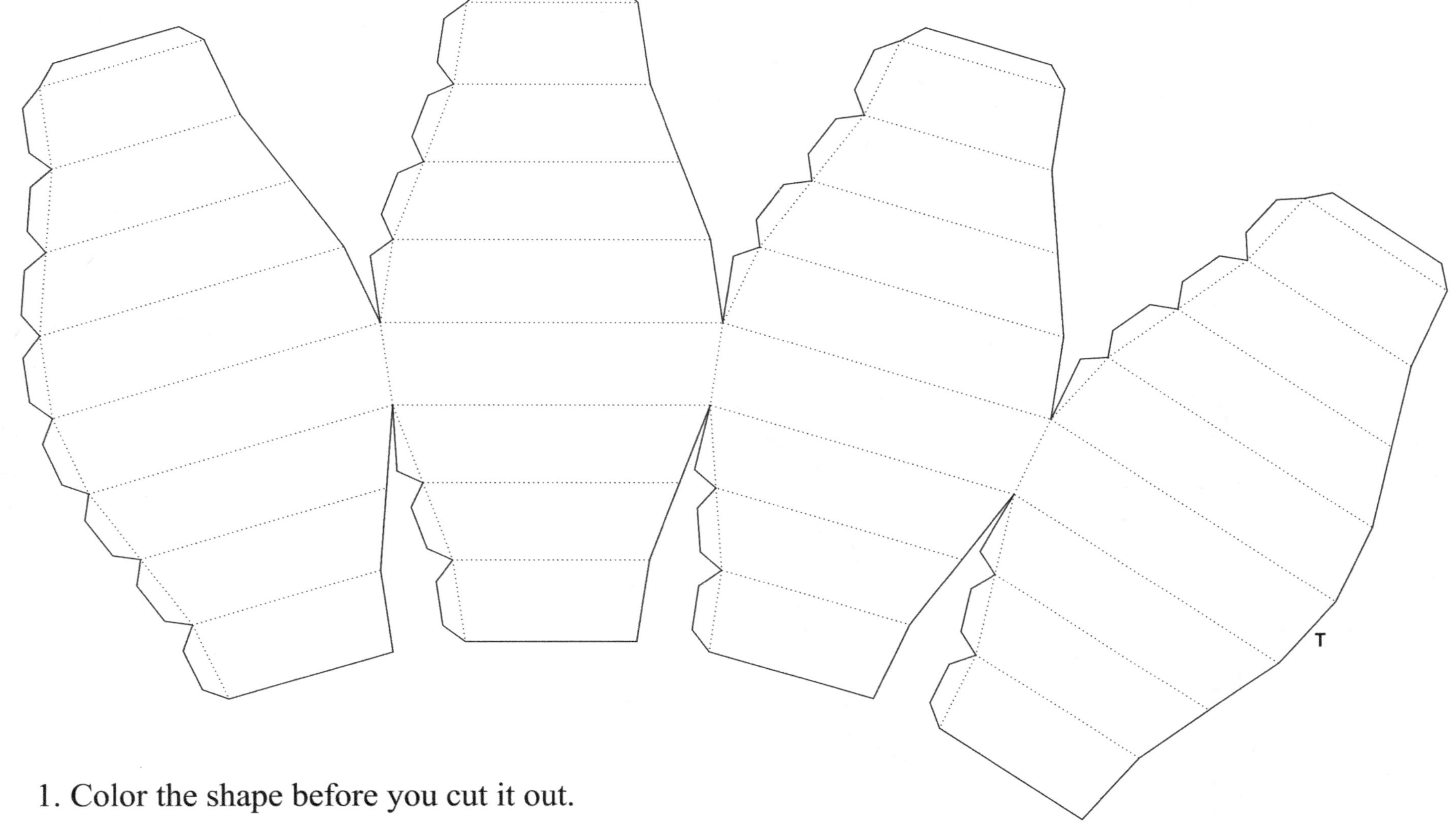

1. Color the shape before you cut it out.
2. Cut out along the solid lines.
3. Fold along all of the dotted lines.
4. Dab a little glue on each tab and attach it.
5. Once the glue has dried, attach any decorations.

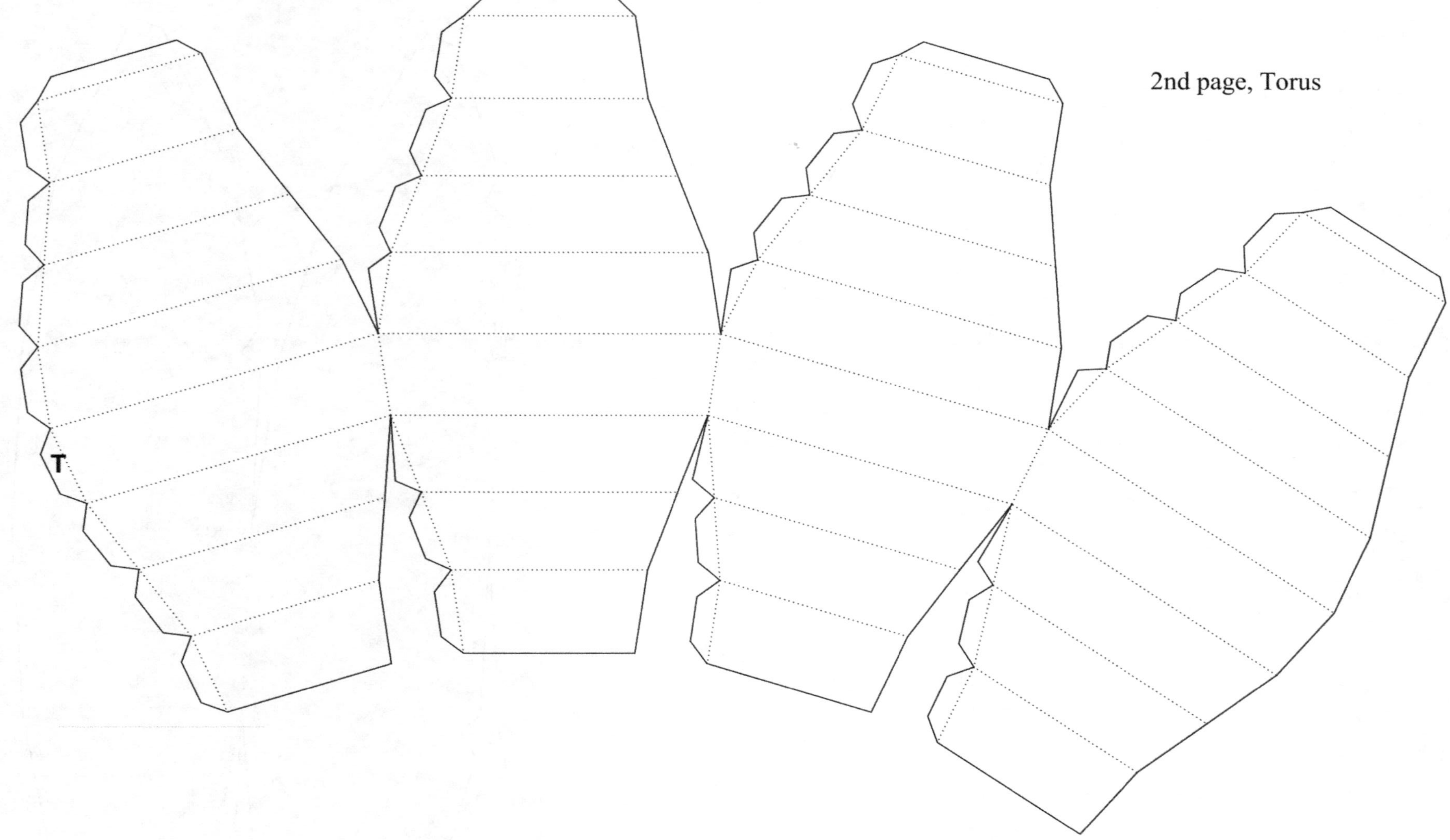

2nd page, Torus

Faceted oblate ellipsoid

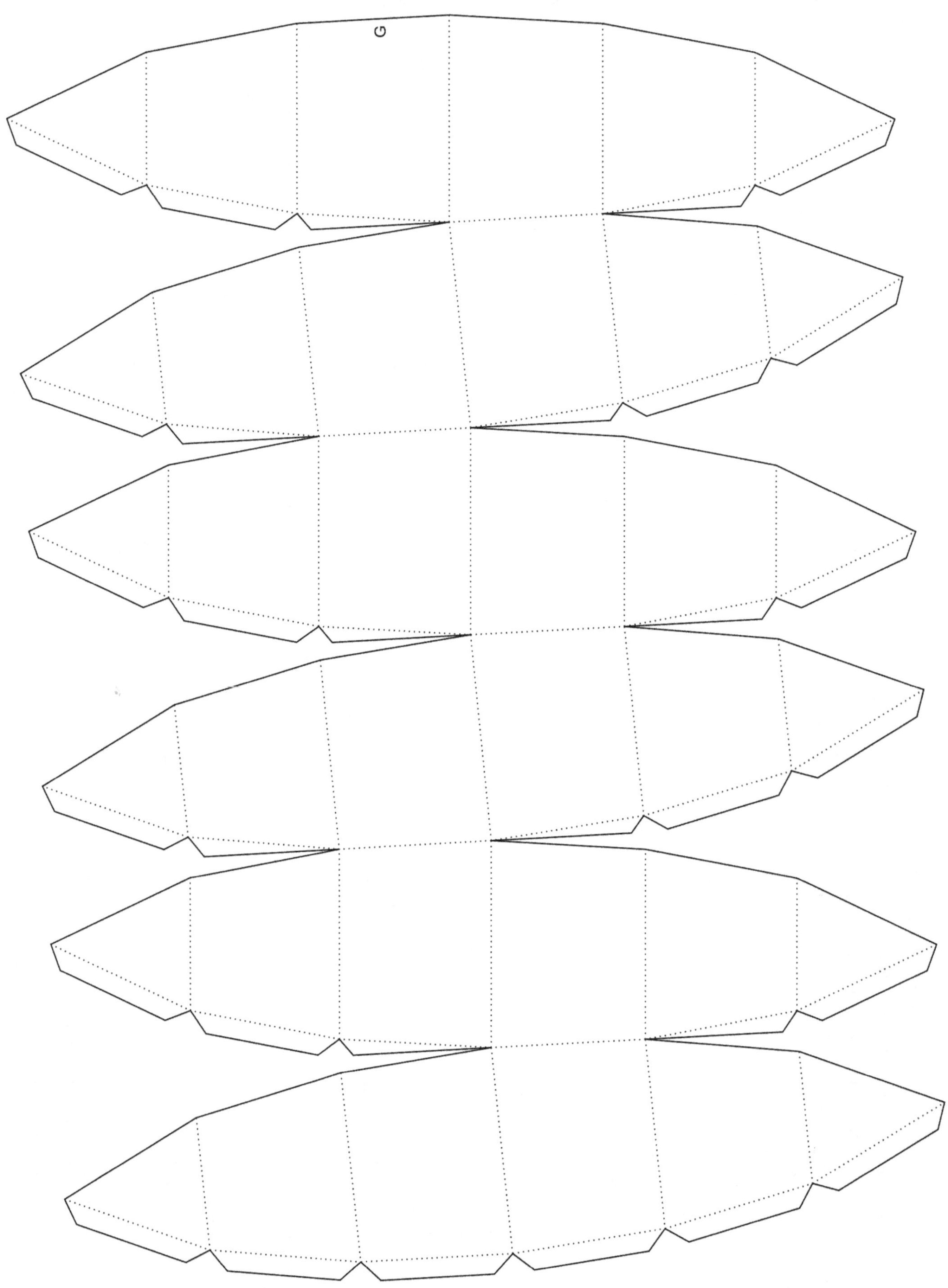

Geometric Nets Mega Project Book, Tabbed By David E. McAdams

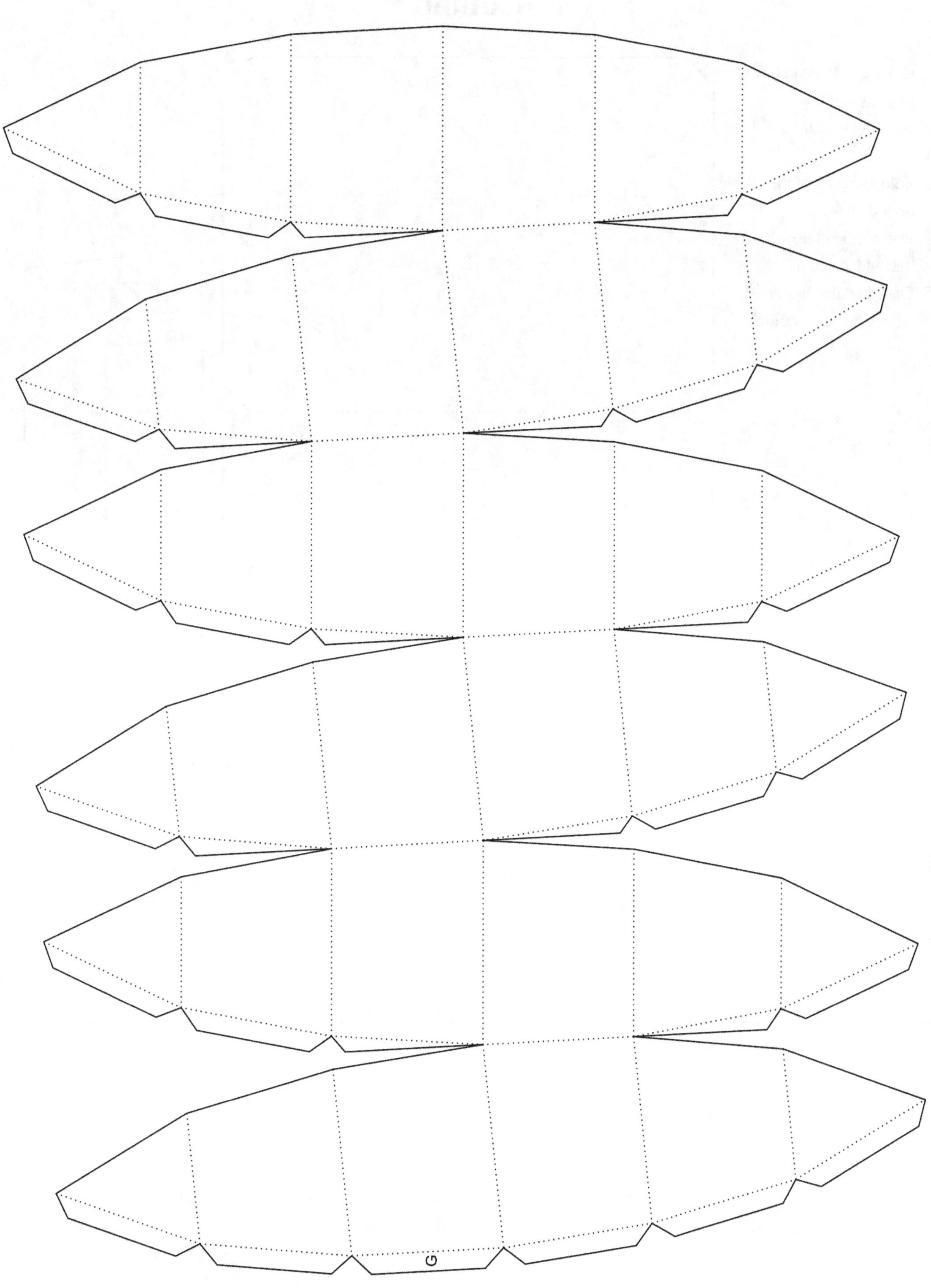

Cuboid

1. Color the shape before you cut it out.
2. Cut out along the solid lines.
3. Fold along all of the dotted lines.
4. Dab a little glue on each tab and attach it.
5. Once the glue has dried, attach any decorations.

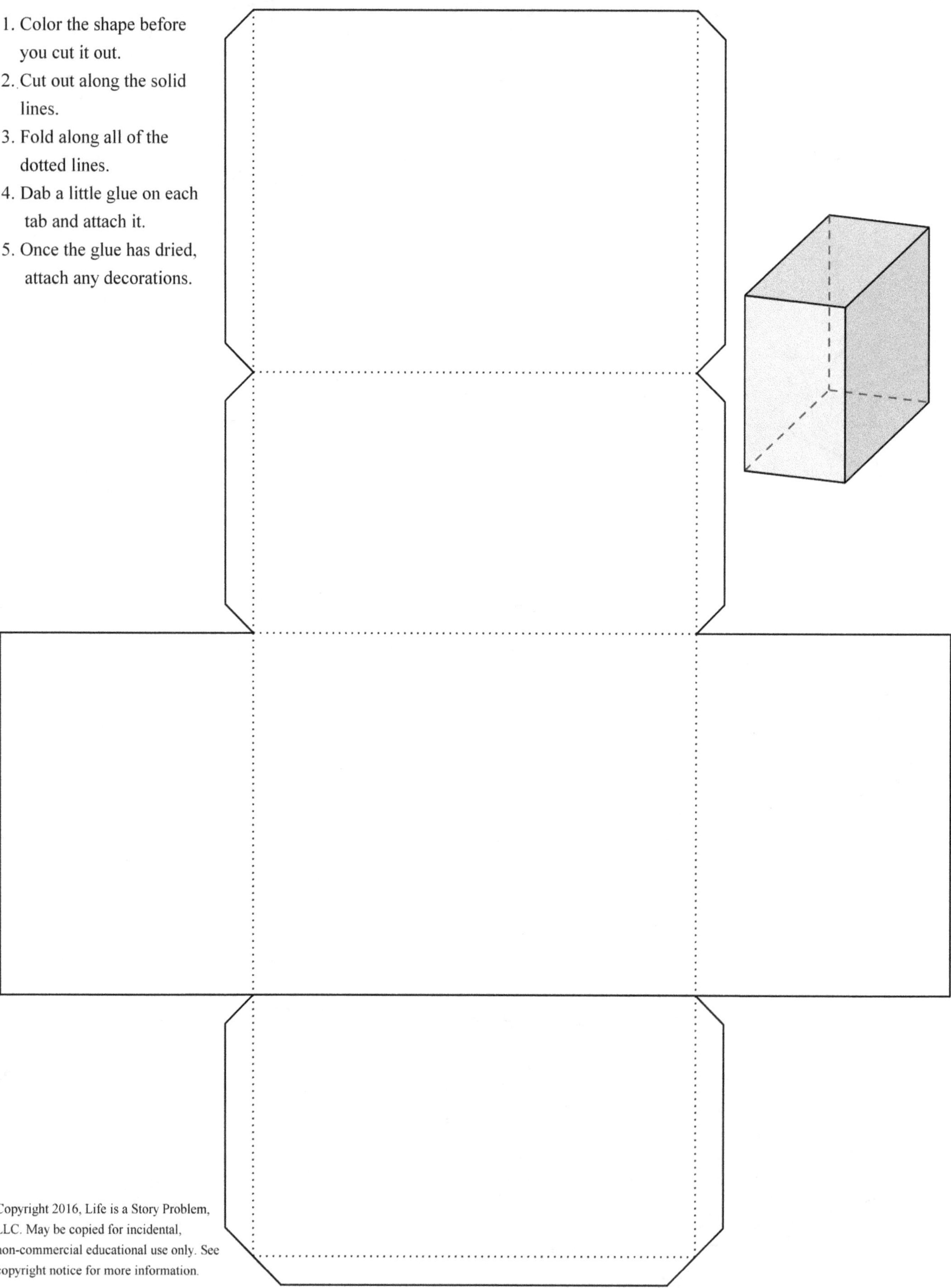

 Geometric Nets Mega Project Book, Tabbed By David E. McAdams

Cuboid augmented cube

1. Color the shape before you cut
   it out.

2. Cut out along the solid
   lines.

3. Fold forward along all of
   the dotted lines.

4. Fold backward along all
   of the dashed lines.

5. Dab a little glue on each
   tab and attach it.

6. Once the glue has dried,
   attach any decorations.

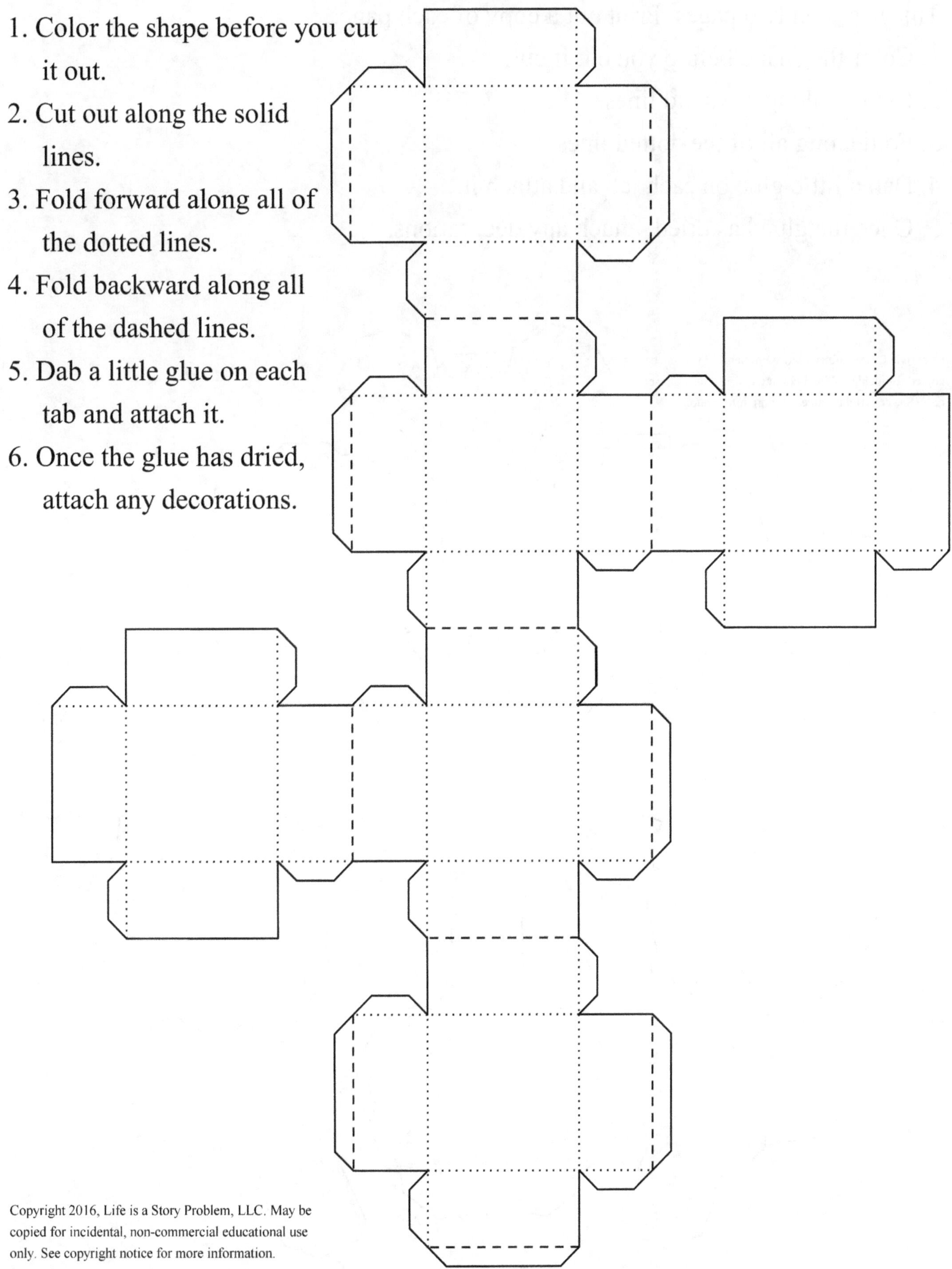

Truncated icosidodechedon

This net is on two pages. Print out a copy of each page.

1. Color the shape before you cut it out.

2. Cut out along the solid lines.

3. Fold along all of the dotted lines.

4. Dab a little glue on each tab and attach it.

5. Once the glue has dried, attach any decorations.

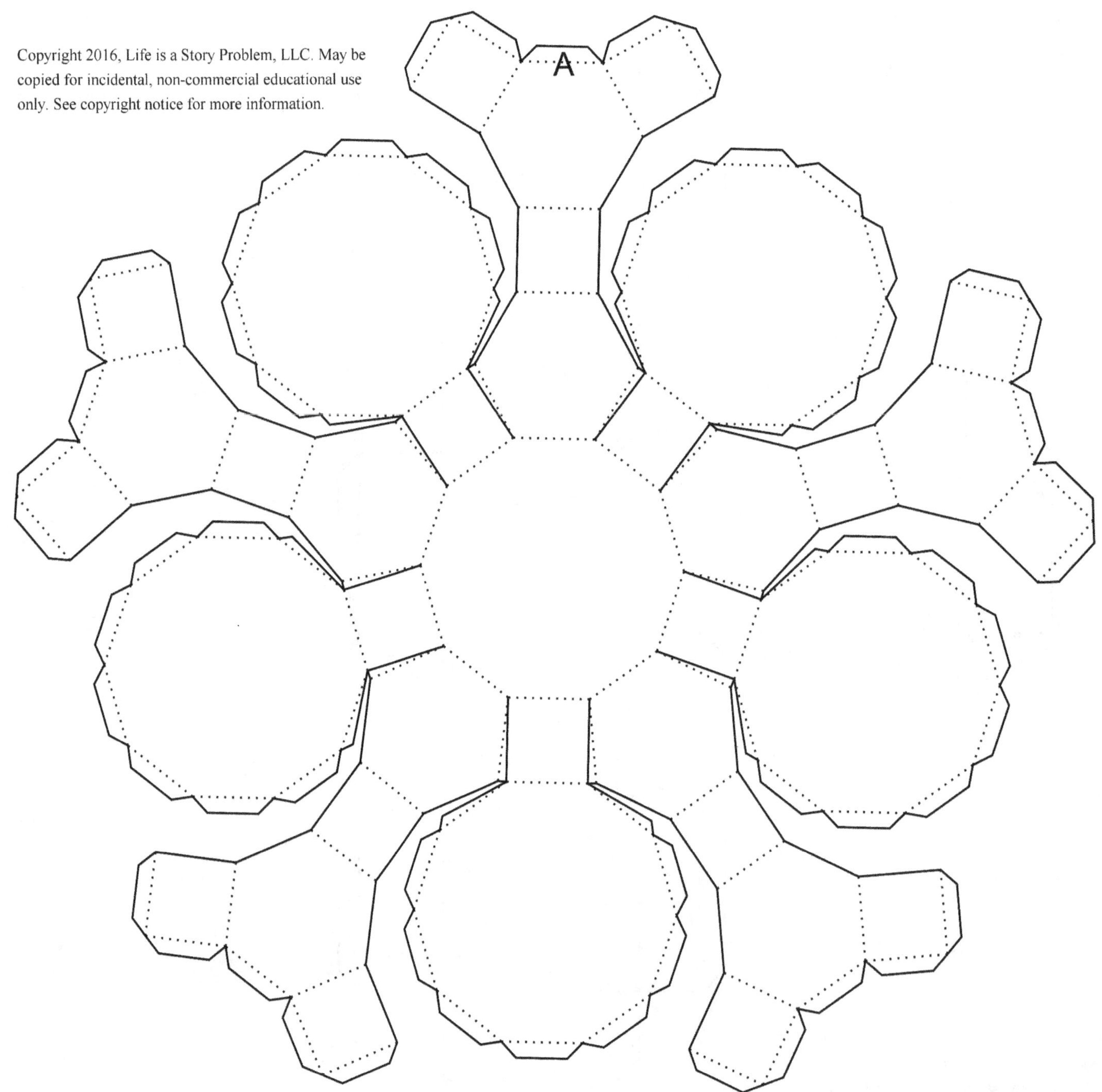

2nd page, truncated icosidodechedon

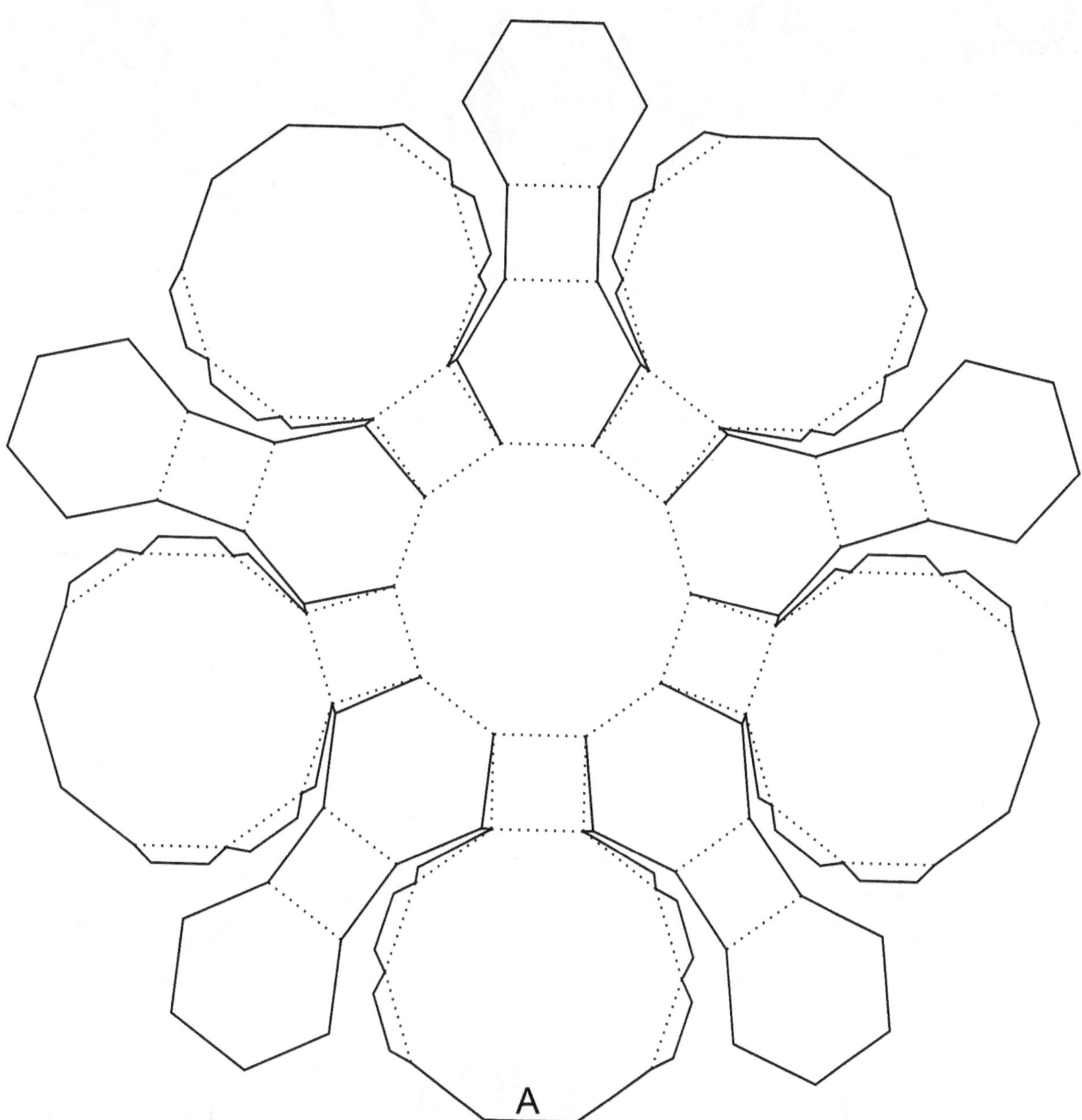

Reduced cube

1. Color the shape before you cut it out.

2. Cut out along the solid lines.

3. Fold forward along all of the dotted lines.

4. Fold backward along all of the dashed lines.

5. Dab a little glue on each tab and attach it.

6. Once the glue has dried, attach any decorations.

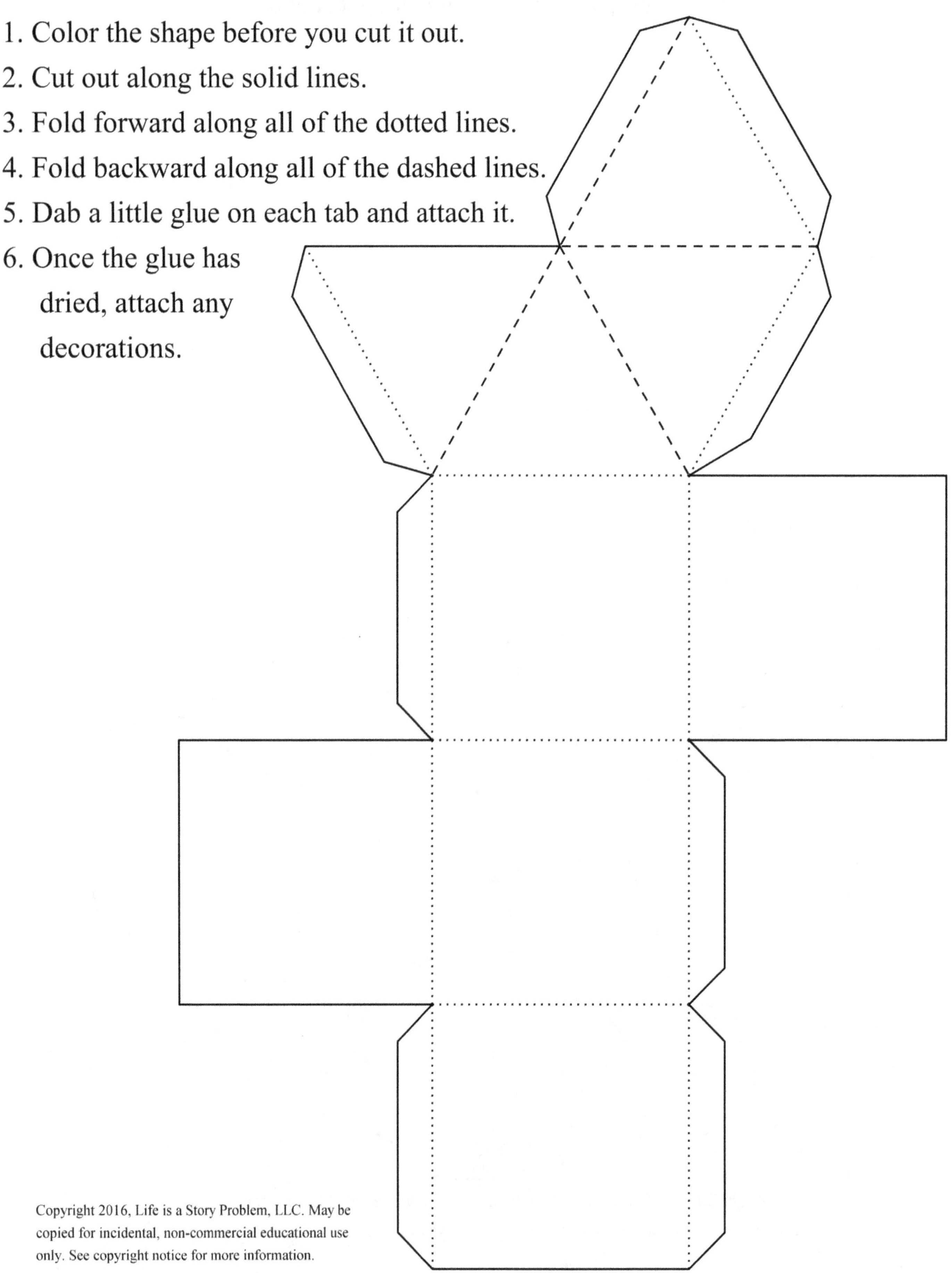

 Geometric Nets Mega Project Book, Tabbed By David E. McAdams

Bireduced cube

1. Color the shape before you cut it out.
2. Cut out along the solid lines.
3. Fold forward along all of the dotted lines.
4. Fold backward along all of the dashed lines.
5. Dab a little glue on each tab and attach it.
6. Once the glue has dried, attach any decorations.

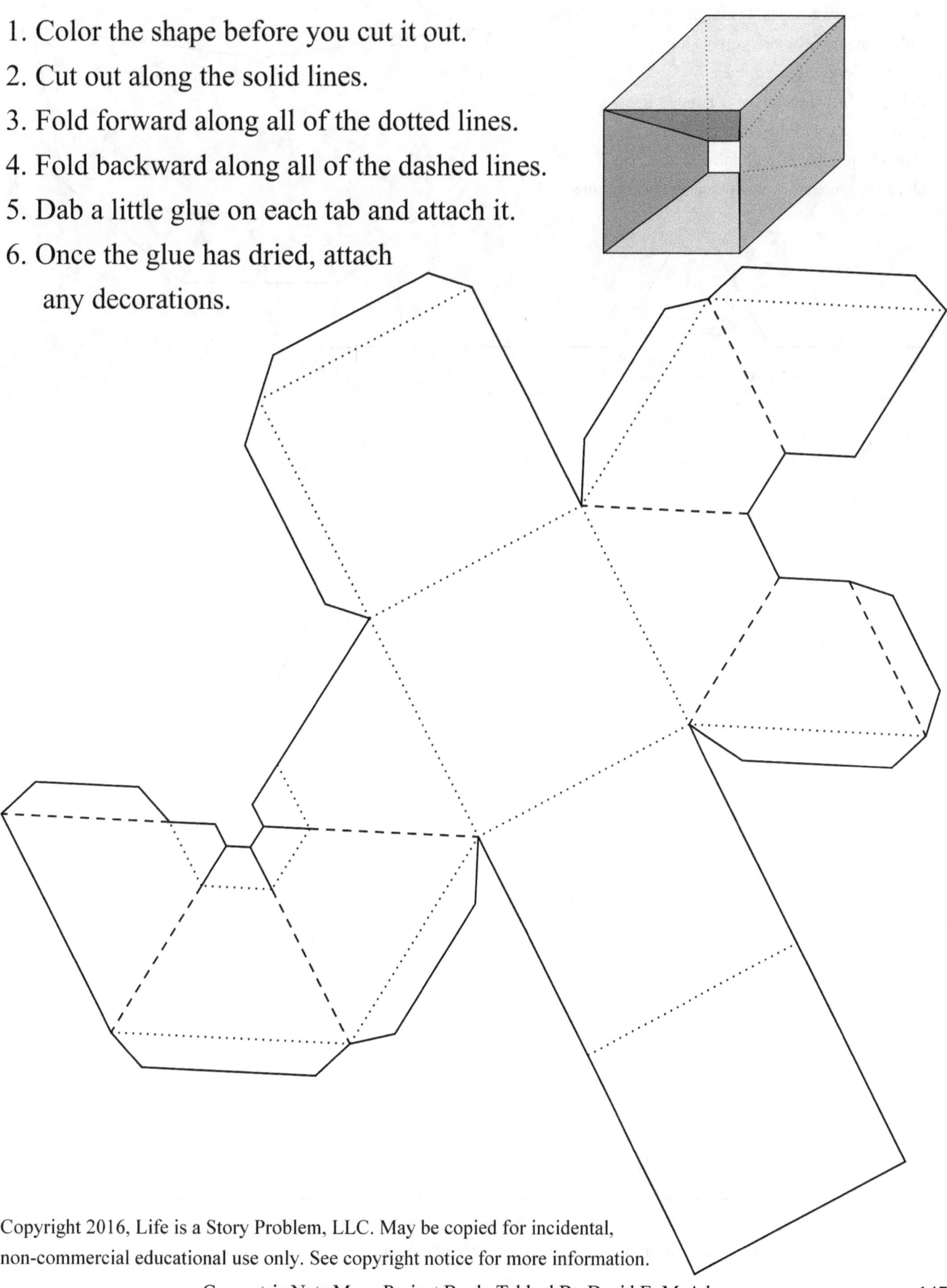

Twist augmented countertwist pentagonal prism

This net is on two pages. Print out a copy of each page.

1. Color the shape before you cut it out.
2. Cut out along the solid lines.
3. Fold forward along all of the dotted lines.
4. Fold backward along all of the dashed lines.
5. Dab a little glue on each tab and attach it.
6. Once the glue has dried, attach any decorations.

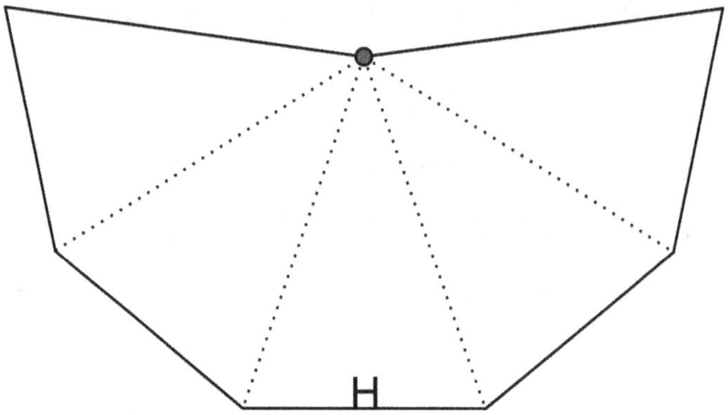

H

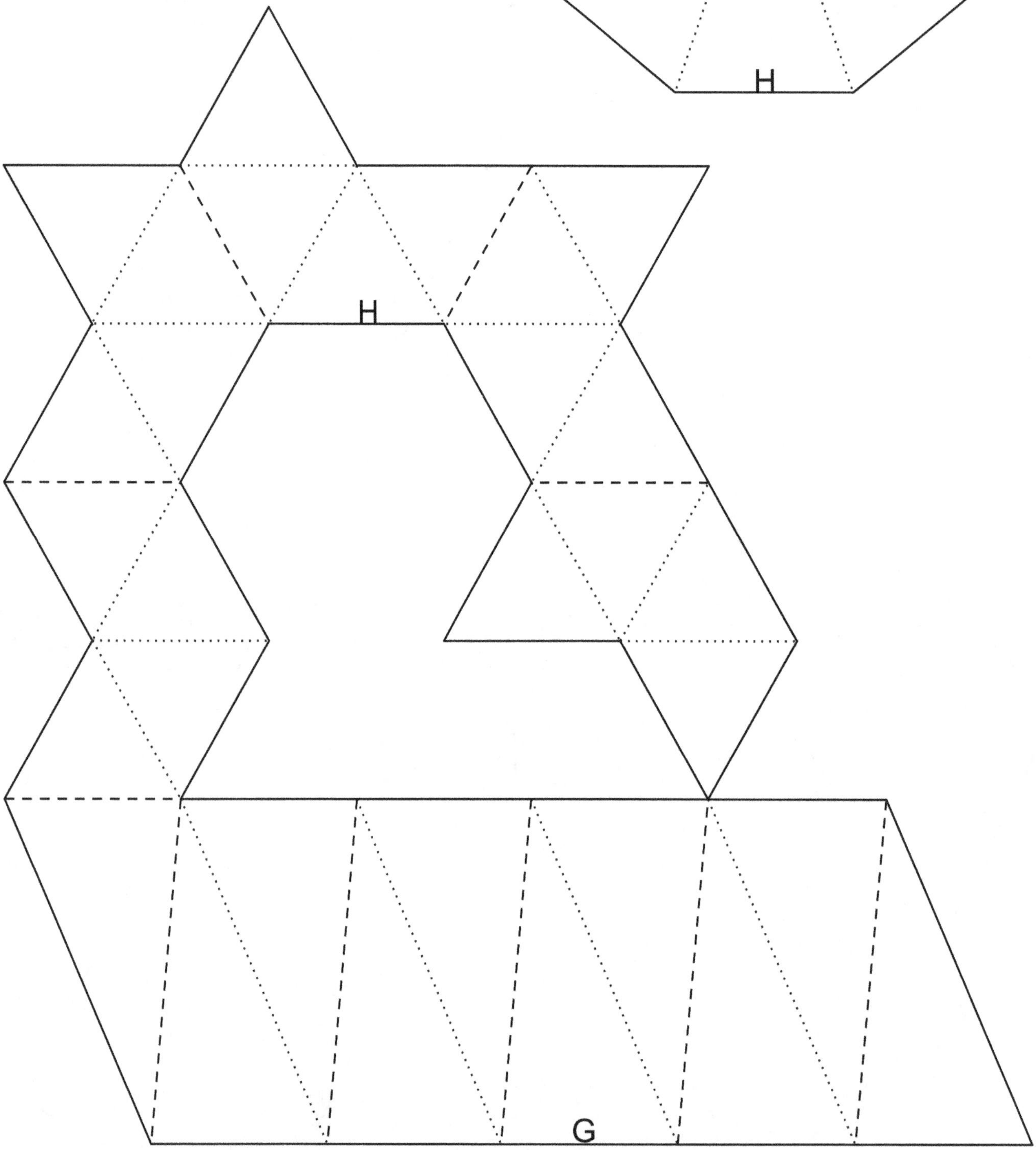

H

G

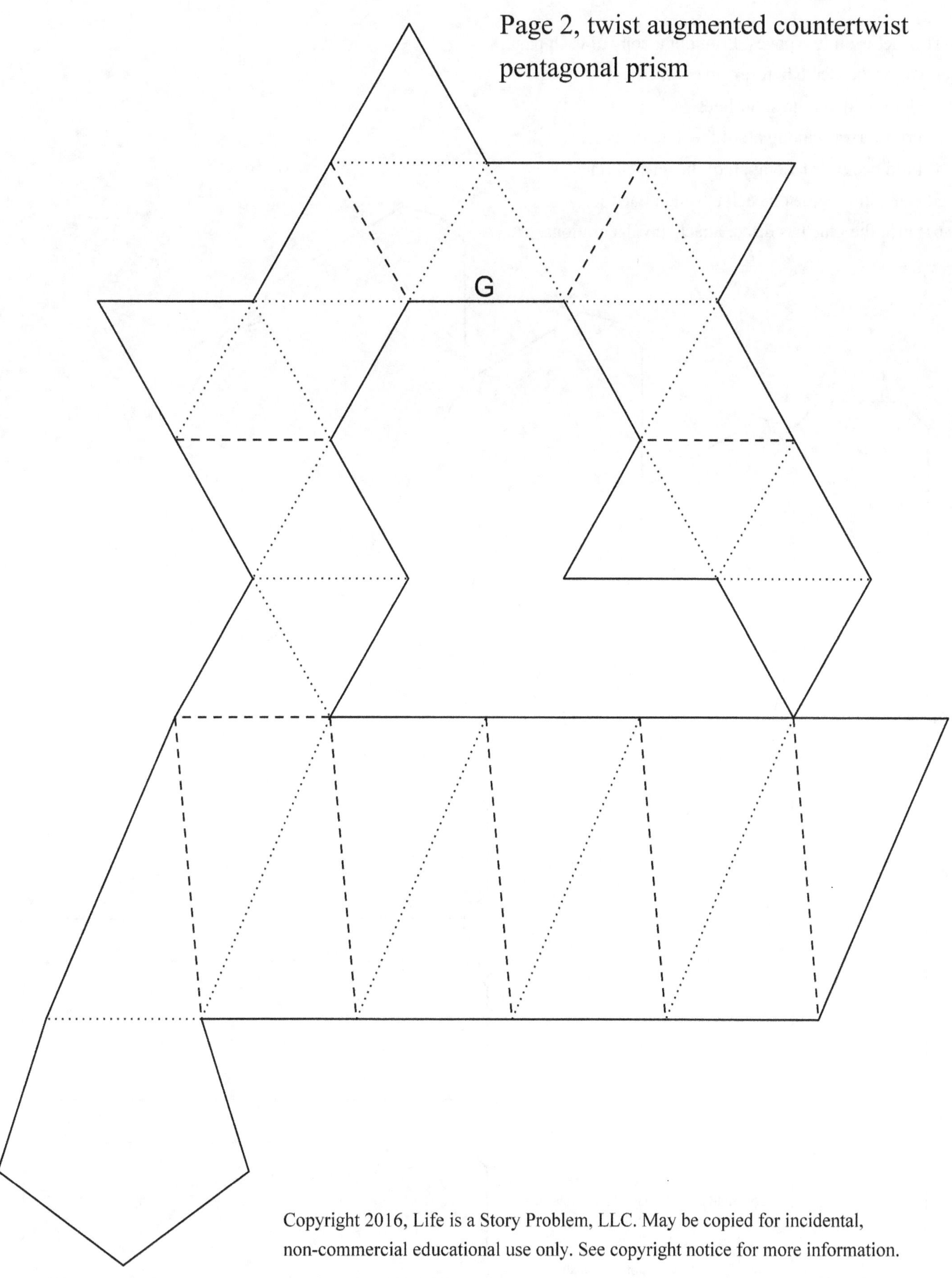

Page 2, twist augmented countertwist pentagonal prism
G

Sierpinski Triangle, M2

This net is on two pages. Print out a copy of each page.

1. Color the shape before you cut it out.

2. Cut out along the solid lines.

3. Fold forward along all of the dotted lines.

4. Fold backward along all of the dashed lines.

5. Dab a little glue on each tab and attach it.

6. Once the glue has dried, attach any decorations.

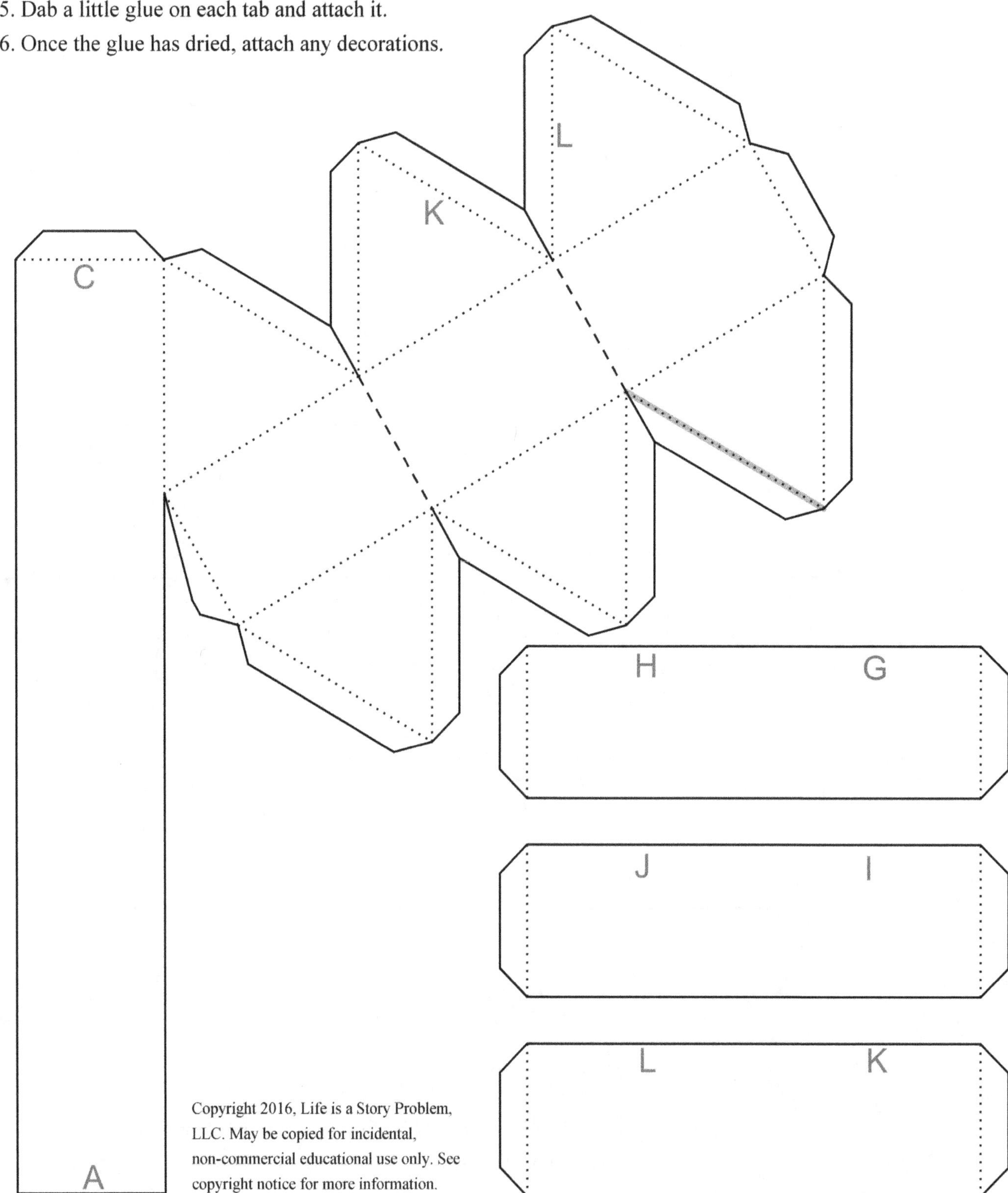

Geometric Nets Mega Project Book, Tabbed By David E. McAdams

Page 2, Sierpinski triangle

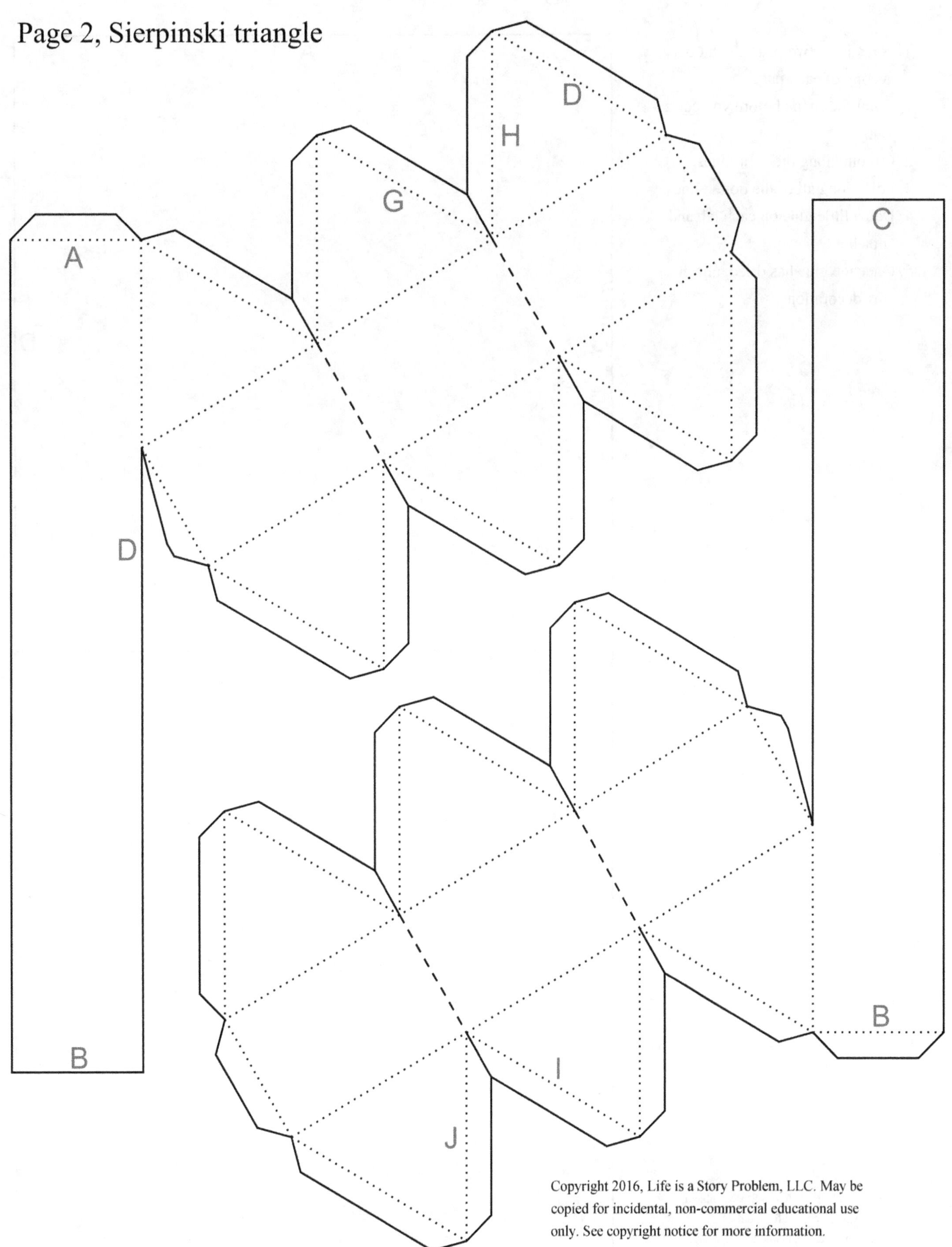

Menger Sponge, M0

This net is on two pages. Print out a copy of each page.

1. Color the shape before you cut it out.
2. Cut out along the solid lines.
3. Fold along all of the dotted lines.
4. Dab a little glue on each tab and attach it.
5. Once the glue has dried, attach any decorations.

A

C

D

E

F

B

Geometric Nets Mega Project Book, Tabbed By David E. McAdams

B
G
I
H
J
A

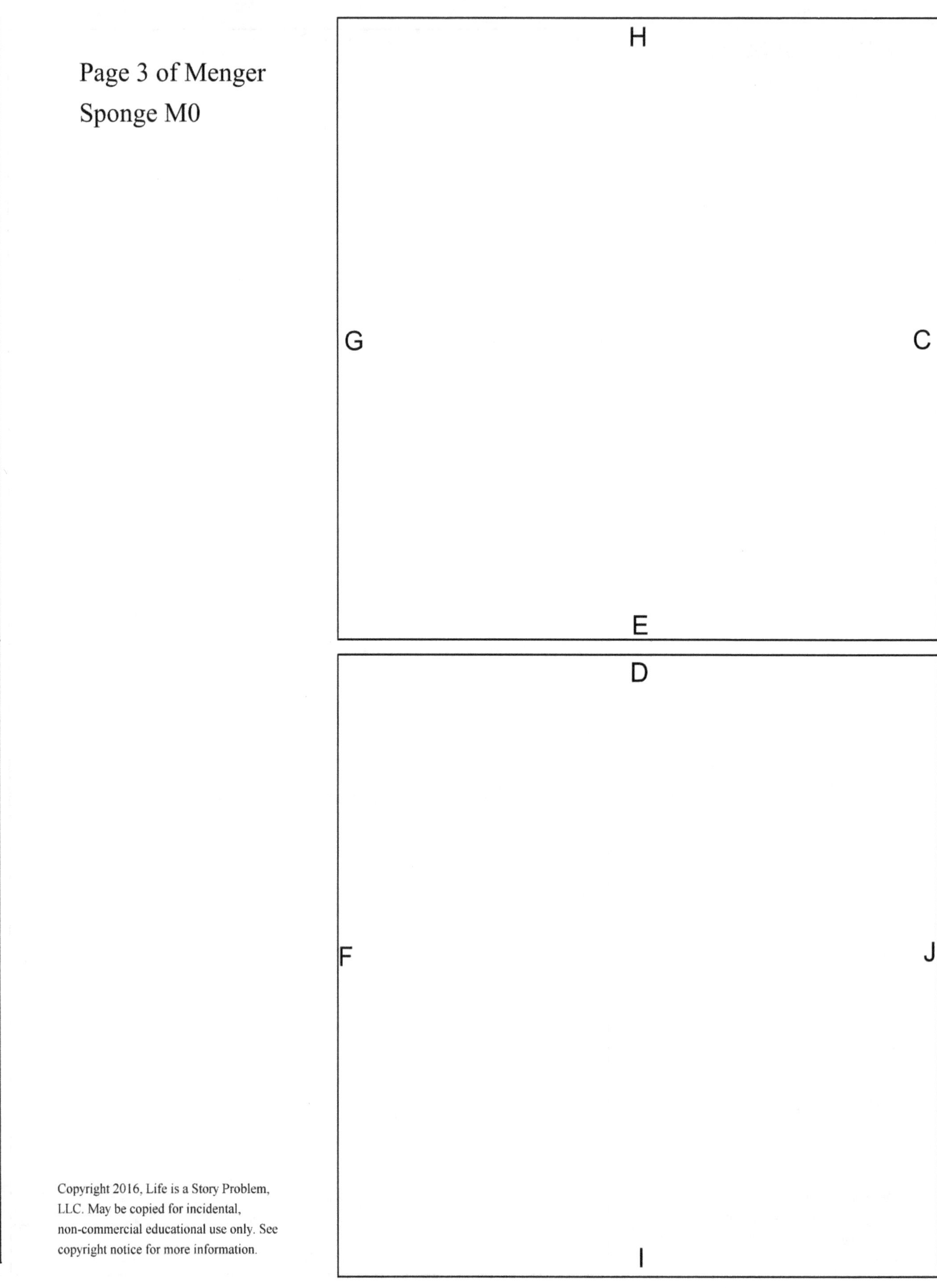
H
G
C
E
D
F
J
I

Menger Sponge, M1

This net is on two pages. Print out a copy of each page.

1. Color the shape before you cut it out.
2. Cut out along the solid lines.
3. Fold along all of the dotted lines.
4. Dab a little glue on each tab and attach it.
5. Once the glue has dried, attach any decorations.

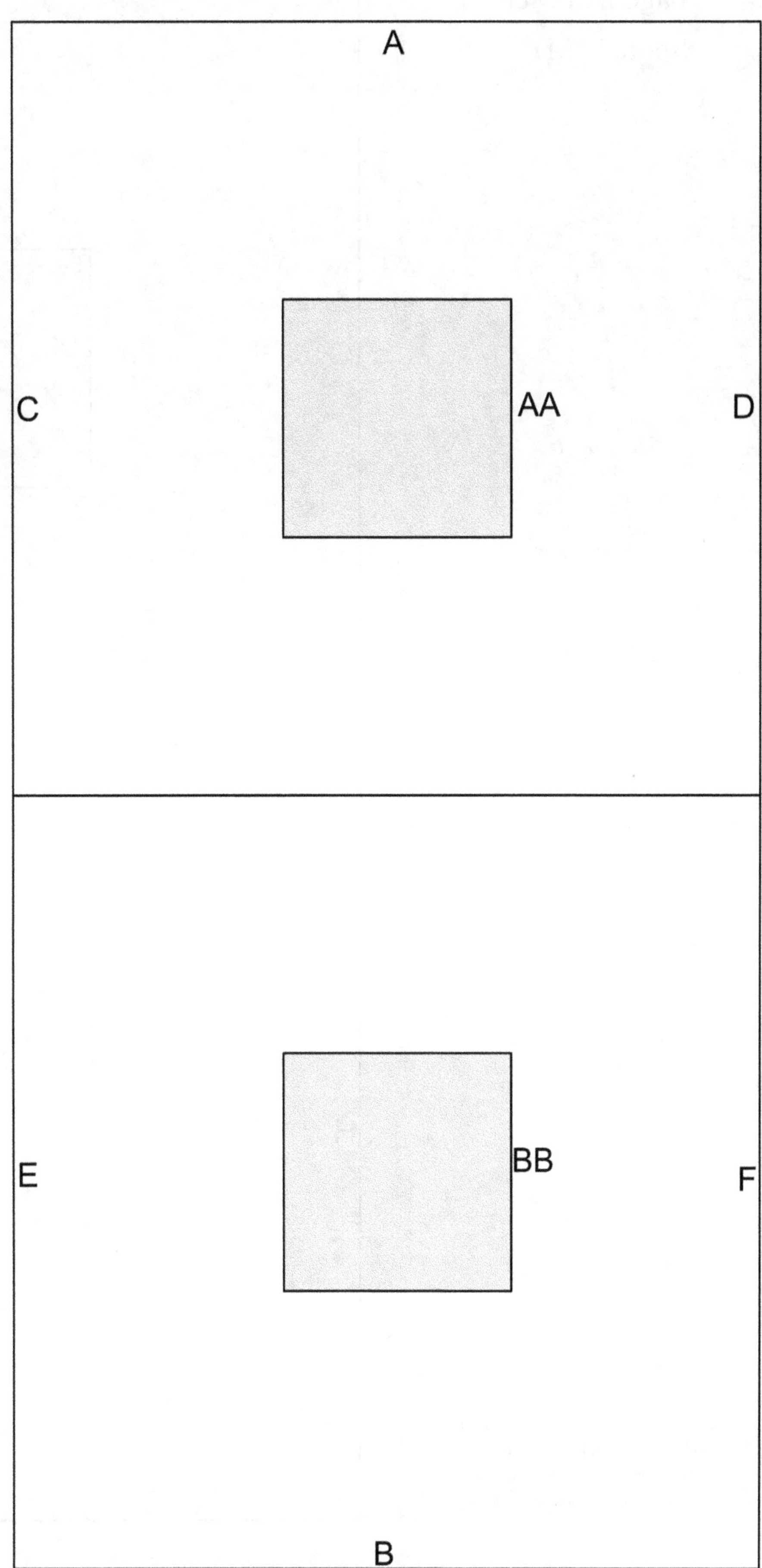

B

G

I

H

J

A

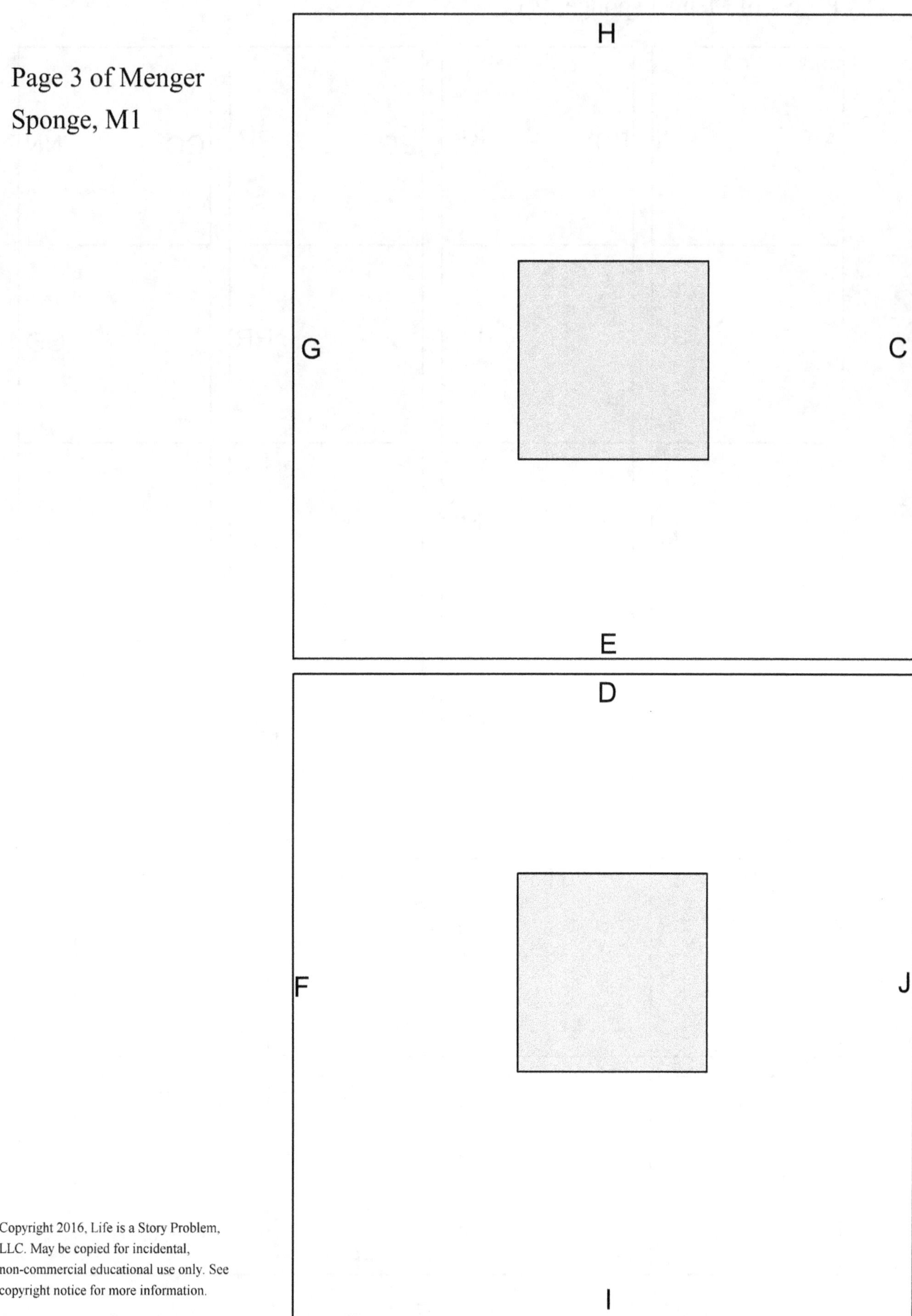
H
G
C
E
D
F
J
I

AA II	BB KK	EE PP	CC NN
GG	LL	RR	OO
JJ	MM	JJ	PP
HH	OO	MM	LL

QQ	HH	RR	OO
DD			

KK	II	QQ	NN
		FF	

Menger Sponge, M2

This net is on two pages. Print out a copy of each page.

1. Color the shape before you cut it out.
2. Cut out along the solid lines.
3. Fold along all of the dotted lines.
4. Dab a little glue on each tab and attach it.
5. Once the glue has dried, attach any decorations.

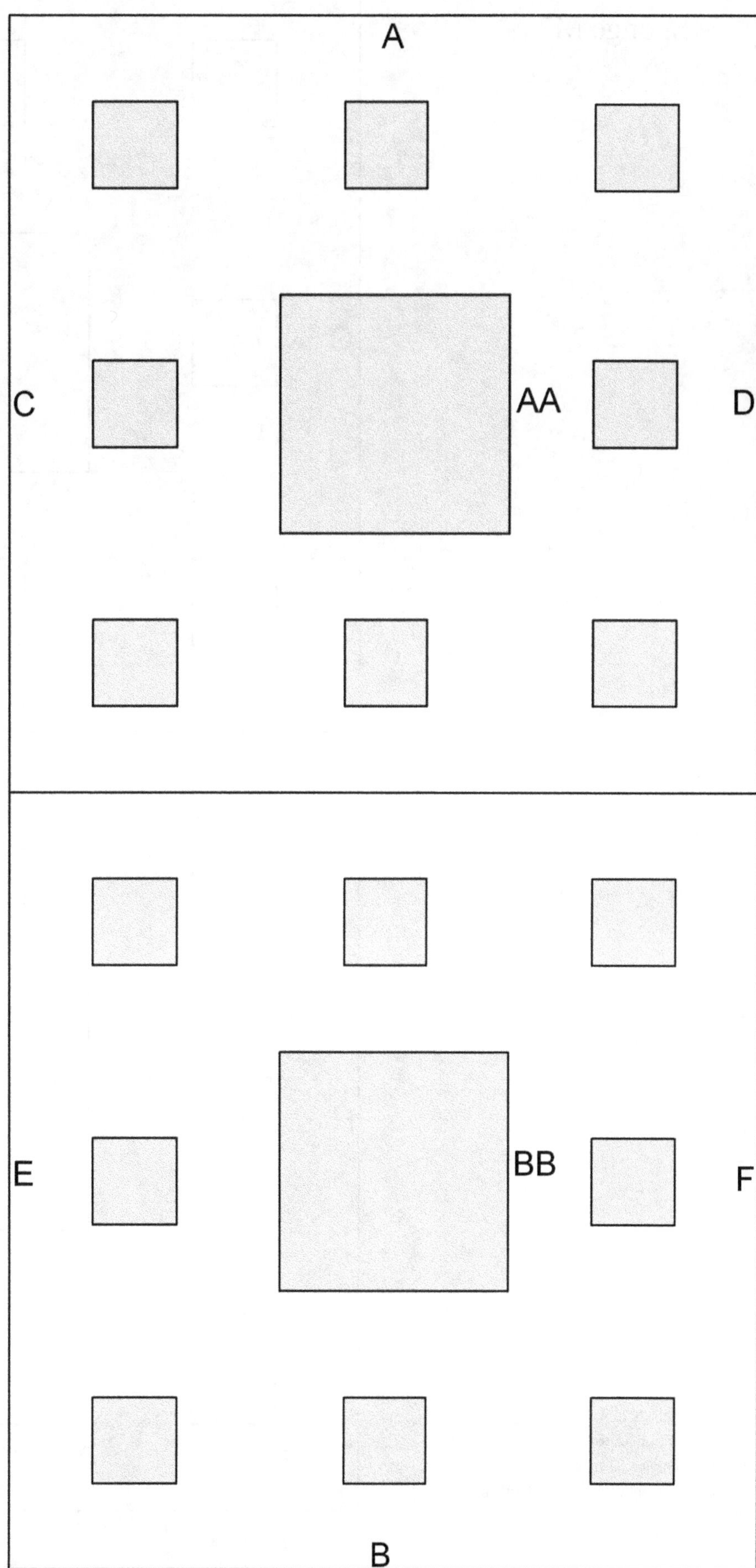

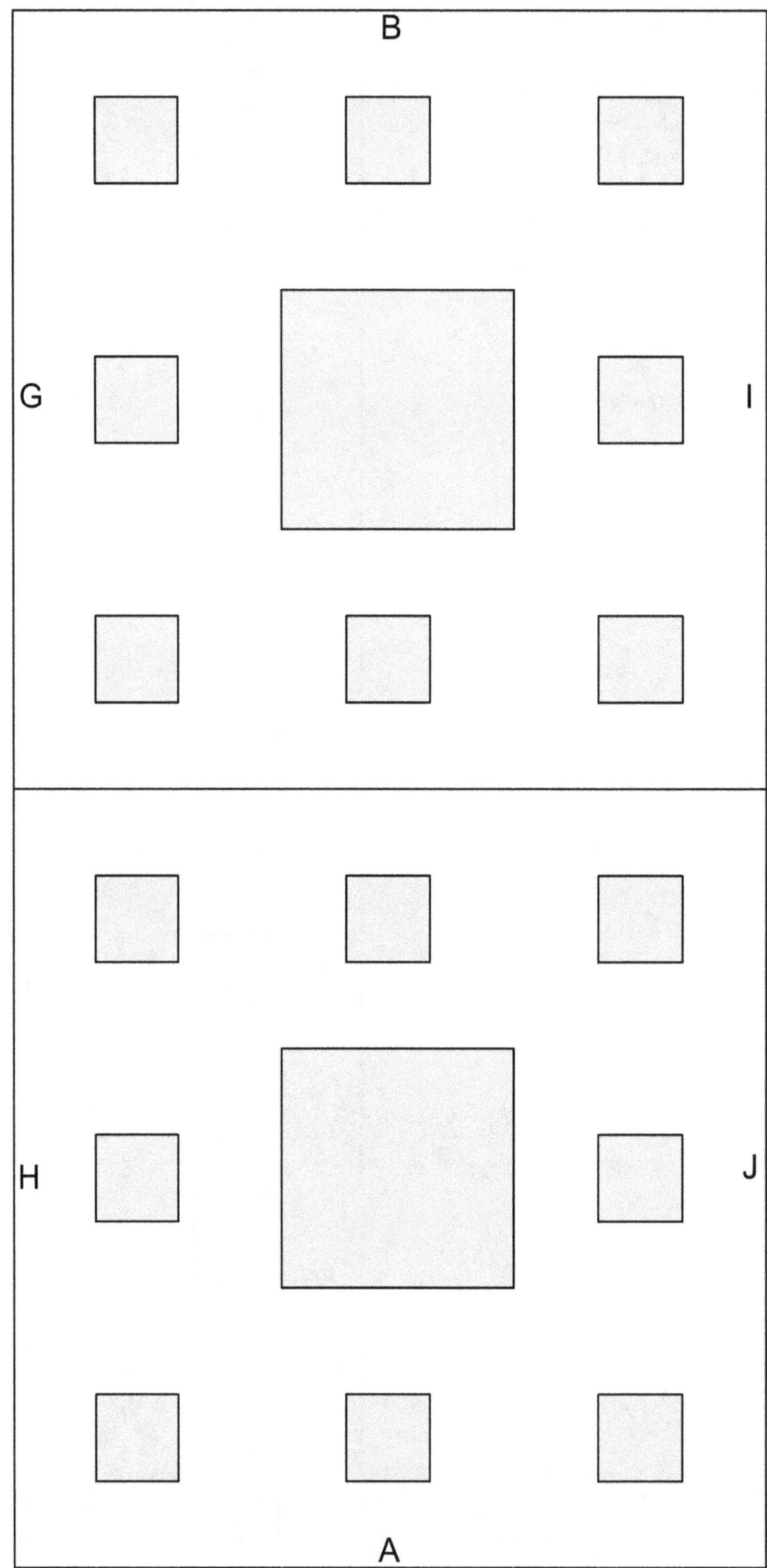

Geometric Nets Mega Project Book, Tabbed By David E. McAdams

Page 3 of Menger
Sponge M2

AA II
BB KK
EE PP
CC NN

GG
LL
RR
OO

JJ
MM
JJ
PP

HH
OO
MM
LL

QQ
HH
RR
OO

DD

KK
II
QQ
NN

FF

Geometric Nets Mega Project Book, Tabbed By David E. McAdams

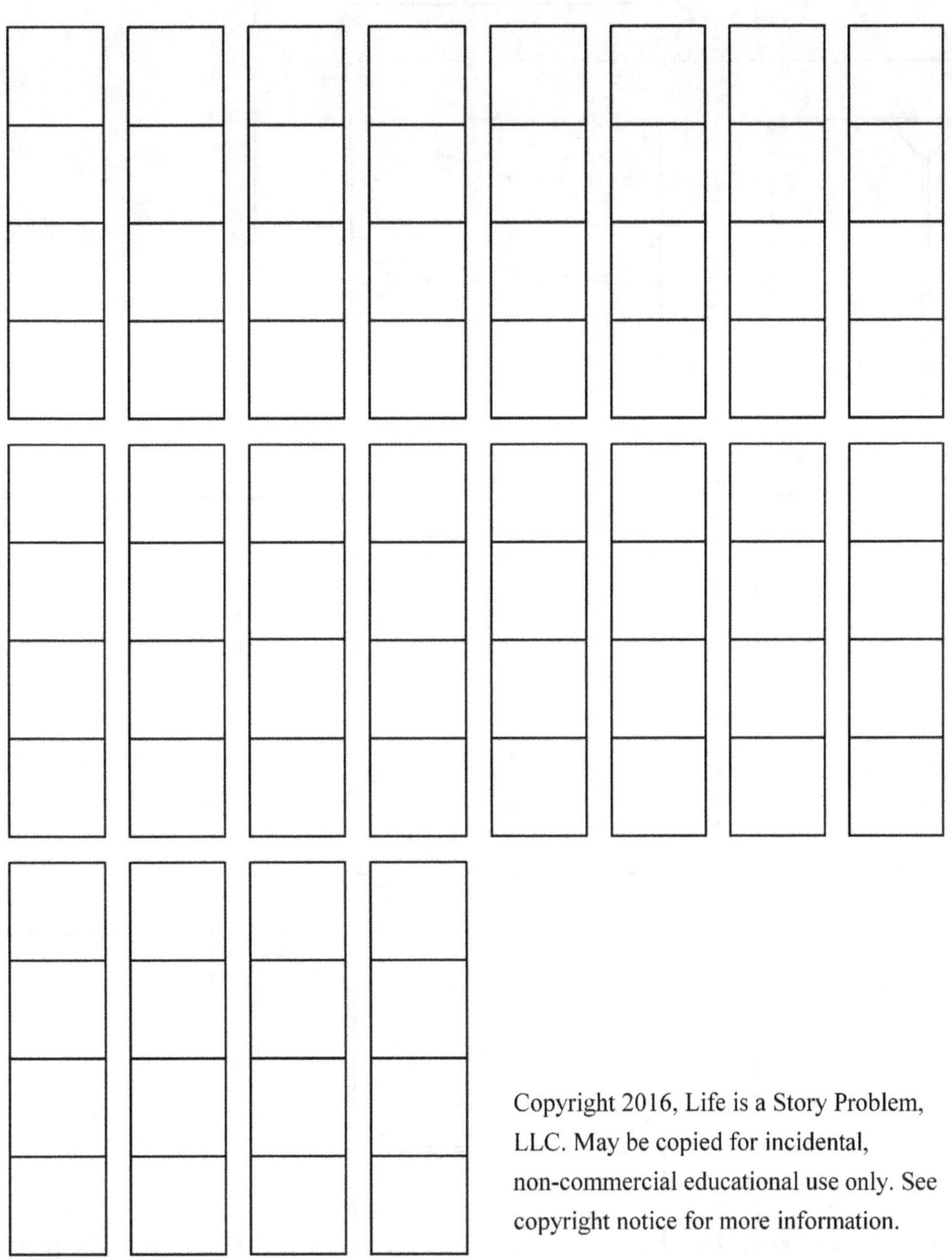

Square outline of a cube

1. Color the shape before you cut it out.
2. Cut out along the solid lines.
3. Fold along all of the dotted lines.
4. Dab a little glue on each tab and attach it.
5. Once the glue has dried, attach any decorations.

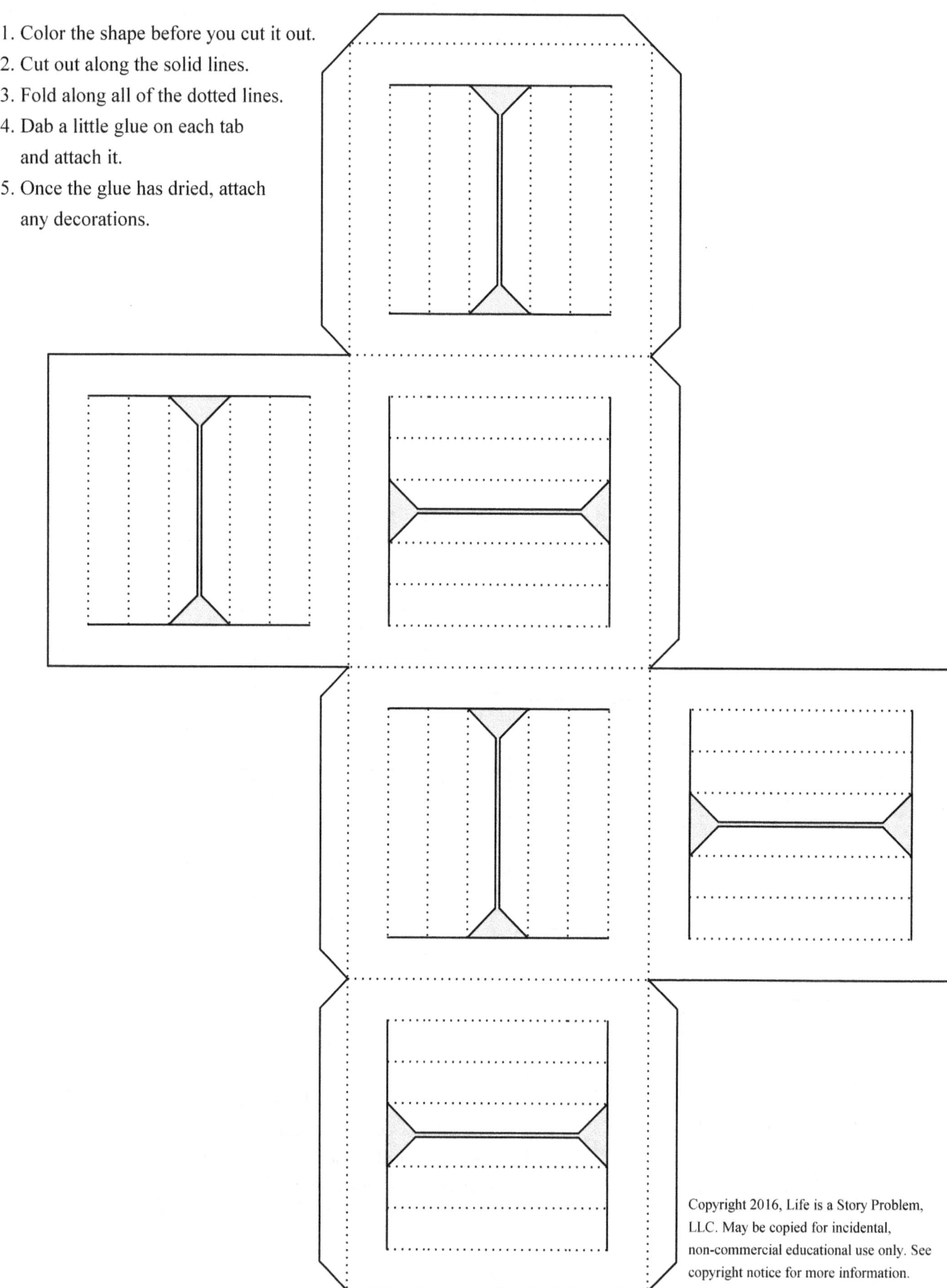

     Geometric Nets Mega Project Book, Tabbed By David E. McAdams

Cube impaled on a square pyramid

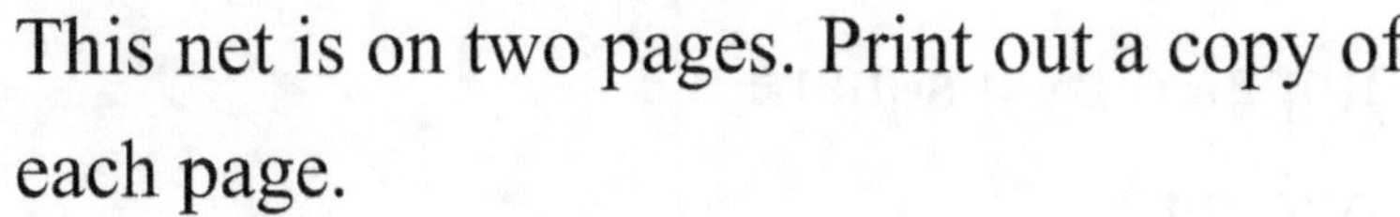

This net is on two pages. Print out a copy of each page.

1. Color the shape before you cut it out.
2. Cut out along the solid lines.
3. Fold forward along all of the dotted lines.
4. Fold backward along all of the dashed lin(
5. Dab a little glue on each tab and attach it.
6. Once the glue has dried, attach any decorations.

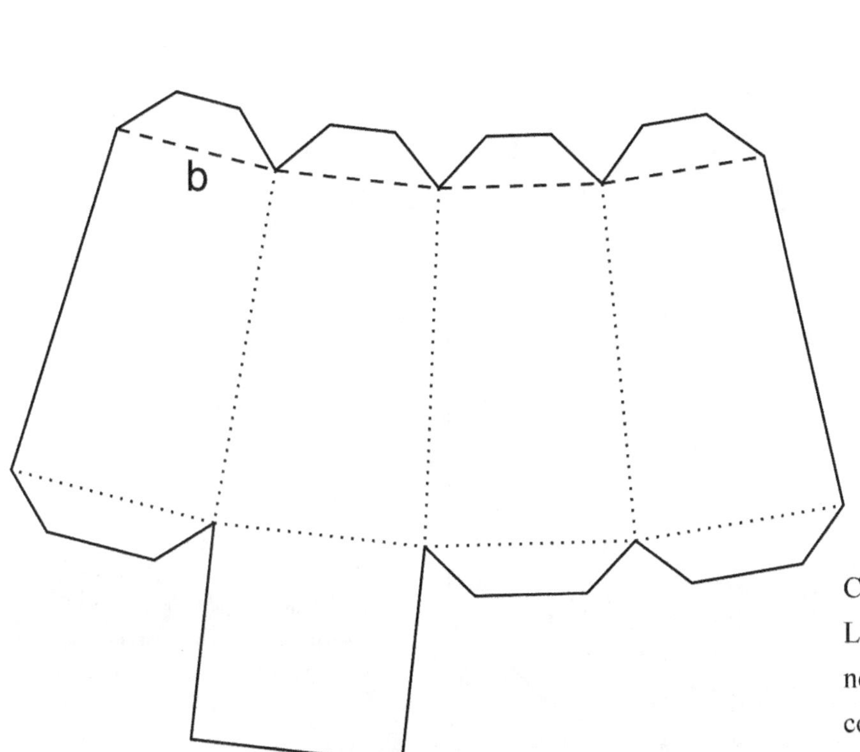

2nd page, Cube
impaled by a square
pyramid

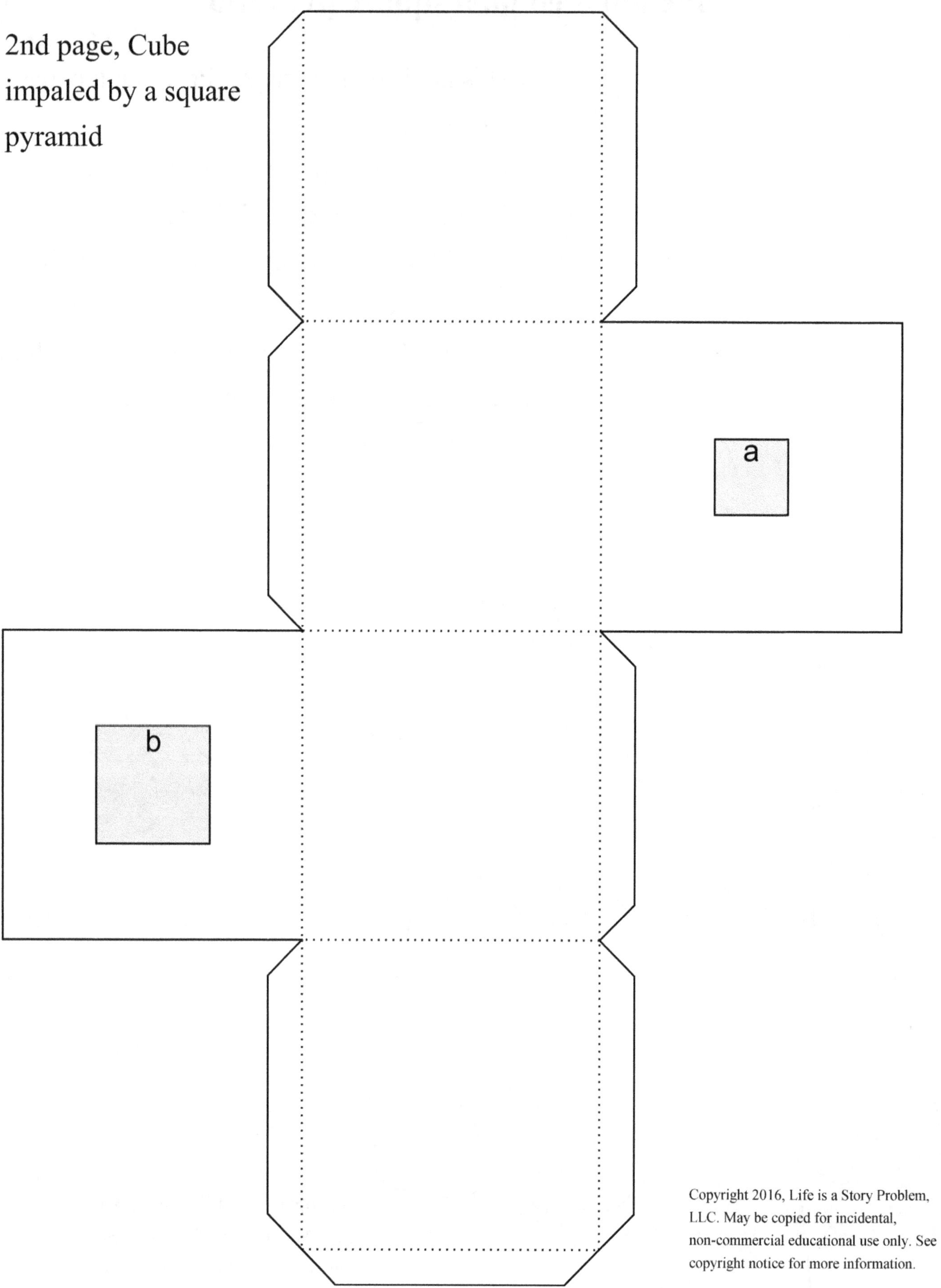

Geometric Nets Mega Project Book, Tabbed By David E. McAdams

Triangular prisms joined by square prisms

This net is on two pages. Print out a copy of each page.

1. Color the shape before you cut it out. 2. Cut out along the solid lines.

3. Fold along all of the dotted lines. 4. Dab a little glue on each tab and attach it.

5. Once the glue has dried, attach any decorations.

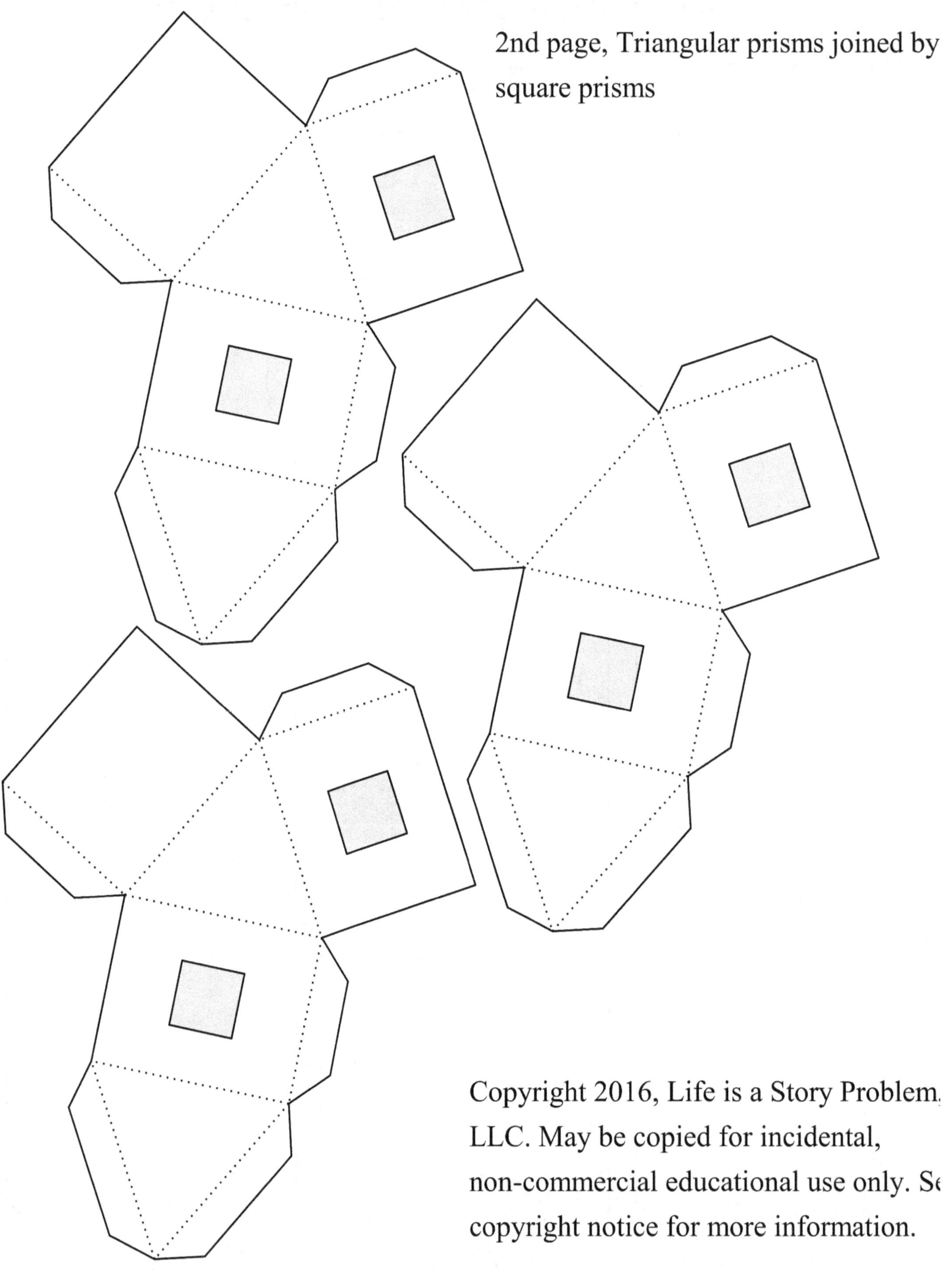

Geometric Nets Mega Project Book, Tabbed By David E. McAdams

Slanted spire

Slanted spire

1. Color the shape before you cut it out.
2. Cut out along the solid lines.
3. Fold along all of the dotted lines.
4. Dab a little glue on each tab and attach it.
5. Once the glue has dried, attach any decorations.

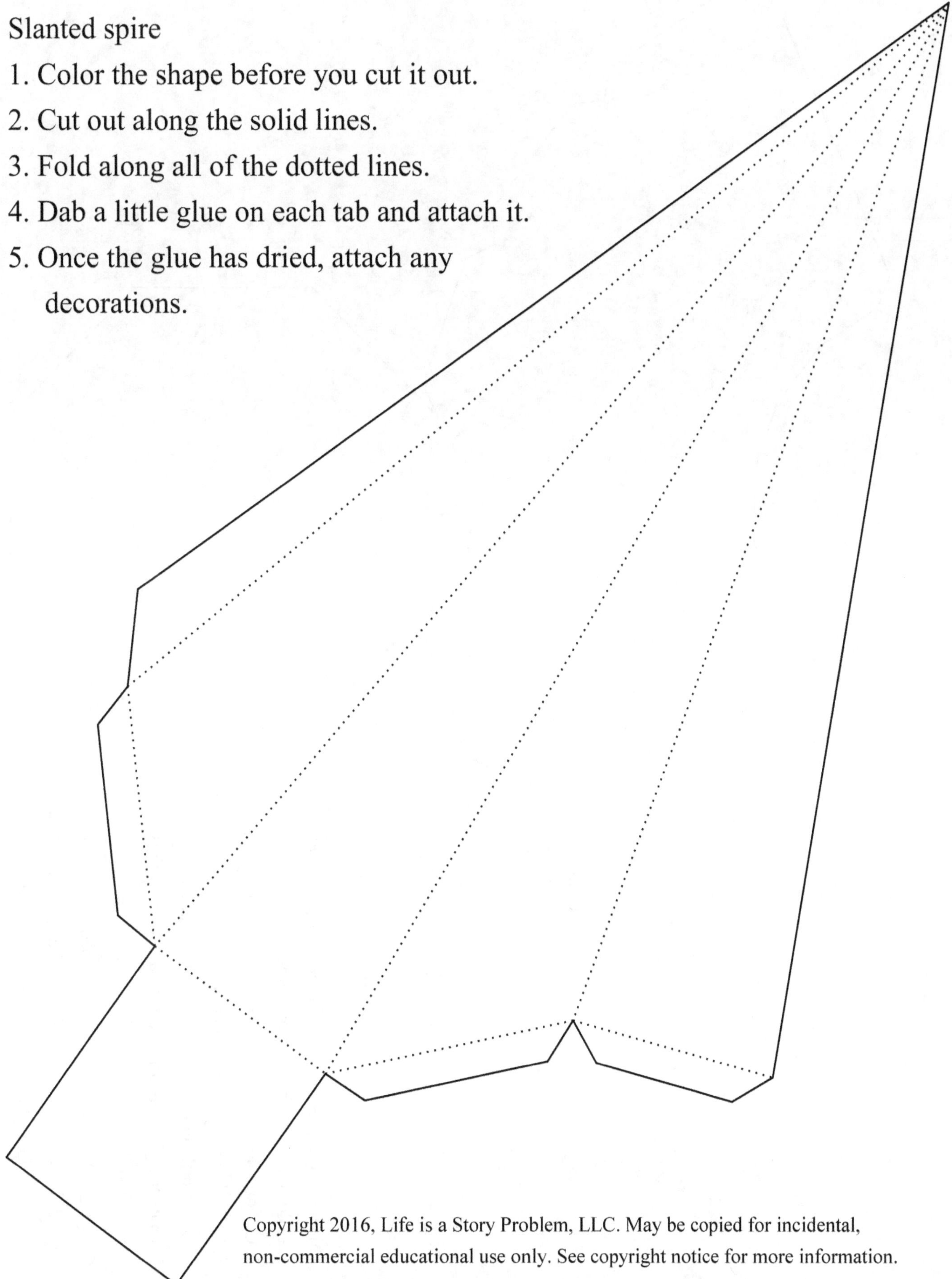

Guard tower

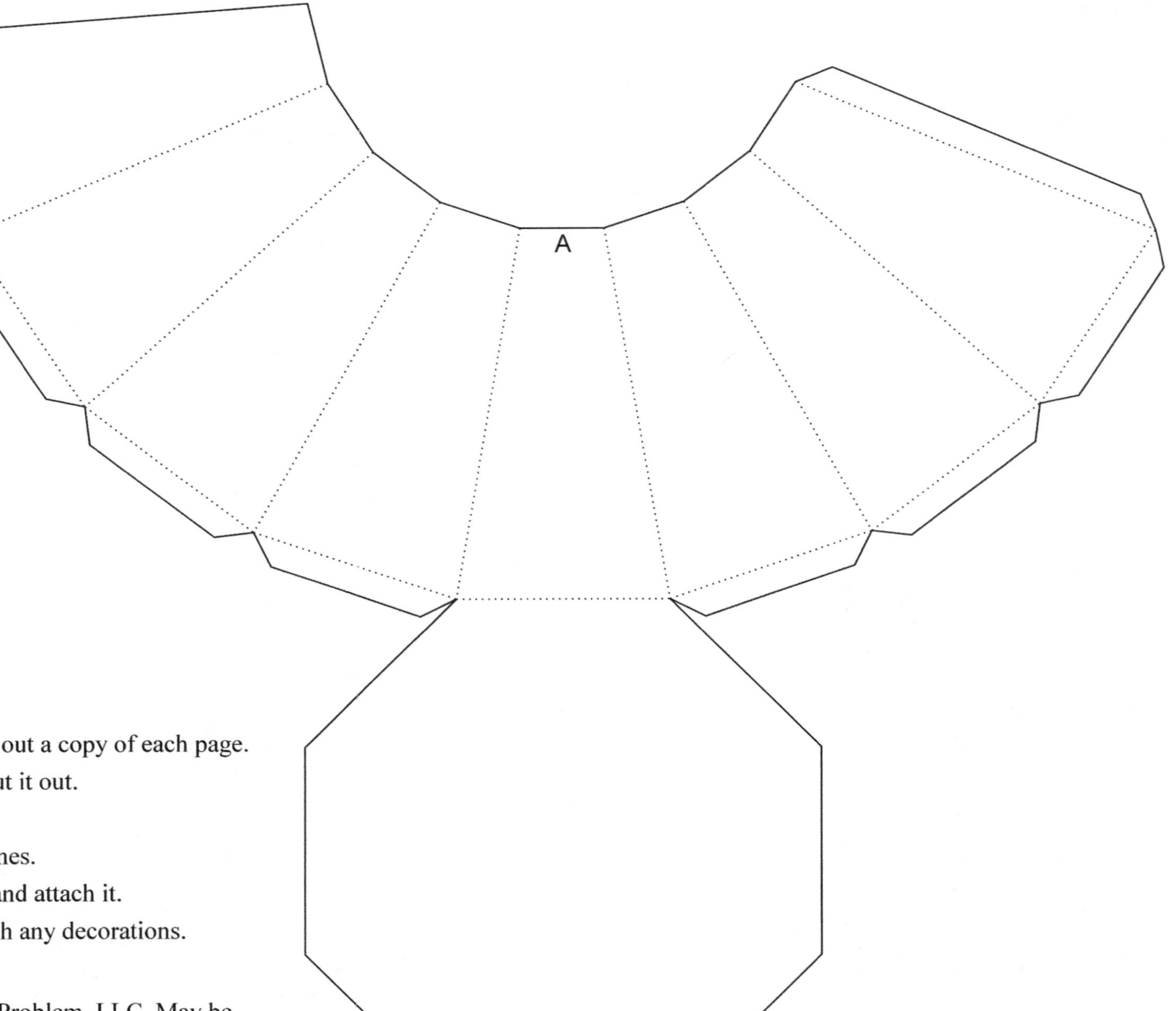

This net is on three pages. Print out a copy of each page.

1. Color the shape before you cut it out.
2. Cut out along the solid lines.
3. Fold along all of the dotted lines.
4. Dab a little glue on each tab and attach it.
5. Once the glue has dried, attach any decorations.

A

B

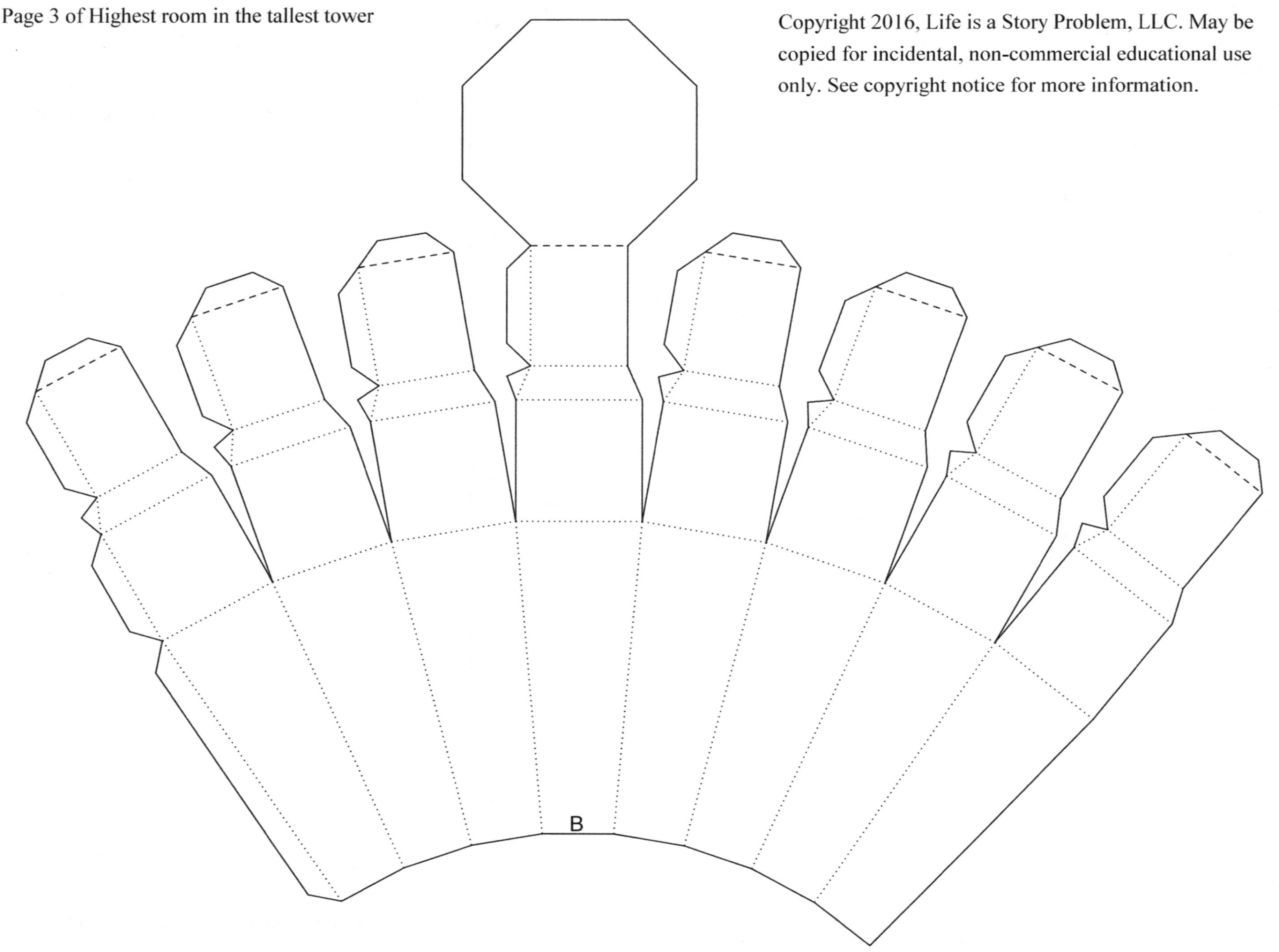
B

Kaleidocycle N=6 closed

1. Color the shape before you cut it out.
2. Cut out along the solid lines.
3. Fold along all of the dotted lines.

4. Dab a little glue on
   each tab and attach it.
5. Once the glue has dried,
   attach any decorations.

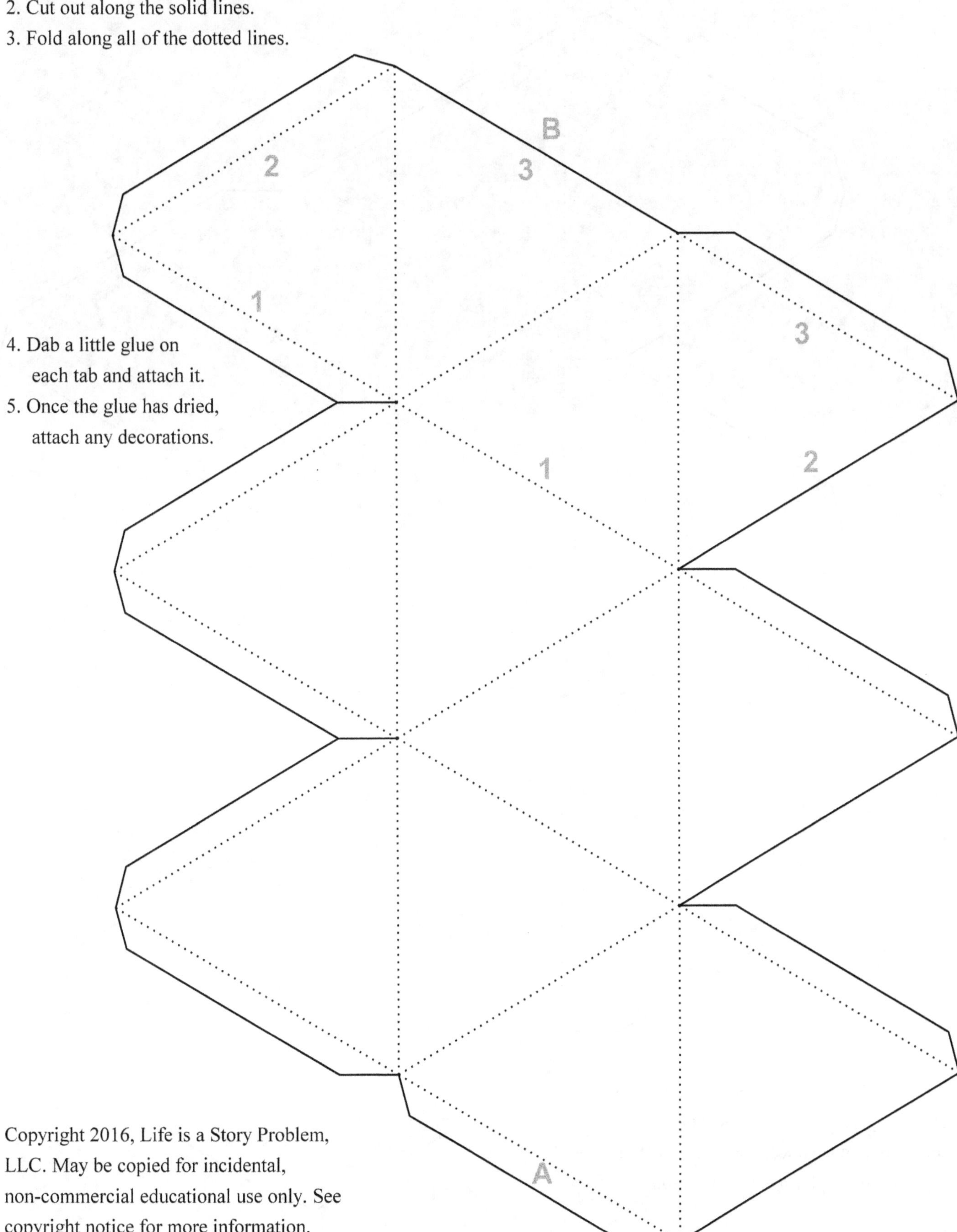

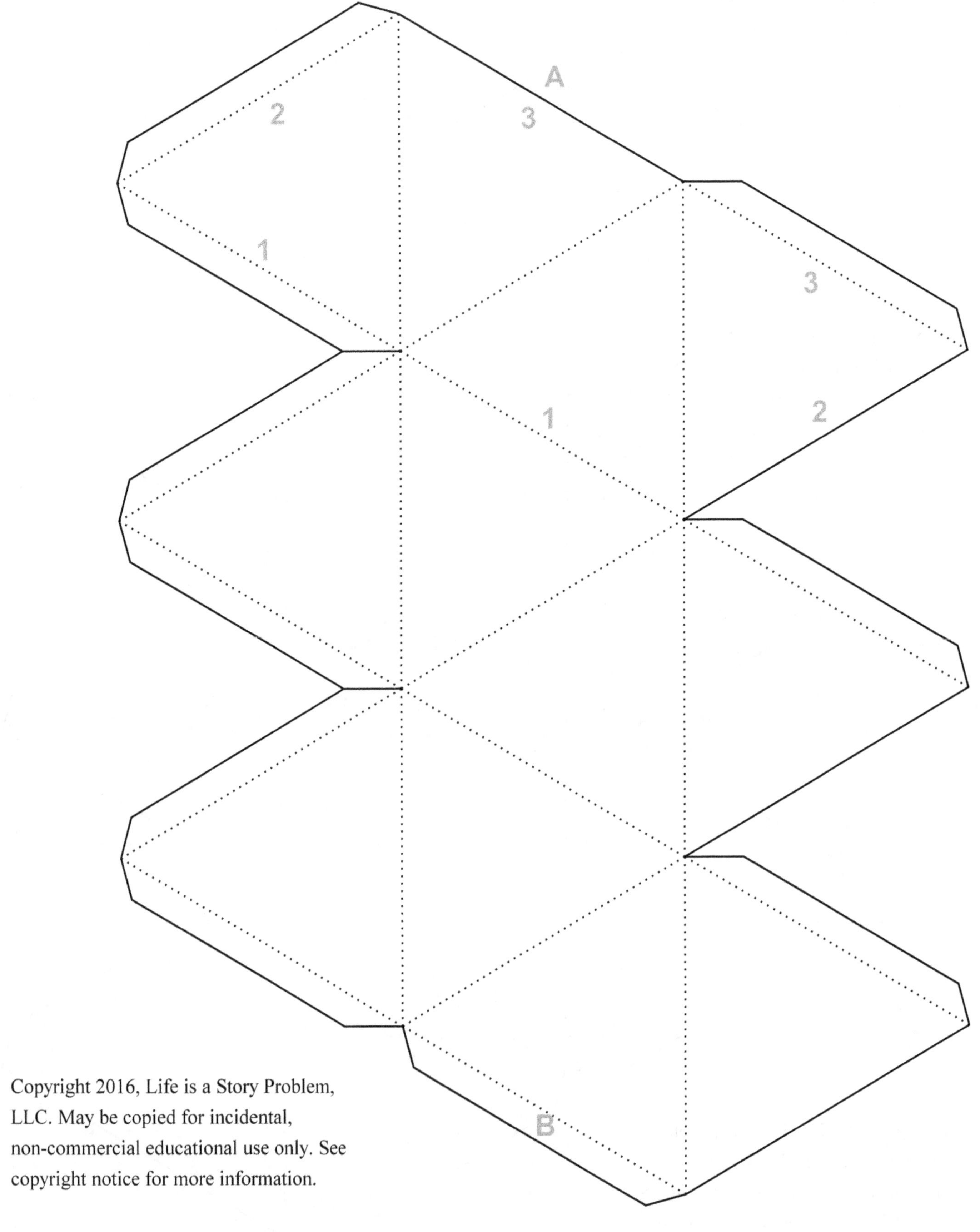

 Geometric Nets Mega Project Book, Tabbed By David E. McAdams

Kaleidocycle N=8 regular

1. Color the shape before you cut it out.
2. Cut out along the solid lines.
3. Fold along all of the dotted lines.
4. Dab a little glue on each tab and attach it.
5. Once the glue has dried, attach any decorations.

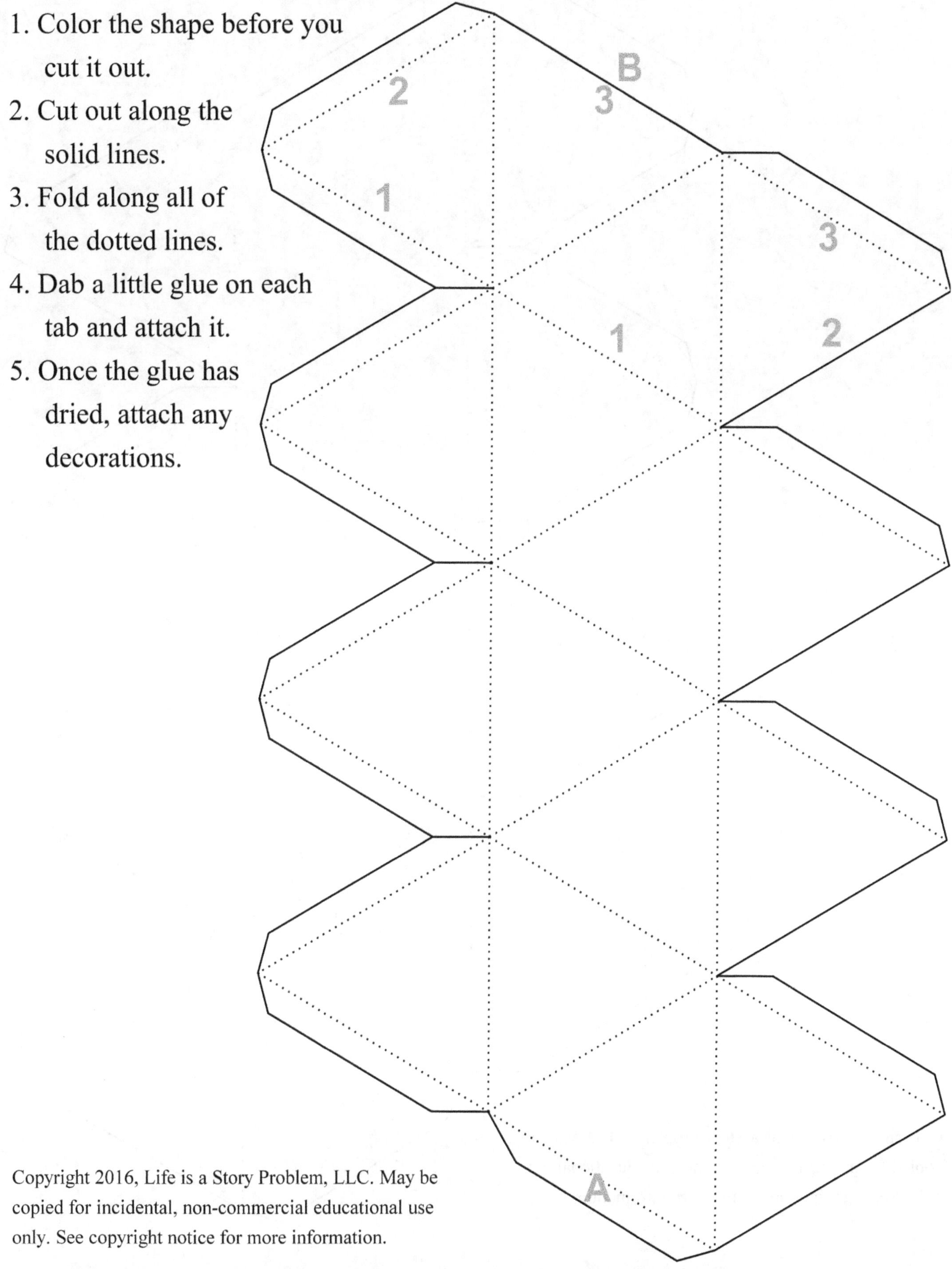

2nd page Kaleidocycle 8 regular

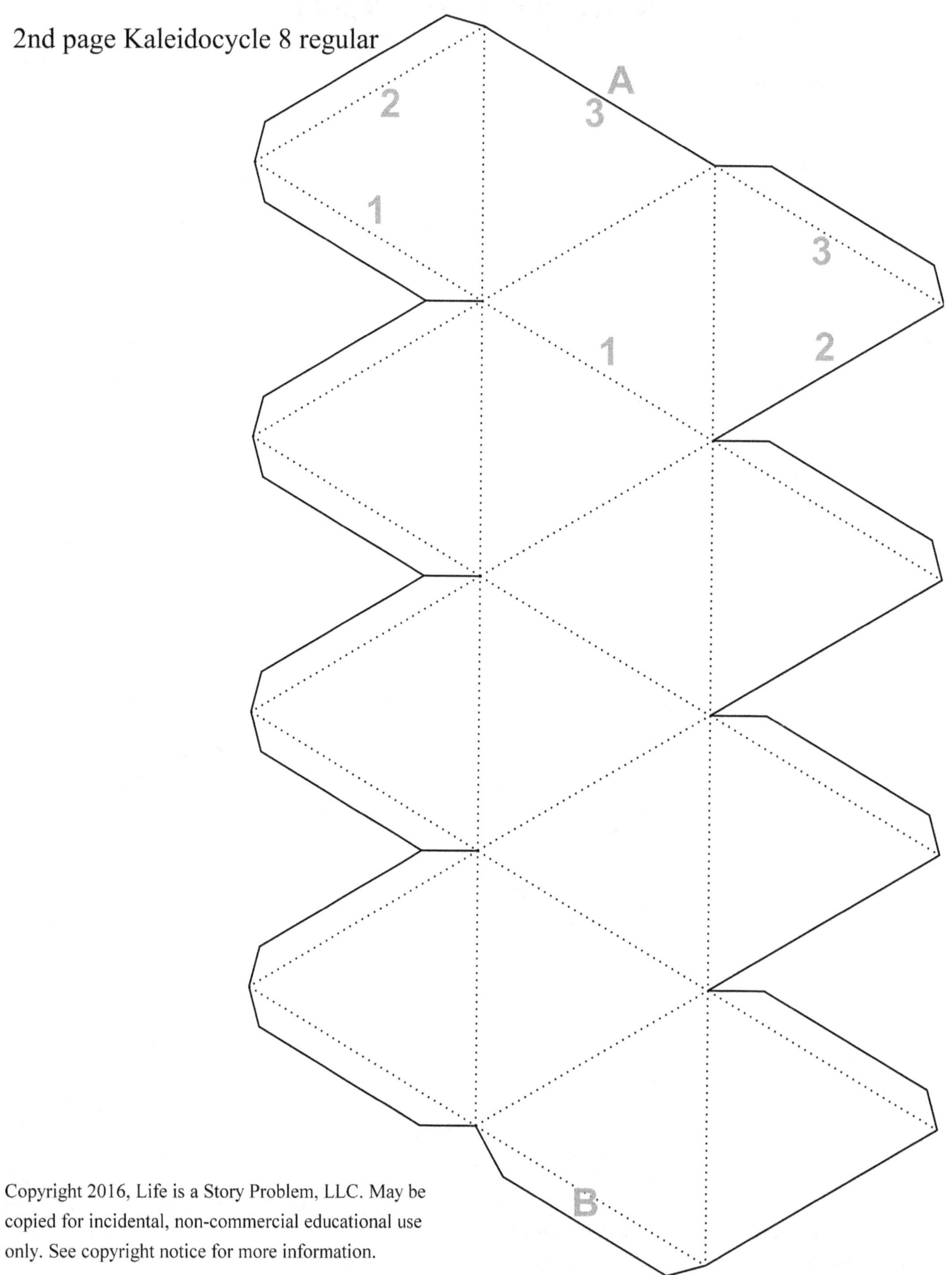

Geometric Nets Mega Project Book, Tabbed By David E. McAdams

Cone

1. Color the shape before you cut it out.
2. Cut out along the solid lines.
3. Fold along all of the dotted lines.
4. Dab a little glue on each tab and attach it.
5. Once the glue has dried, attach any decorations.

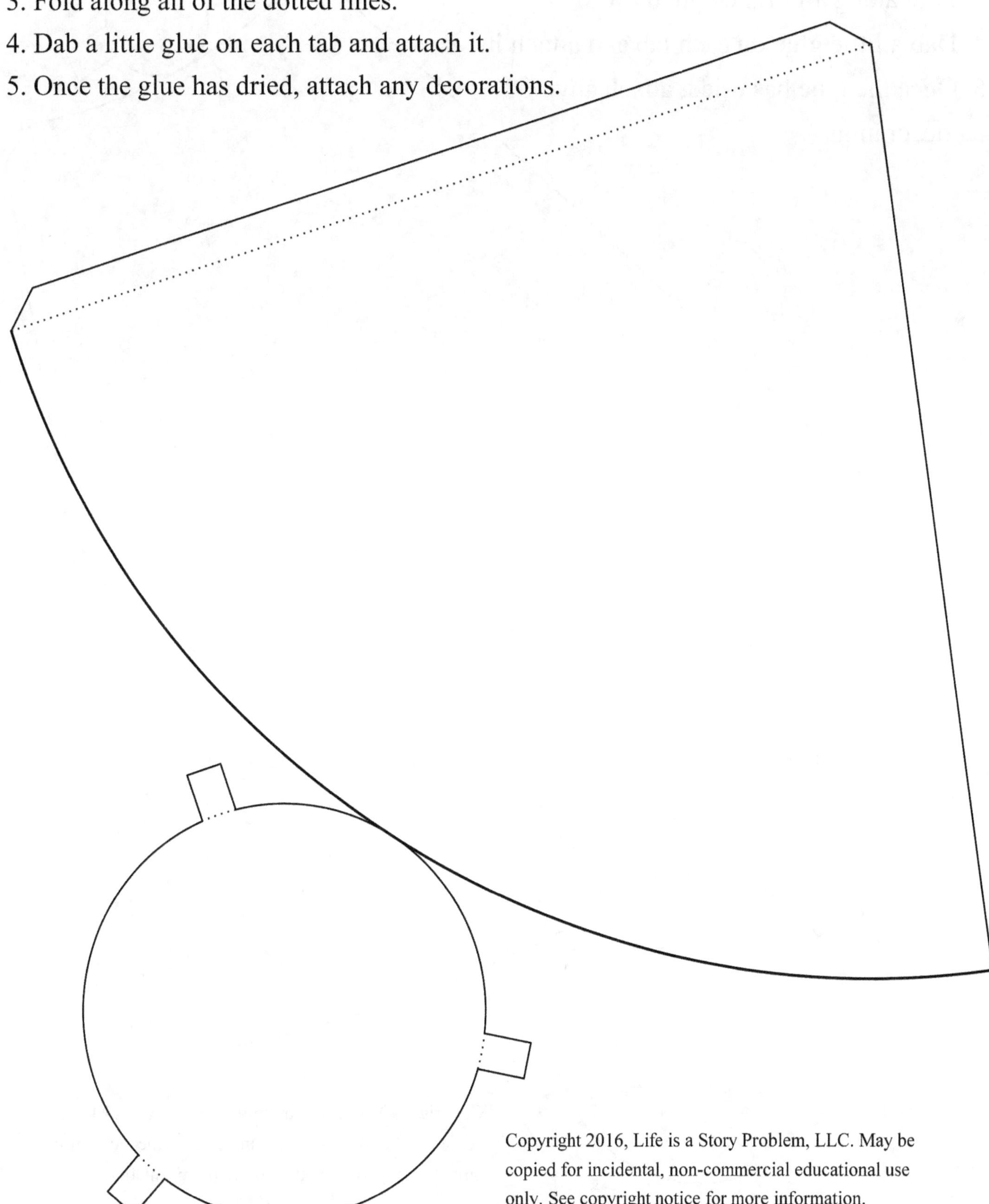

Truncated cone

1. Color the shape before you cut it out.
2. Cut out along the solid lines.
3. Fold along all of the dotted lines.
4. Dab a little glue on each tab and attach it.
5. Once the glue has dried, attach any
   decorations.

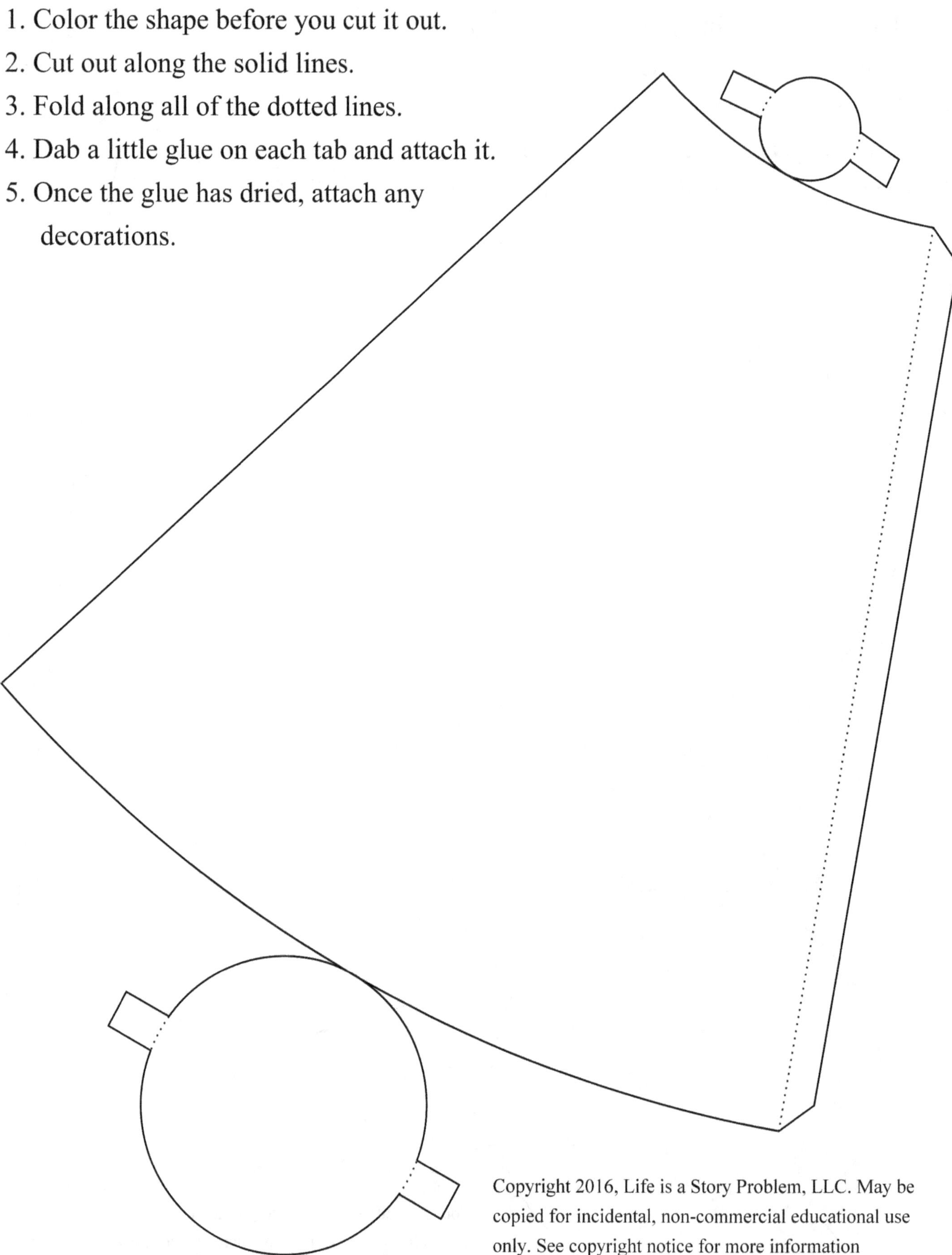

 Geometric Nets Mega Project Book, Tabbed By David E. McAdams

Cylinder

1. Color the shape before you cut it out.

2. Cut out along the solid lines.

3. Fold along all of the dotted lines.

4. Dab a little glue on each tab and attach it.

5. Once the glue has dried, attach any decorations.

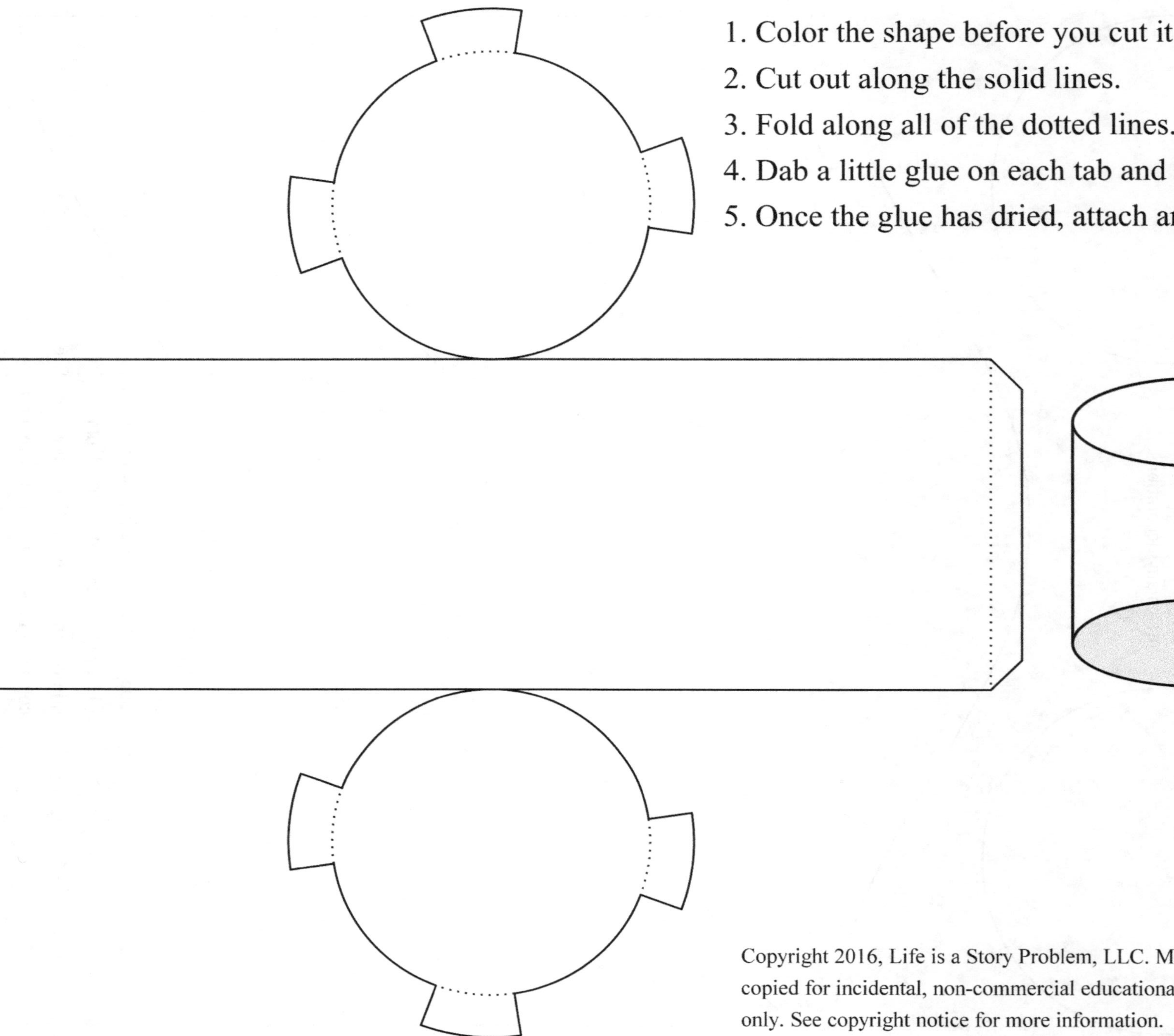

Oblique cylinder

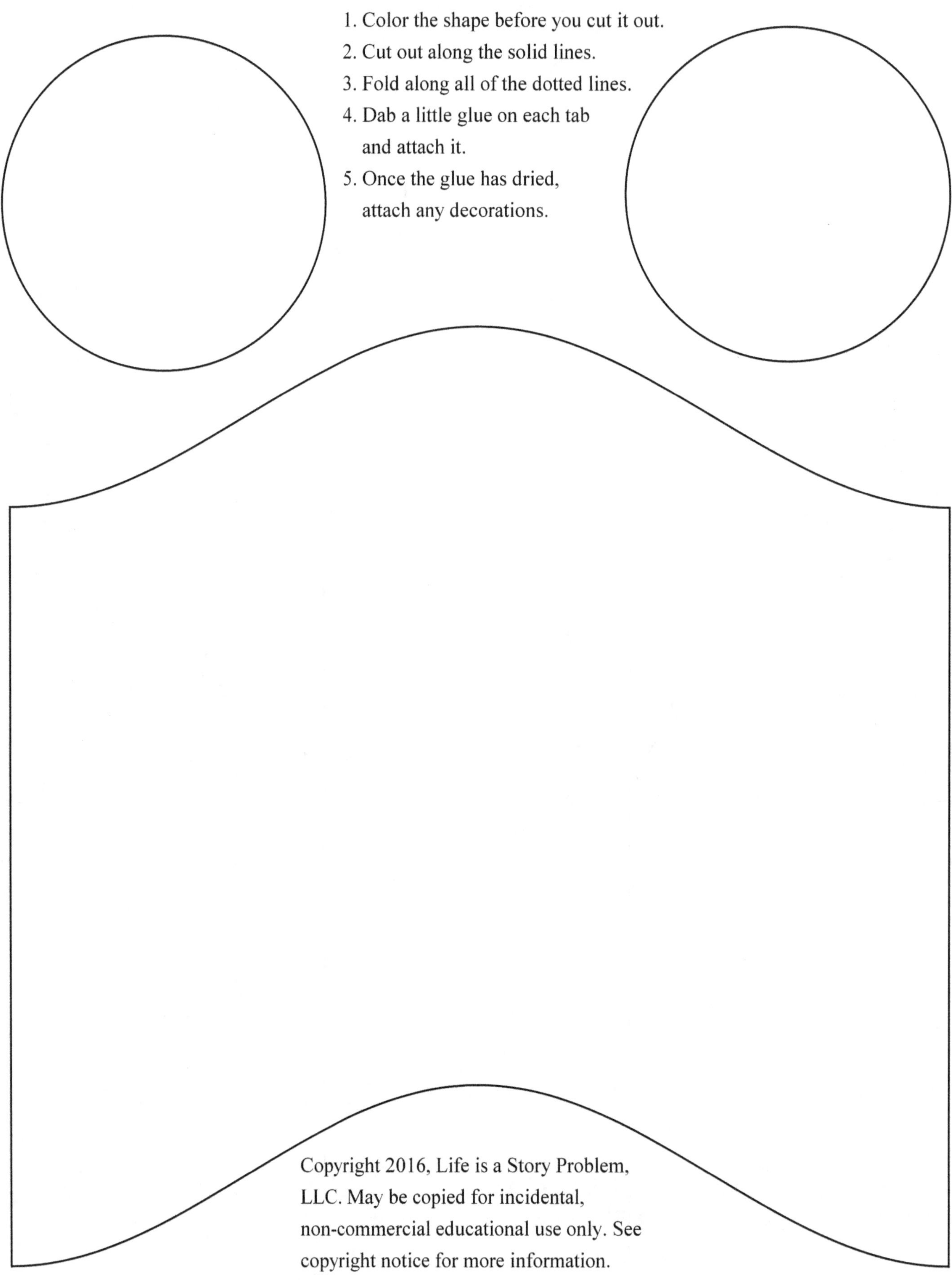

1. Color the shape before you cut it out.
2. Cut out along the solid lines.
3. Fold along all of the dotted lines.
4. Dab a little glue on each tab and attach it.
5. Once the glue has dried, attach any decorations.

 Geometric Nets Mega Project Book, Tabbed By David E. McAdams

Elliptical cylinder

1. Color the shape before you cut it out.
2. Cut out along the solid lines.
3. Fold along all of the dotted lines.
4. Dab a little glue on each tab and attach it.
5. Once the glue has dried, attach any decorations.

Square pyramid

1. Color the shape before you cut it out.

2. Cut out along the solid lines.

3. Fold along all of the dotted lines.

4. Dab a little glue on each tab
   and attach it.

5. Once the glue has dried,
   attach any decorations.

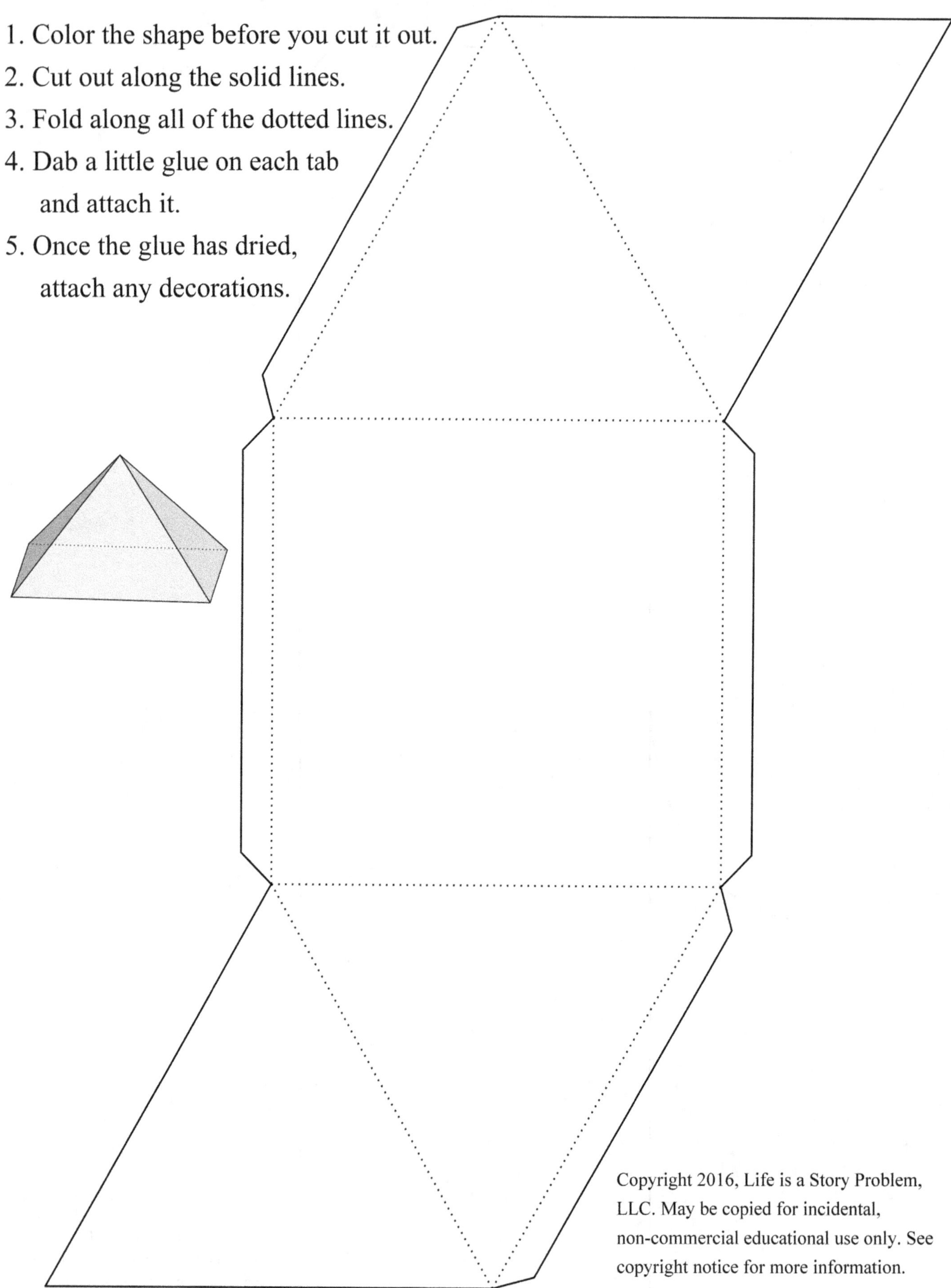

 Geometric Nets Mega Project Book, Tabbed By David E. McAdams

Square antipyramid

1. Color the shape before you cut it out.
2. Cut out along the solid lines.
3. Fold along all of the dotted lines.
4. Dab a little glue on each tab and attach it.
5. Once the glue has dried, attach any decorations.

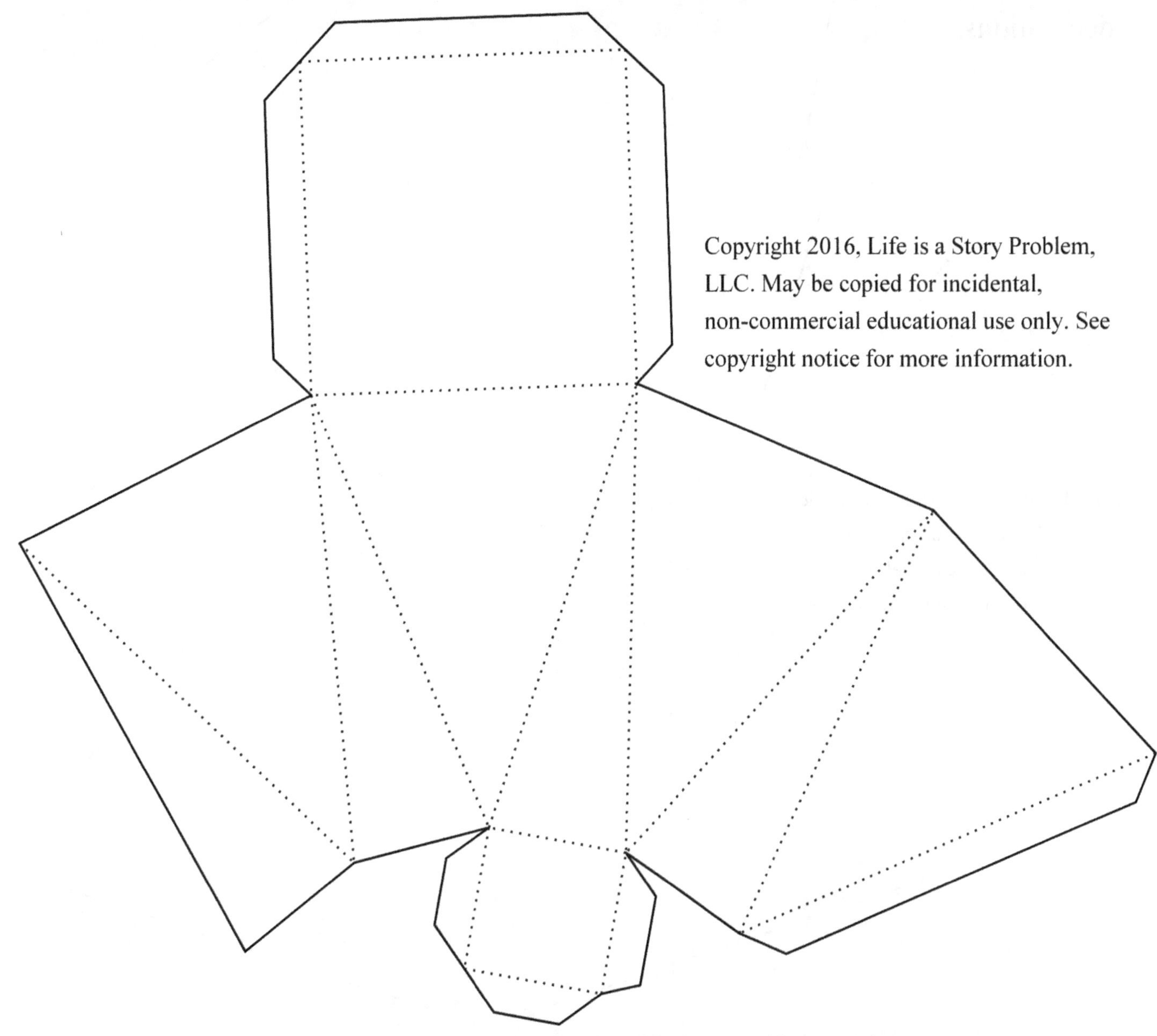

Pentagonal pyramid

1. Color the shape before you cut it out.
2. Cut out along the solid lines.
3. Fold along all of the dotted lines.
4. Dab a little glue on each tab and attach it.
5. Once the glue has dried, attach any decorations.

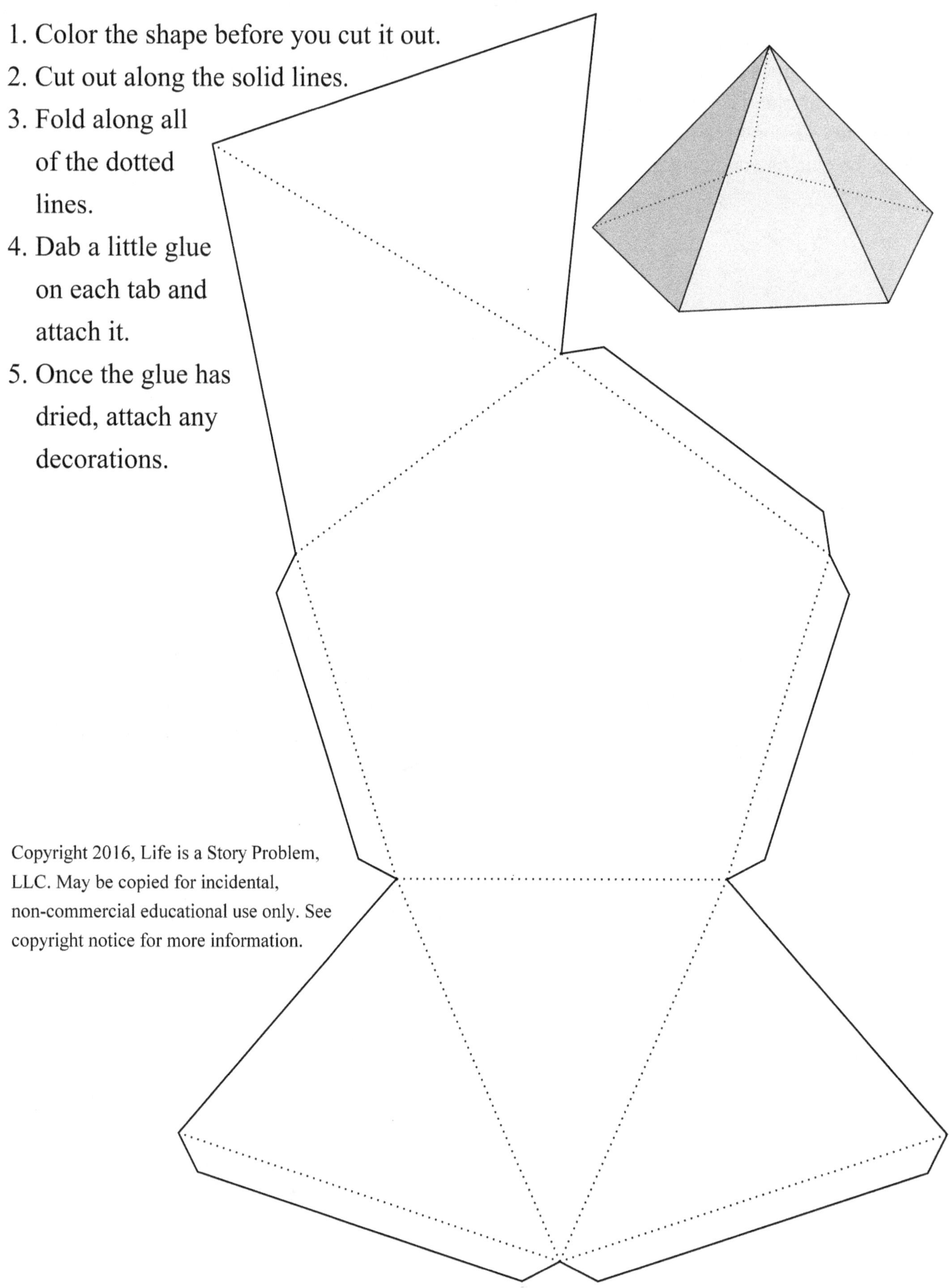

 Geometric Nets Mega Project Book, Tabbed By David E. McAdams

Pentagonal step pyramid

1. Color the shape before you cut it out.
2. Cut out along the solid lines.
3. Fold forward along all of the dotted lines.
4. Fold backward along all of the dashed lines.
5. Dab a little glue on each tab and attach it.
6. Once the glue has dried, attach any decorations.

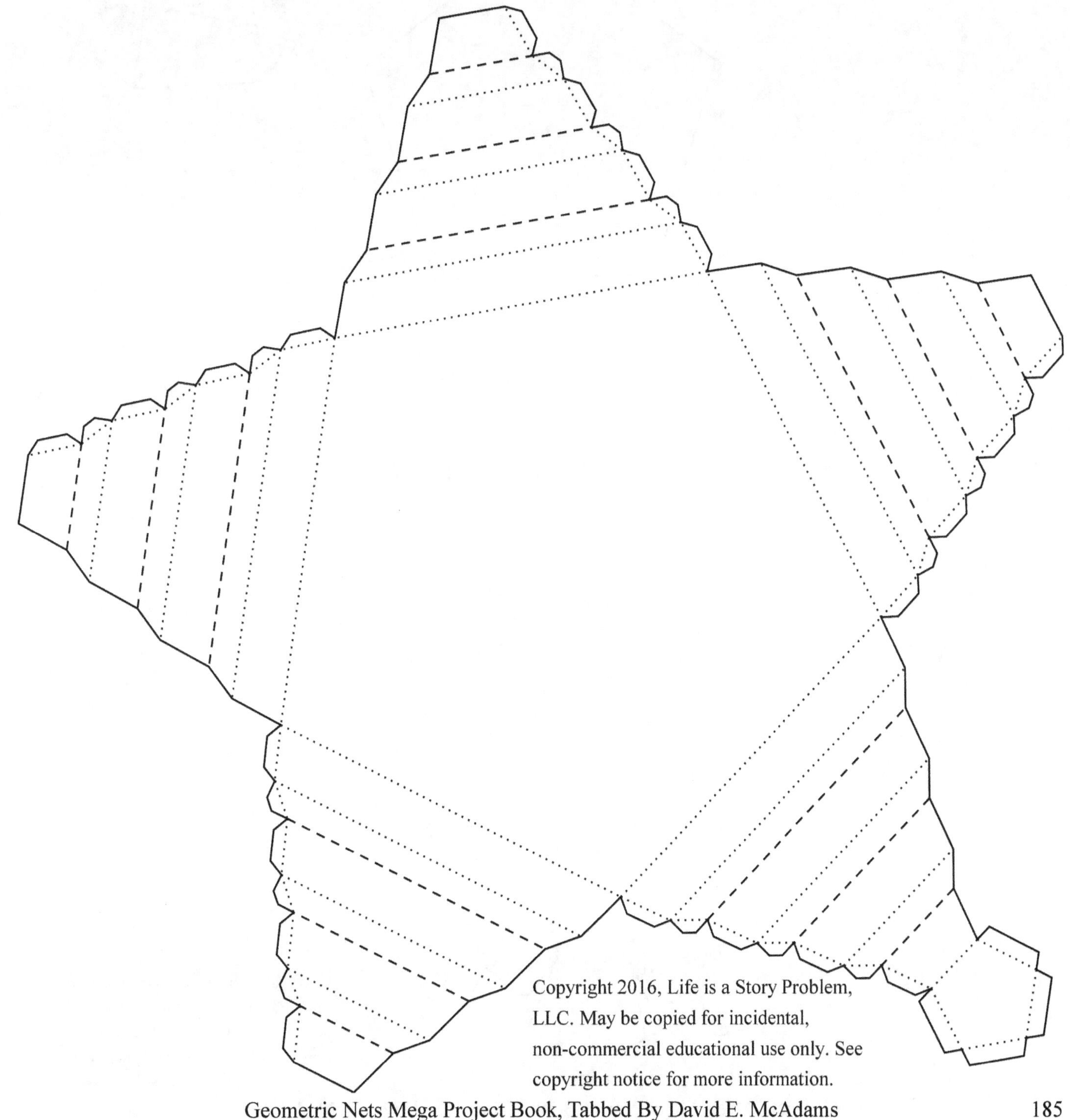

Pentagonal antipyramid

1. Color the shape before you cut it out.

2. Cut out along the solid lines.

3. Fold along all of the dotted lines.

4. Dab a little glue on each tab and attach it.

5. Once the glue has dried, attach any decorations.

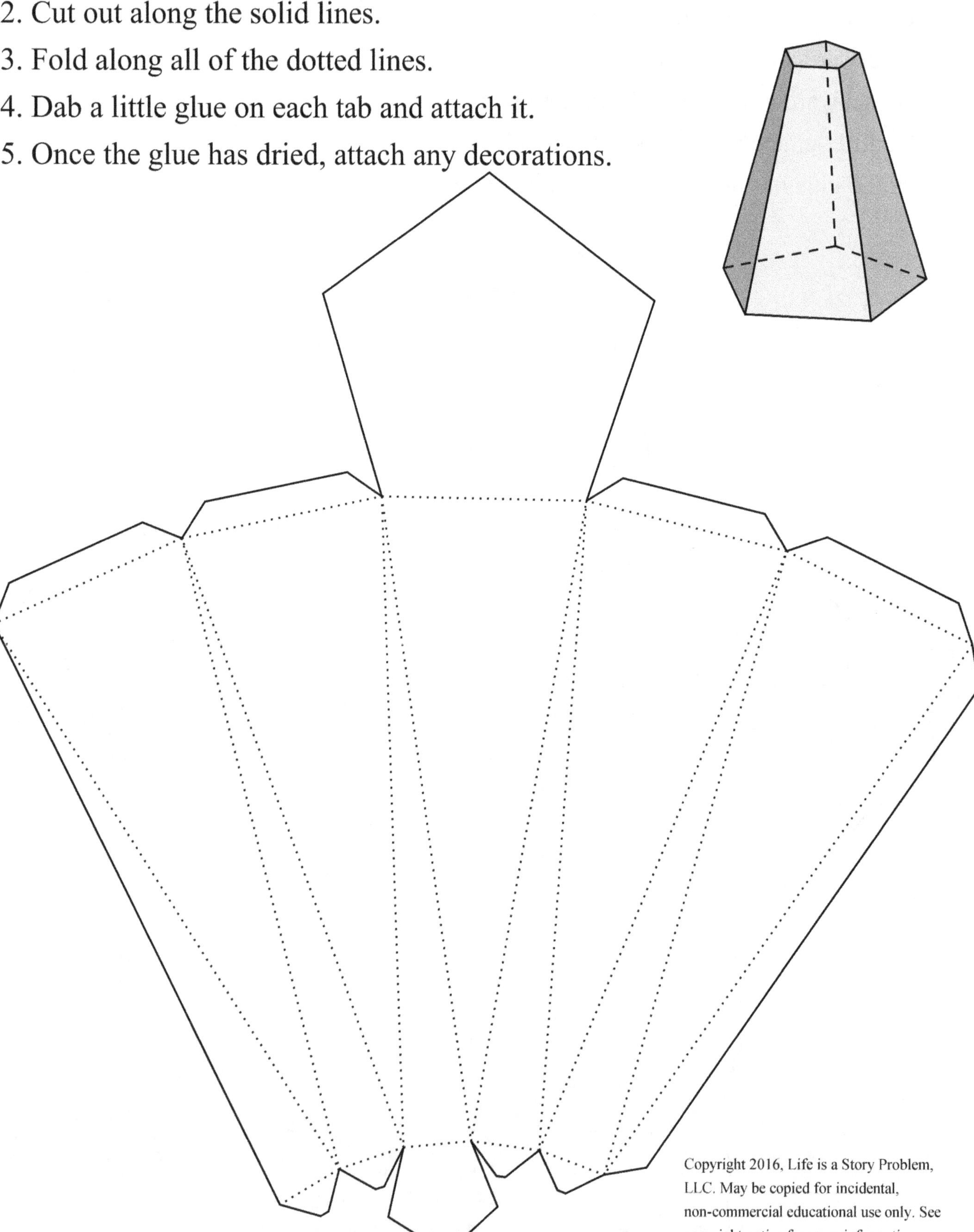

 Geometric Nets Mega Project Book, Tabbed By David E. McAdams

Triangular cupola

1. Color the shape before you cut it out.
2. Cut out along the solid lines.
3. Fold along all of the dotted lines.
4. Dab a little glue on each tab and attach it.
5. Once the glue has dried, attach any decorations.

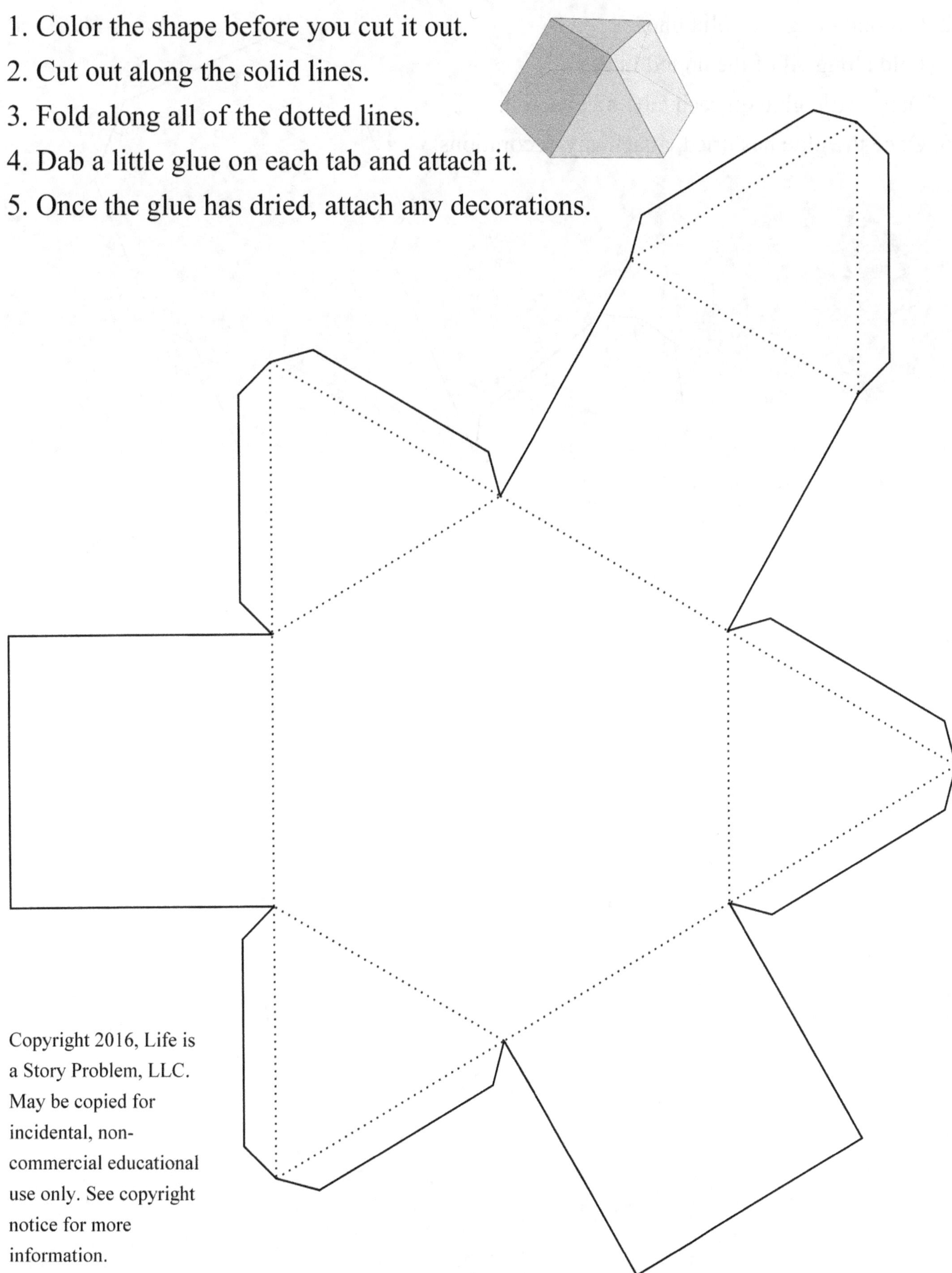

Square cupola

1. Color the shape before you cut it out.

2. Cut out along the solid lines.

3. Fold along all of the dotted lines.

4. Dab a little glue on each tab and attach it.

5. Once the glue has dried, attach any decorations.

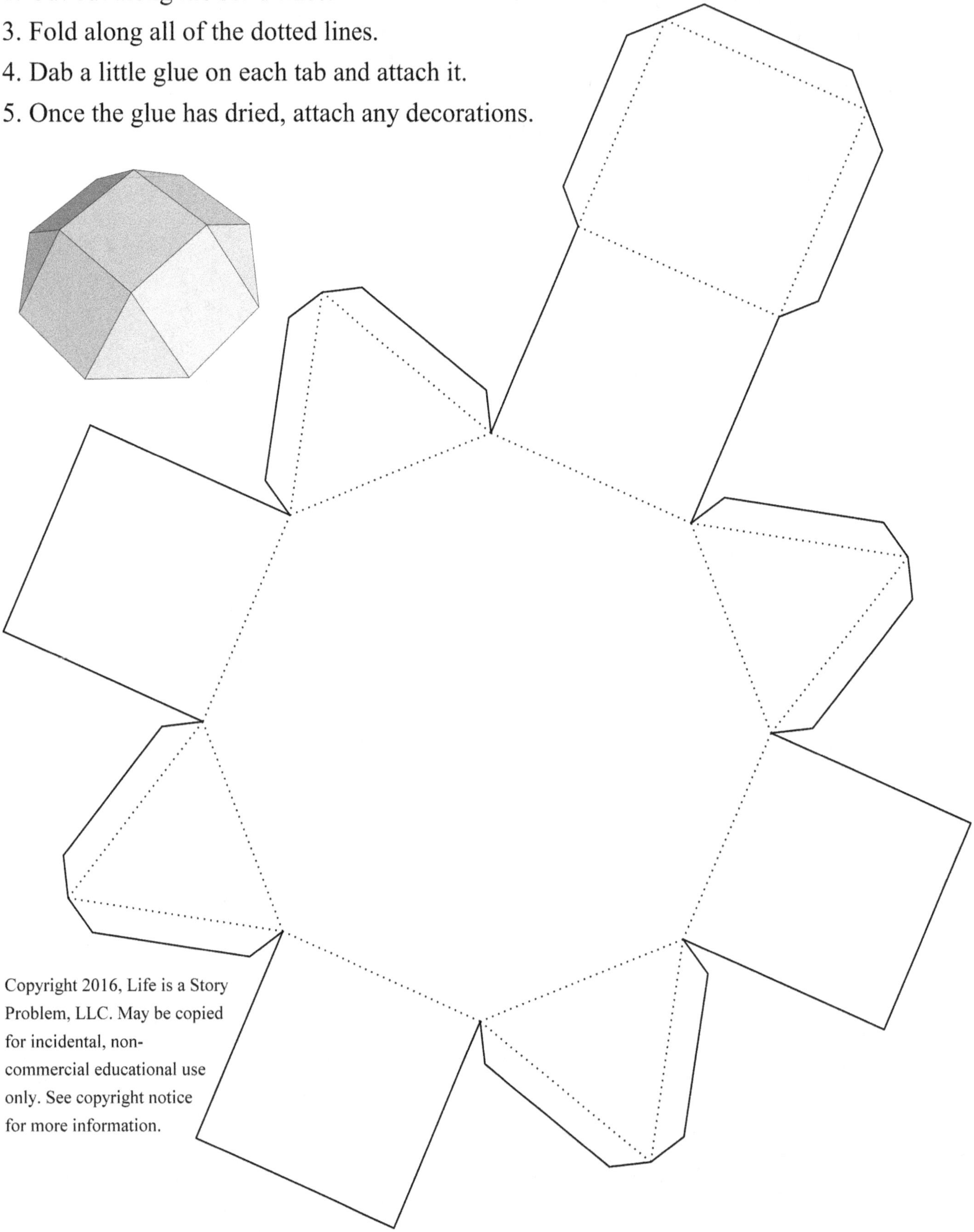

Pentagonal cupola

1. Color the shape before you cut it out.
2. Cut out along the solid lines.
3. Fold along all of the dotted lines.
4. Dab a little glue on each tab and attach it.
5. Once the glue has dried, attach any decorations.

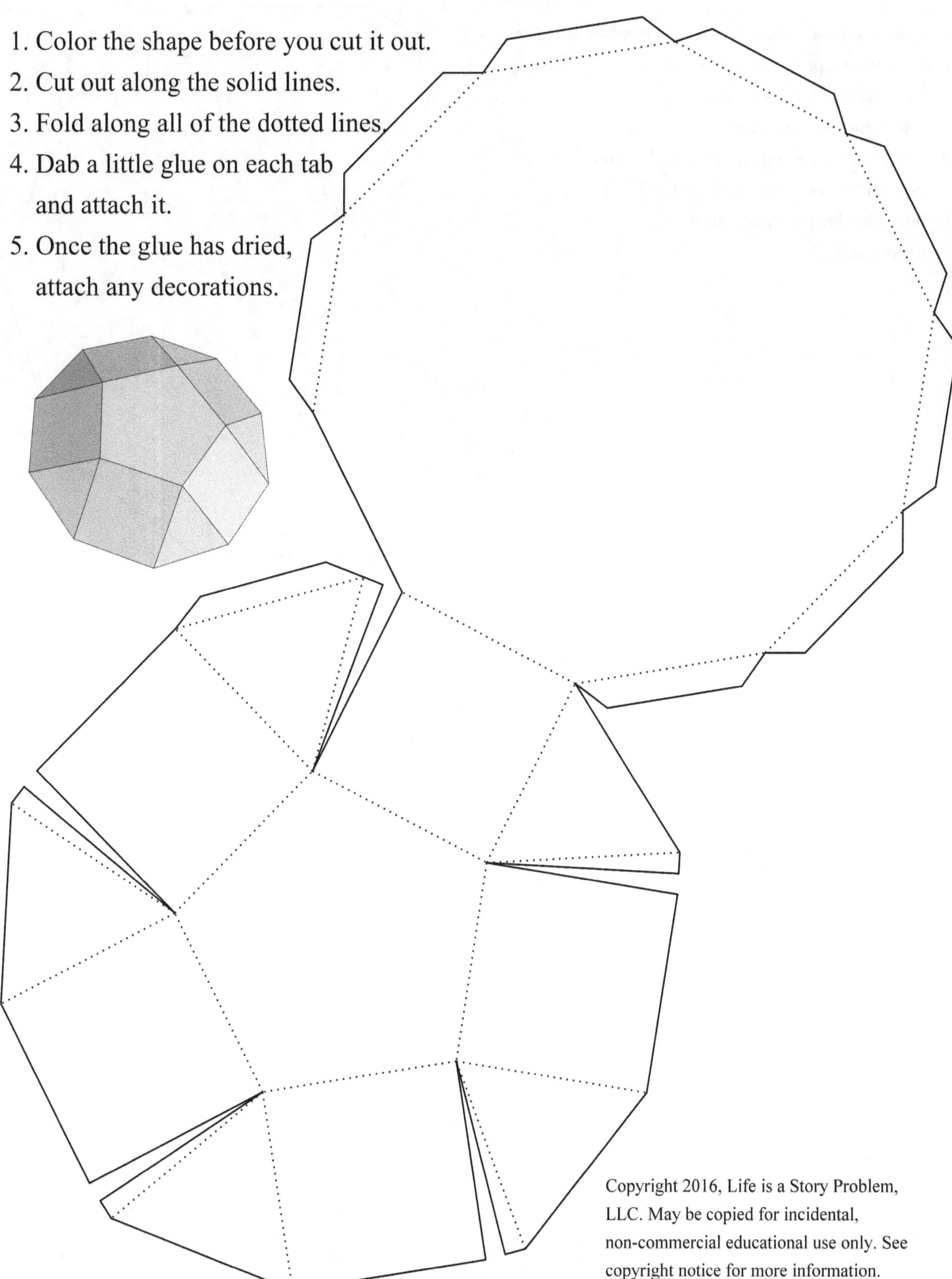

Twist elongated pentagonal cupola

This net is on two pages. Print out a copy of each page.

1. Color the shape before you cut it out.
2. Cut out along the solid lines.
3. Fold forward along all of the dotted lines.
4. Fold backward along all of the dashed lines.
5. Dab a little glue on each tab and attach it.
6. Once the glue has dried, attach any decorations.

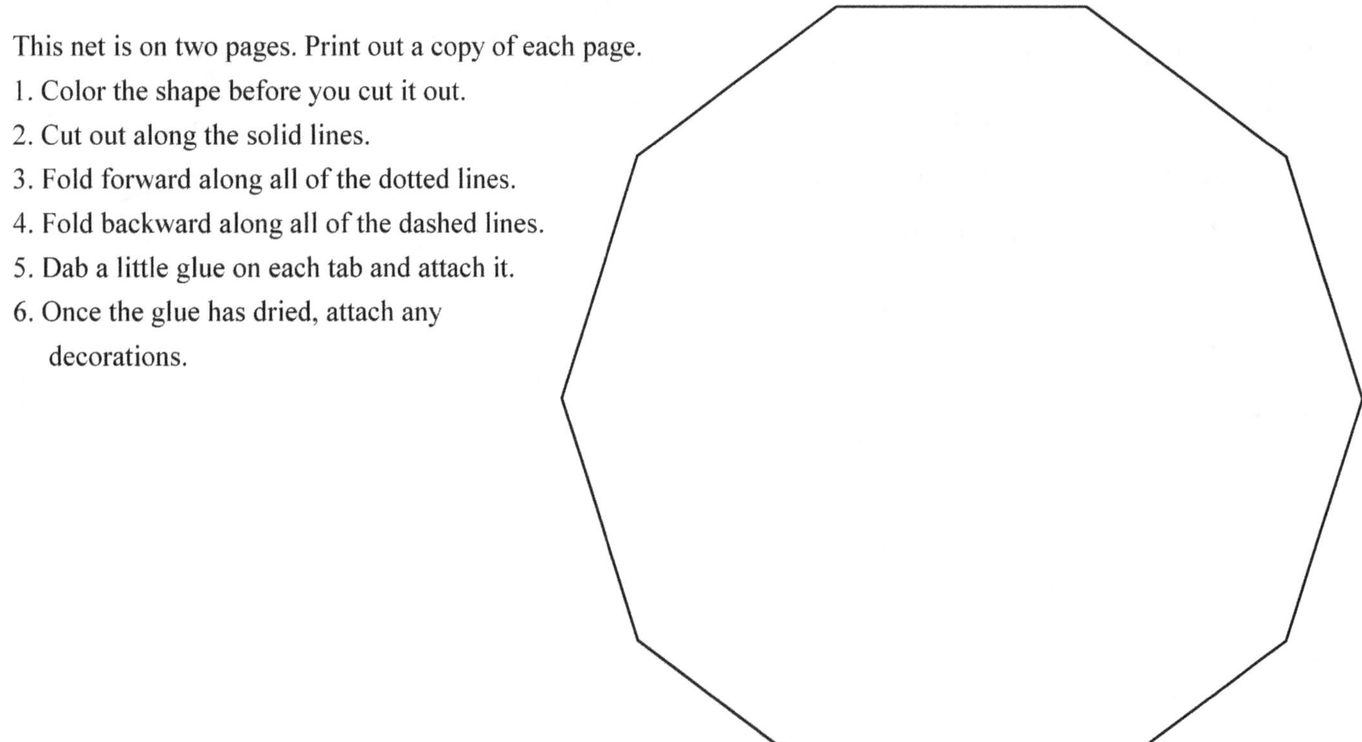

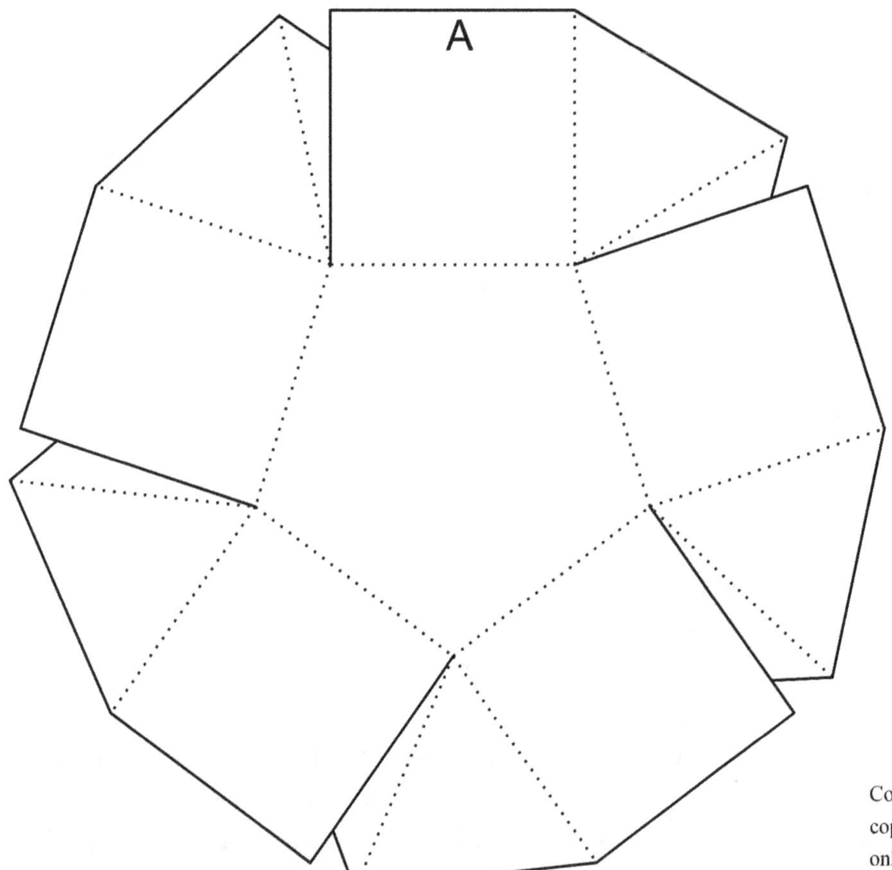

Geometric Nets Mega Project Book, Tabbed By David E. McAdams

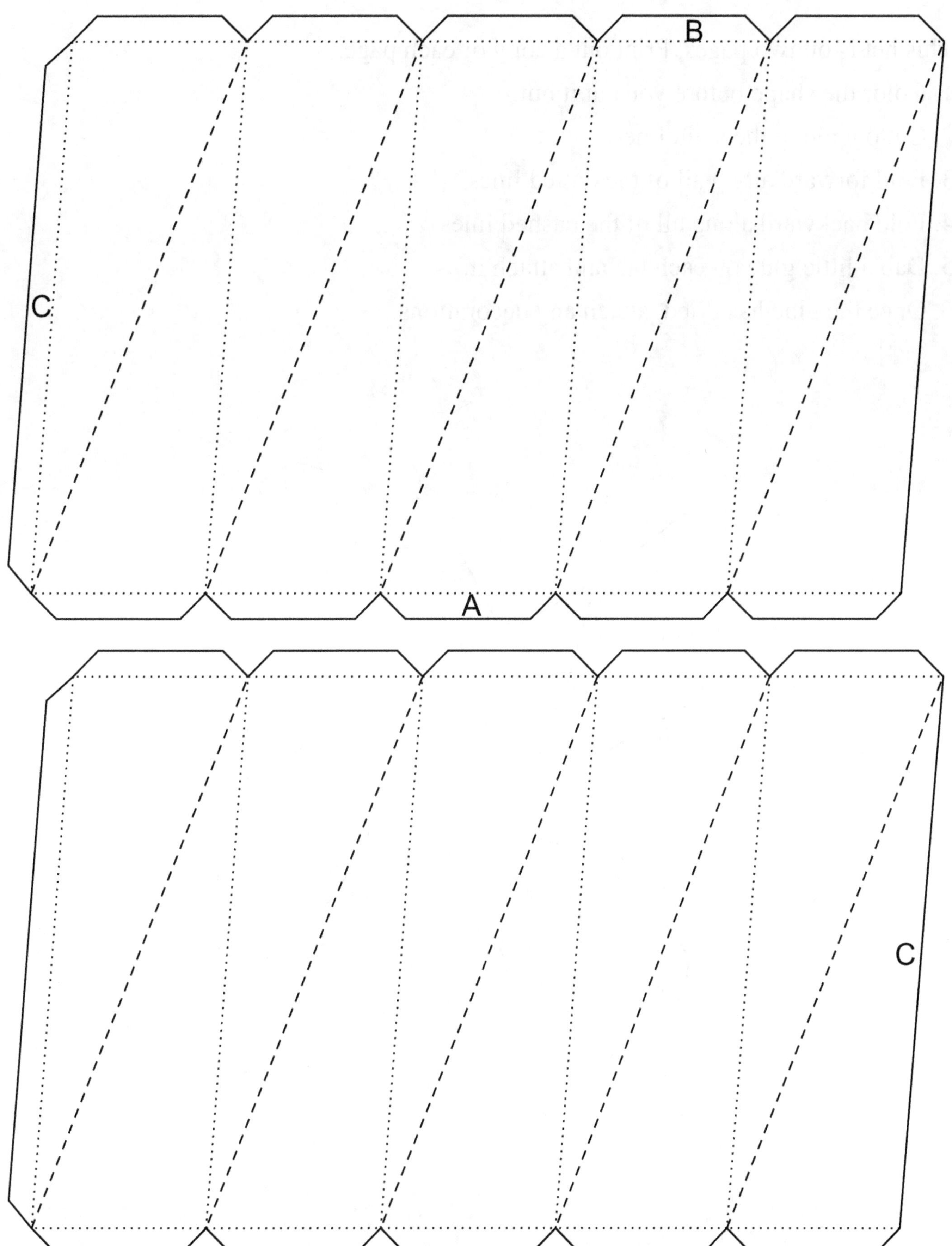

B
C
A
C

Crossed square cupola

This net is on two pages. Print out a copy of each page.

1. Color the shape before you cut it out.
2. Cut out along the solid lines.
3. Fold forward along all of the dotted lines.
4. Fold backward along all of the dashed lines.
5. Dab a little glue on each tab and attach it.
6. Once the glue has dried, attach any decorations.

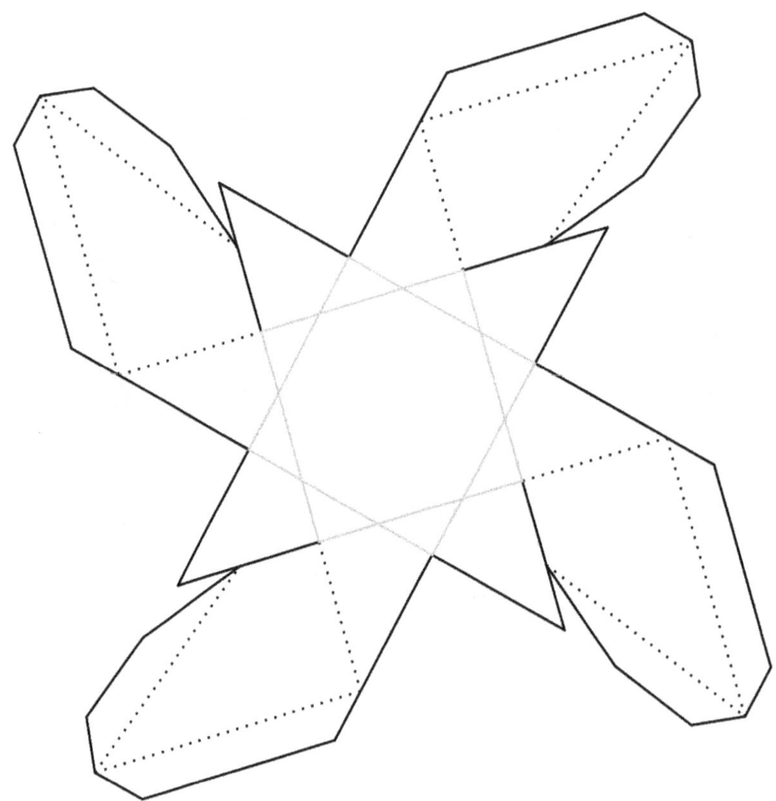

 Geometric Nets Mega Project Book, Tabbed By David E. McAdams

2nd page, Crossed square cupola

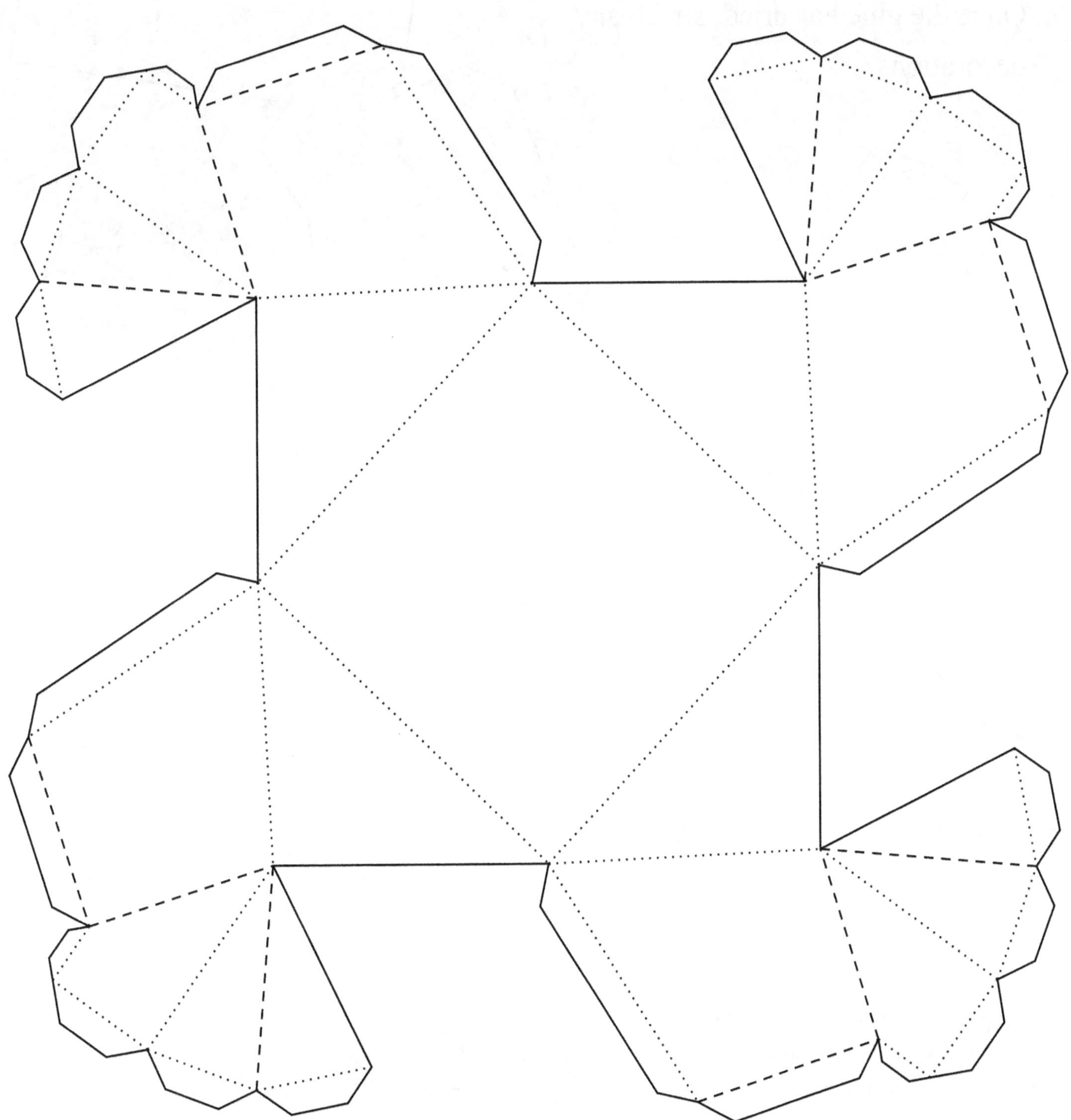

Pentagonal star cupoloid

1. Color the shape before you cut it out.
2. Cut out along the solid lines.
3. Fold forward along all of the dotted lines.
4. Fold backward along all of the dashed lines.
5. Dab a little glue on each tab and attach it.
6. Once the glue has dried, attach any decorations.

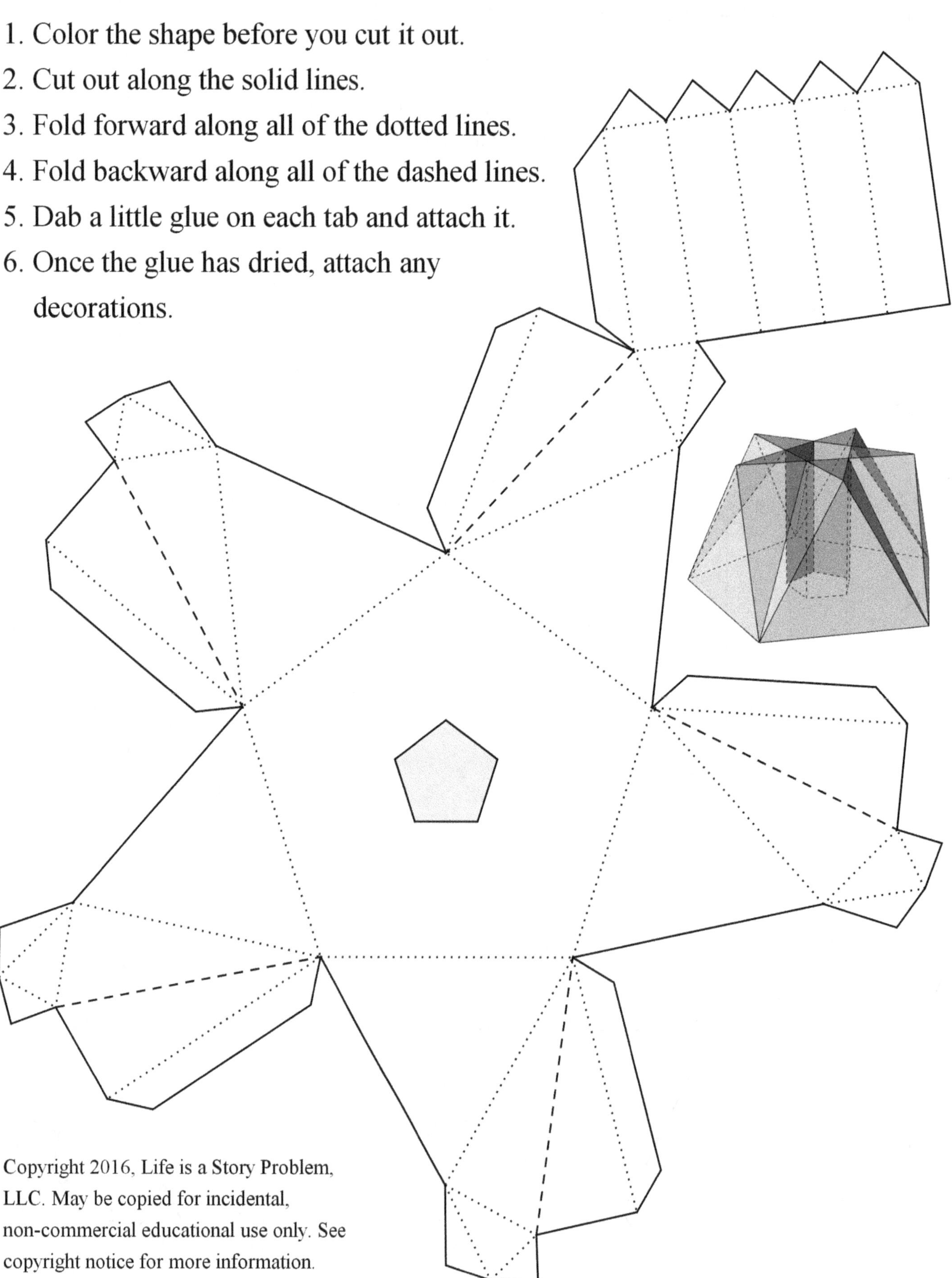

Octagonal star cupola

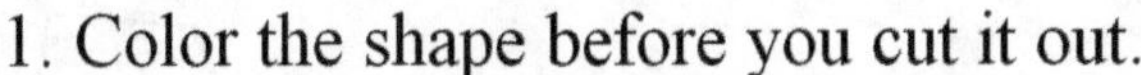

This net is on five pages. Print out a copy of each page.

1. Color the shape before you cut it out.

2. Cut out along the solid lines.

3. Fold forward along all of the dotted lines.

4. Fold backward along all of the dashed lines.

5. Dab a little glue on each tab and attach it.

6. Once the glue has dried, attach any decorations.

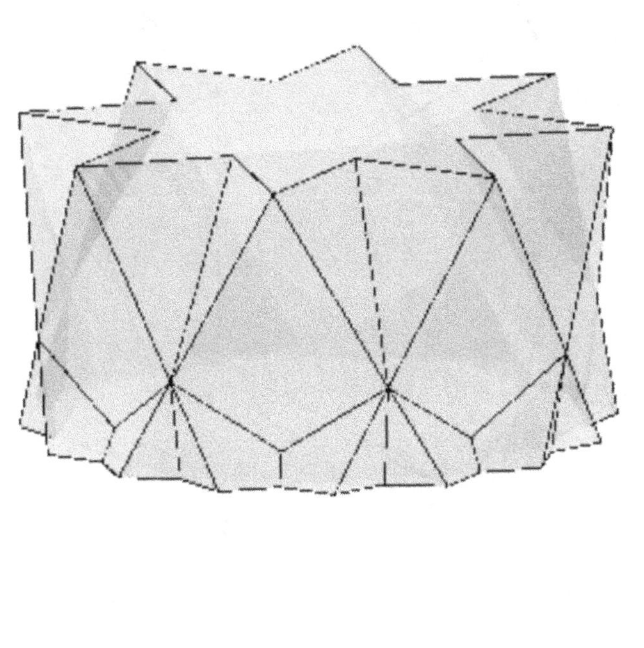

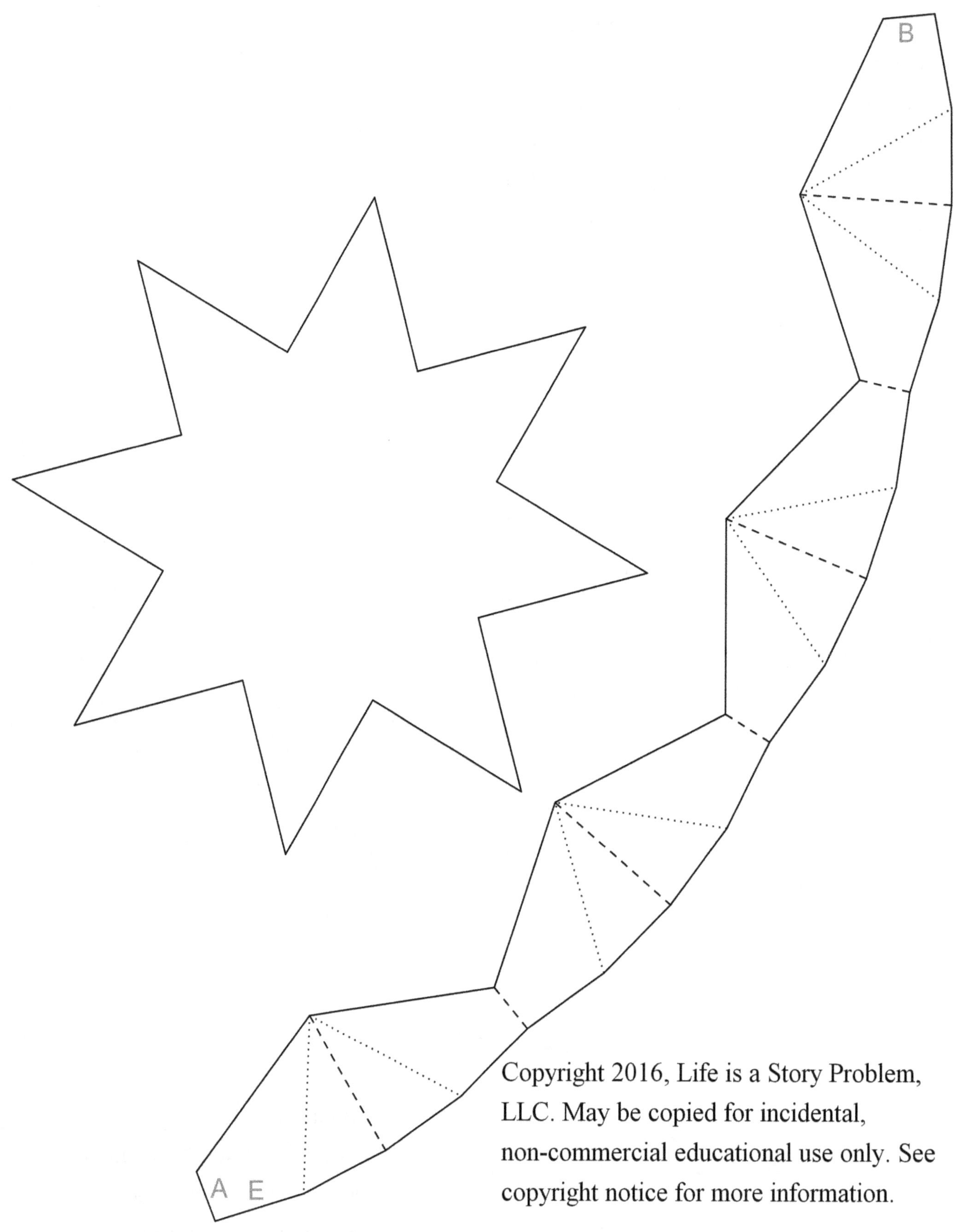

Geometric Nets Mega Project Book, Tabbed By David E. McAdams

E

4th page, Octagrammic cupola

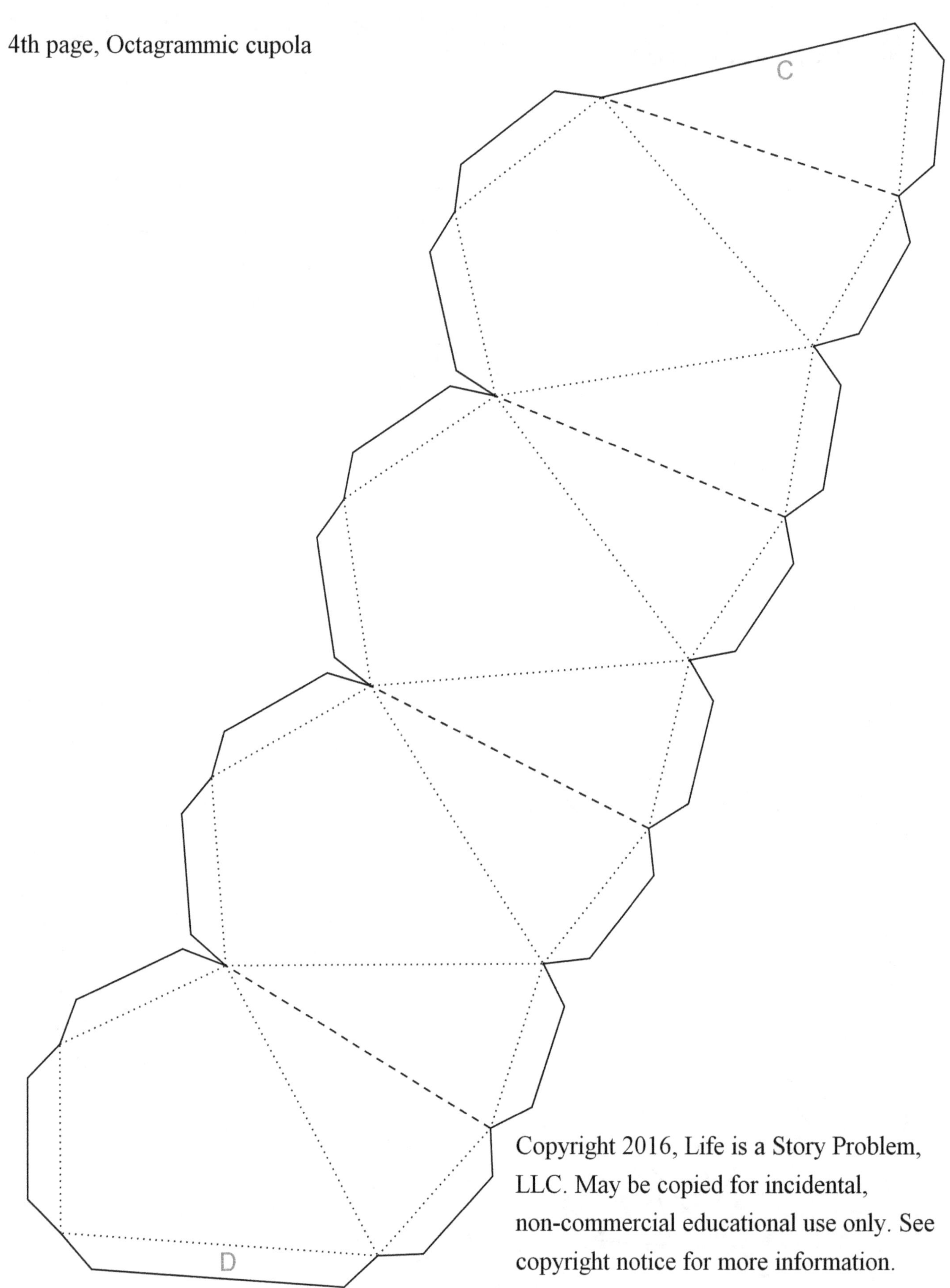

Geometric Nets Mega Project Book, Tabbed By David E. McAdams

5th page, Octagrammic cupola

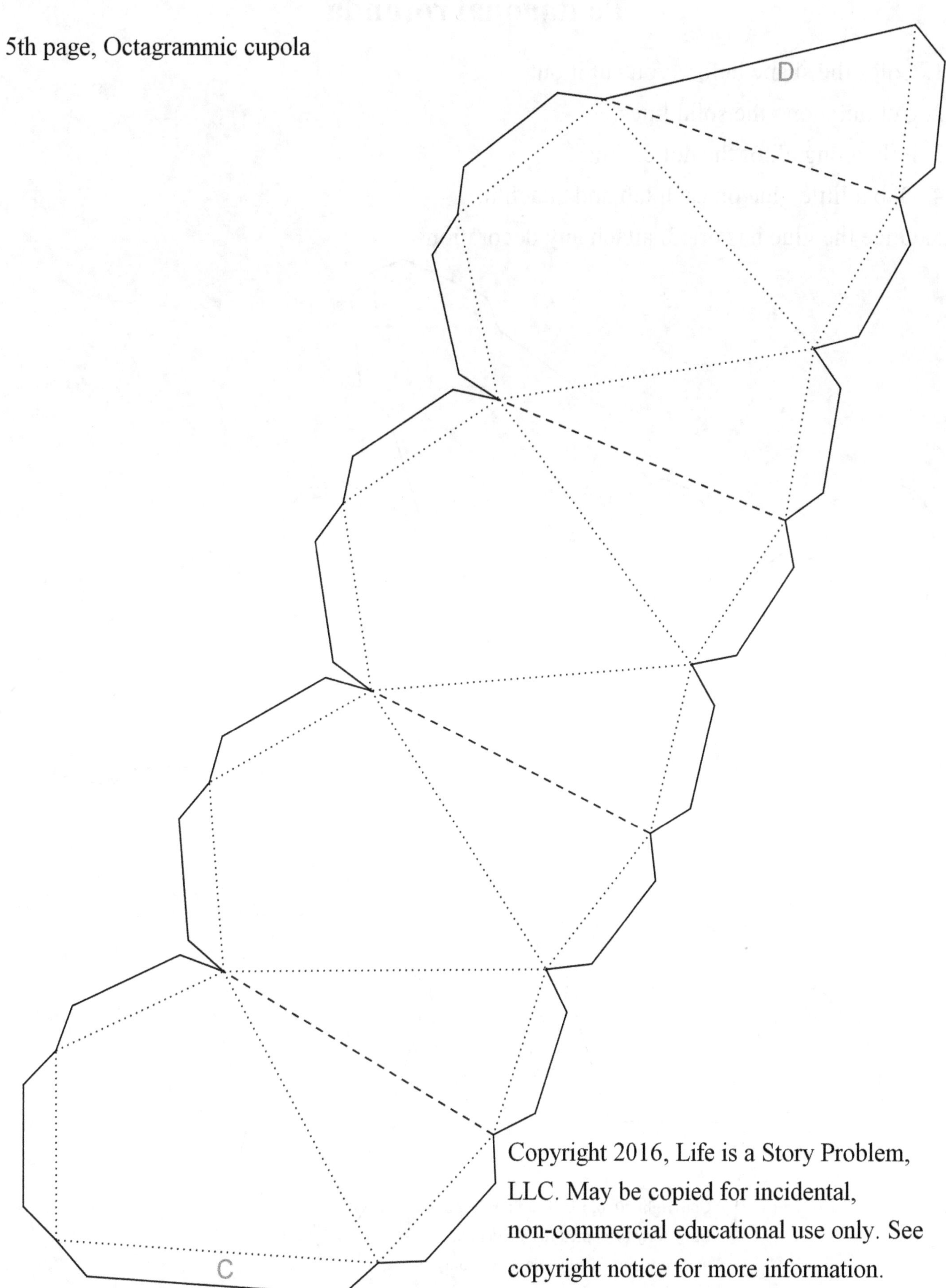

Pentagonal rotunda

1. Color the shape before you cut it out.

2. Cut out along the solid lines.

3. Fold along all of the dotted lines.

4. Dab a little glue on each tab and attach it.

5. Once the glue has dried, attach any decorations.

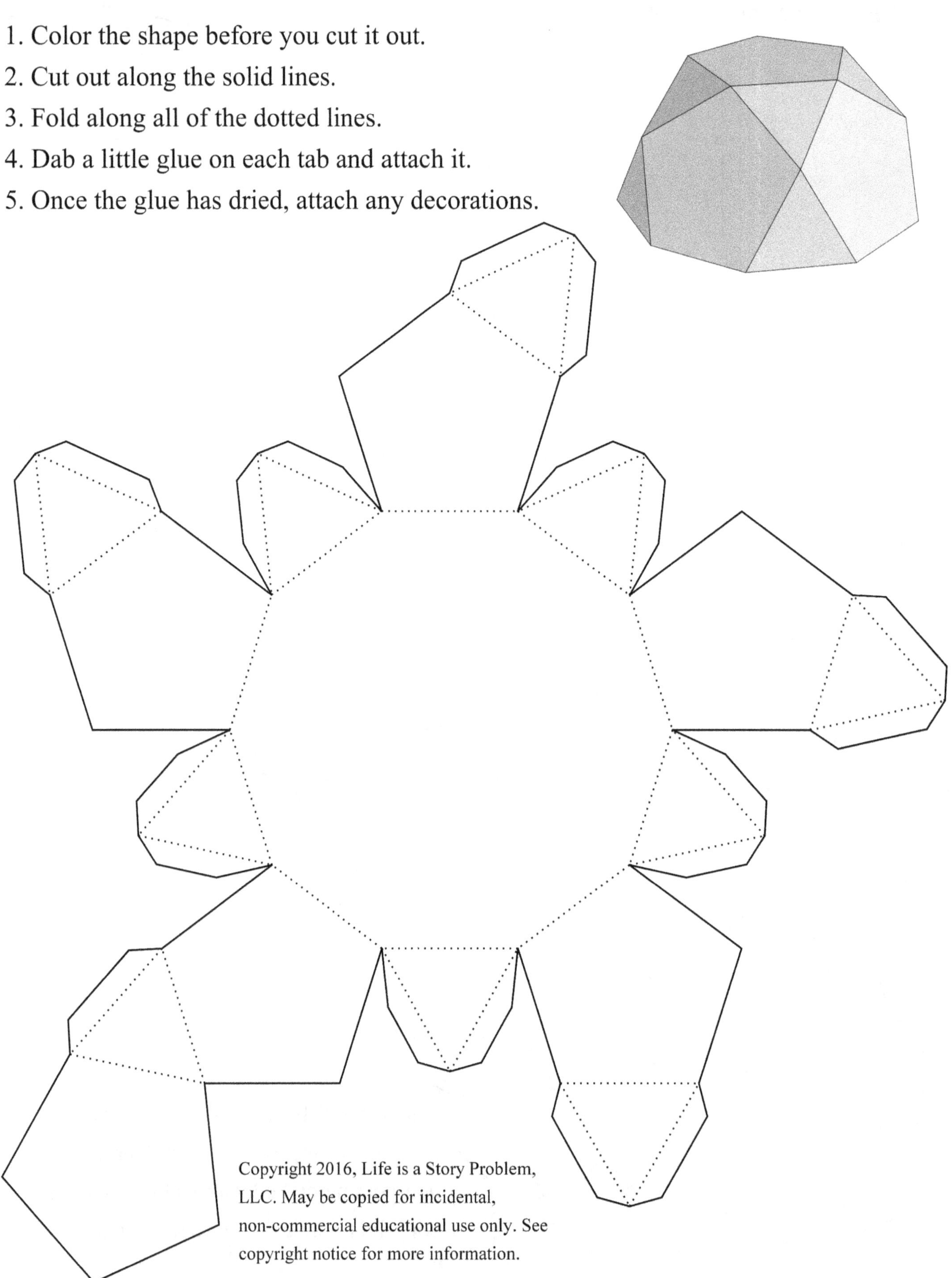

  Geometric Nets Mega Project Book, Tabbed By David E. McAdams

Elongated triangular pyramid

1. Color the shape before you cut it out.
2. Cut out along the solid lines.
3. Fold along all of the dotted lines.
4. Dab a little glue on each tab and attach it.
5. Once the glue has dried, attach any decorations.

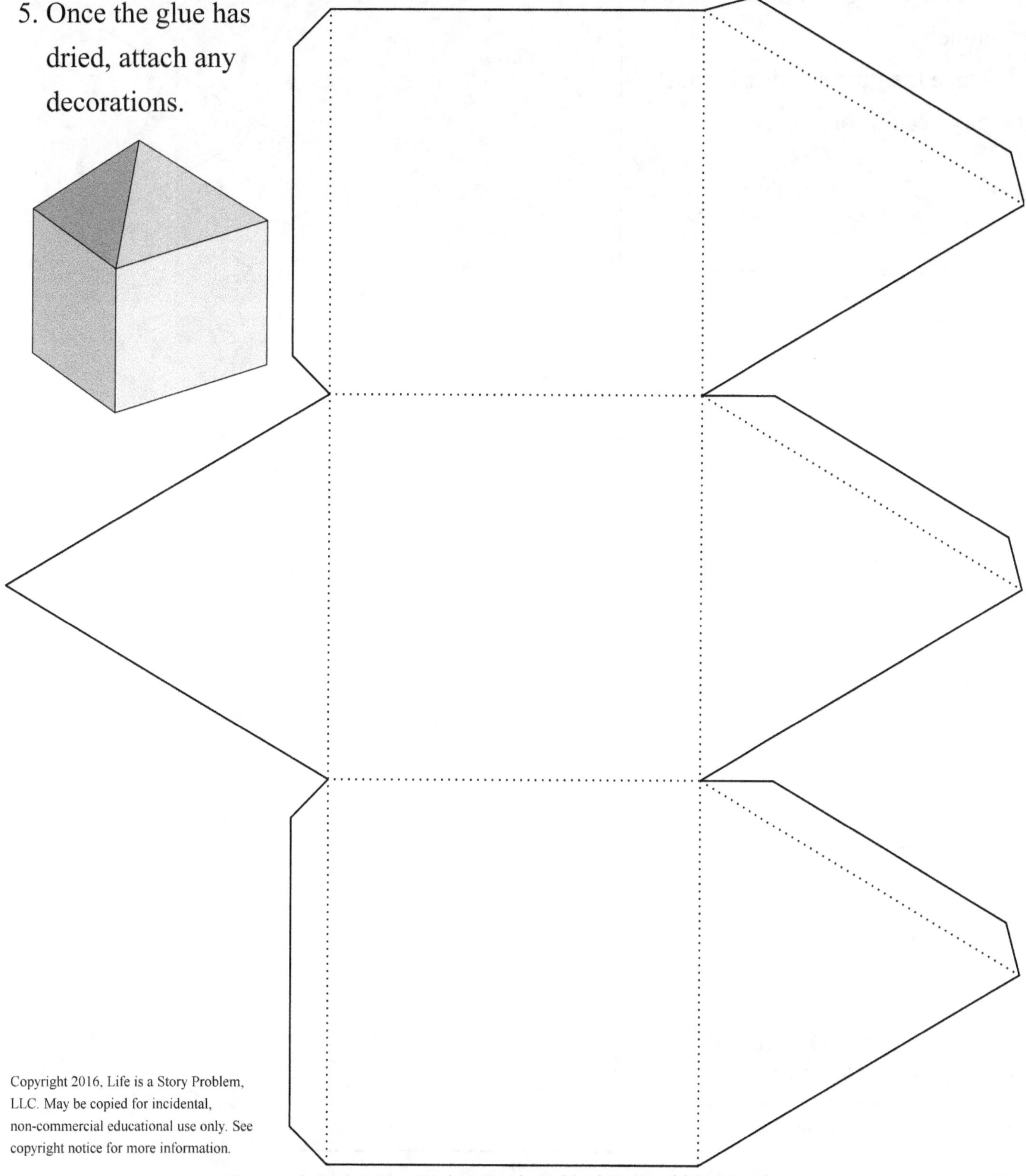

Elongated square pyramid

1. Color the shape before you cut it out.
2. Cut out along the solid lines.
3. Fold along all of the dotted lines.
4. Dab a little glue on each tab and attach it.
5. Once the glue has dried, attach any decorations.

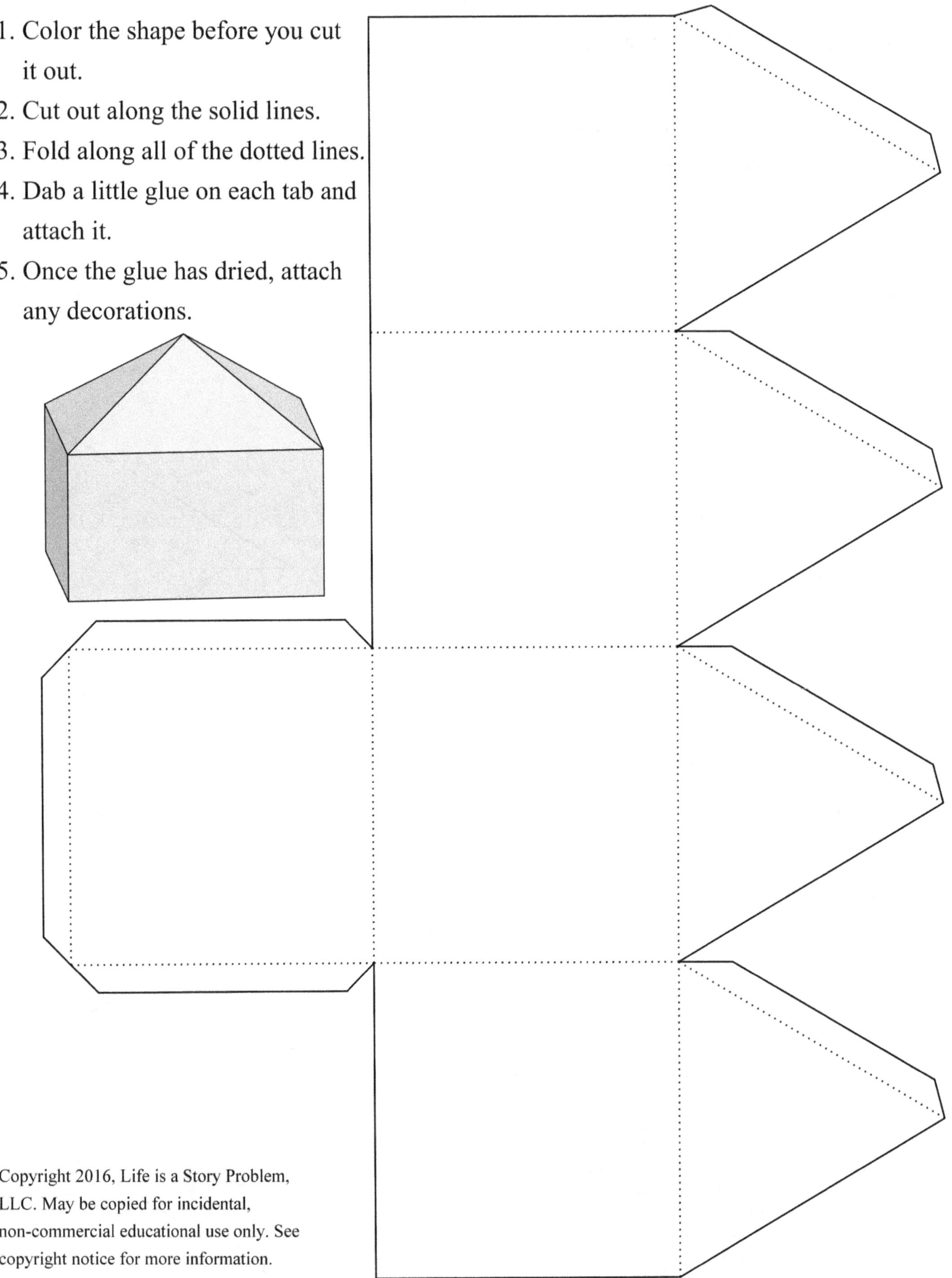

Elongated pentagonal pyramid

1. Color the shape before you cut it out.

2. Cut out along the solid lines.

3. Fold along all of the dotted lines.

4. Dab a little glue on each tab and attach it.

5. Once the glue has dried, attach
   any decorations.

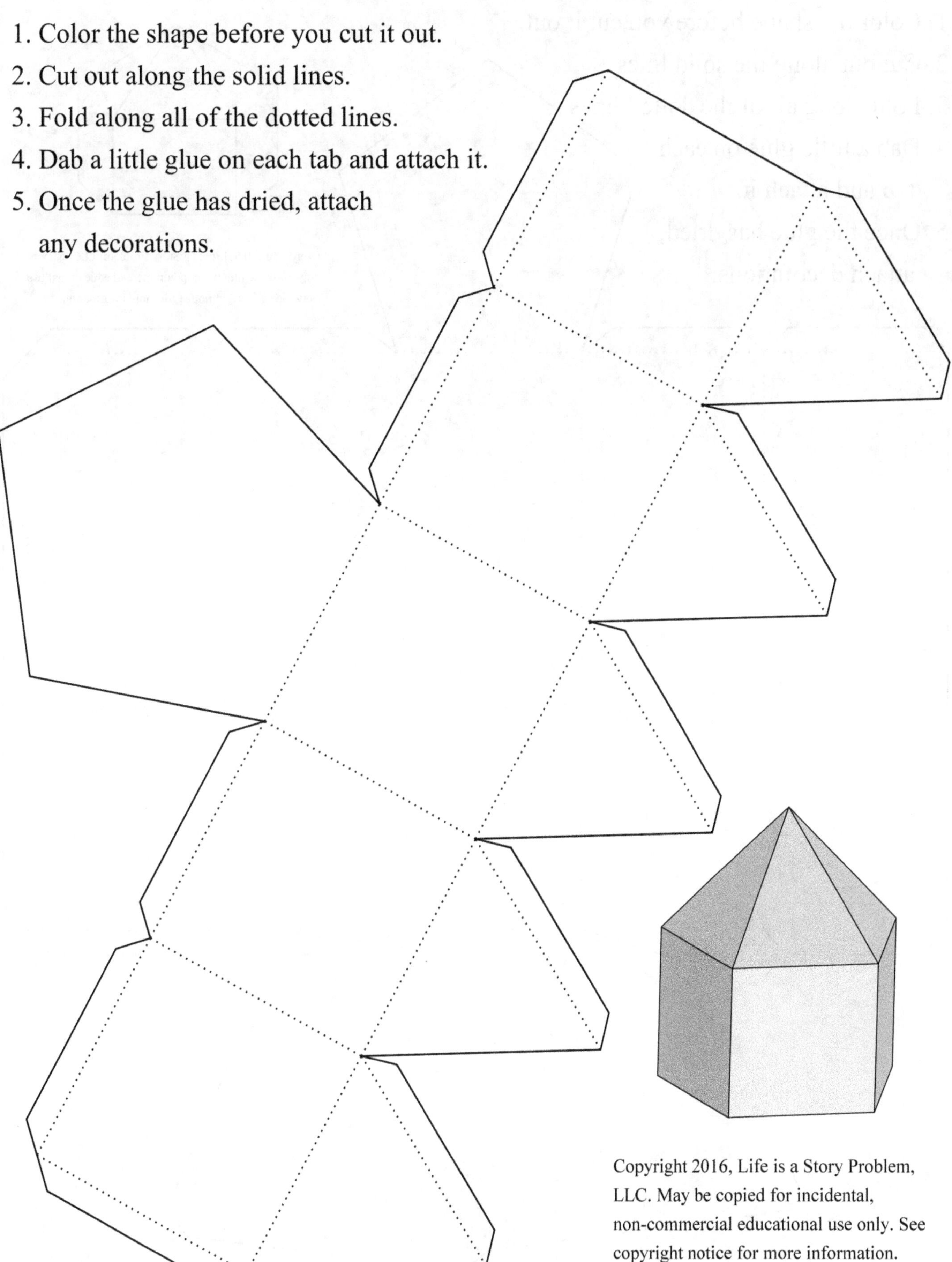

Elongated pentagonal pyramid 2

1. Color the shape before you cut it out.
2. Cut out along the solid lines.
3. Fold along all of the dotted lines.
4. Dab a little glue on each tab and attach it.
5. Once the glue has dried, attach decorations.

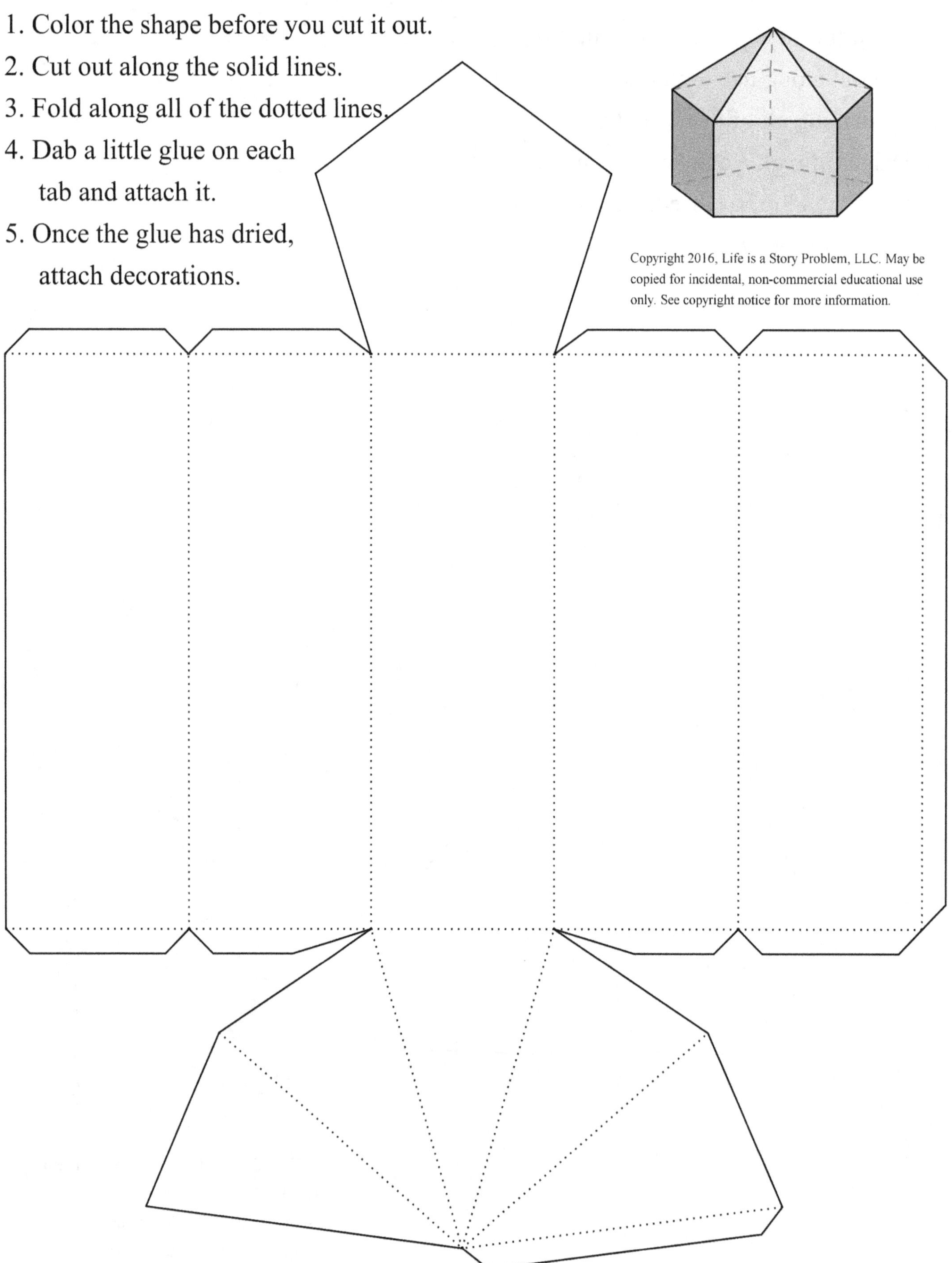

     Geometric Nets Mega Project Book, Tabbed By David E. McAdams

Twist elongated pentagonal pyramid

1. Color the shape before you cut it out.

2. Cut out along the solid lines.

3. Fold forward along all of the dotted lines.

4. Fold backward along all of the dashed lines.

5. Dab a little glue on each tab and attach it.

6. Once the glue has dried, attach any decorations.

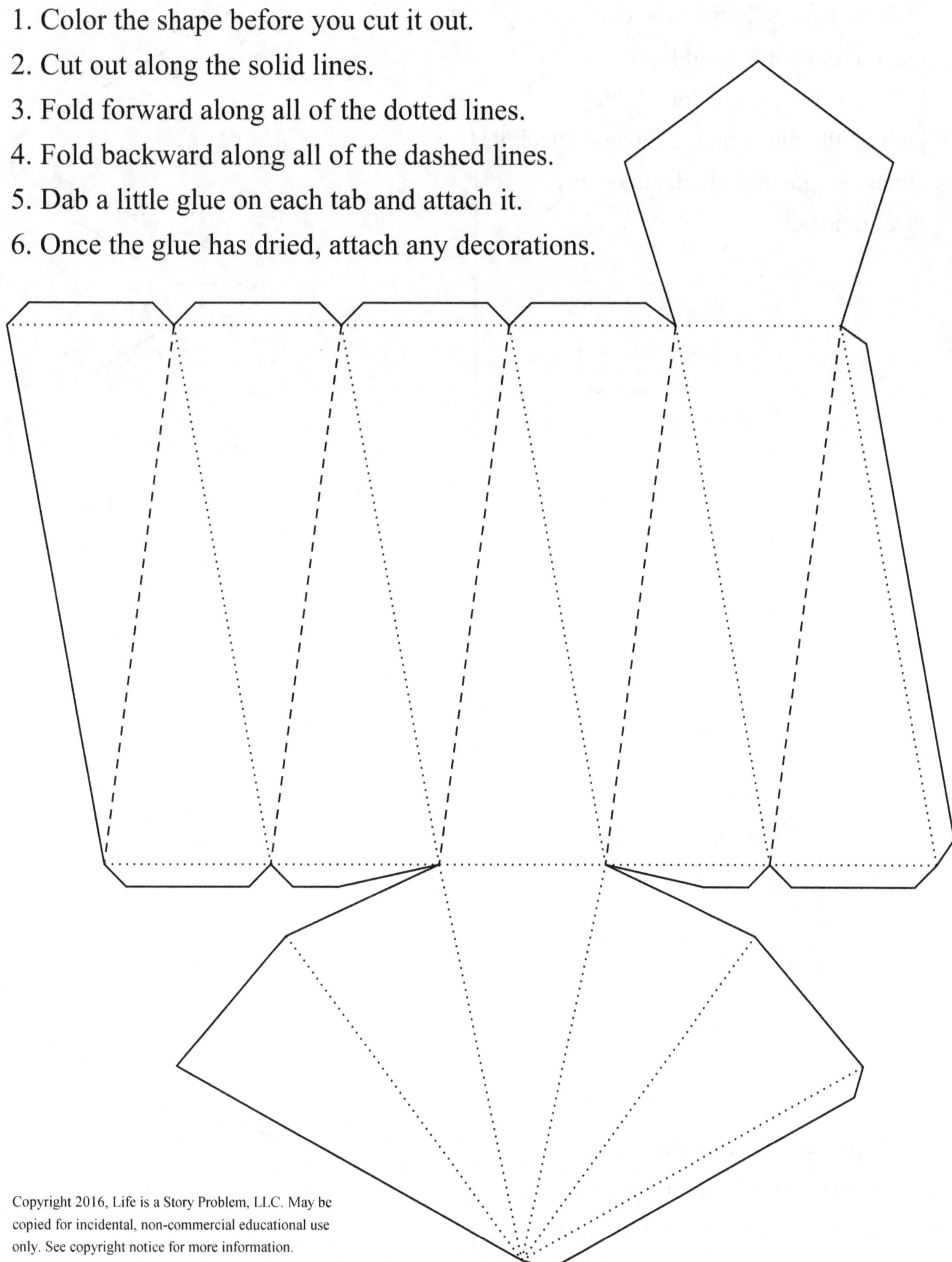

Gyroelongated square pyramid

1. Color the shape before you cut it out.

2. Cut out along the solid lines.

3. Fold along all of the dotted lines.

4. Dab a little glue on each tab and attach it.

5. Once the glue has dried, attach any
   decorations.

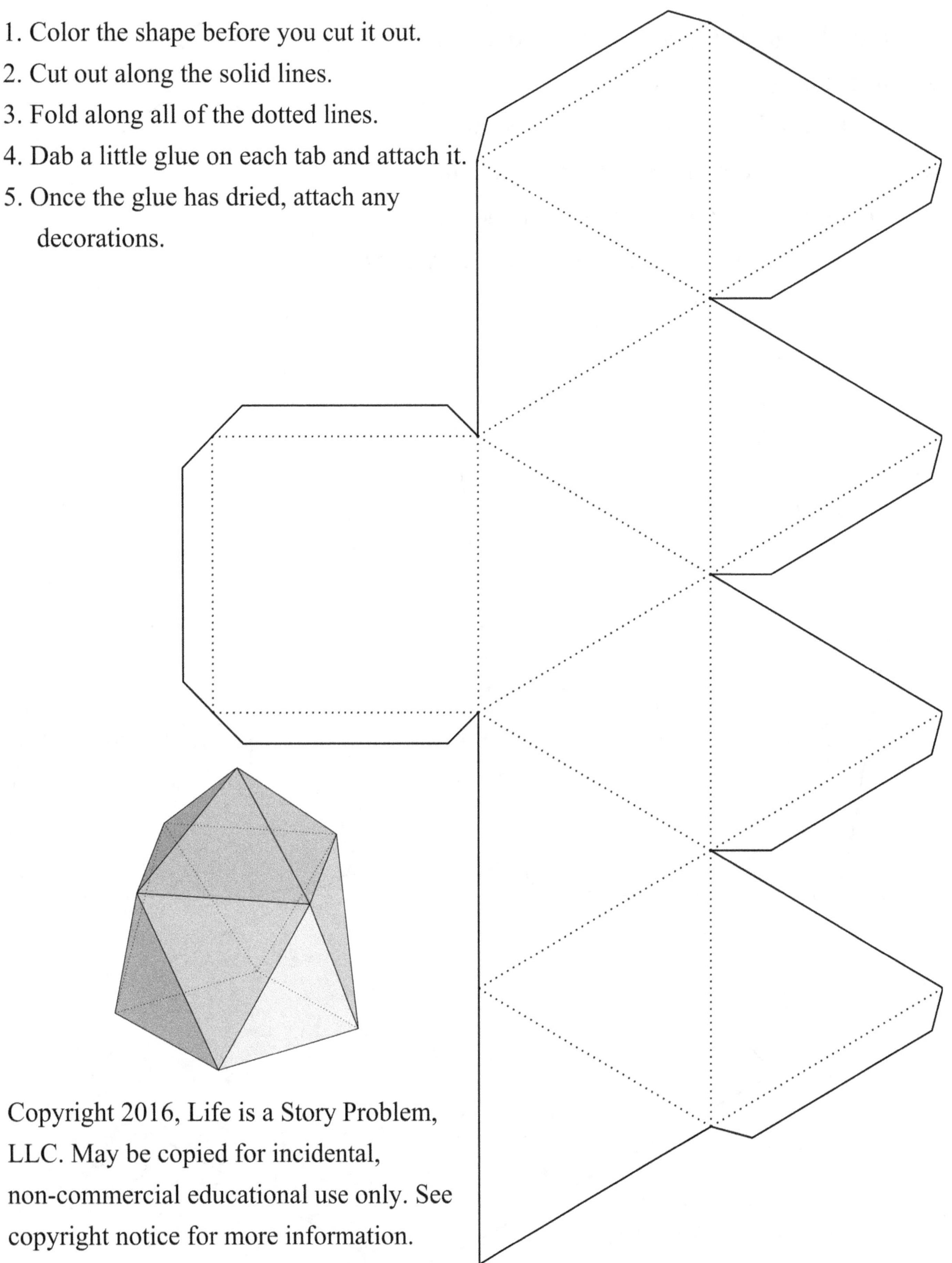

 Geometric Nets Mega Project Book, Tabbed By David E. McAdams

Gyroelongated pentagonal pyramid

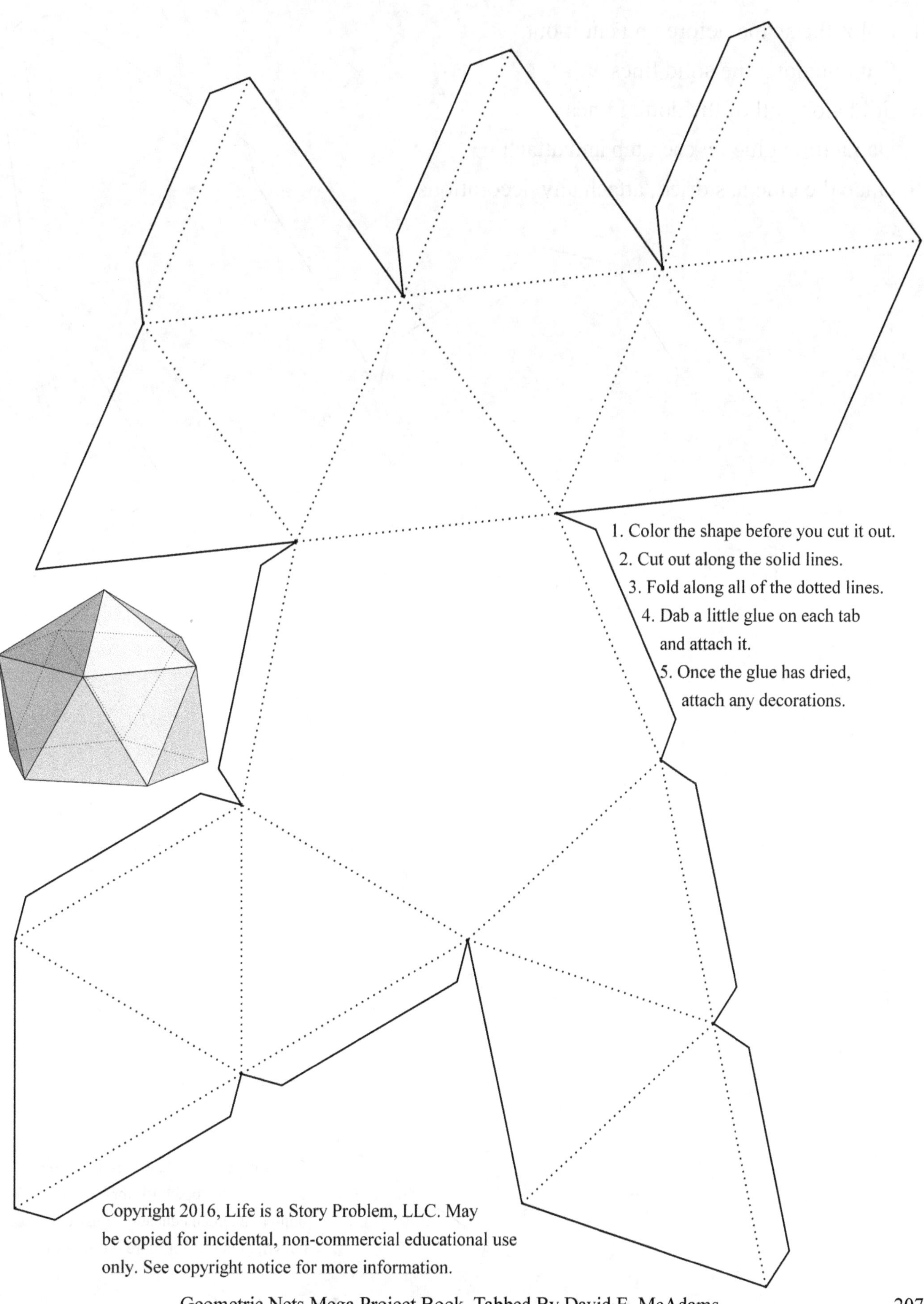

1. Color the shape before you cut it out.
2. Cut out along the solid lines.
3. Fold along all of the dotted lines.
4. Dab a little glue on each tab
   and attach it.
5. Once the glue has dried,
   attach any decorations.

Elongated digonal cupola

1. Color the shape before you cut it out.
2. Cut out along the solid lines.
3. Fold along all of the dotted lines.
4. Dab a little glue on each tab and attach it.
5. Once the glue has dried, attach any decorations.

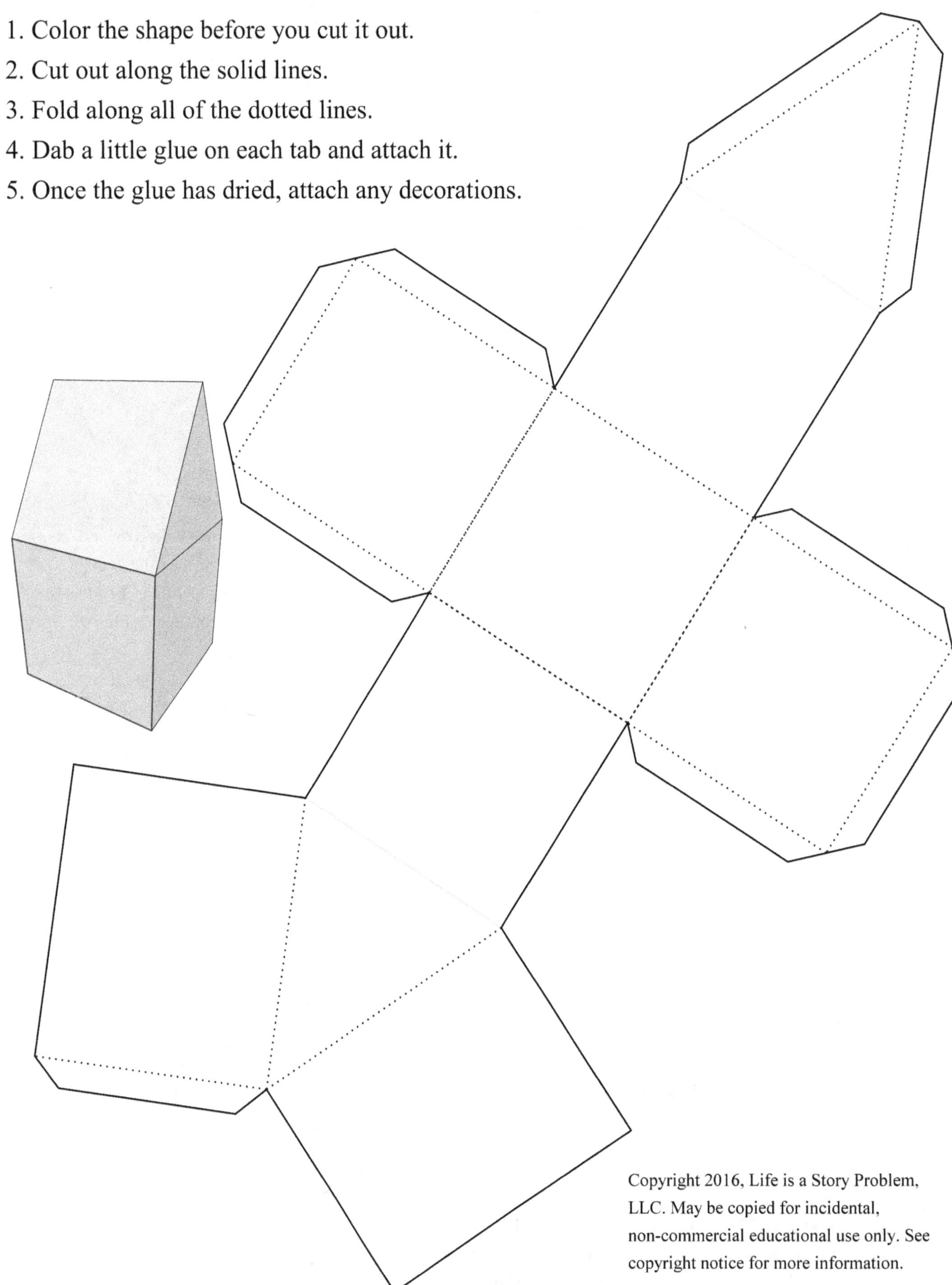

 Geometric Nets Mega Project Book, Tabbed By David E. McAdams

Triangular bipyramid

1. Color the shape before you cut it out.

2. Cut out along the solid lines.

3. Fold along all of the dotted lines.

4. Dab a little glue on each tab and attach it.

5. Once the glue has dried, attach any decorations.

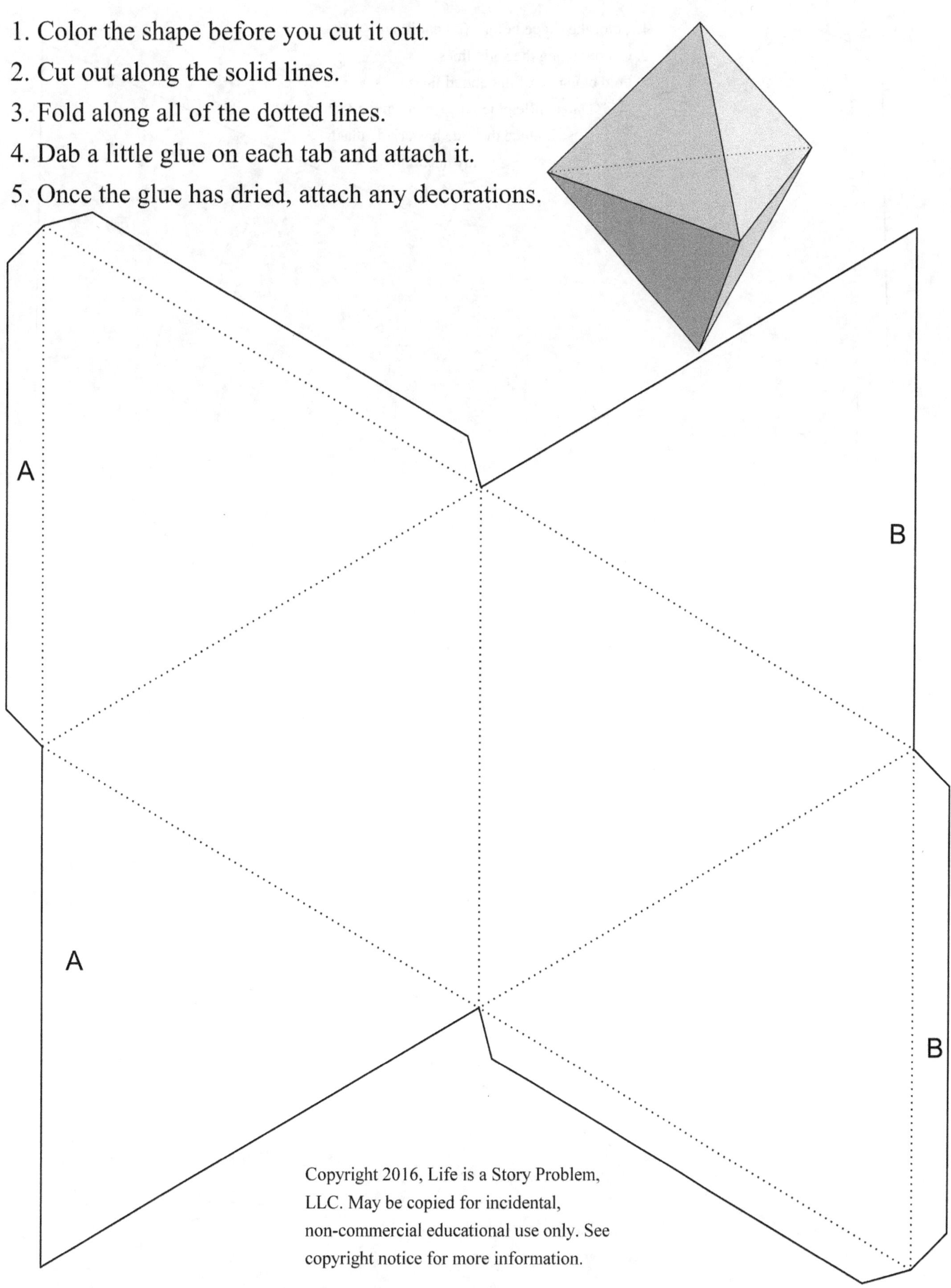

Pentagonal bipyramid

1. Color the shape before you cut it out.
2. Cut out along the solid lines.
3. Fold along all of the dotted lines.
4. Dab a little glue on each tab and attach it.
5. Once the glue has dried, attach any decorations.

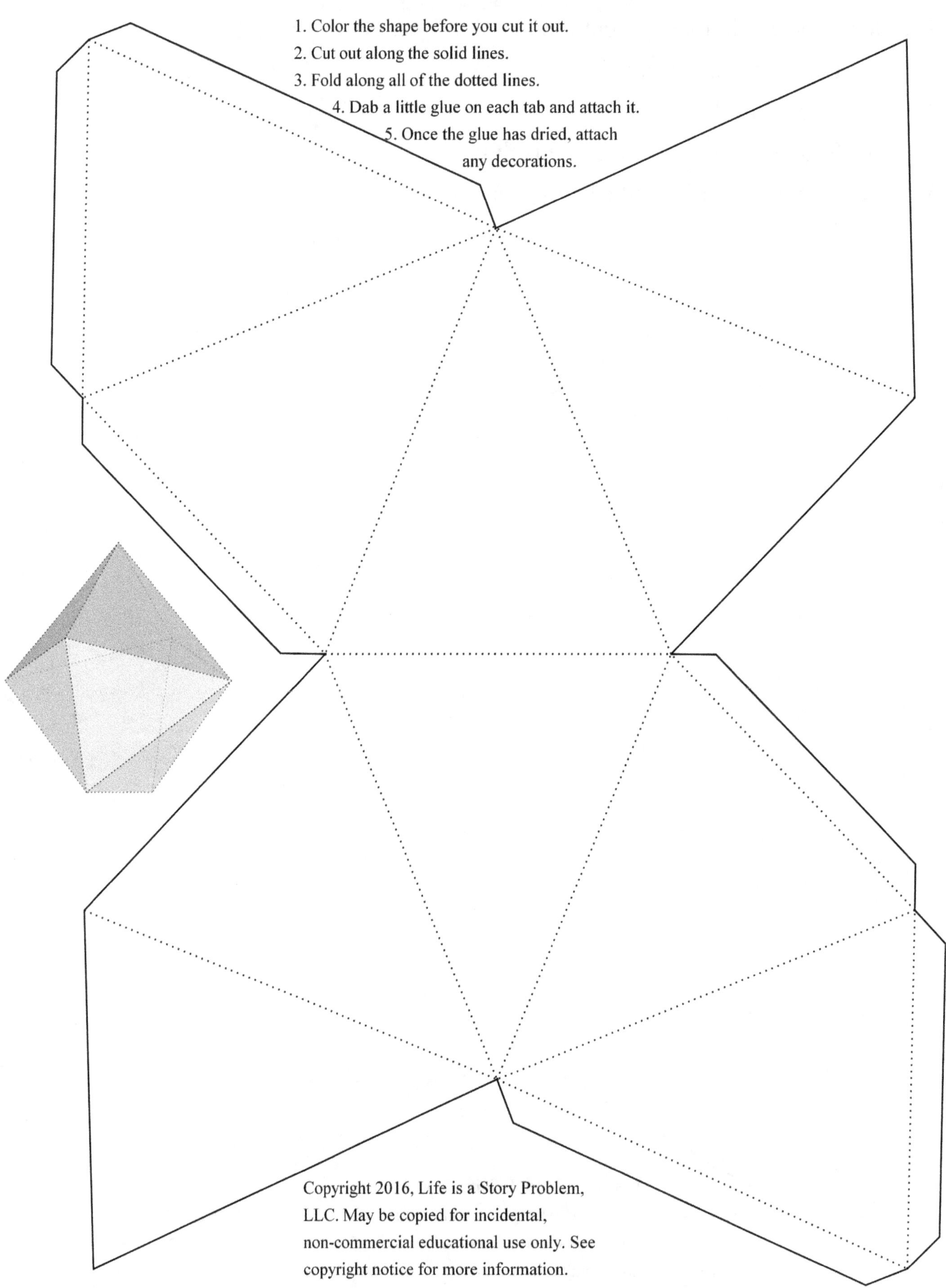

Bitruncated pentagonal bipyramid

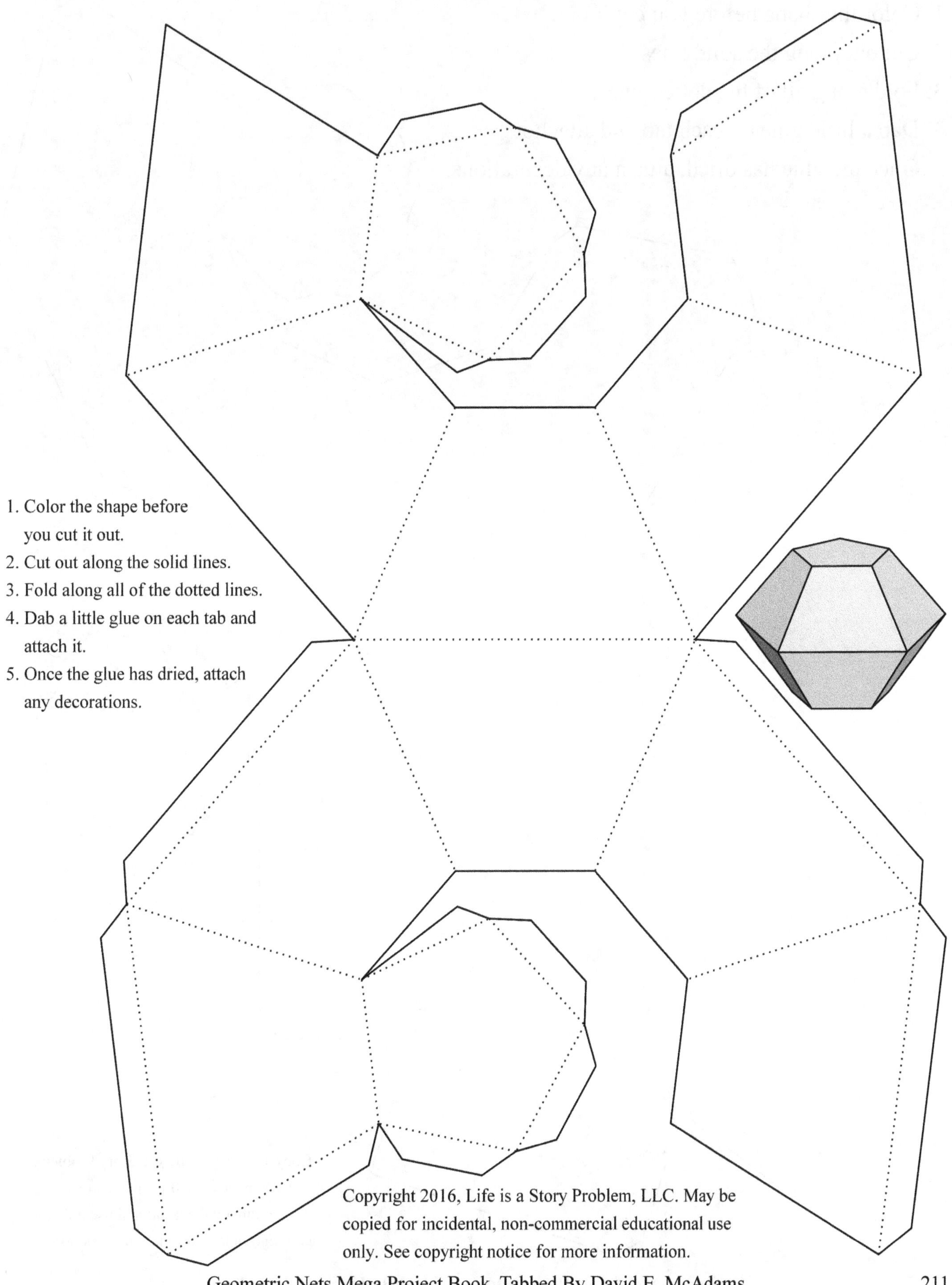

1. Color the shape before you cut it out.
2. Cut out along the solid lines.
3. Fold along all of the dotted lines.
4. Dab a little glue on each tab and attach it.
5. Once the glue has dried, attach any decorations.

Elongated truncated pentagonal pyramid

1. Color the shape before you cut it out.

2. Cut out along the solid lines.

3. Fold along all of the dotted lines.

4. Dab a little glue on each tab and attach it.

5. Once the glue has dried, attach any decorations.

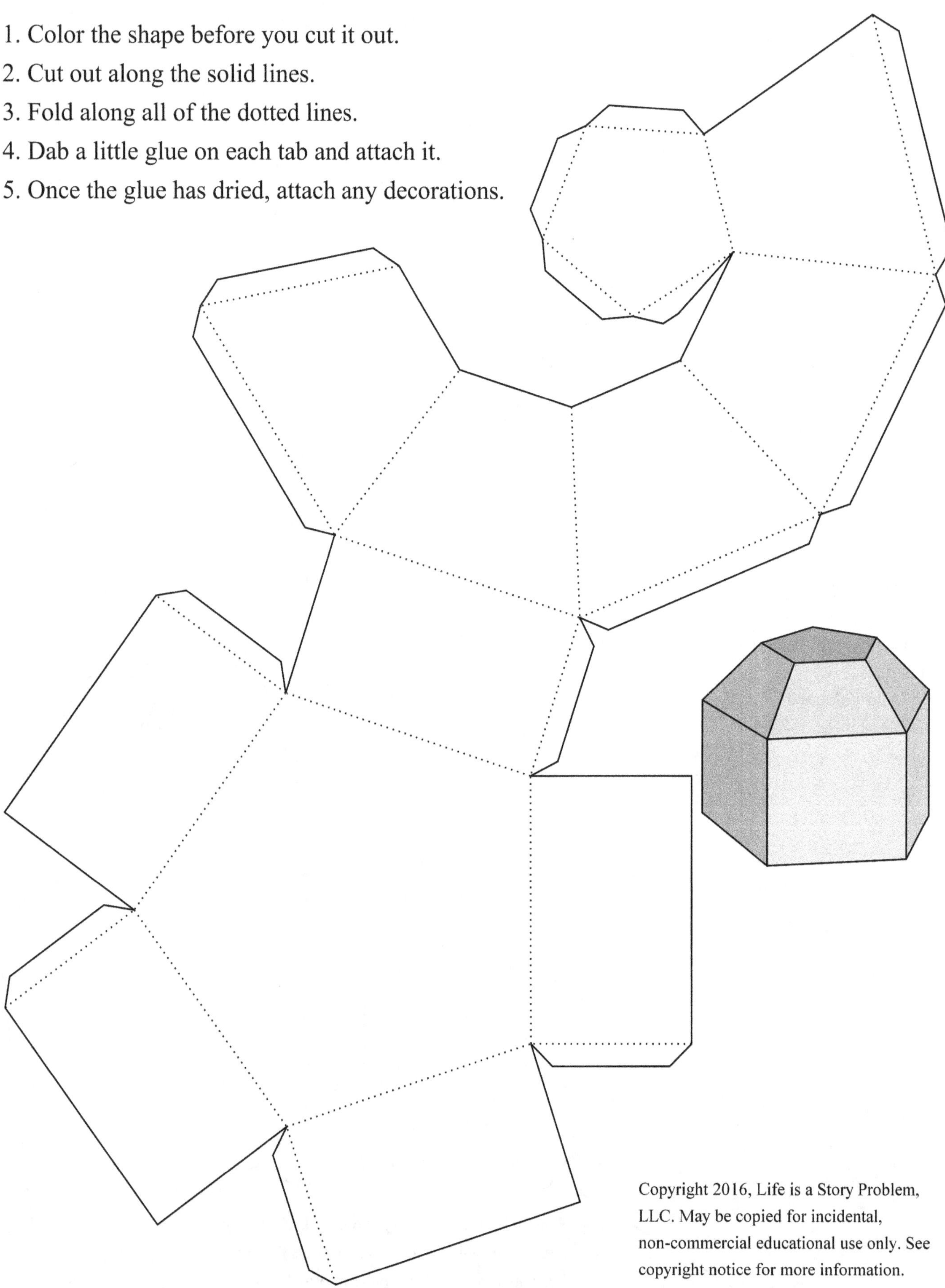

 Geometric Nets Mega Project Book, Tabbed By David E. McAdams

Twisted pentagonal pyramid

1. Color the shape before you cut it out.
2. Cut out along the solid lines.
3. Fold forward along all of the dotted lines.
4. Fold backward along all of the dashed lines.
5. Dab a little glue each tab and attach it.
6. Once the glue has dried, attach decorations.

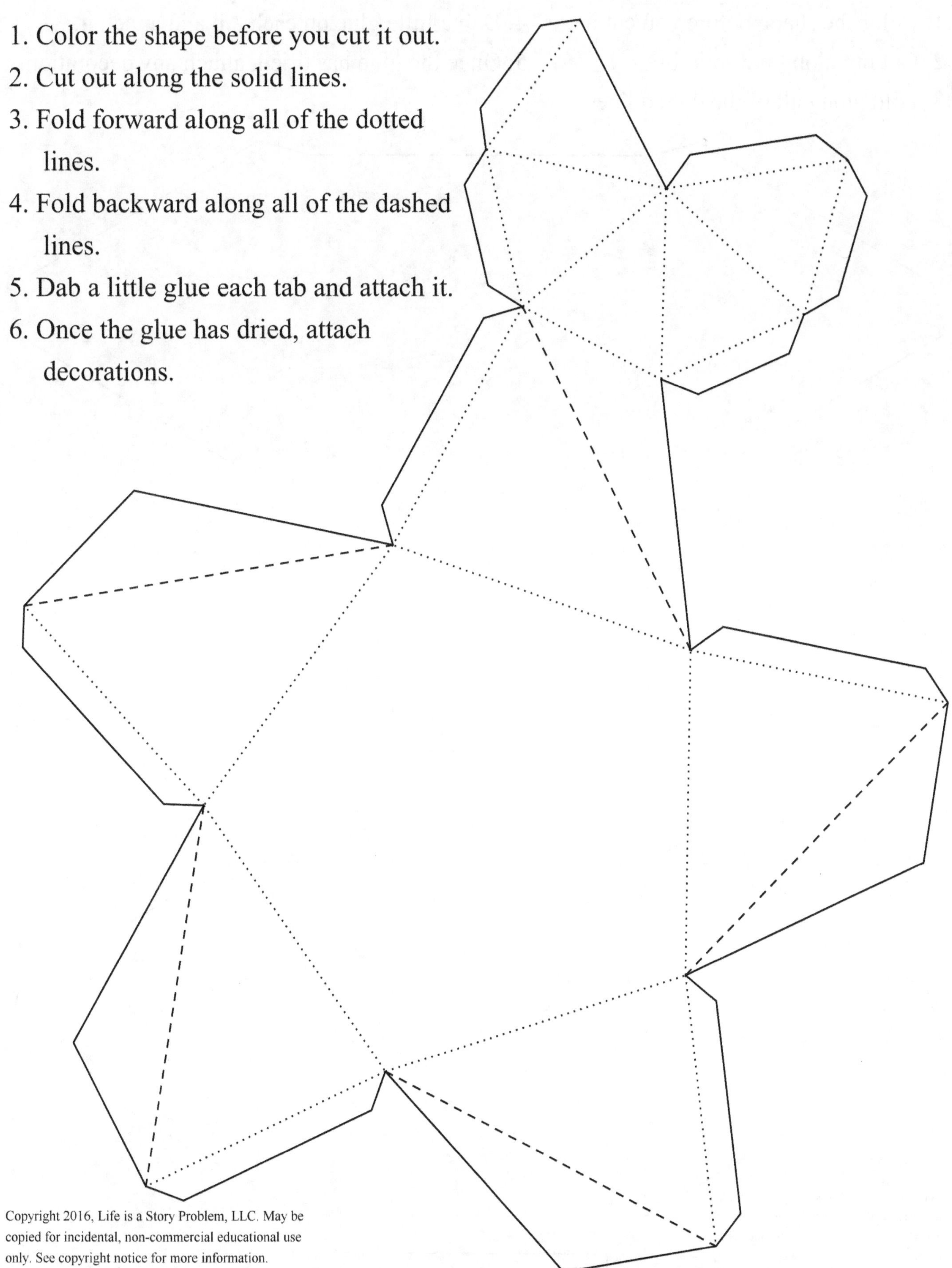

Elongated triangular bipyramid

1. Color the shape before you cut it out.
2. Cut out along the solid lines.
3. Fold along all of the dotted lines.

4. Dab a little glue on each tab and attach it.
5. Once the glue has dried, attach any decorations.

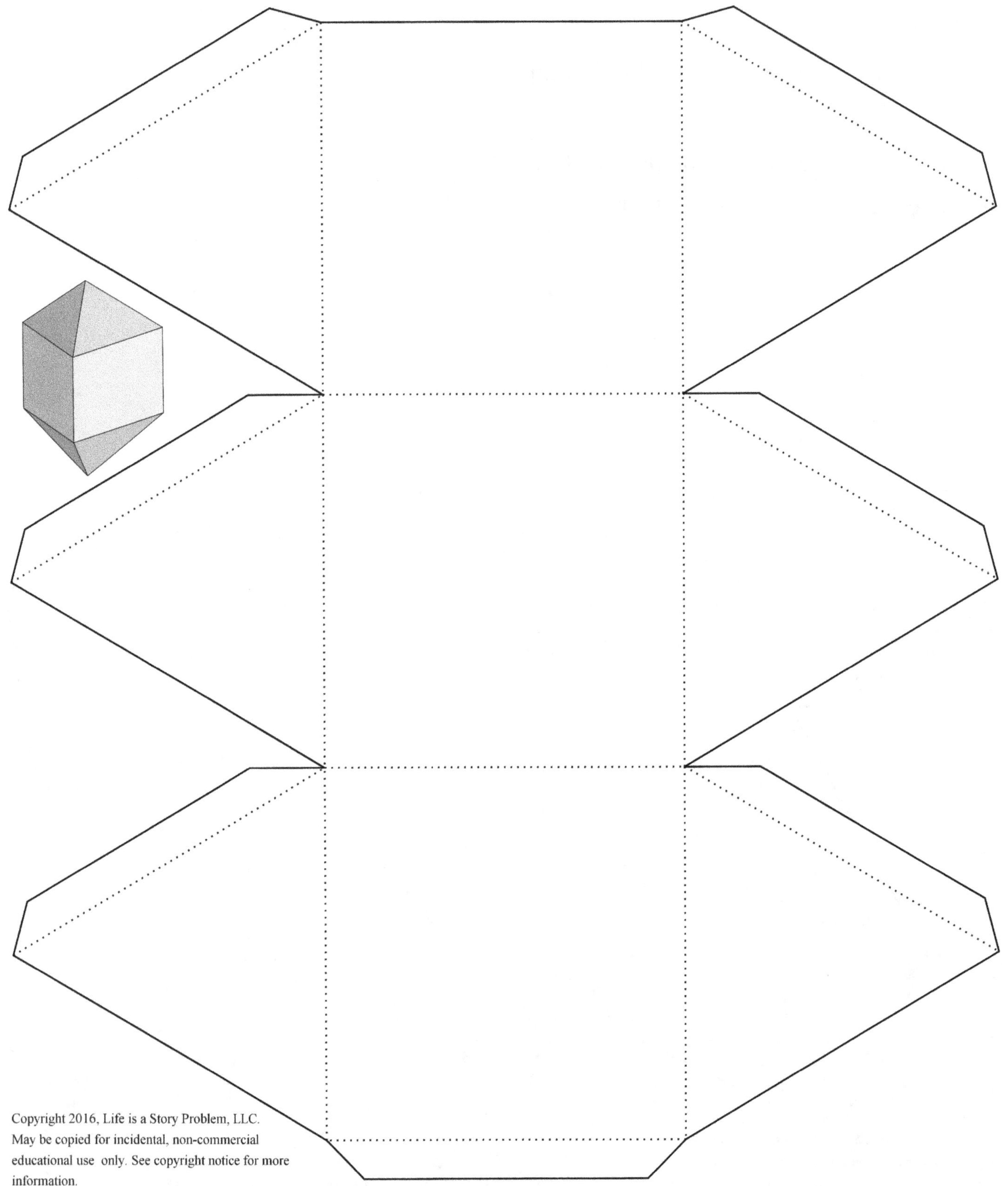

Elongated square bipyramid

1. Color the shape before you cut it out. 2. Cut out along the solid lines. 3. Fold along all of the dotted lines. 4. Dab a little glue on each tab and attach it.

5. Once the glue has dried, attach any decorations.

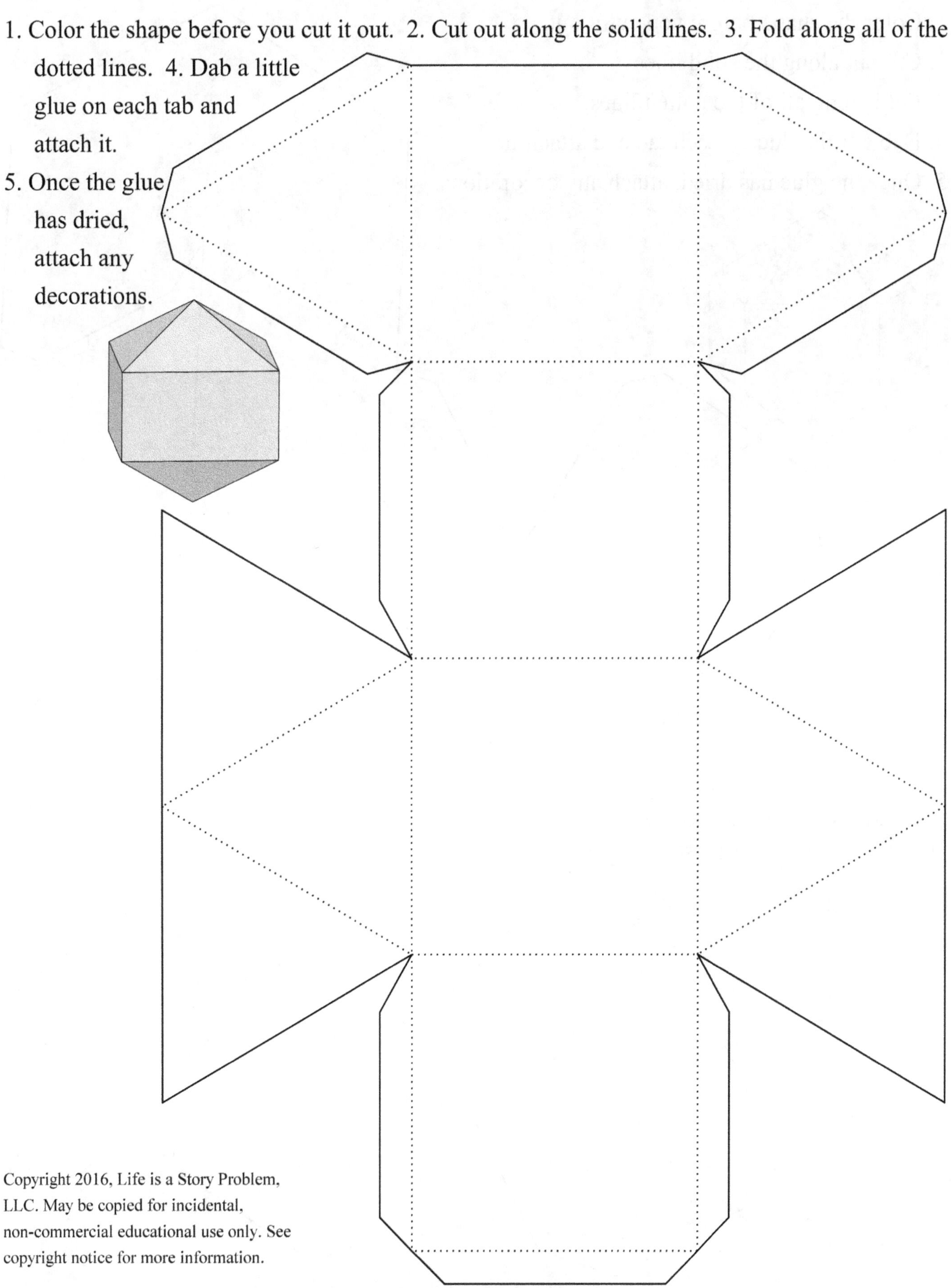

Elongated pentagonal bipyramid

1. Color the shape before you cut it out.

2. Cut out along the solid lines.

3. Fold along all of the dotted lines.

4. Dab a little glue on each tab and attach it.

5. Once the glue has dried, attach any decorations.

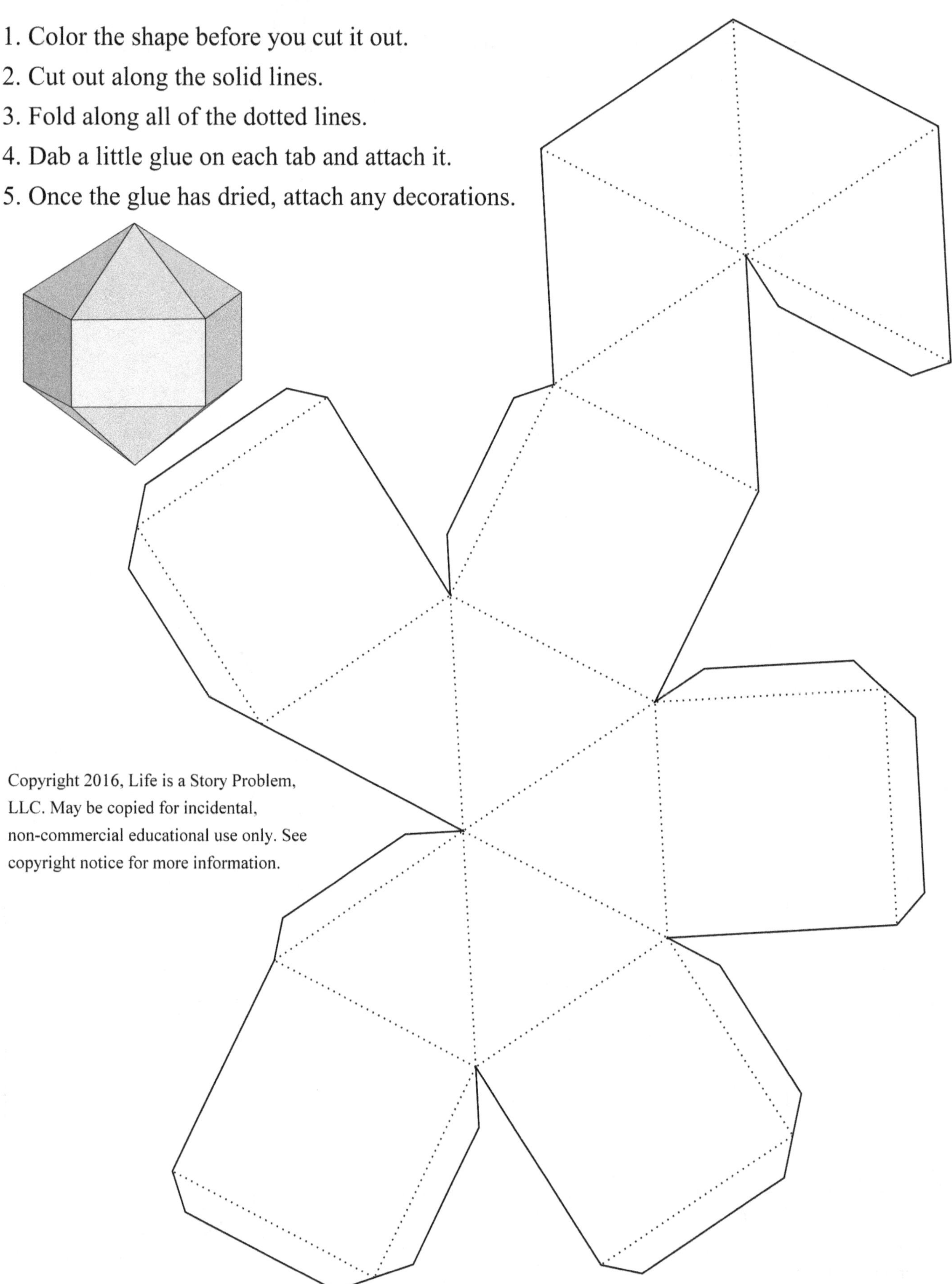

 Geometric Nets Mega Project Book, Tabbed By David E. McAdams

Gyroelongated square bipyramid

1. Color the shape before you cut it out.
2. Cut out along the solid lines.
3. Fold forward along all of the dotted lines.
4. Fold backward along all of the dashed lines.
5. Dab a little glue on each tab and attach it.
6. Once the glue has dried, attach any decorations.

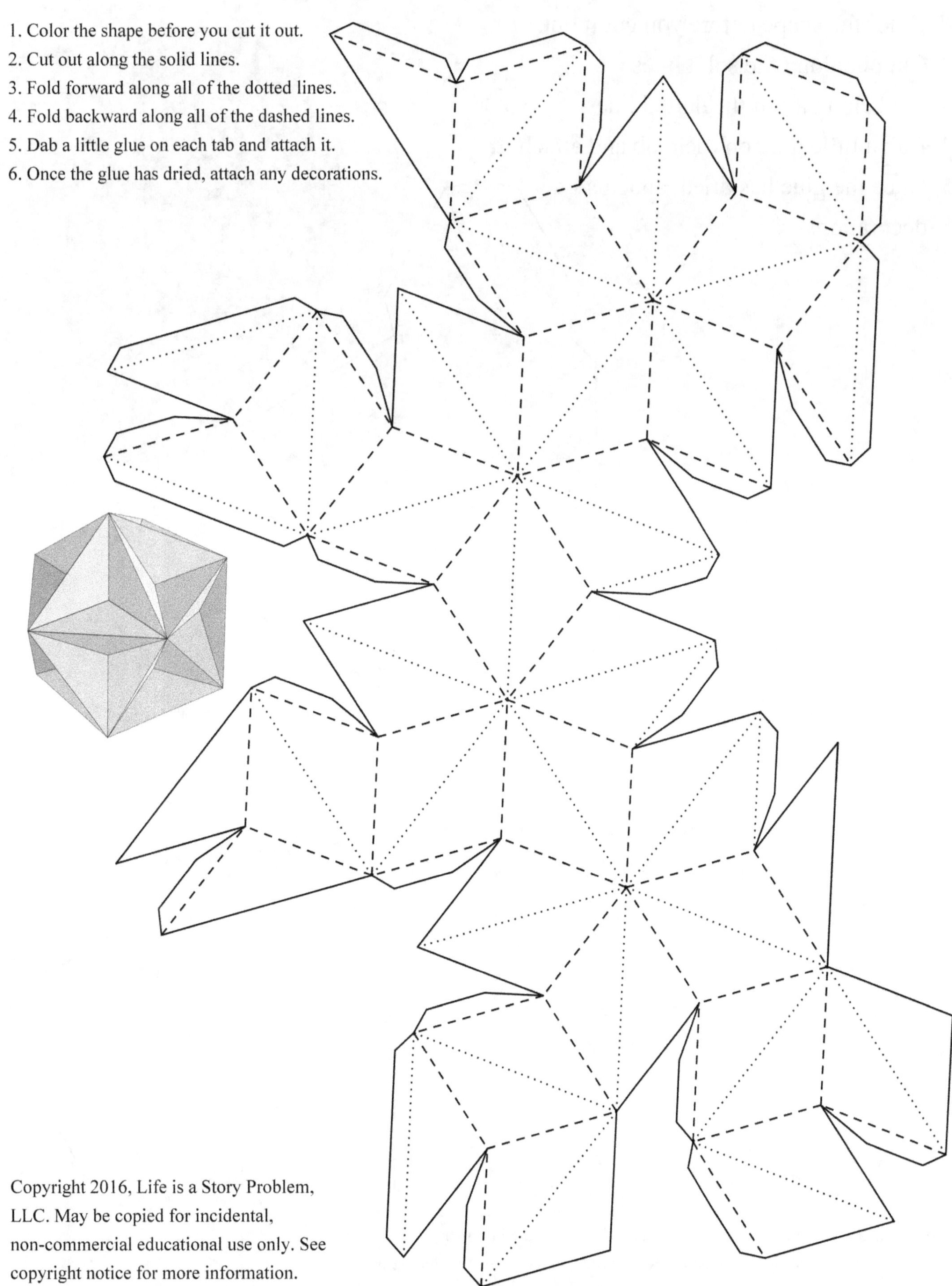

Elongated triangular cupola

1. Color the shape before you cut it out.
2. Cut out along the solid lines.
3. Fold along all of the dotted lines.
4. Dab a little glue on each tab and attach it.
5. Once the glue has dried, attach any decorations.

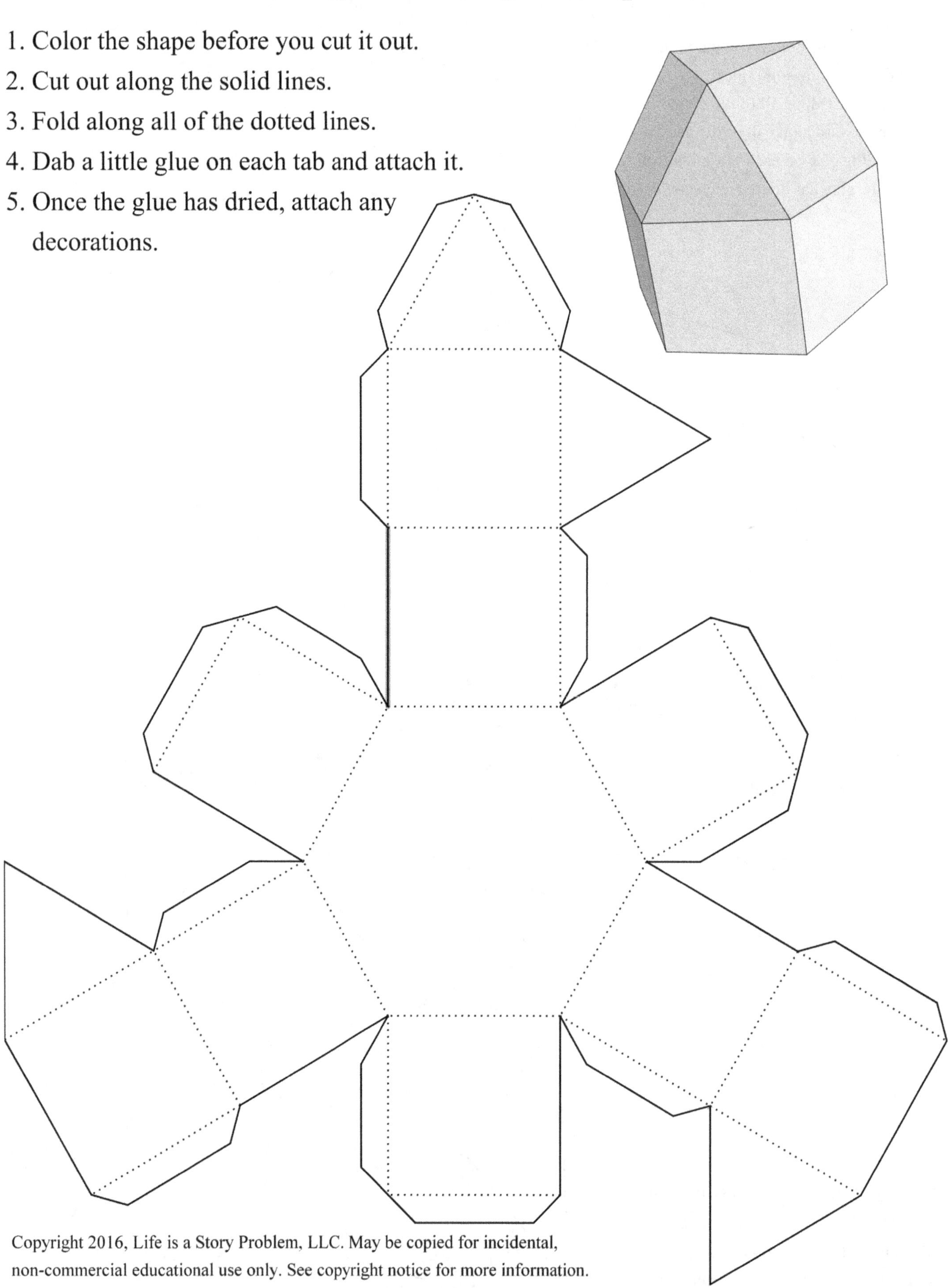

Elongated square cupola

1. Color the shape before you cut it out.
2. Cut out along the solid lines.
3. Fold along all of the dotted lines.
4. Dab a little glue on each tab and attach it.
5. Once the glue has dried, attach any
   decorations.

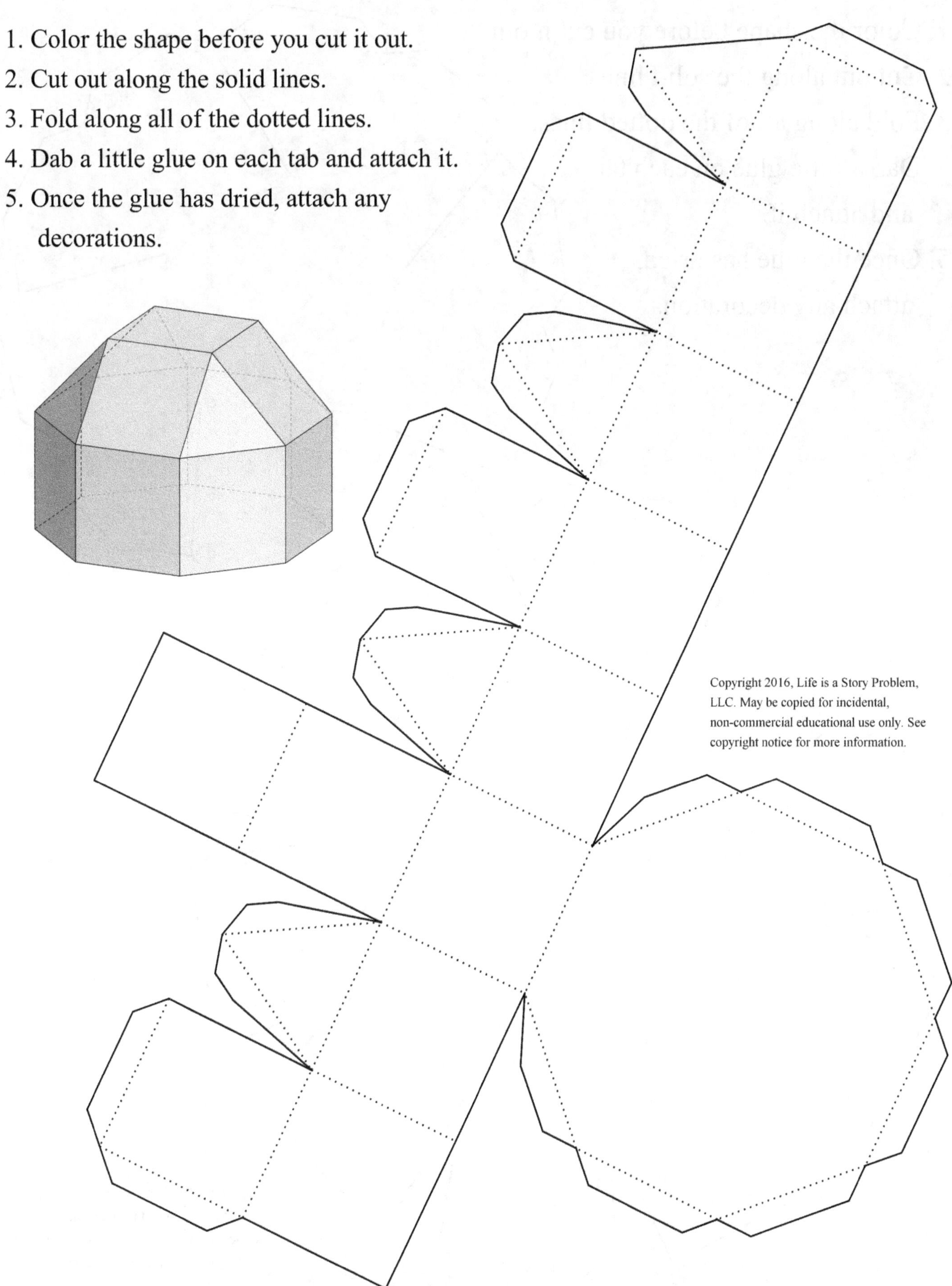

Elongated pentagonal cupola

1. Color the shape before you cut it out.
2. Cut out along the solid lines.
3. Fold along all of the dotted lines.
4. Dab a little glue on each tab and attach it.
5. Once the glue has dried, attach any decorations.

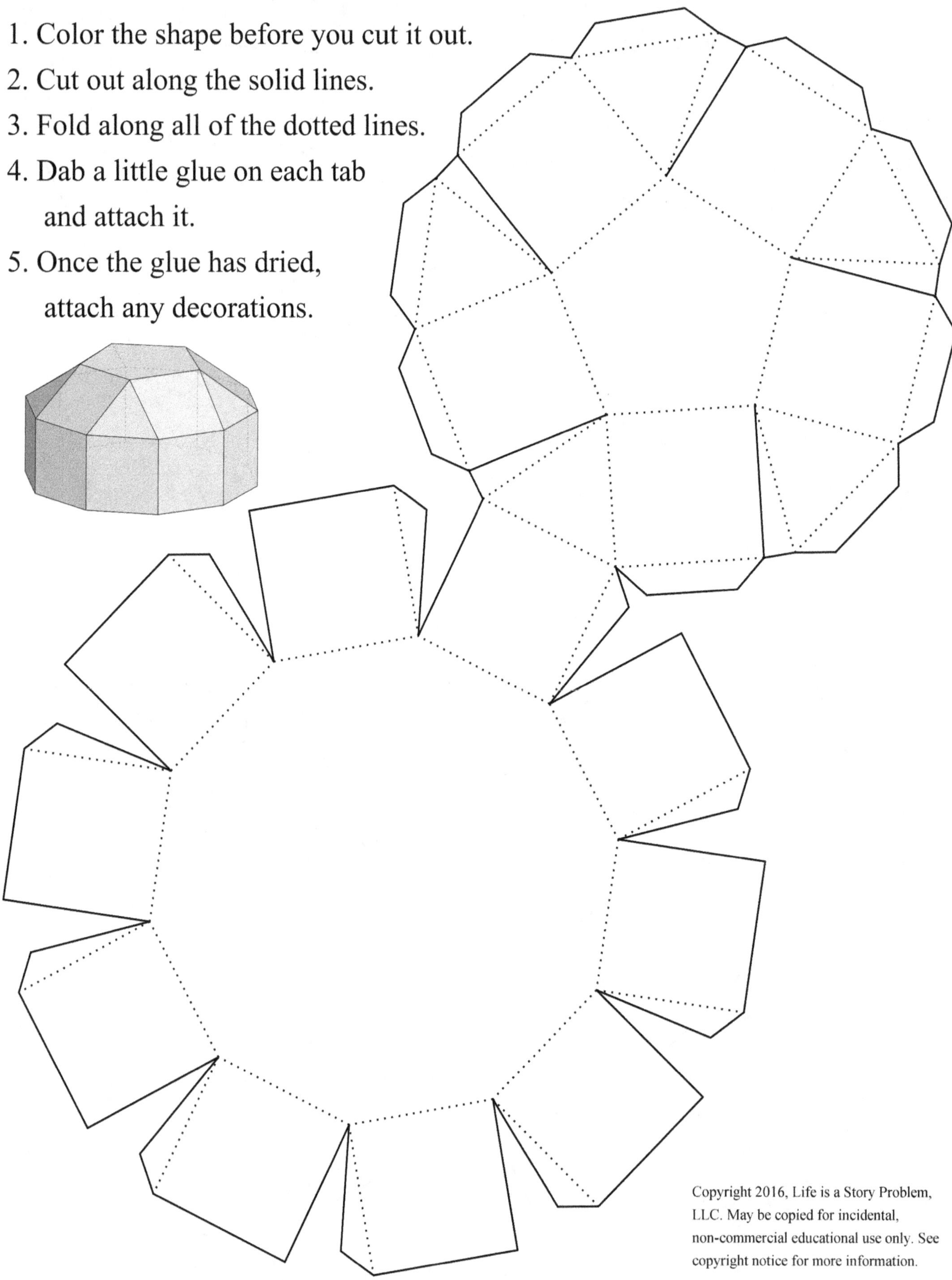

  Geometric Nets Mega Project Book, Tabbed By David E. McAdams

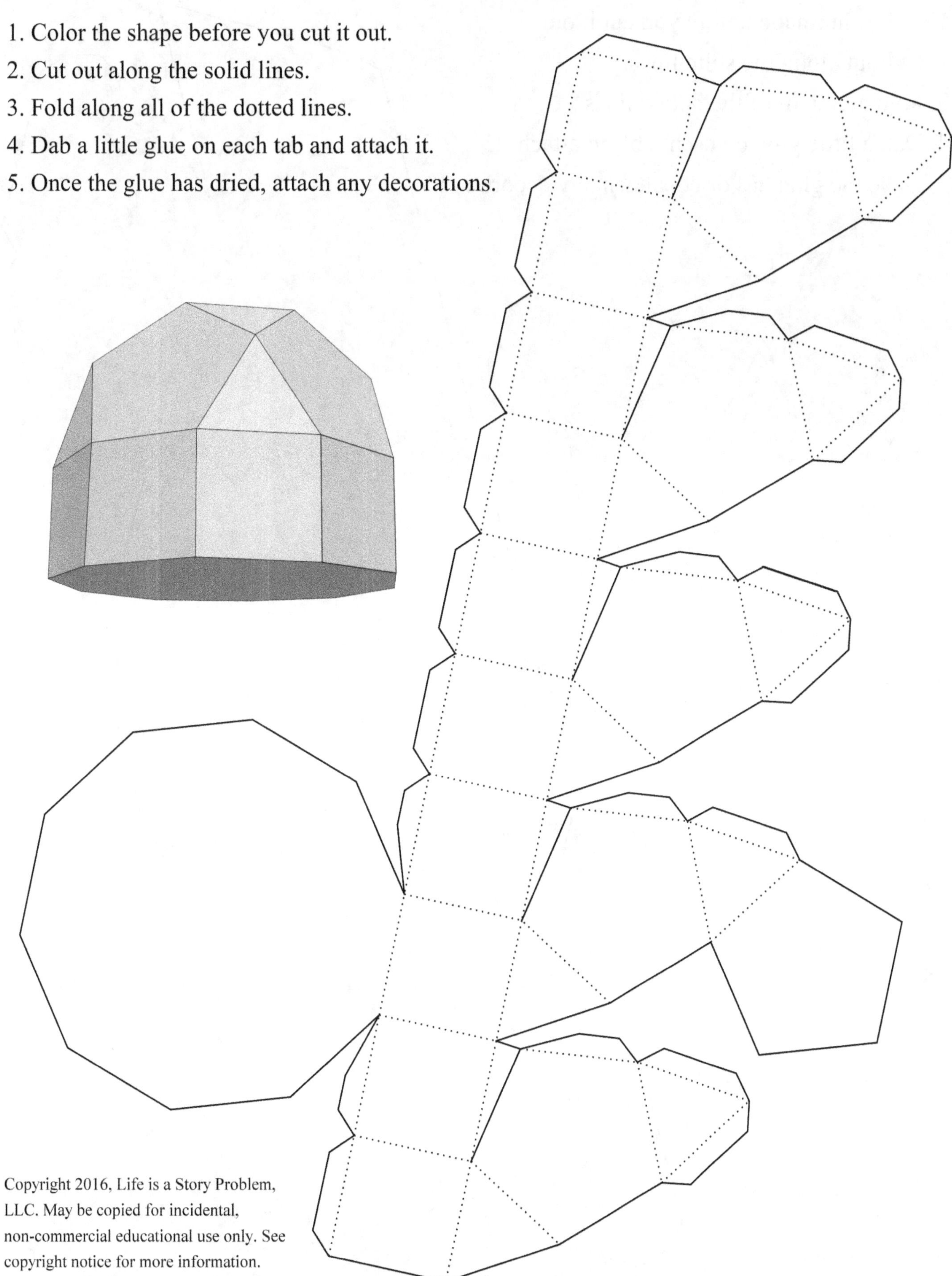

Elongated pentagonal rotunda

1. Color the shape before you cut it out.
2. Cut out along the solid lines.
3. Fold along all of the dotted lines.
4. Dab a little glue on each tab and attach it.
5. Once the glue has dried, attach any decorations.

Gyroelongated triangular cupola

1. Color the shape before you cut it out.

2. Cut out along the solid lines.

3. Fold along all of the dotted lines.

4. Dab a little glue on each tab and attach it.

5. Once the glue has dried, attach any decorations.

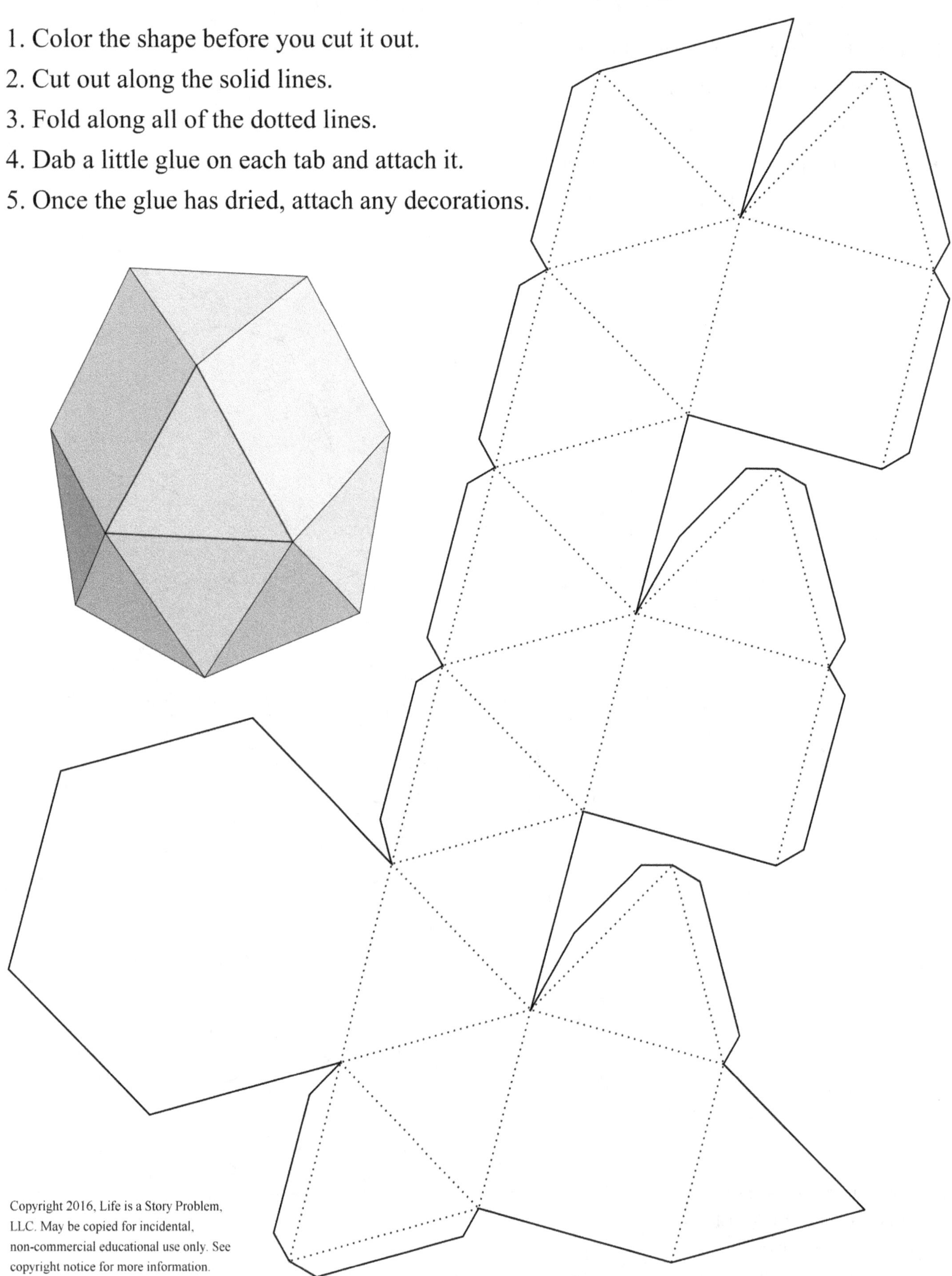

 Geometric Nets Mega Project Book, Tabbed By David E. McAdams

Gyroelongated square cupola

1. Color the shape before you cut it out.
2. Cut out along the solid lines.
3. Fold along all of the dotted lines.
4. Dab a little glue on each tab and attach it.
5. Once the glue has dried, attach any decorations.

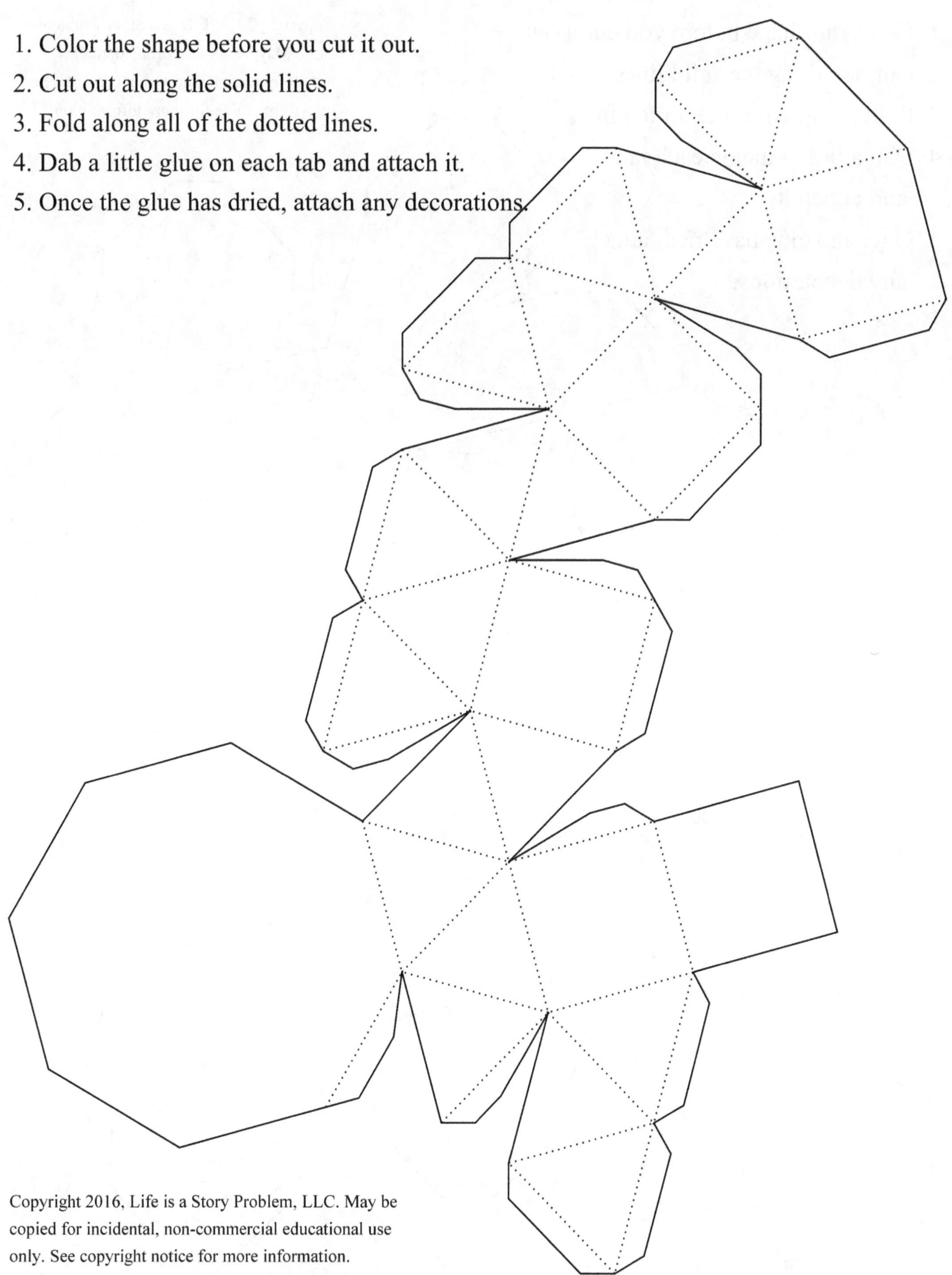

Gyroelongated pentagonal cupola

1. Color the shape before you cut it out.
2. Cut out along the solid lines.
3. Fold along all of the dotted lines.
4. Dab a little glue on each tab
   and attach it.
5. Once the glue has dried, attach
   any decorations.

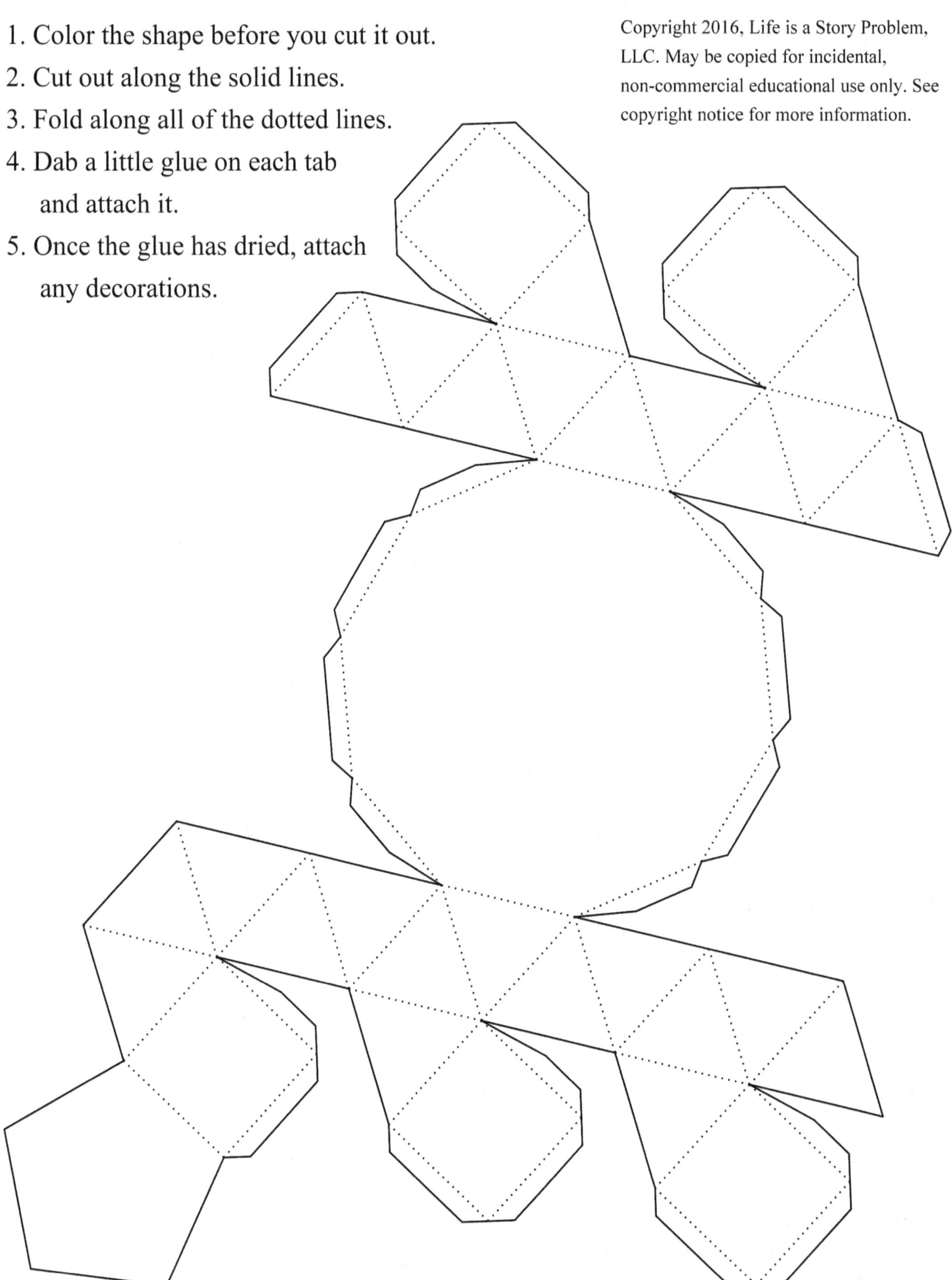

Geometric Nets Mega Project Book, Tabbed By David E. McAdams

Gyroelongated pentagonal rotunda

1. Color the shape before you cut it out.

2. Cut out along the solid lines.

3. Fold along all of the dotted lines.

4. Dab a little glue on each tab and attach it.

5. Once the glue has dried, attach any decorations.

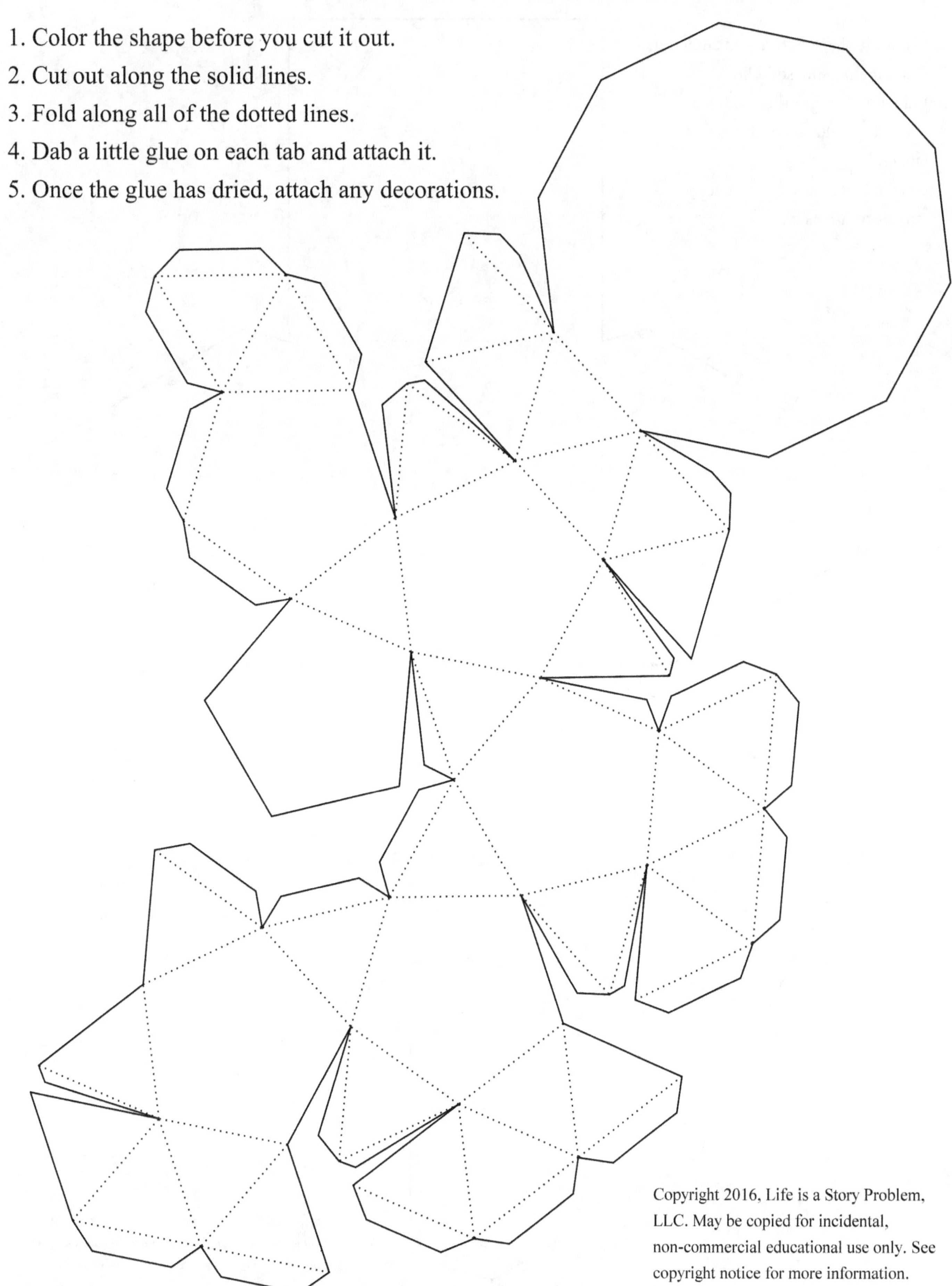

Gyrobifastigium

1. Color the shape before you cut it out.
2. Cut out along the solid lines.
3. Fold along all of the dotted lines.
4. Dab a little glue on each tab and attach it.
5. Once the glue has dried, attach any decorations.

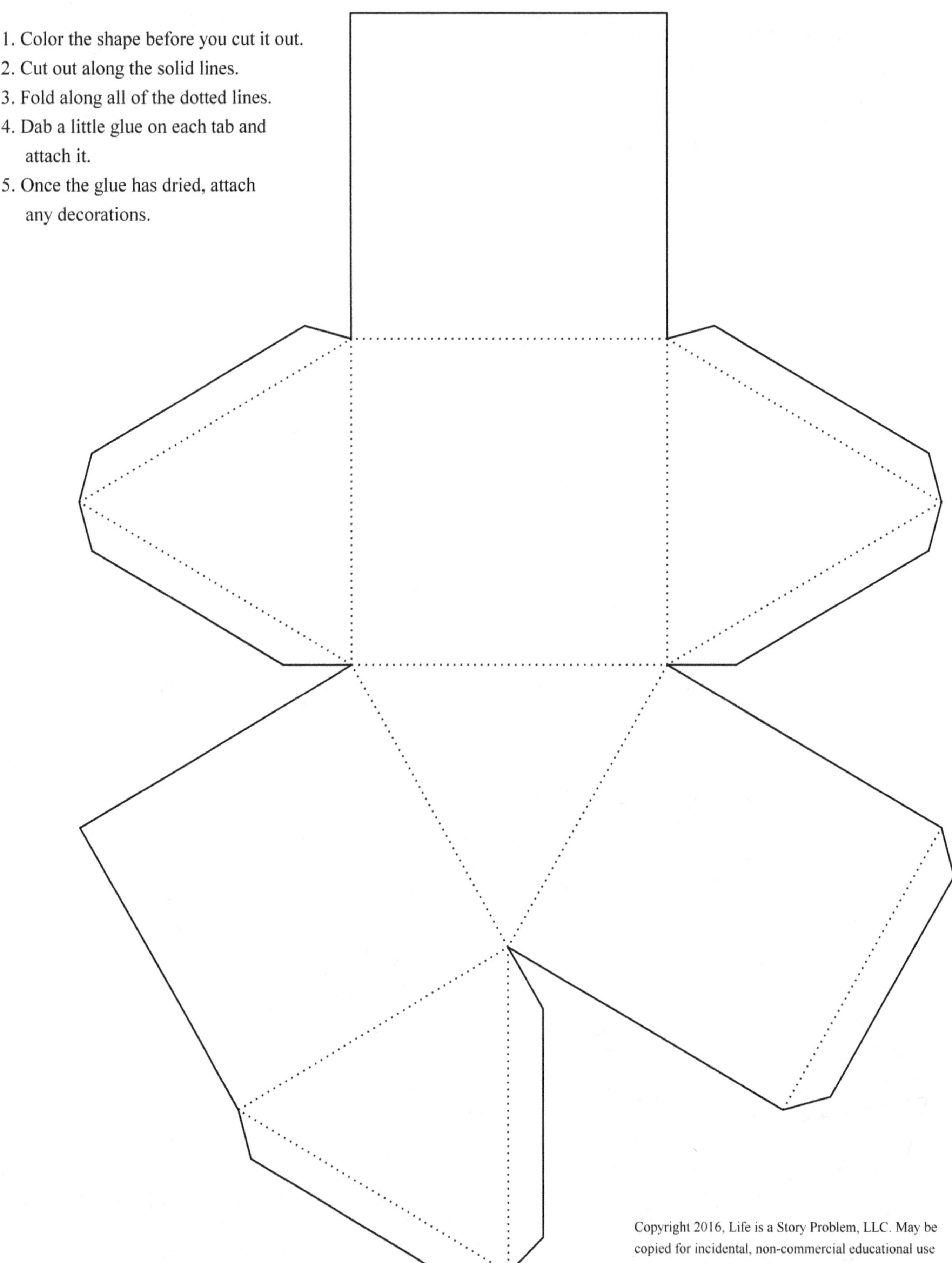

Geometric Nets Mega Project Book, Tabbed By David E. McAdams

Triangular orthobicupola

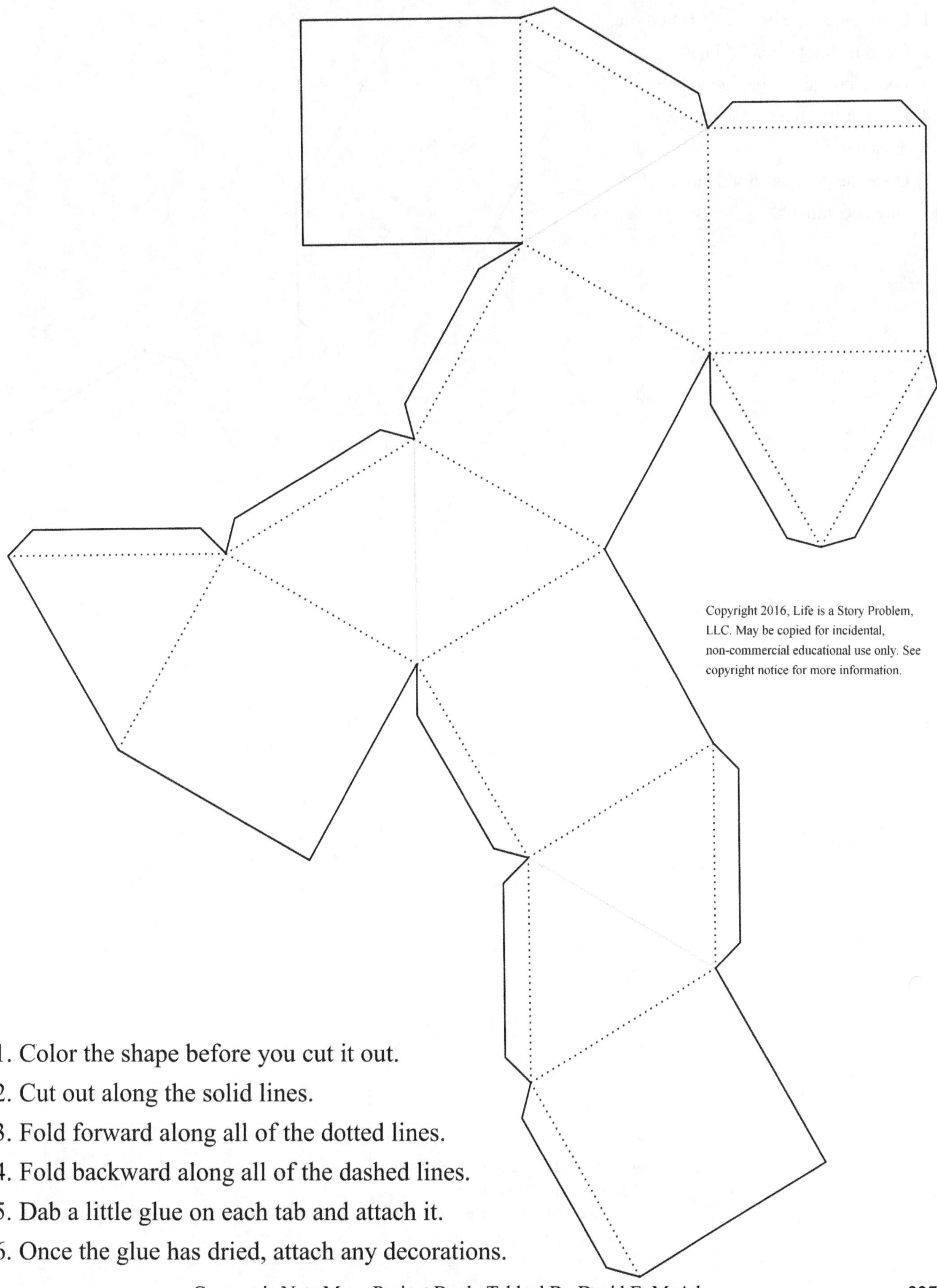

1. Color the shape before you cut it out.

2. Cut out along the solid lines.

3. Fold forward along all of the dotted lines.

4. Fold backward along all of the dashed lines.

5. Dab a little glue on each tab and attach it.

6. Once the glue has dried, attach any decorations.

Square orthobicupola

1. Color the shape before you cut it out.
2. Cut out along the solid lines.
3. Fold along all of the dotted lines.
4. Dab a little glue on each tab and
   attach it.
5. Once the glue has dried, attach
   any decorations.

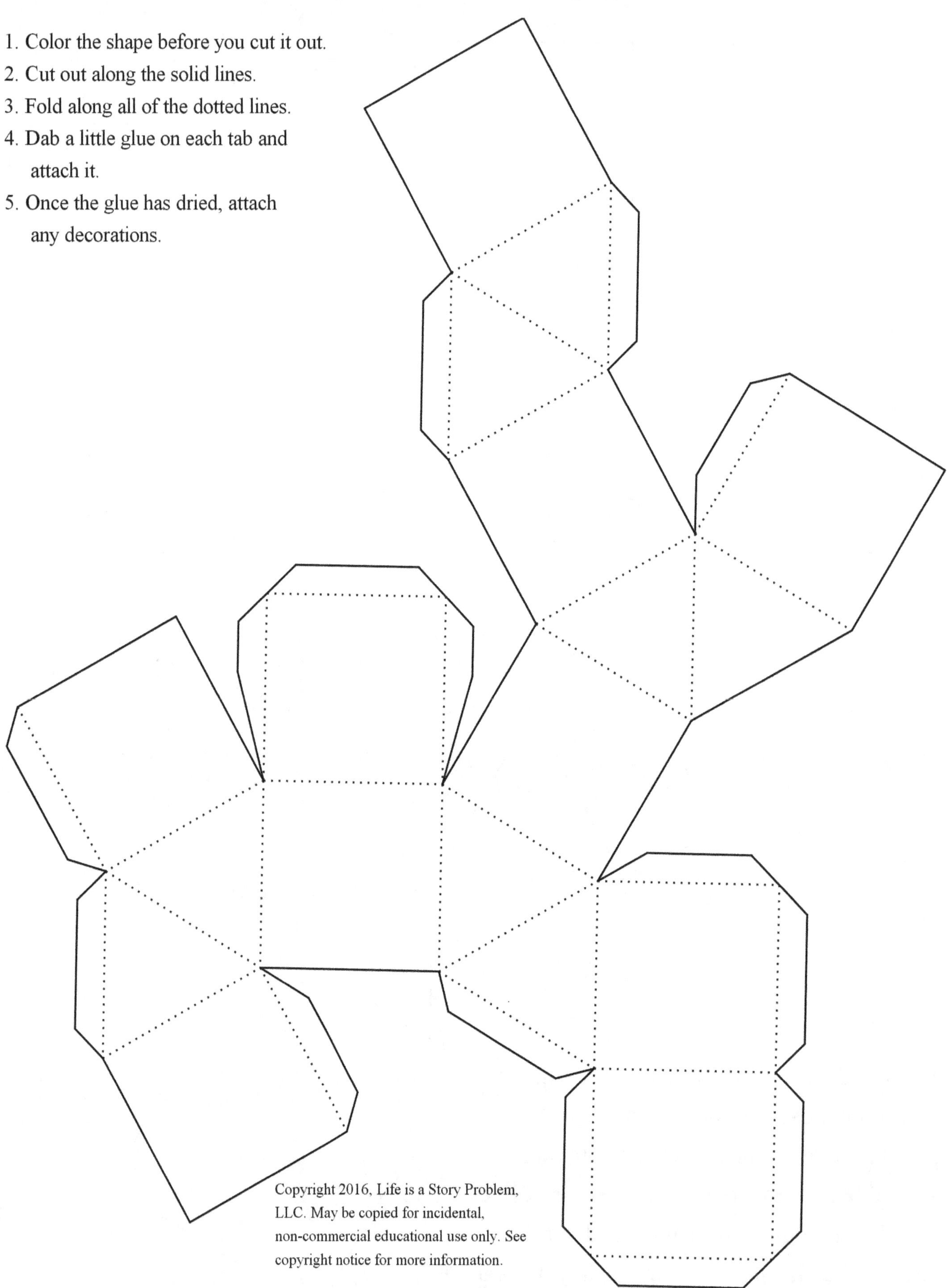

 Geometric Nets Mega Project Book, Tabbed By David E. McAdams

Square gyrobicupola

1. Color the shape before you cut it out.
2. Cut out along the solid lines.
3. Fold along all of the dotted lines.
4. Dab a little glue on each tab and attach it.
5. Once the glue has dried, attach any decorations.

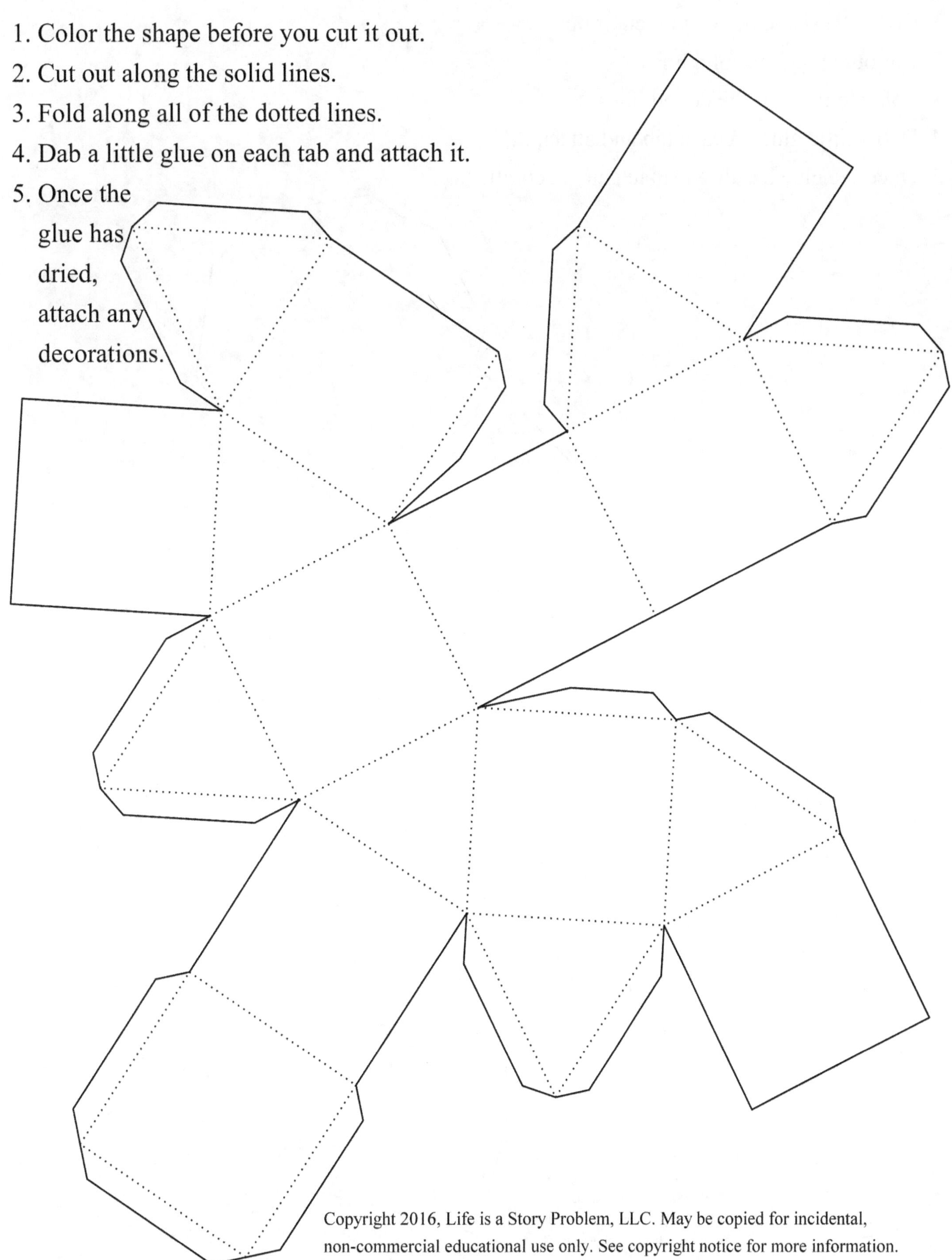

Pentagonal orthobicupola

1. Color the shape before you cut it out.
2. Cut out along the solid lines.
3. Fold along all of the dotted lines.
4. Dab a little glue on each tab and attach it.
5. Once the glue has dried, attach any decorations.

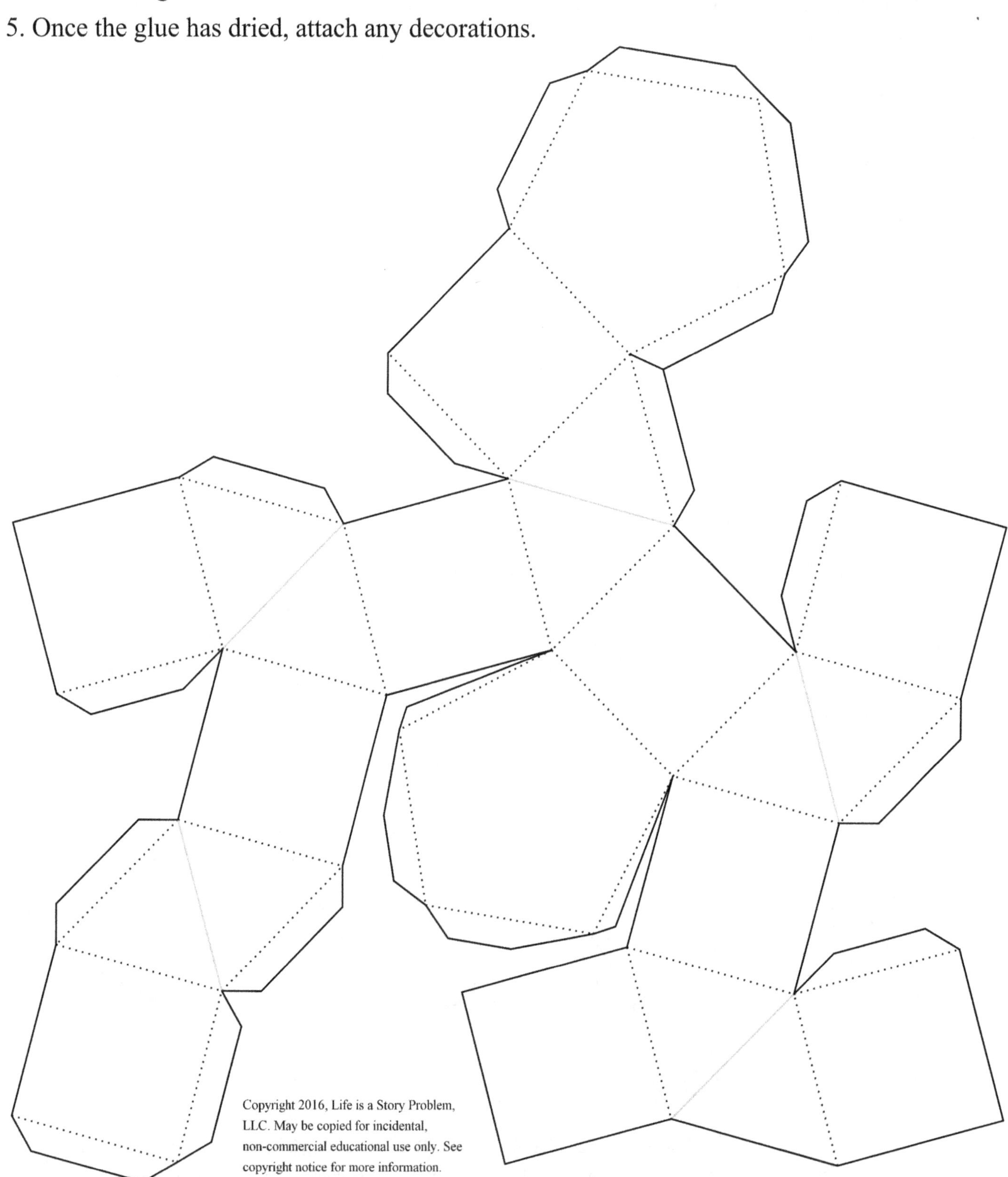

   Geometric Nets Mega Project Book, Tabbed By David E. McAdams

Pentagonal gyrobicupola

1. Color the shape before you cut it out.
2. Cut out along the solid lines.
3. Fold along all of the dotted lines.
4. Dab a little glue on each tab and attach it.
5. Once the glue has dried, attach any decorations.

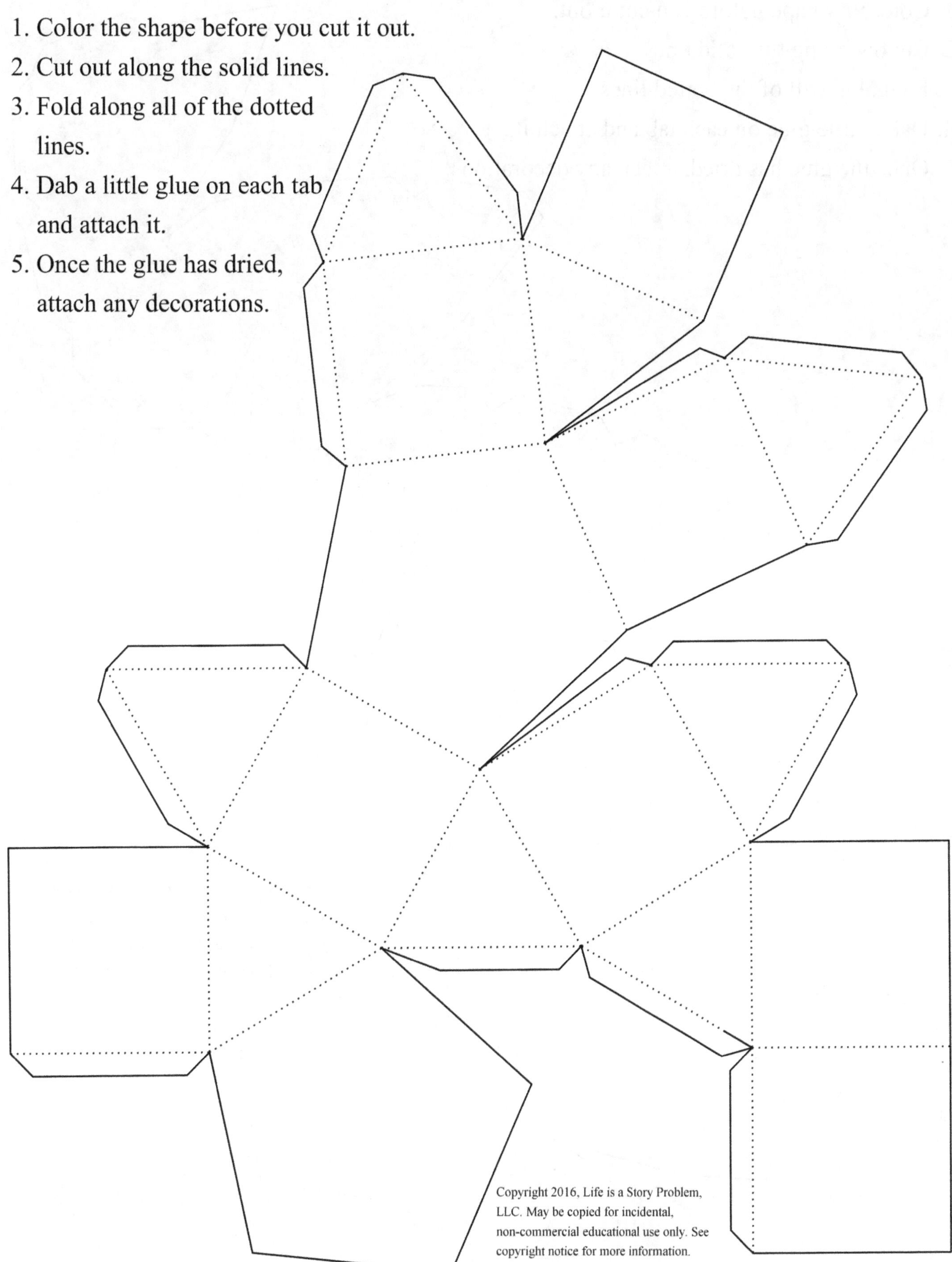

Pentagonal orthocupolarotunda

1. Color the shape before you cut it out.
2. Cut out along the solid lines.
3. Fold along all of the dotted lines.
4. Dab a little glue on each tab and attach it.
5. Once the glue has dried, attach any decorations.

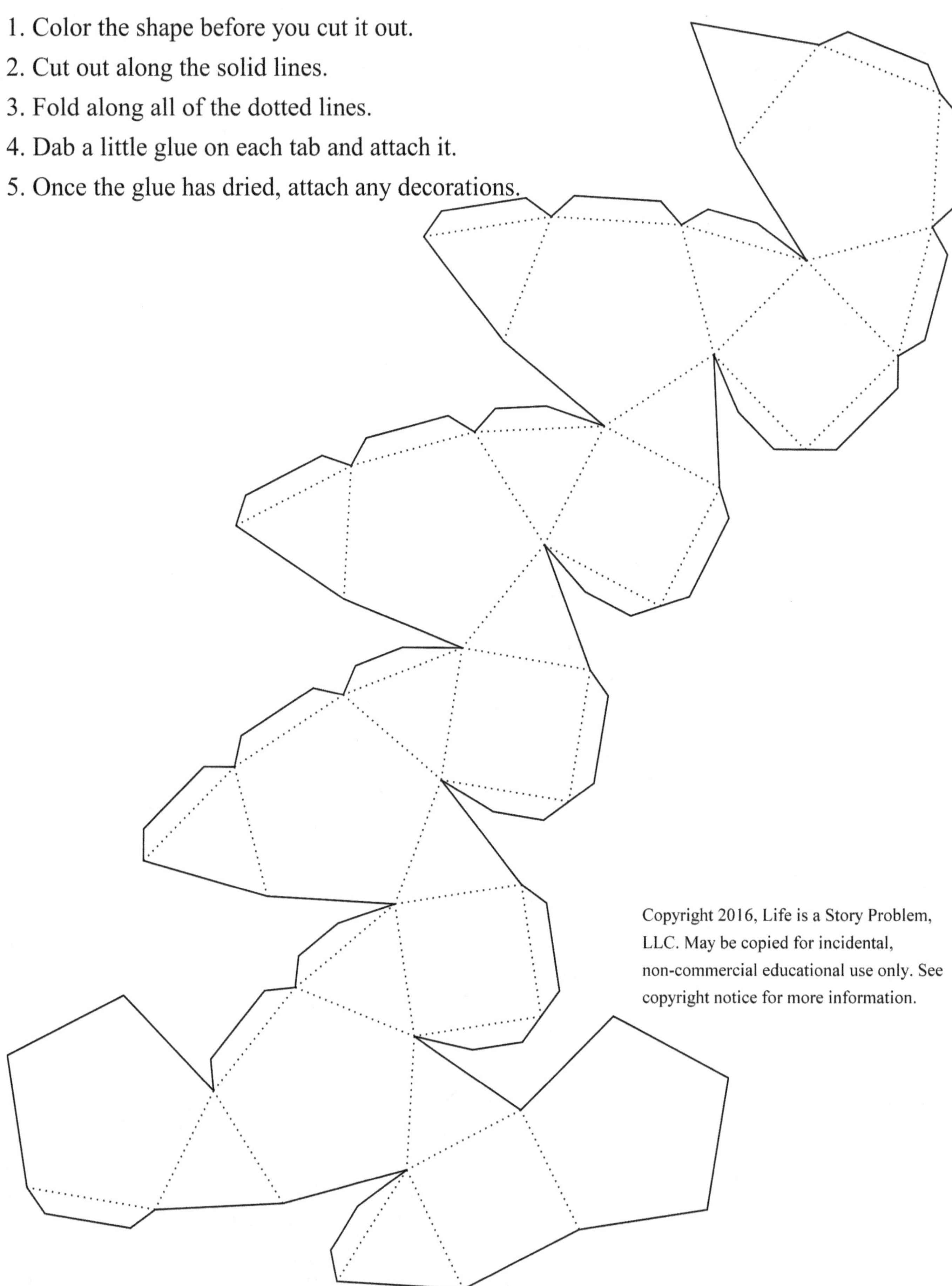

     Geometric Nets Mega Project Book, Tabbed By David E. McAdams

Pentagonal gyrocupolarotunda

1. Color the shape before you cut it out.
2. Cut out along the solid lines.
3. Fold along all of the dotted lines.
4. Dab a little glue on each tab and attach it.
5. Once the glue has dried, attach
   any decorations.

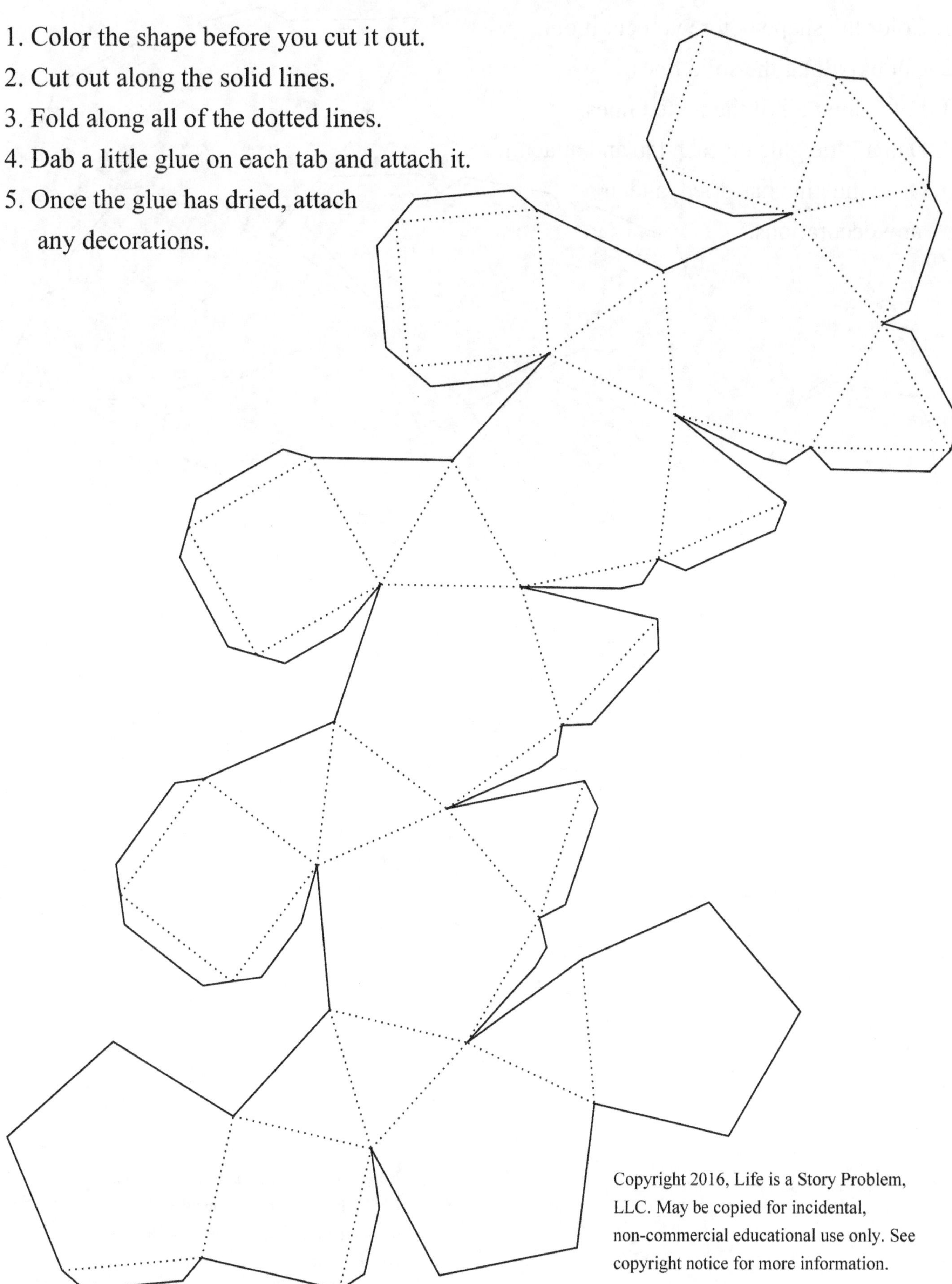

Pentagonal orthobirotunda

1. Color the shape before you cut it out.
2. Cut out along the solid lines.
3. Fold along all of the dotted lines.
4. Dab a little glue on each tab and attach it.
5. Once the glue has dried, attach any decorations.

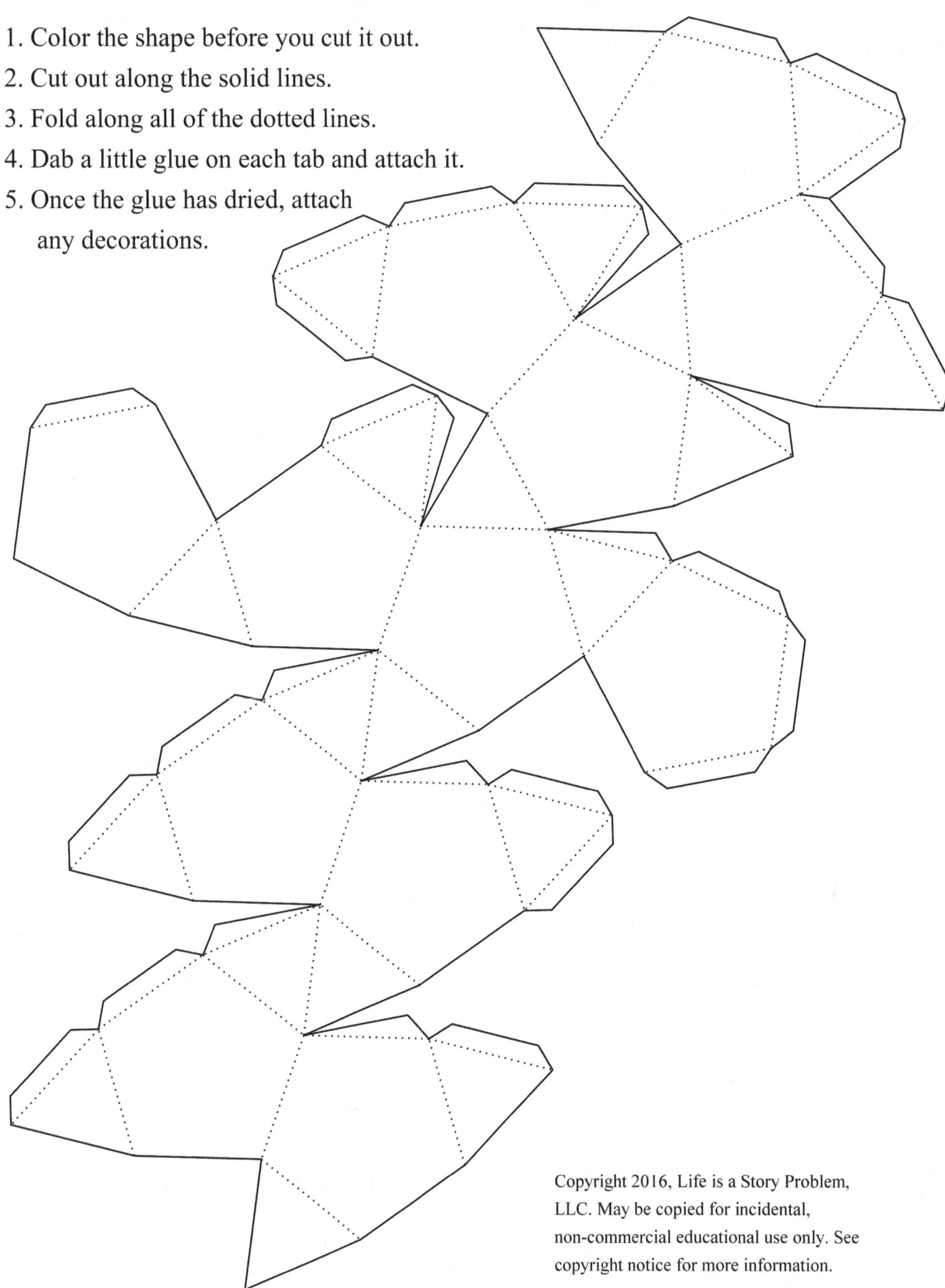

Elongated triangular orthobicupola

1. Color the shape before you cut it out.
2. Cut out along the solid lines.
3. Fold along all of the dotted lines.
4. Dab a little glue on each tab and attach it.
5. Once the glue has dried, attach any decorations.

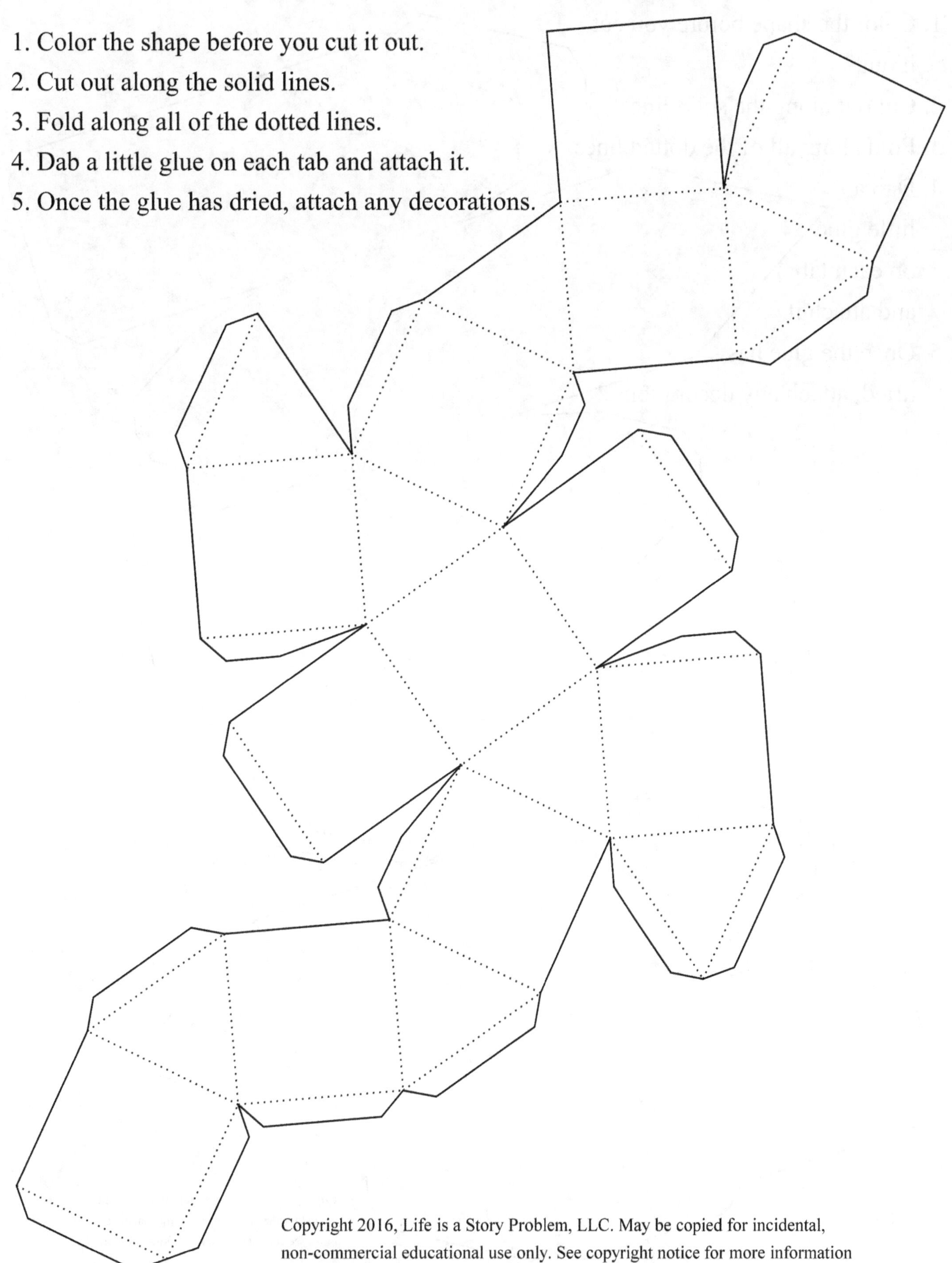

Elongated triangular gyrobicupola

1. Color the shape before you cut it out.
2. Cut out along the solid lines.
3. Fold along all of the dotted lines.
4. Dab a little glue on each tab and attach it.
5. Once the glue has dried, attach any decorations.

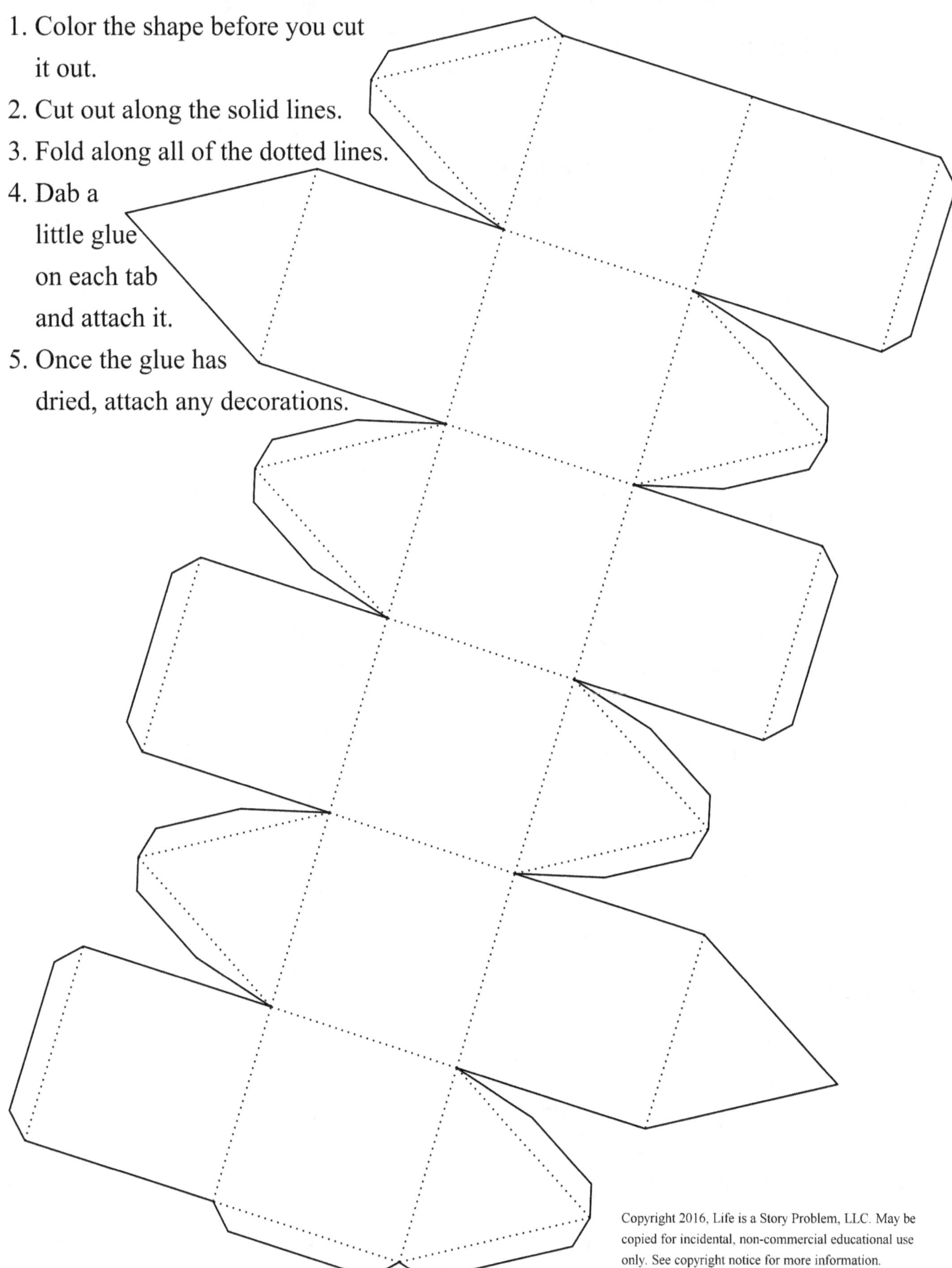

 Geometric Nets Mega Project Book, Tabbed By David E. McAdams

Elongated square gyrobicupola

1. Color the shape before you cut it out.
2. Cut out along the solid lines.
3. Fold along all of the dotted lines.
4. Dab a little glue on each tab and attach it.
5. Once the glue has dried, attach any decorations.

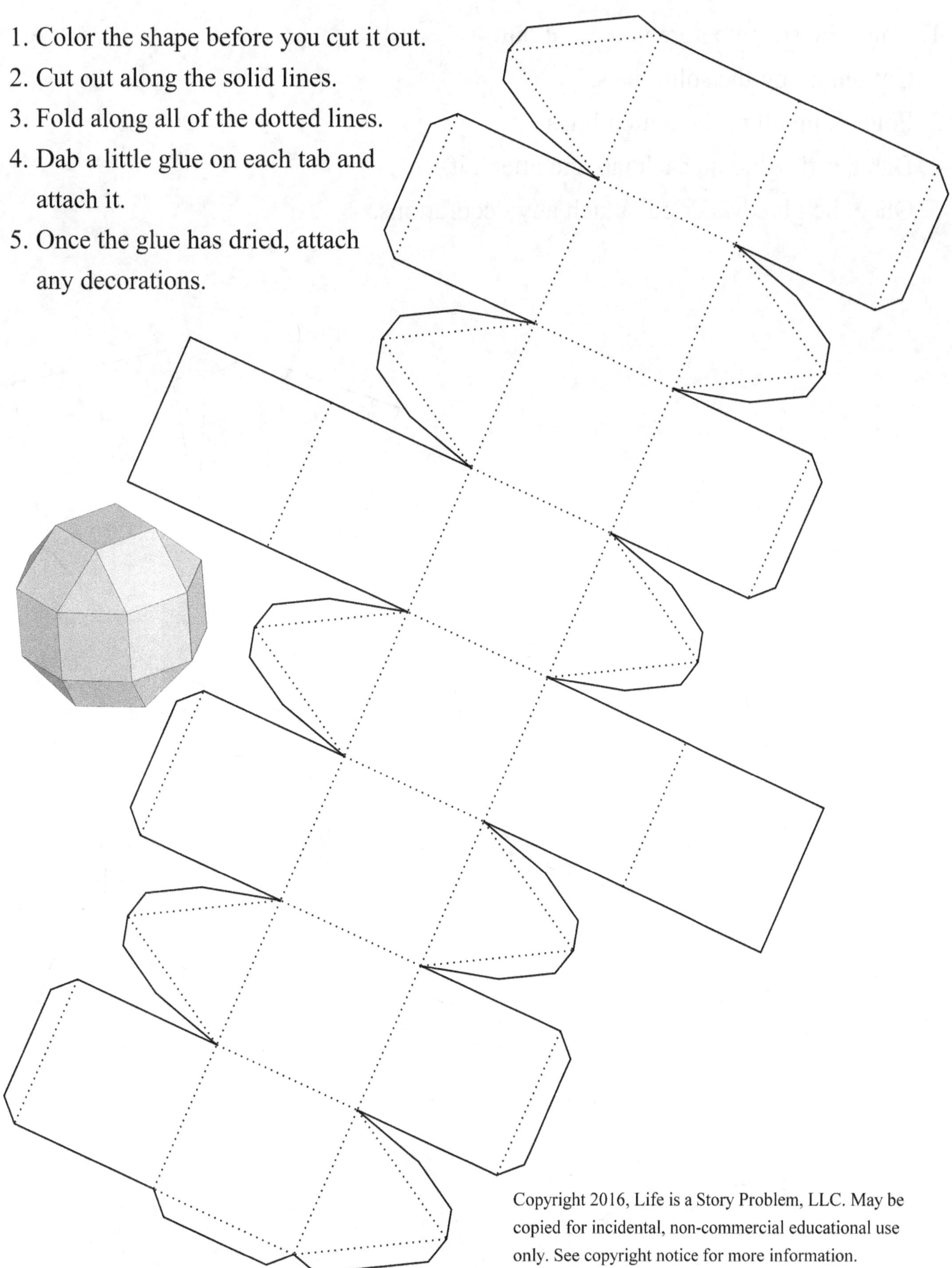

Elongated pentagonal orthobicupola

1. Color the shape before you cut it out.

2. Cut out along the solid lines.

3. Fold along all of the dotted lines.

4. Dab a little glue on each tab and attach it.

5. Once the glue has dried, attach any decorations.

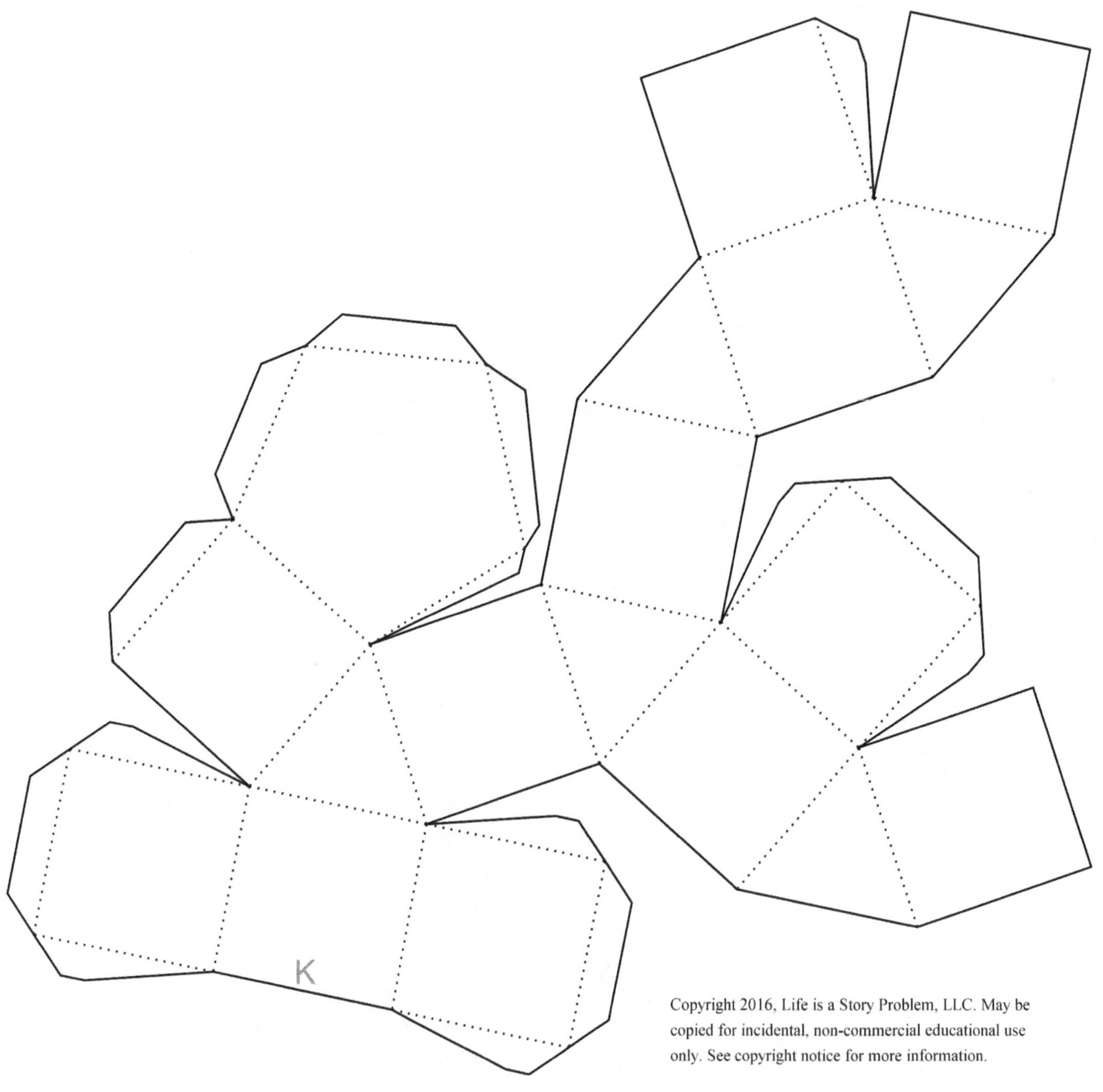

Geometric Nets Mega Project Book, Tabbed By David E. McAdams

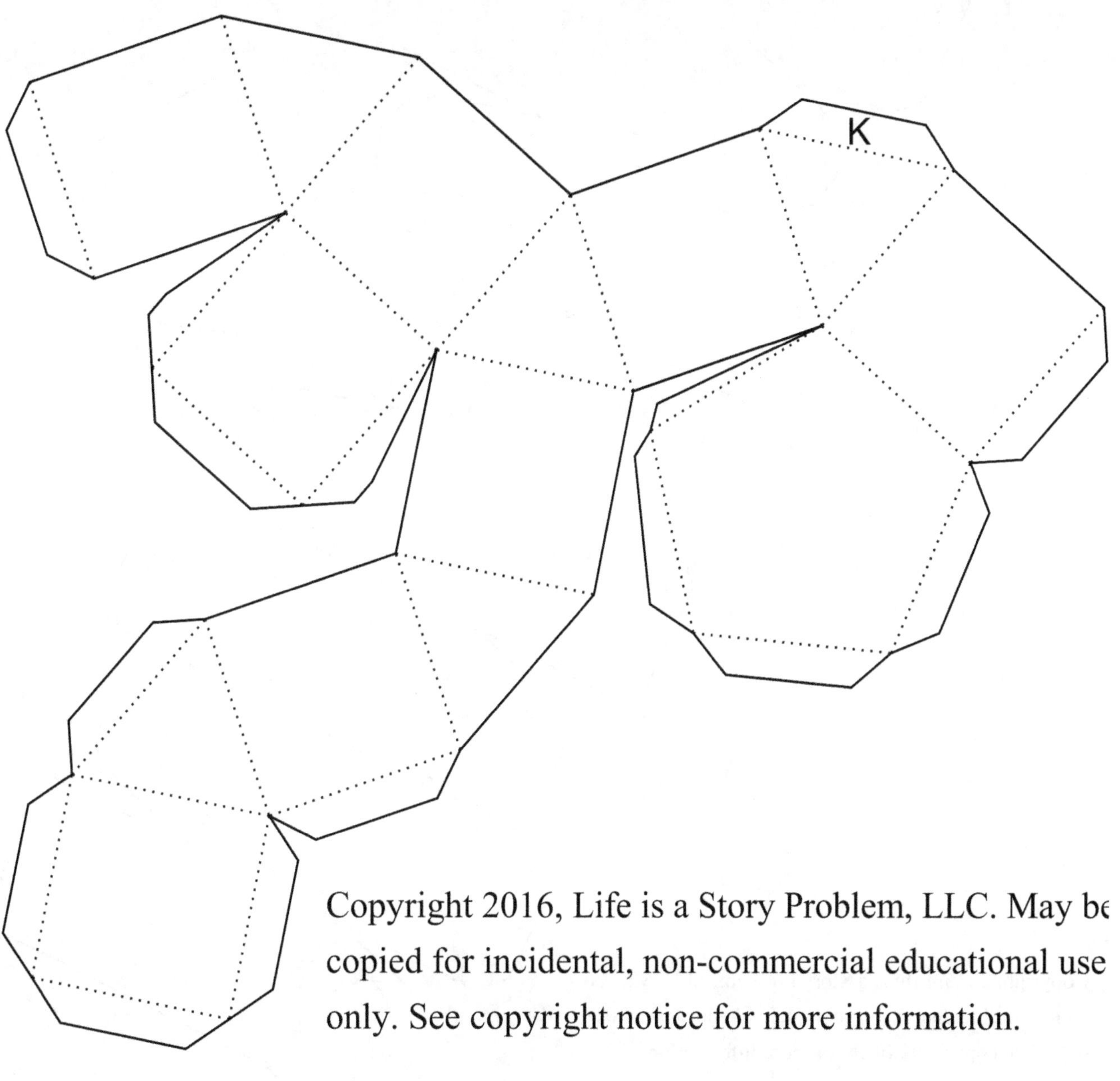

K

Elongated pentagonal gyrobicupola

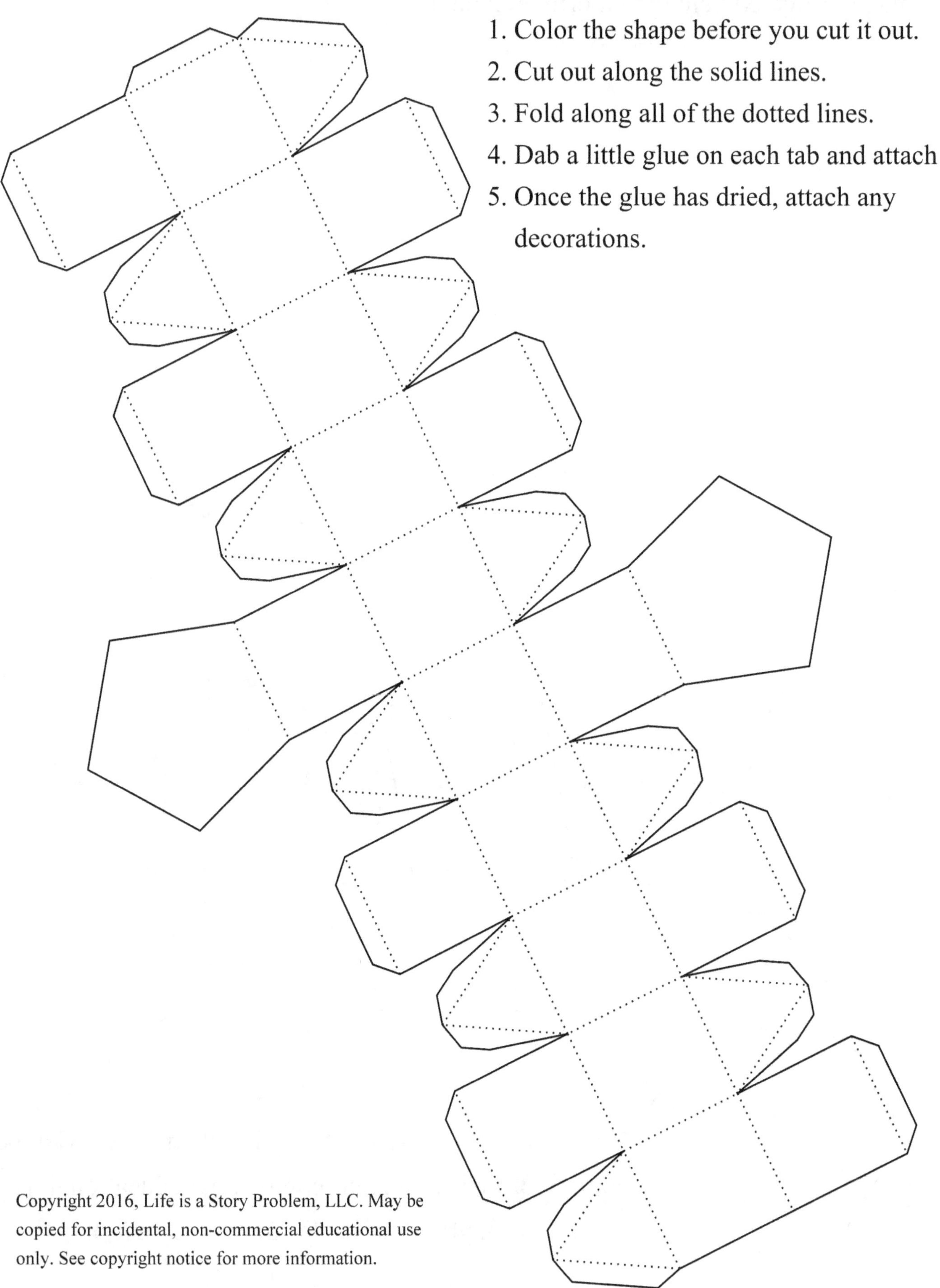

1. Color the shape before you cut it out.
2. Cut out along the solid lines.
3. Fold along all of the dotted lines.
4. Dab a little glue on each tab and attach it.
5. Once the glue has dried, attach any decorations.

Geometric Nets Mega Project Book, Tabbed By David E. McAdams

Elongated pentagonal orthocupolarotunda

1. Color the shape before you cut it out.
2. Cut out along the solid lines.
3. Fold along all of the dotted lines.
4. Dab a little glue on each tab and attach it.
5. Once the glue has dried, attach any decorations.

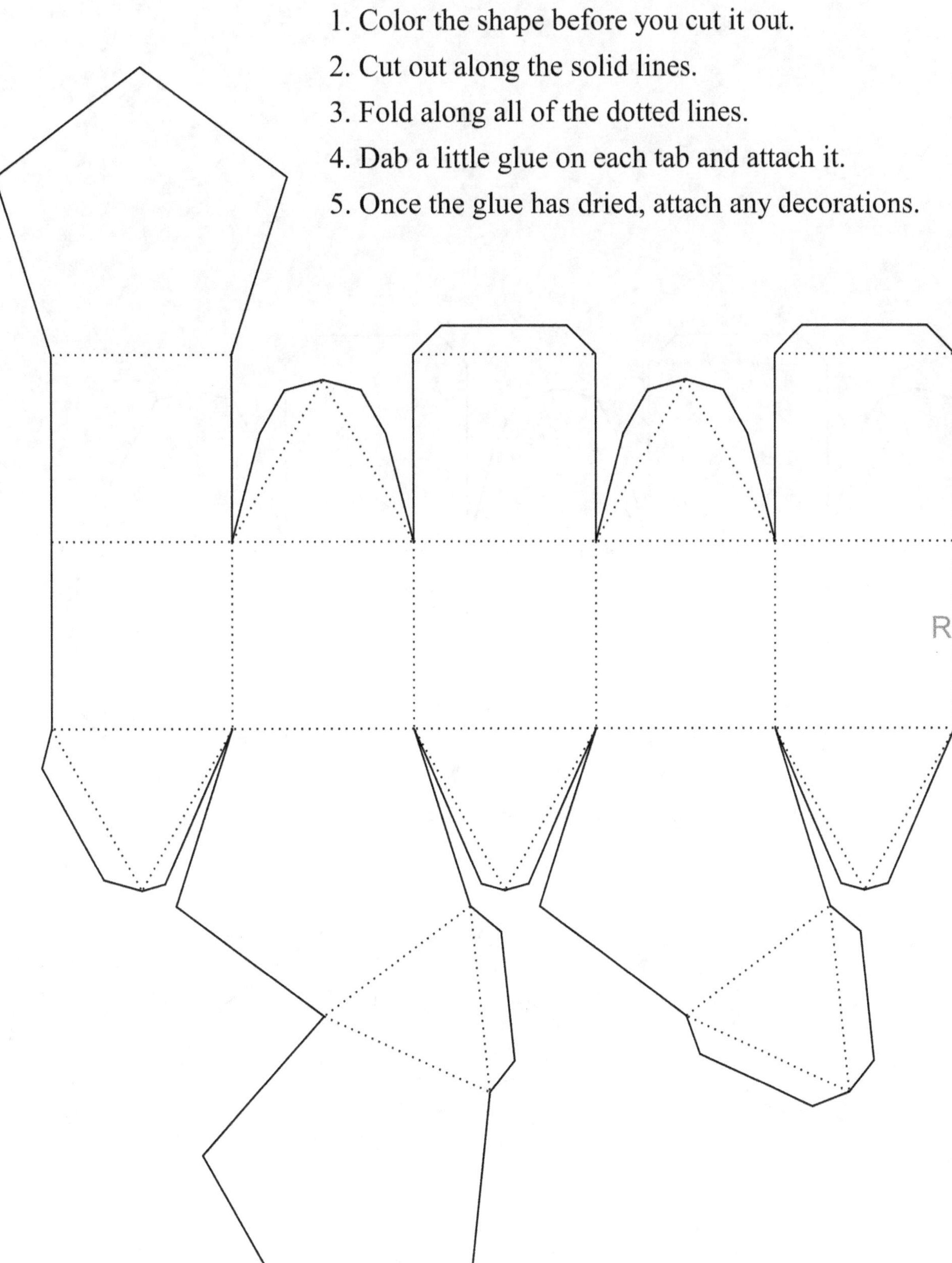

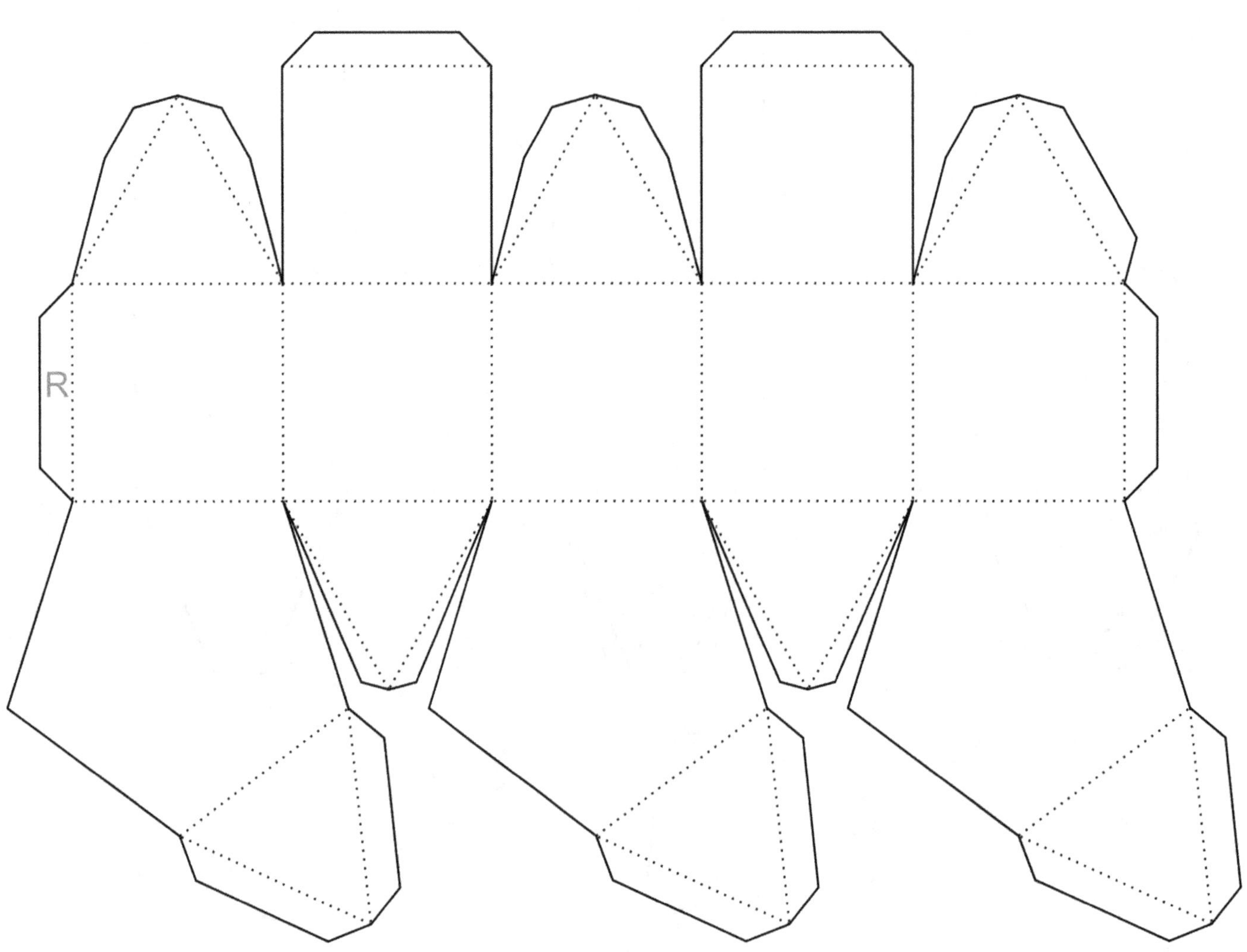

Geometric Nets Mega Project Book, Tabbed By David E. McAdams

Elongated pentagonal gyrocupolarotunda

1. Color the shape before you cut it out.
2. Cut out along the solid lines.
3. Fold along all of the dotted lines.
4. Dab a little glue on each tab and attach it.
5. Once the glue has dried, attach any decorations.

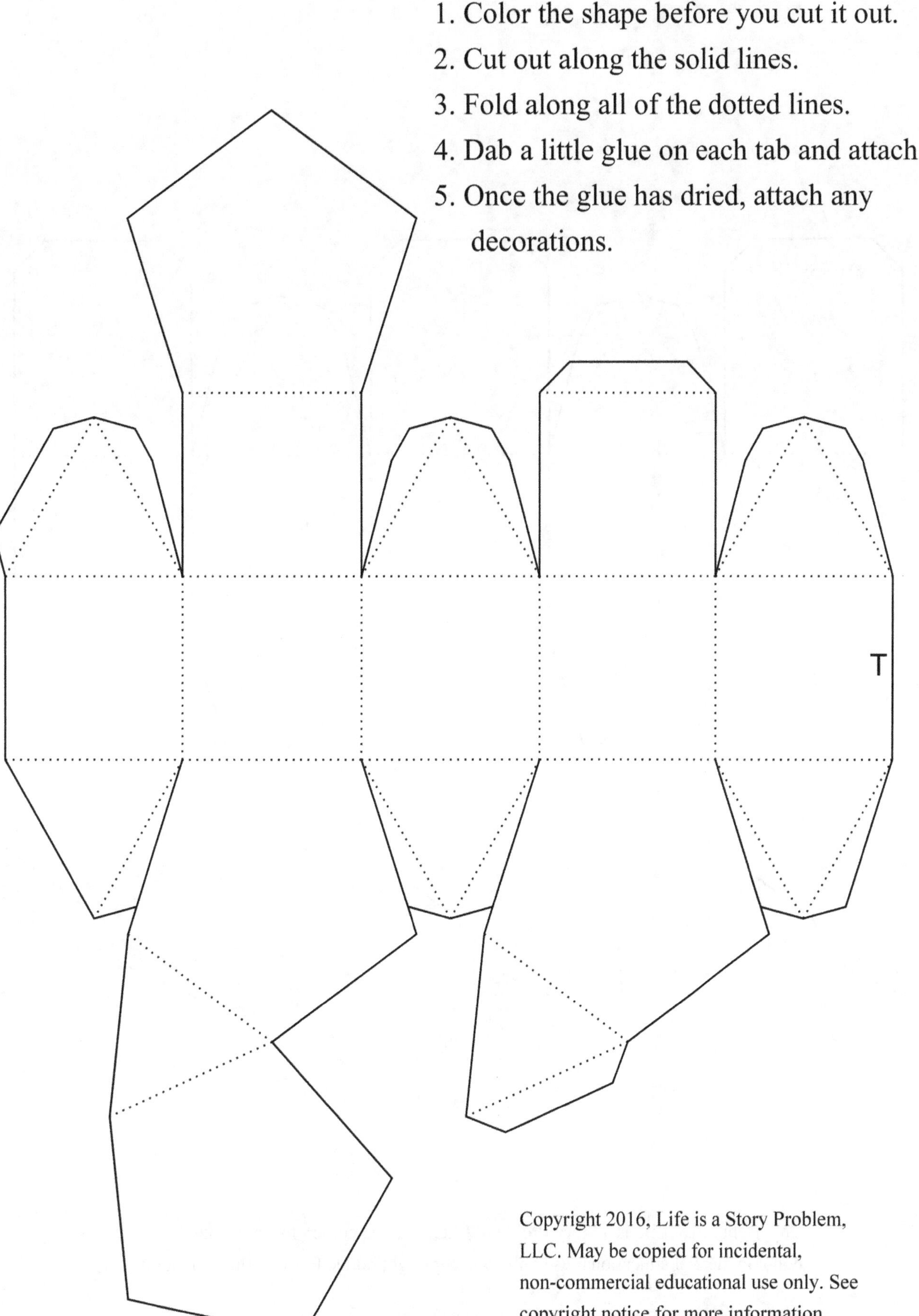

2nd page, Elongated pentagonal gyrocupolaroutunda

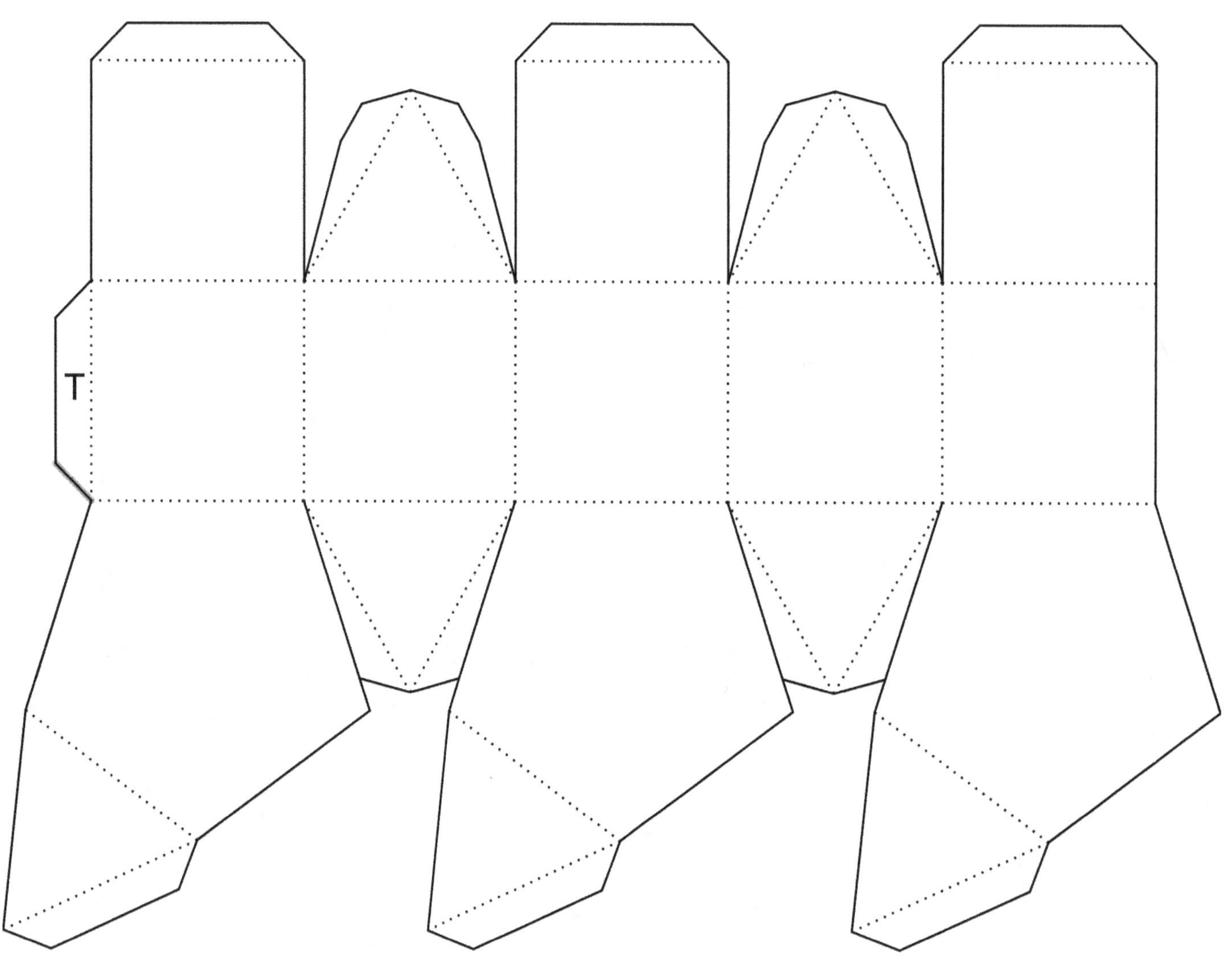

Geometric Nets Mega Project Book, Tabbed By David E. McAdams

Elongated pentagonal orthobirotunda

1. Color the shape before you cut it out.
2. Cut out along the solid lines.
3. Fold along all of the dotted lines.
4. Dab a little glue on each tab and attach it.
5. Once the glue has dried, attach
   any decorations.

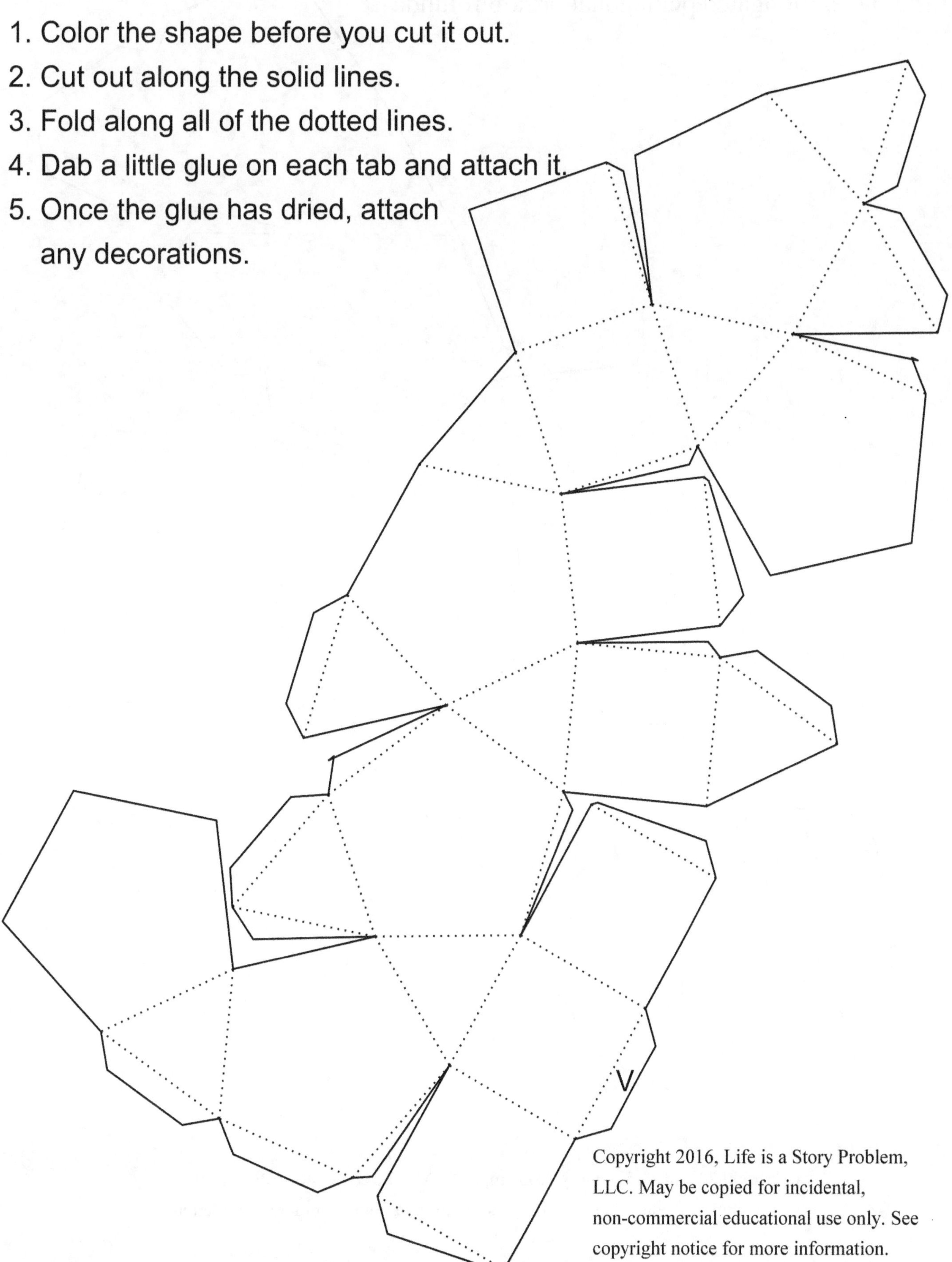

2nd page, Elongated pentagonal orthobirotunda

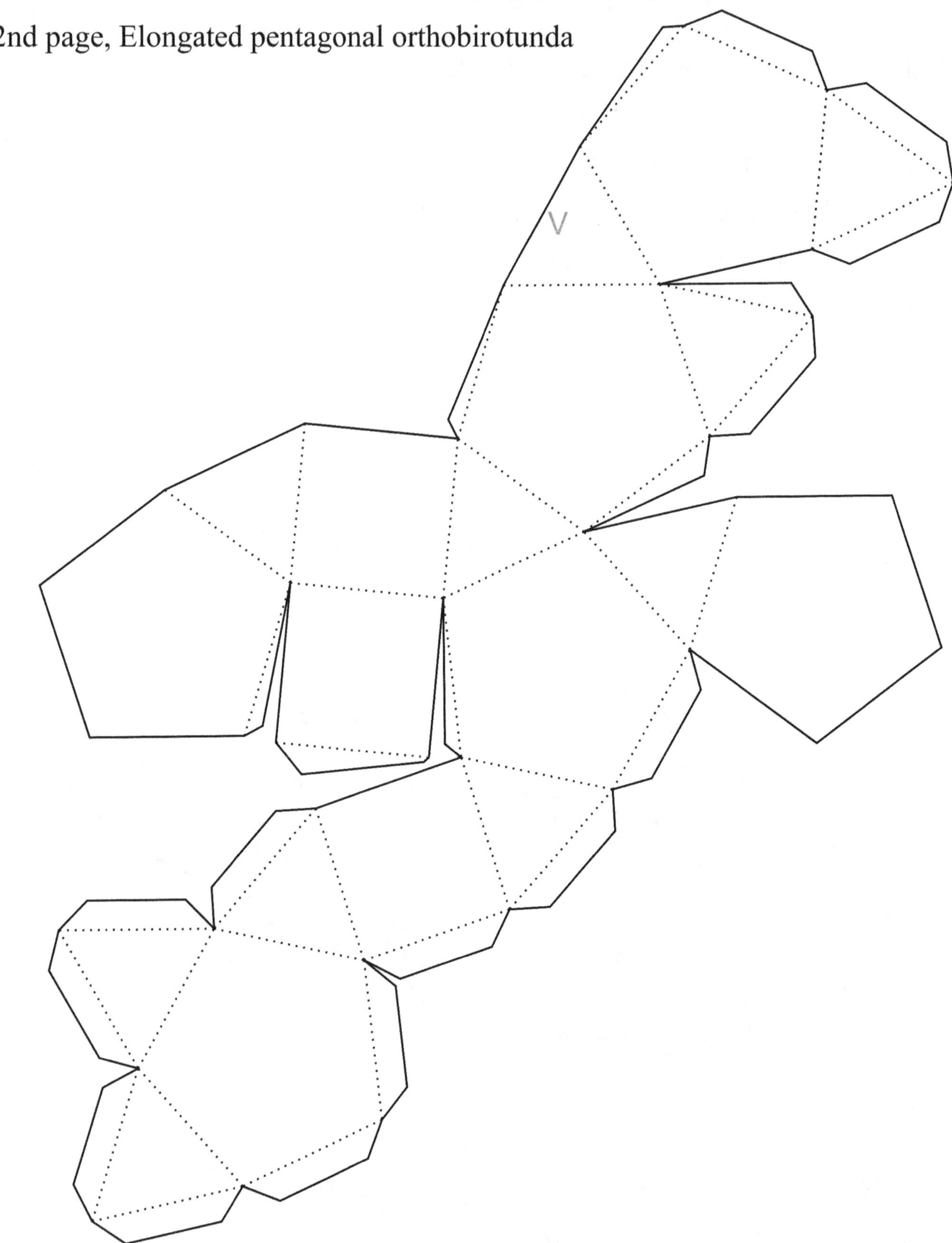

Elongated pentagonal gyrobirotunda

1. Color the shape before you cut it out.
2. Cut out along the solid lines.
3. Fold along all of the dotted lines.
4. Dab a little glue on each tab and attach it.
5. Once the glue has dried, attach any decorations.

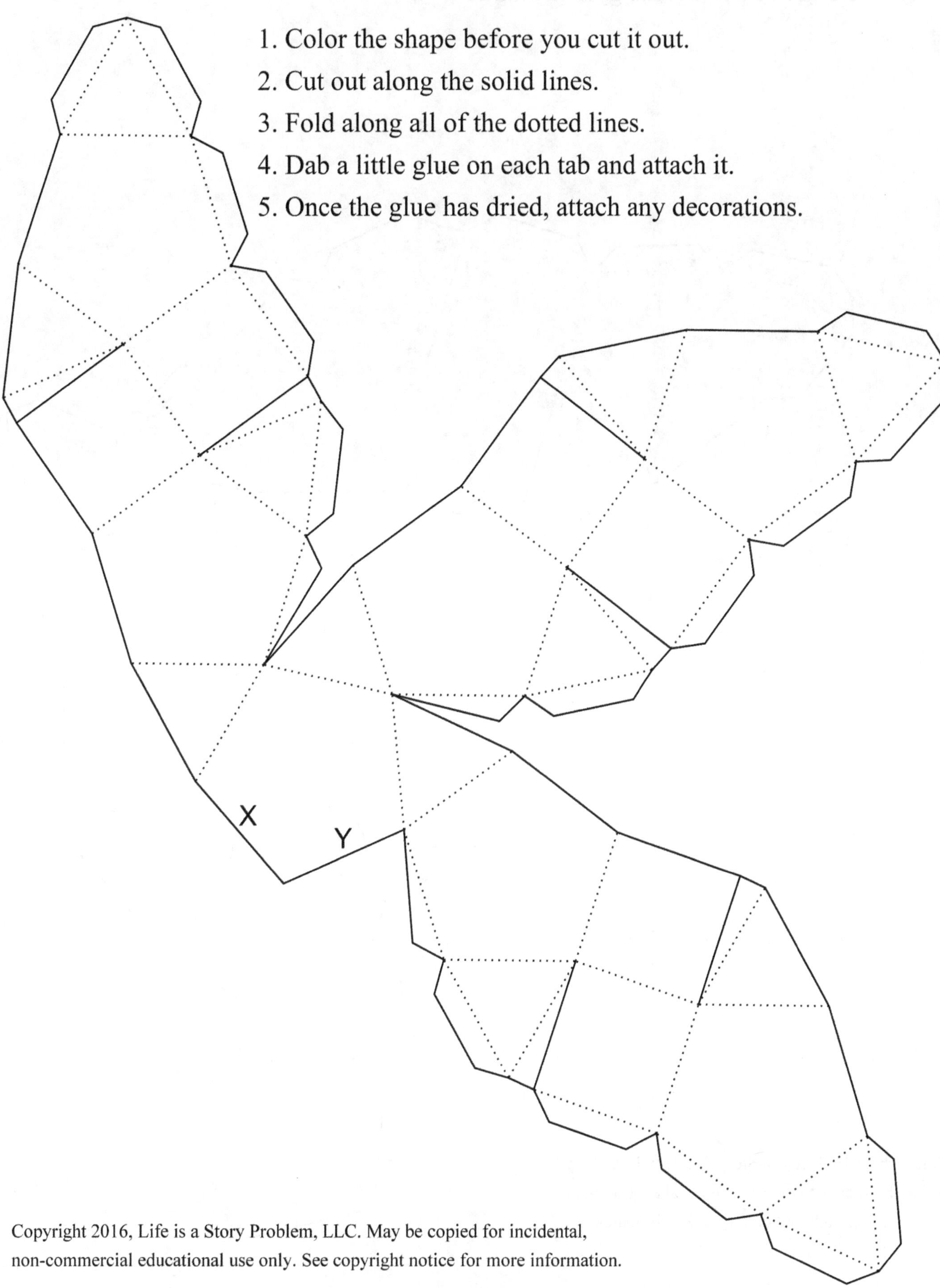

Page 2, Elongated pentagonal gyrobirotunda

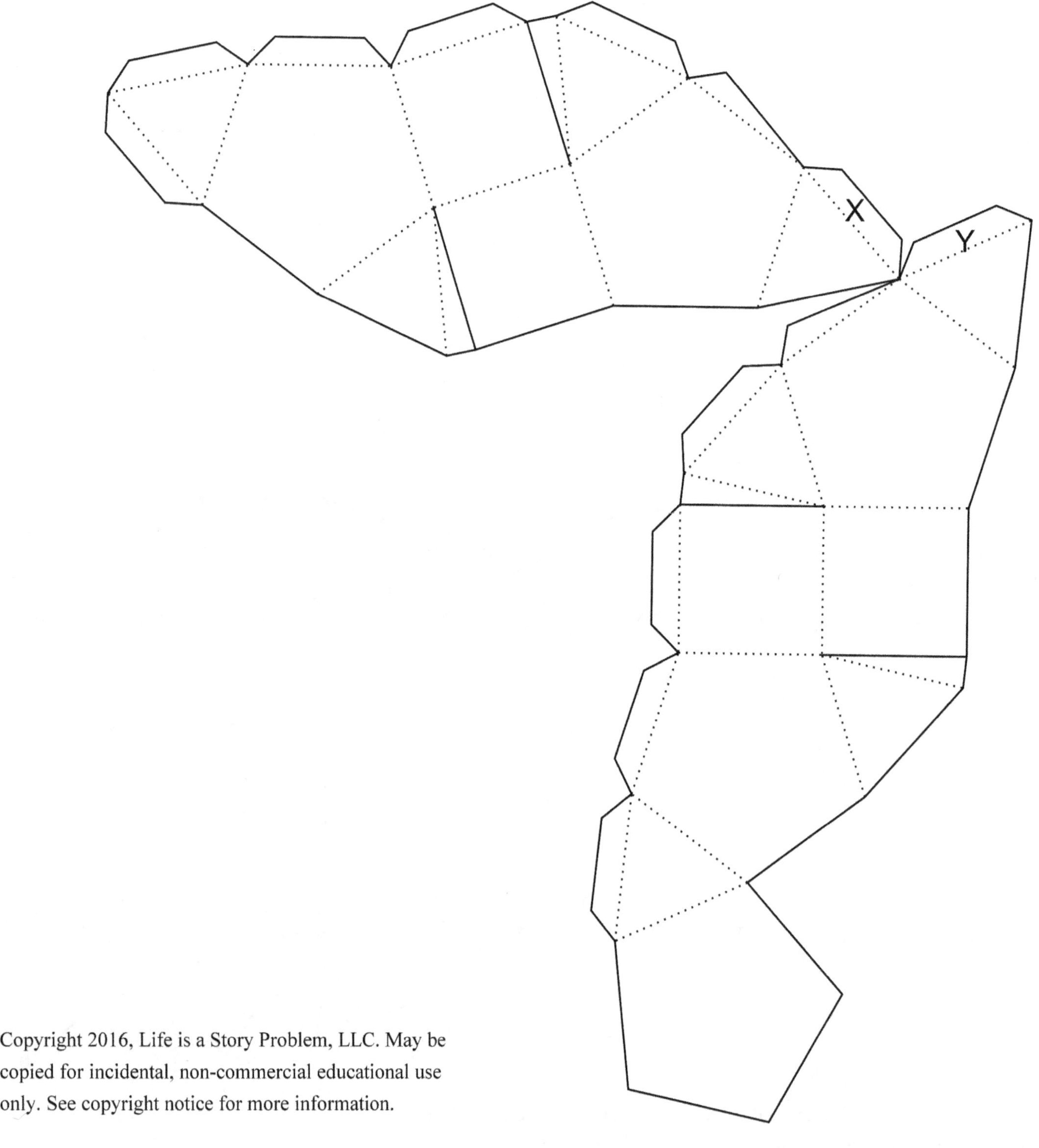

Geometric Nets Mega Project Book, Tabbed By David E. McAdams

Gyroelongated triangular bicupola

1. Color the shape before you cut it out.

2. Cut out along the solid lines.

3. Fold along all of the dotted lines.

4. Dab a little glue on each tab and attach it.

5. Once the glue has dried, attach any decorations.

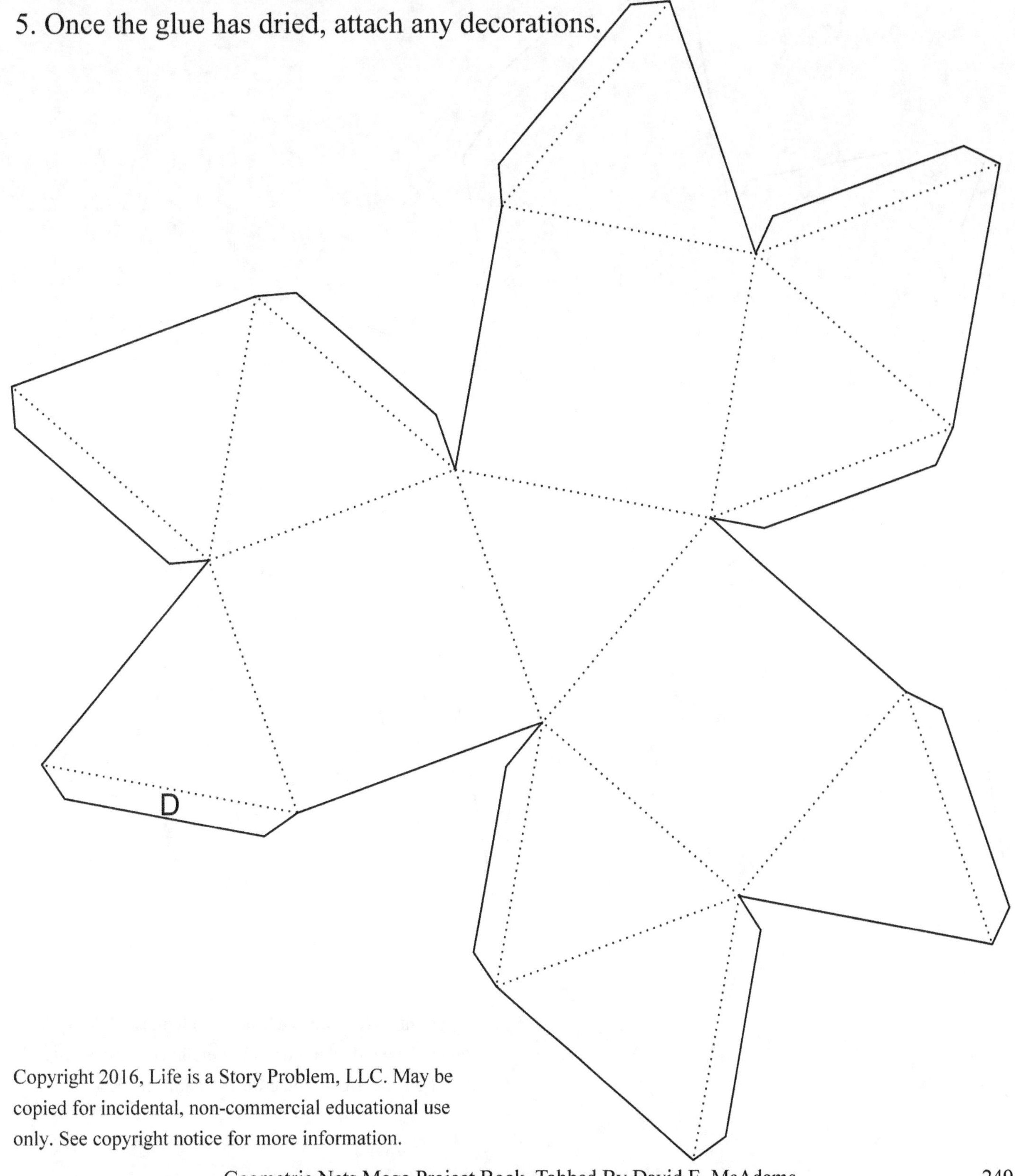

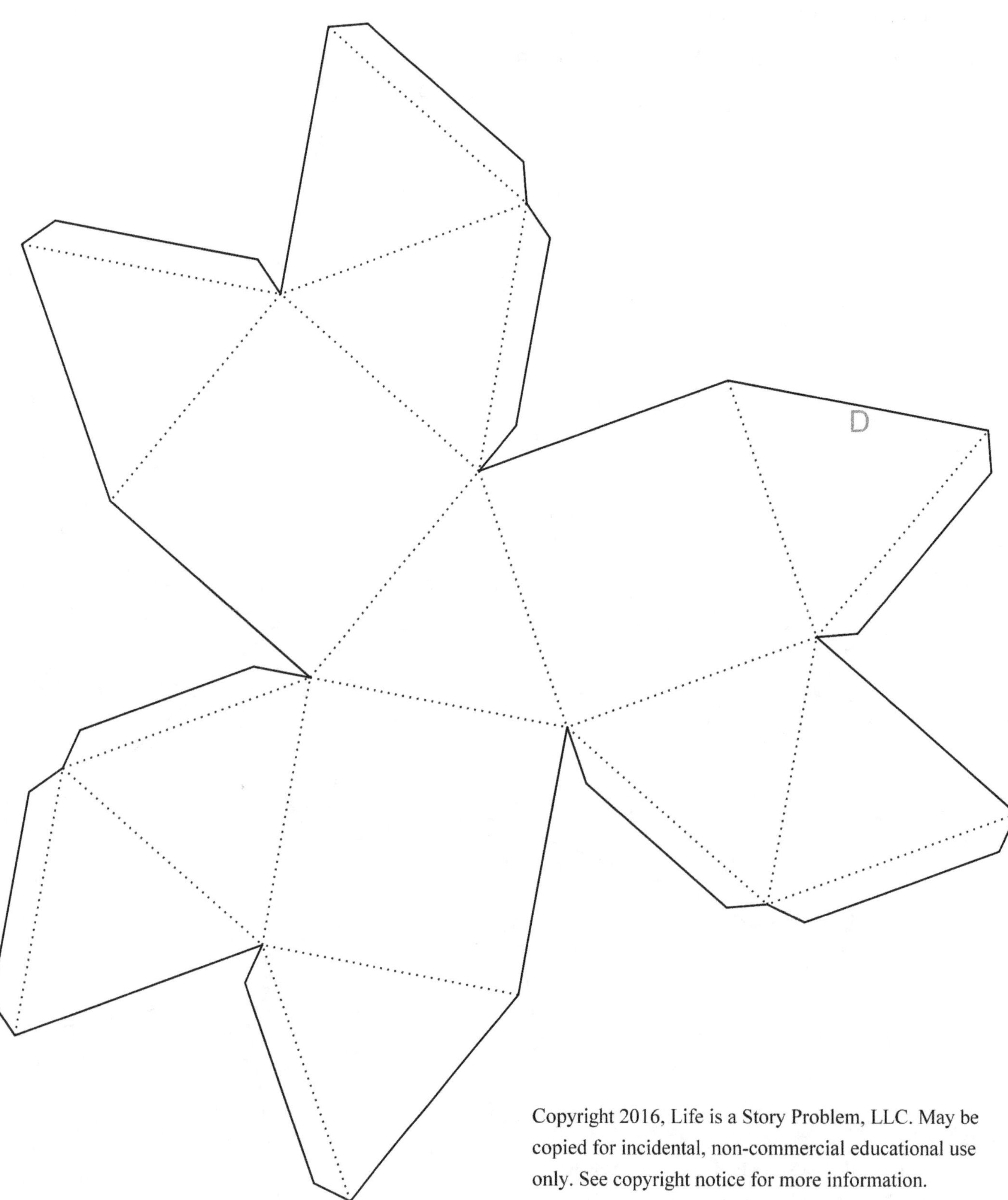

Geometric Nets Mega Project Book, Tabbed By David E. McAdams

Gyroelongated square bicupola

1. Color the shape before you cut it out.

2. Cut out along the solid lines. 3. Fold along all of the dotted lines.

4. Dab a little glue on each tab and attach it.

5. Once the glue has dried, attach any decorations.

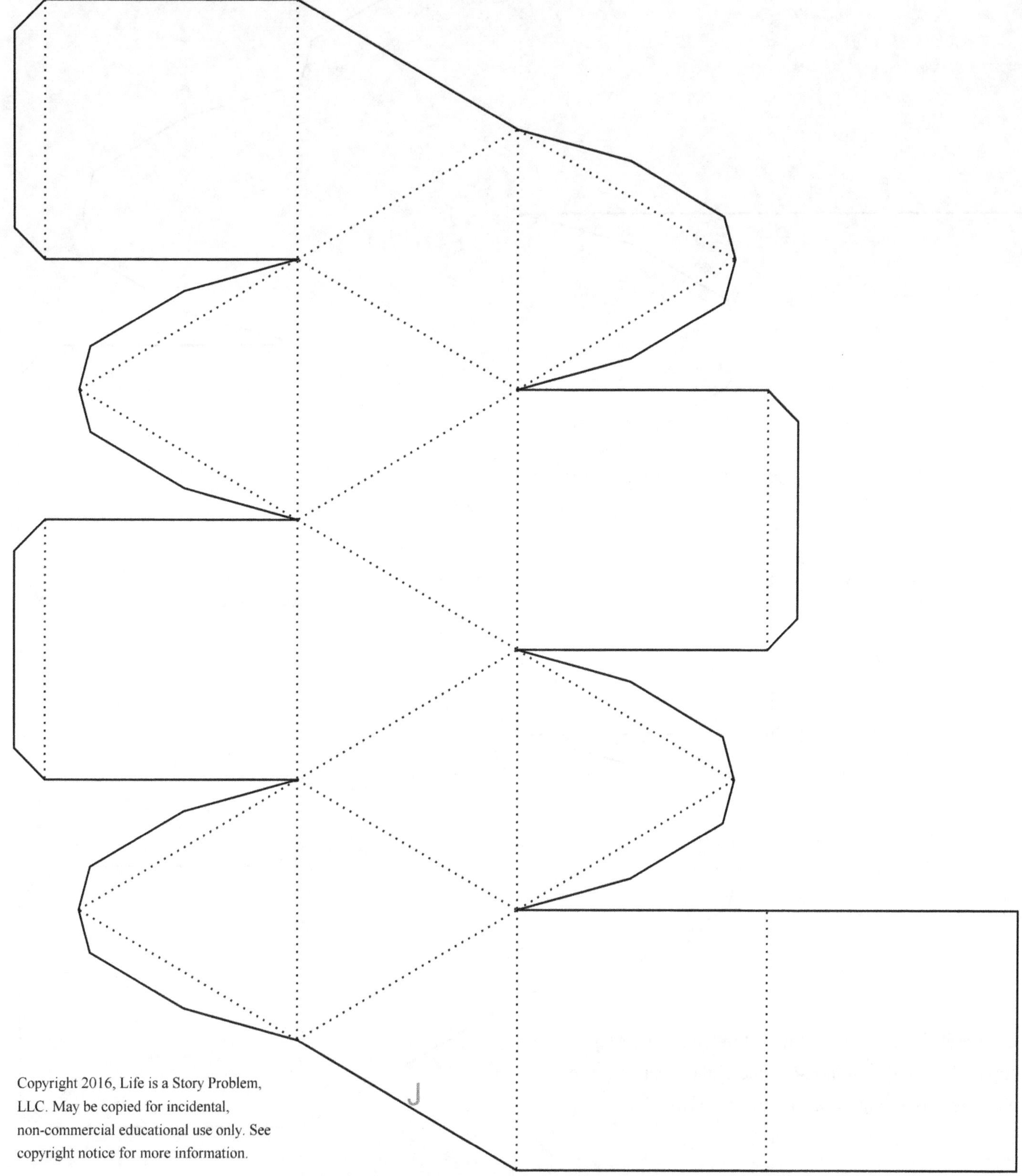

2nd page, Gyroelongated Square Bicupola

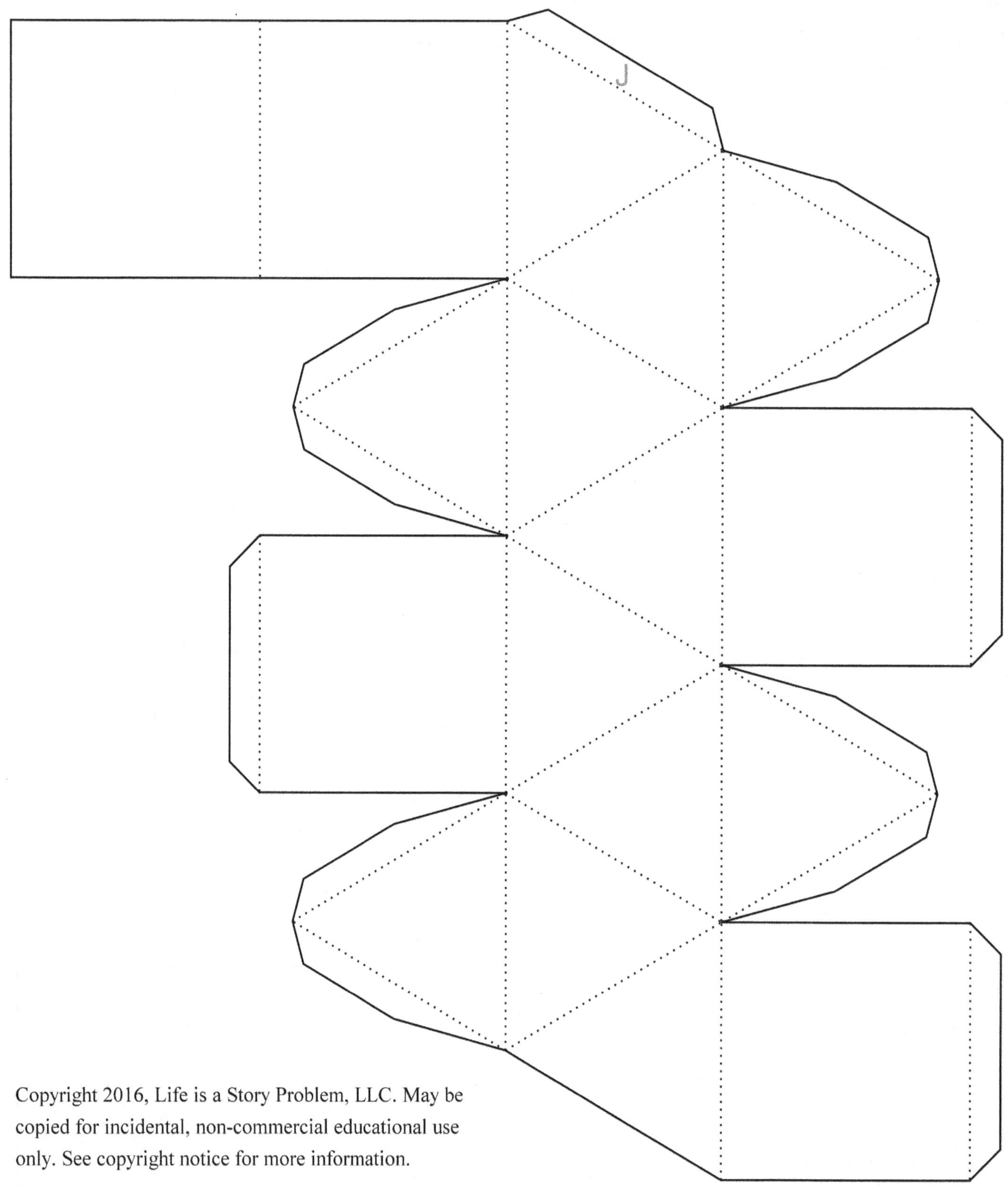

     Geometric Nets Mega Project Book, Tabbed By David E. McAdams

Gyroelongated pentagonal bicupola

1. Color the shape before you cut it out.
2. Cut out along the solid lines.
3. Fold along all of the dotted lines.
4. Dab a little glue on each tab and attach it.
5. Once the glue has dried, attach any decorations.

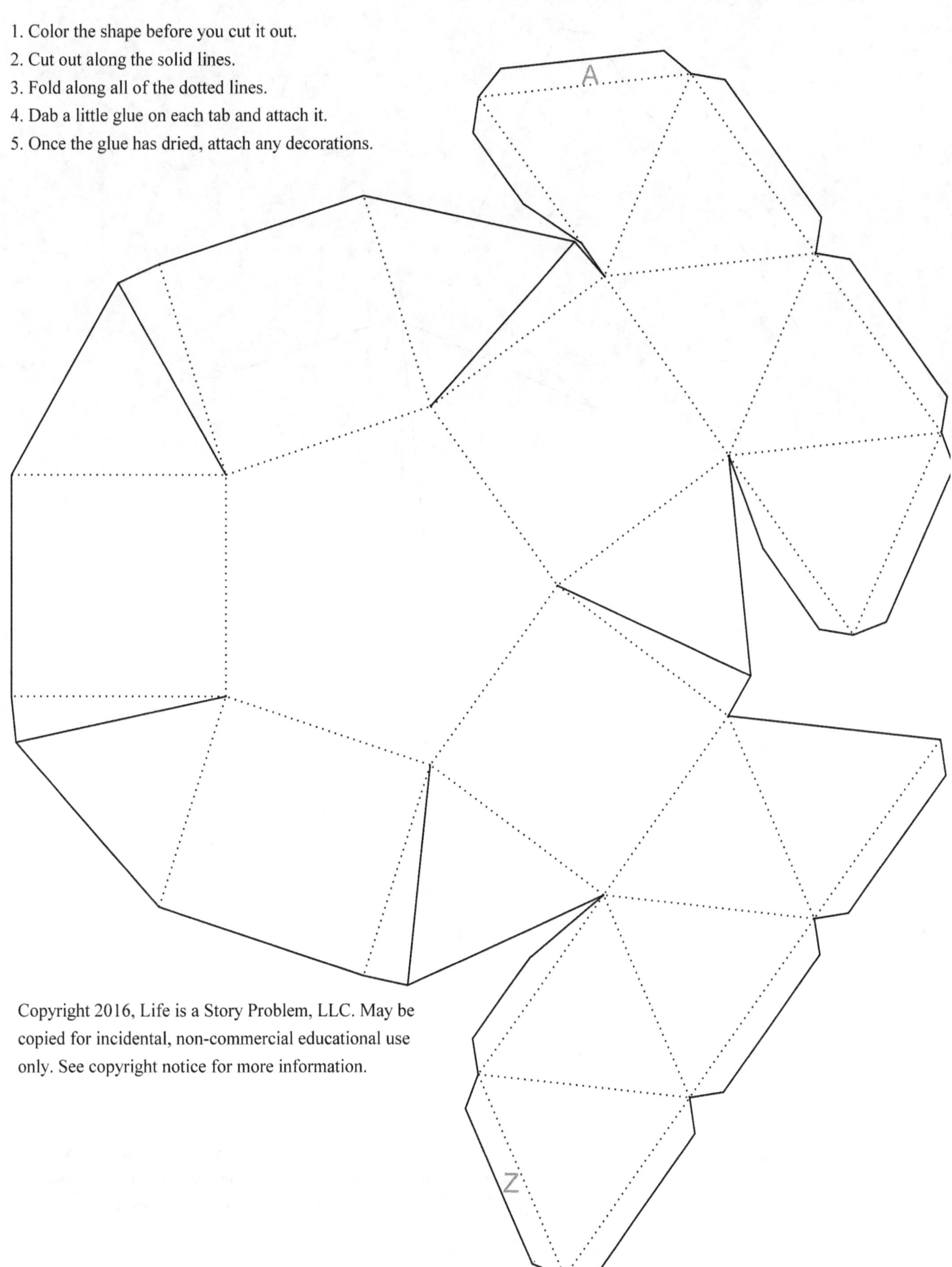

2nd page, Gyroelongated pentagonal bicupola

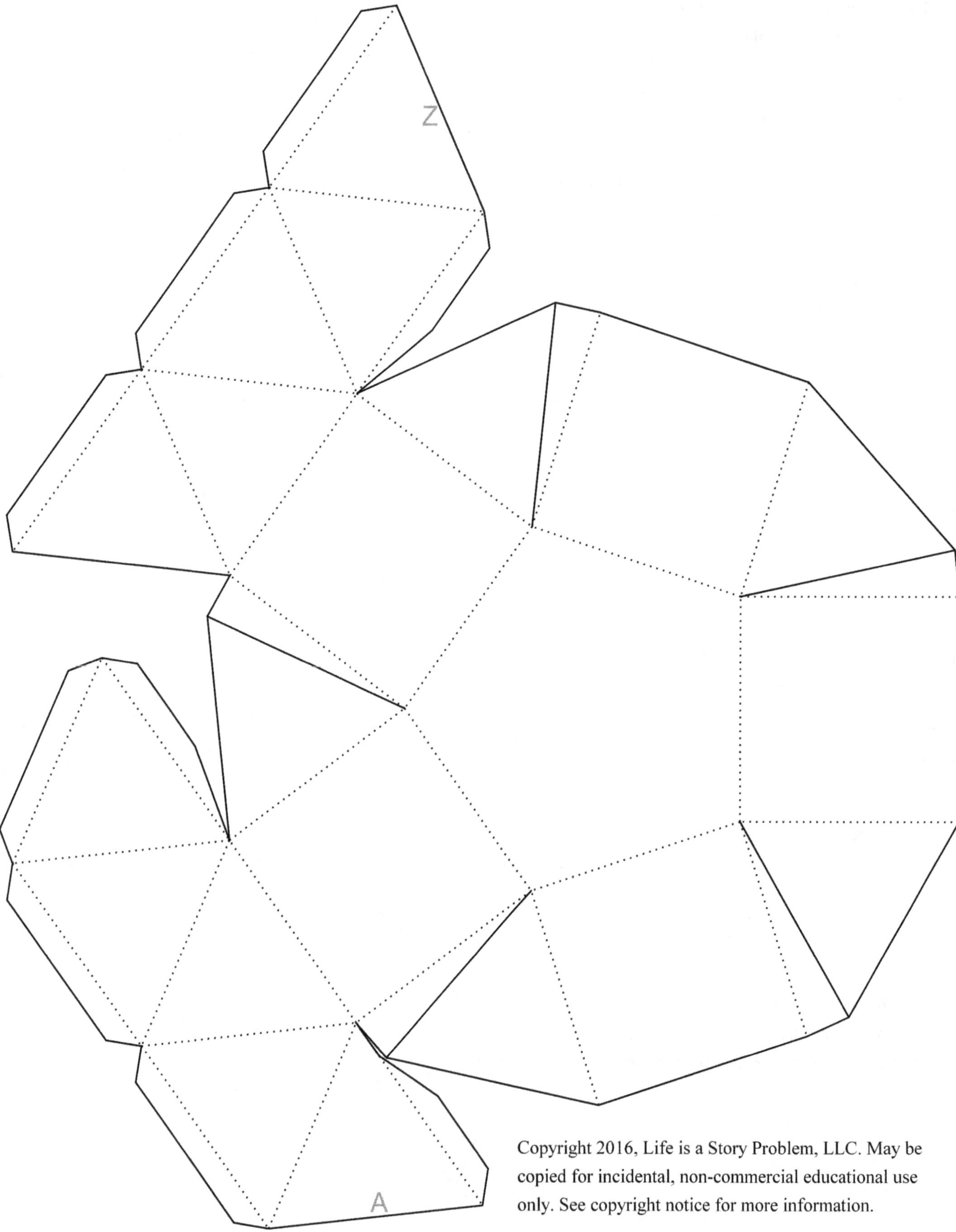

Geometric Nets Mega Project Book, Tabbed By David E. McAdams

Gyroelongated pentagonal cupolarotunda

1. Color the shape before you cut it out.

2. Cut out along the solid lines.

3. Fold along all of the dotted lines.

4. Dab a little glue on each tab and attach it.

5. Once the glue has dried, attach any decorations.

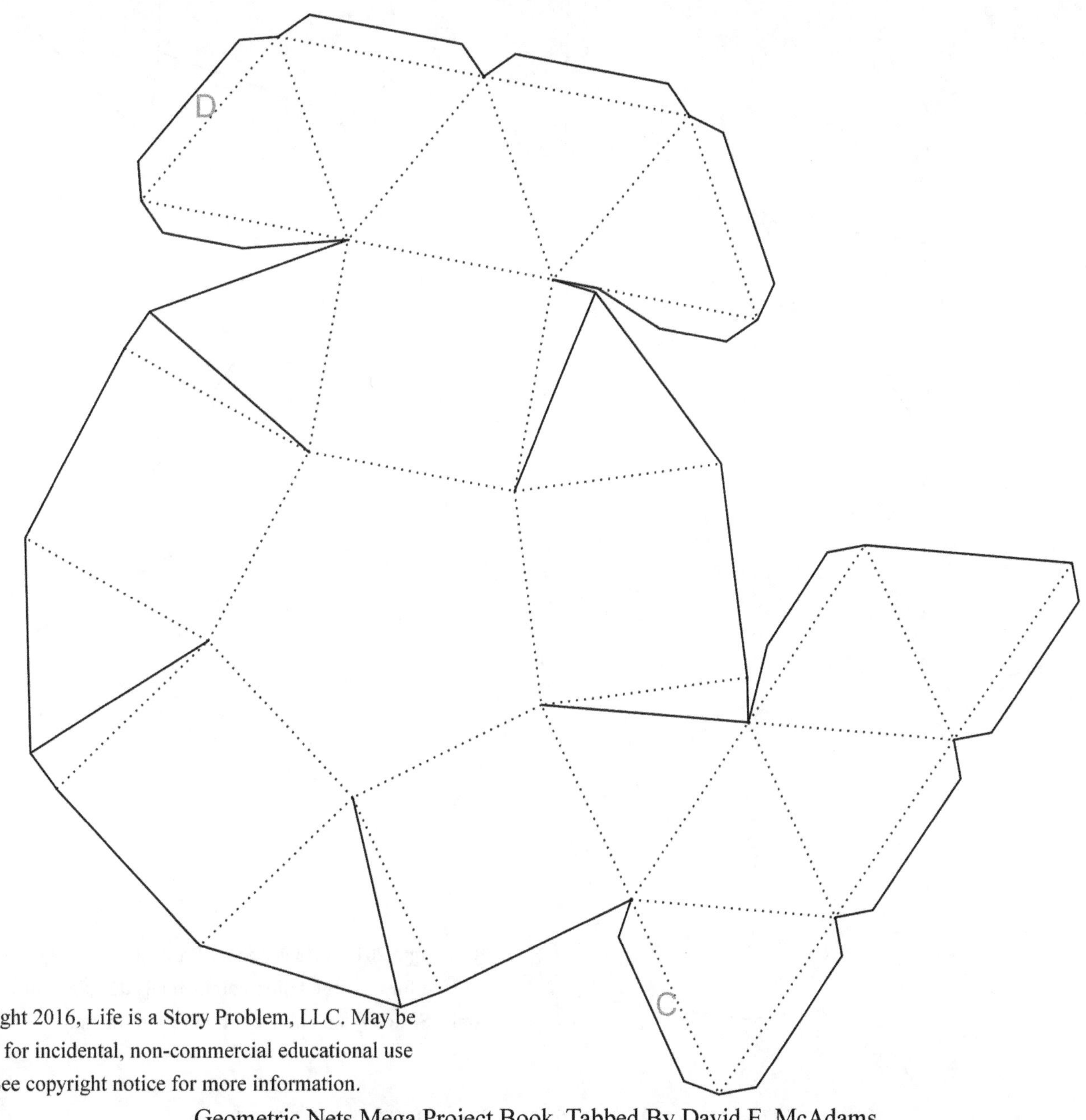

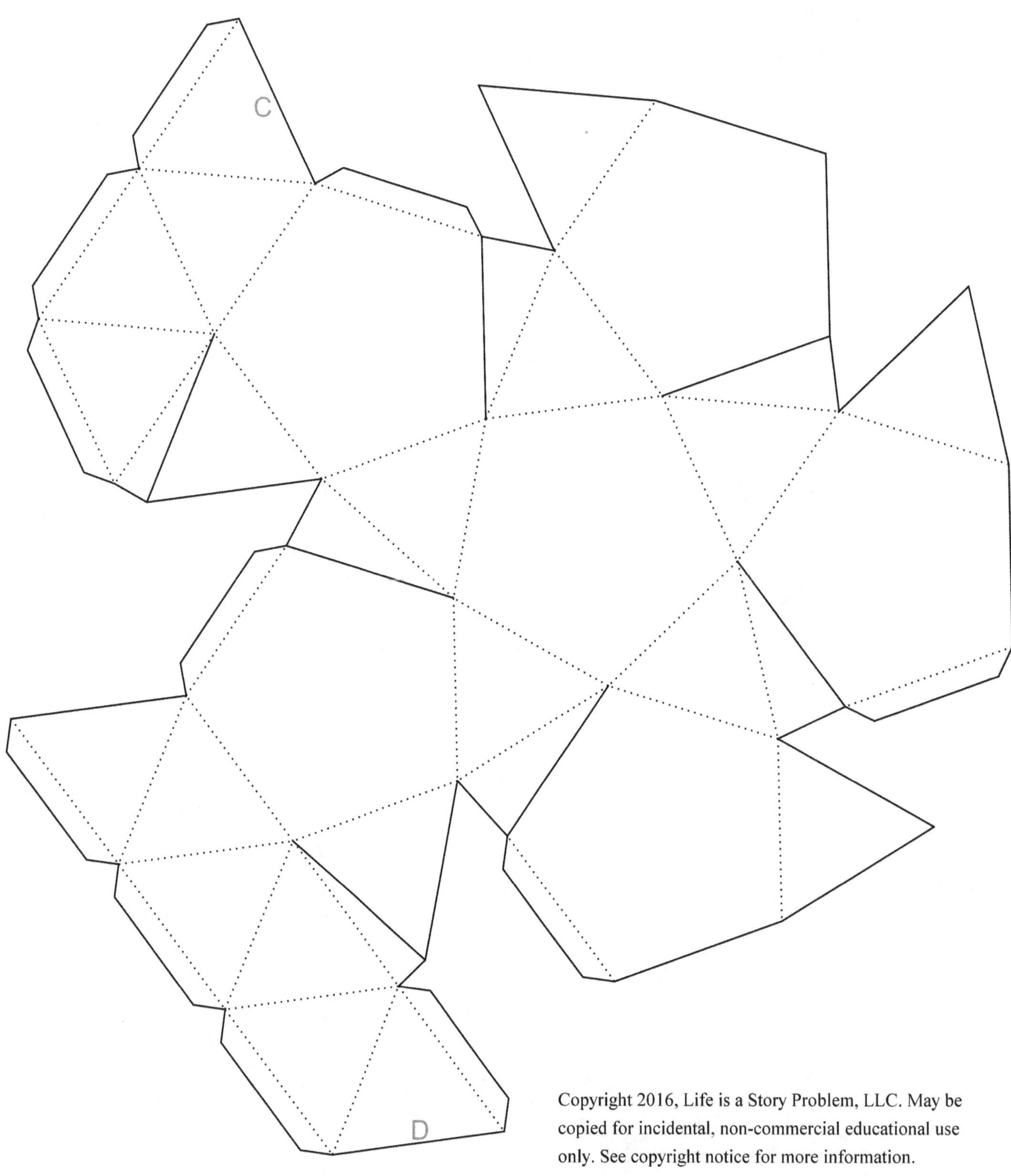

Geometric Nets Mega Project Book, Tabbed By David E. McAdams

Gyroelongated pentagonal birotunda

1. Color the shape before you cut it out.
2. Cut out along the solid lines.
3. Fold along all of the dotted lines.
4. Dab a little glue on each tab and attach it.
5. Once the glue has dried, attach any decorations.

This net is on two pages. Print out a copy of each page.

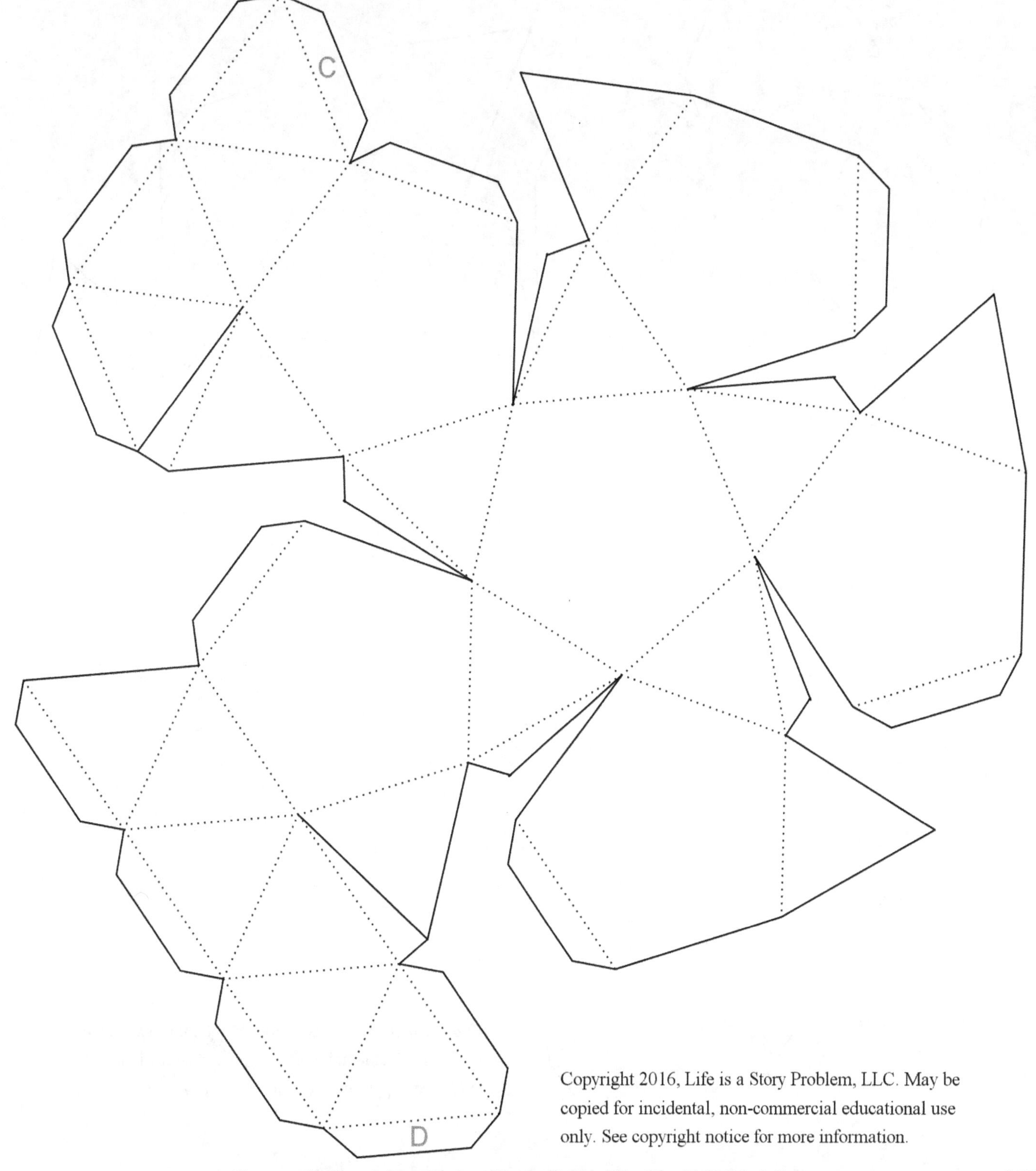

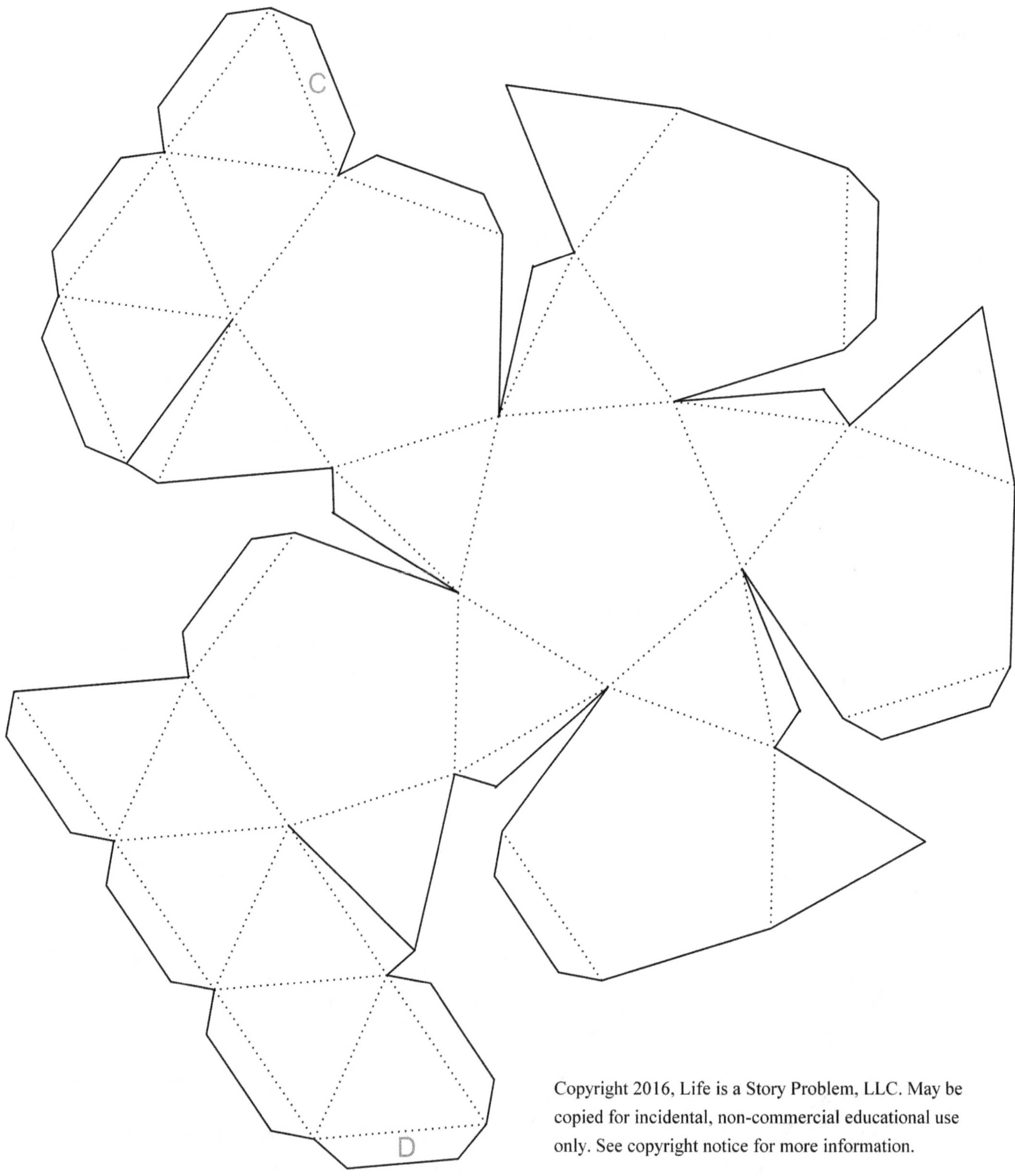

Geometric Nets Mega Project Book, Tabbed By David E. McAdams

Augmented triangular prism

1. Color the shape before you cut it out.
2. Cut out along the solid lines.
3. Fold along all of the dotted lines.
4. Dab a little glue on each tab and attach it.
5. Once the glue has dried, attach any decorations.

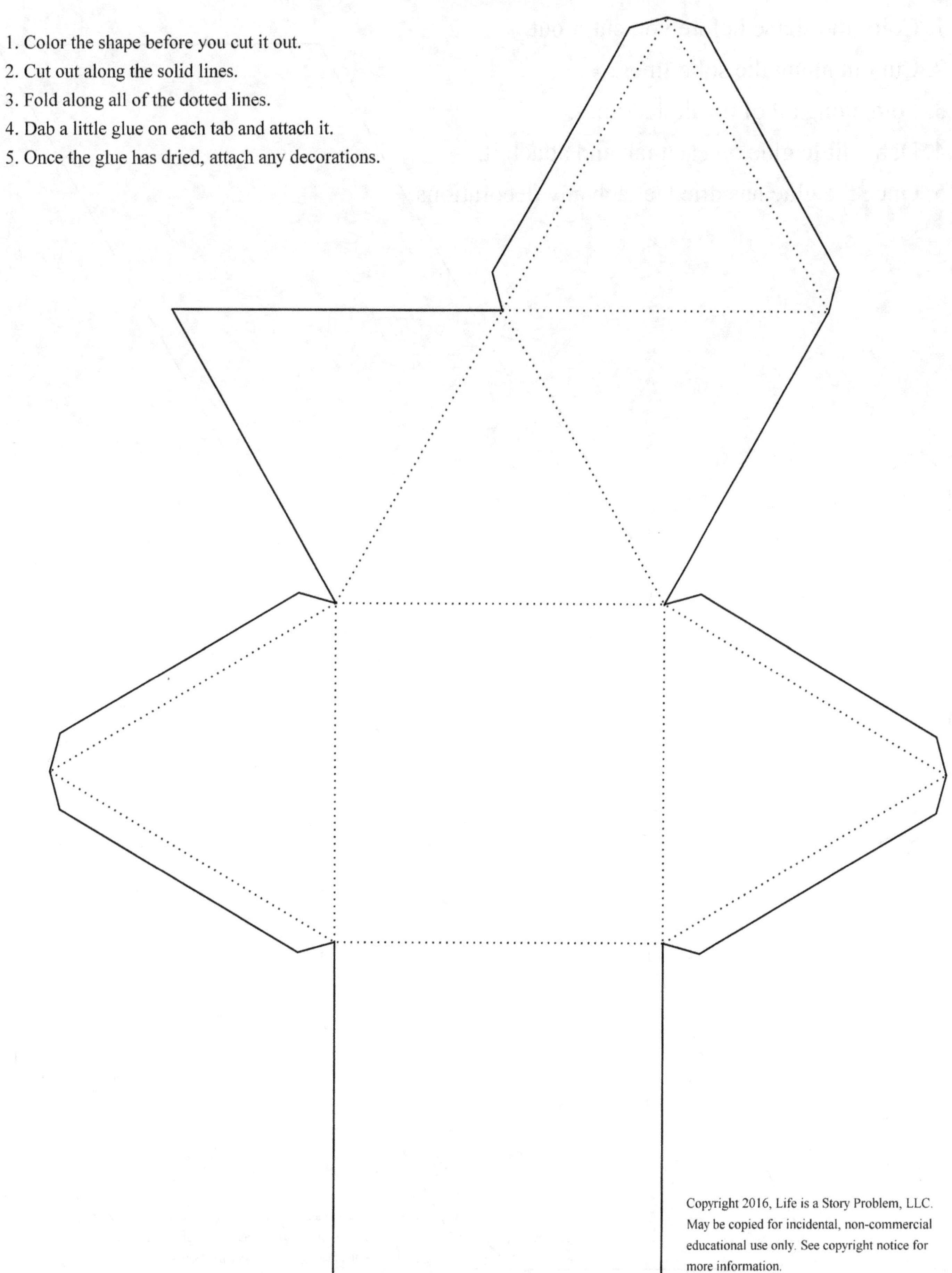

Biaugmented triangular prism

1. Color the shape before you cut it out.

2. Cut out along the solid lines.

3. Fold along all of the dotted lines.

4. Dab a little glue on each tab and attach it.

5. Once the glue has dried, attach any decorations.

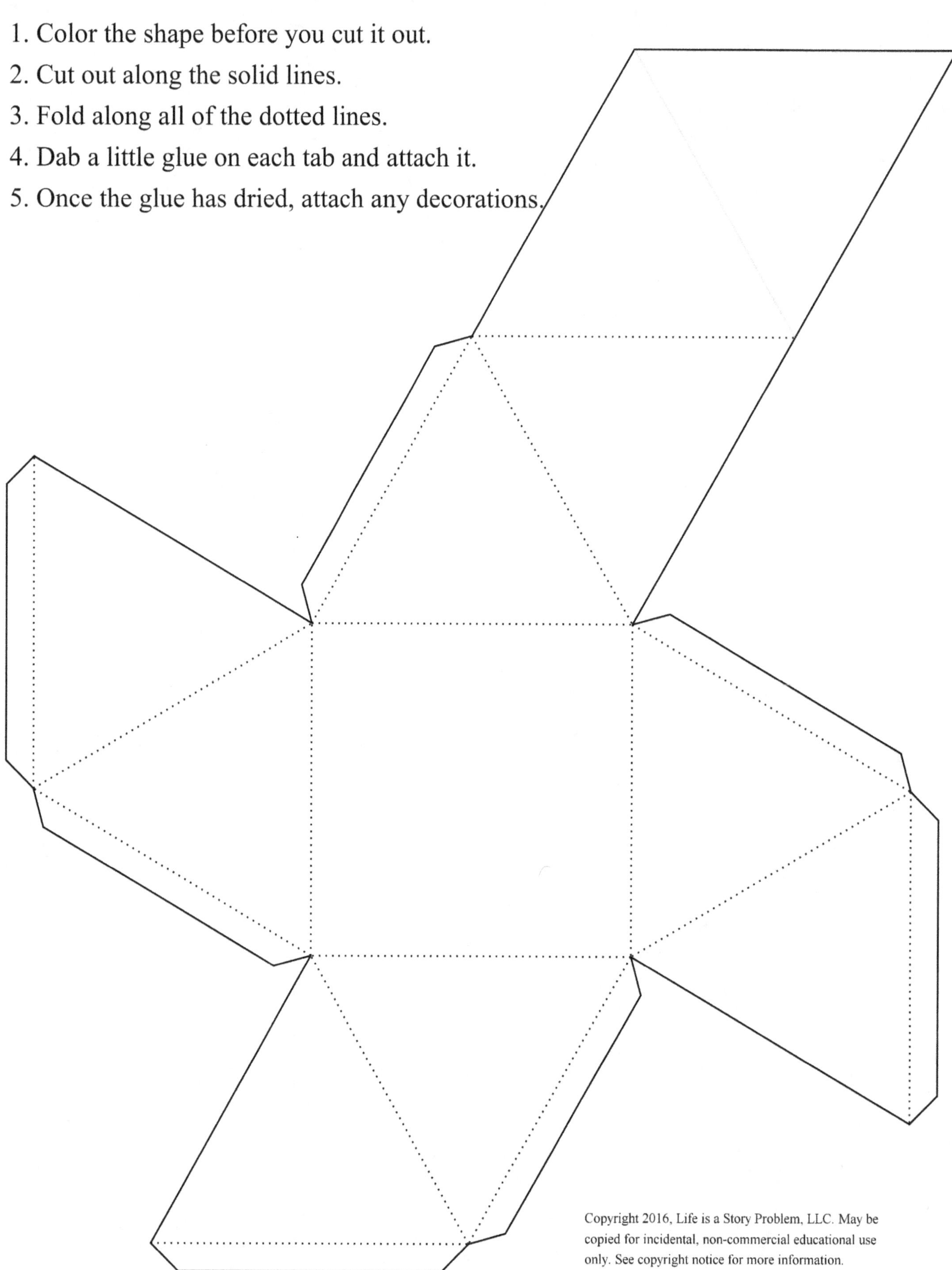

Geometric Nets Mega Project Book, Tabbed By David E. McAdams

Triaugmented triangular prism

1. Color the shape before you cut it out.
2. Cut out along the solid lines.
3. Fold along all of the dotted lines.
4. Dab a little glue on each tab and attach it.
5. Once the glue has dried, attach any decorations.

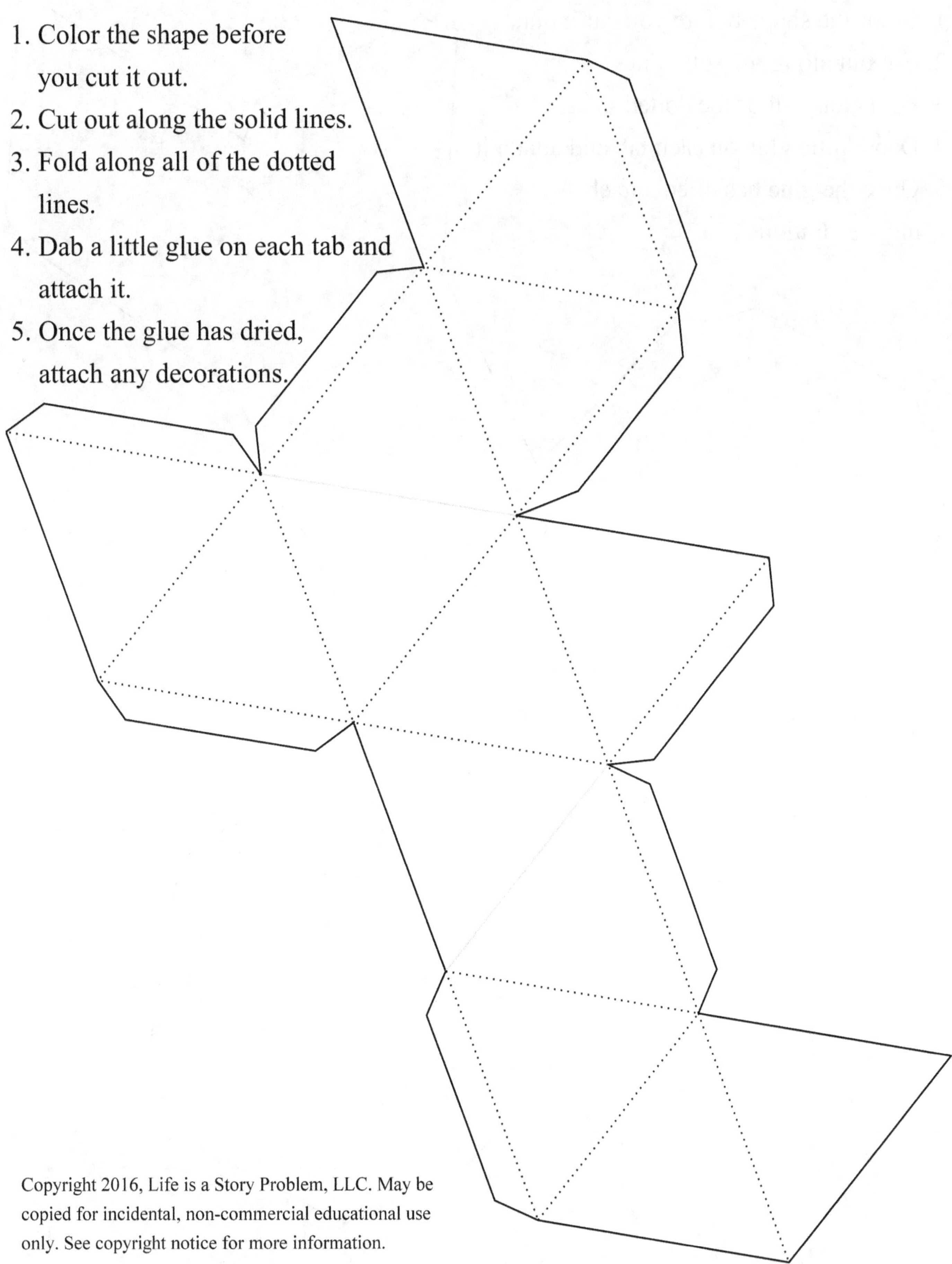

Augmented pentagonal prism

1. Color the shape before you cut it out.
2. Cut out along the solid lines.
3. Fold along all of the dotted lines.
4. Dab a little glue on each tab and attach it.
5. Once the glue has dried, attach
   any decorations.

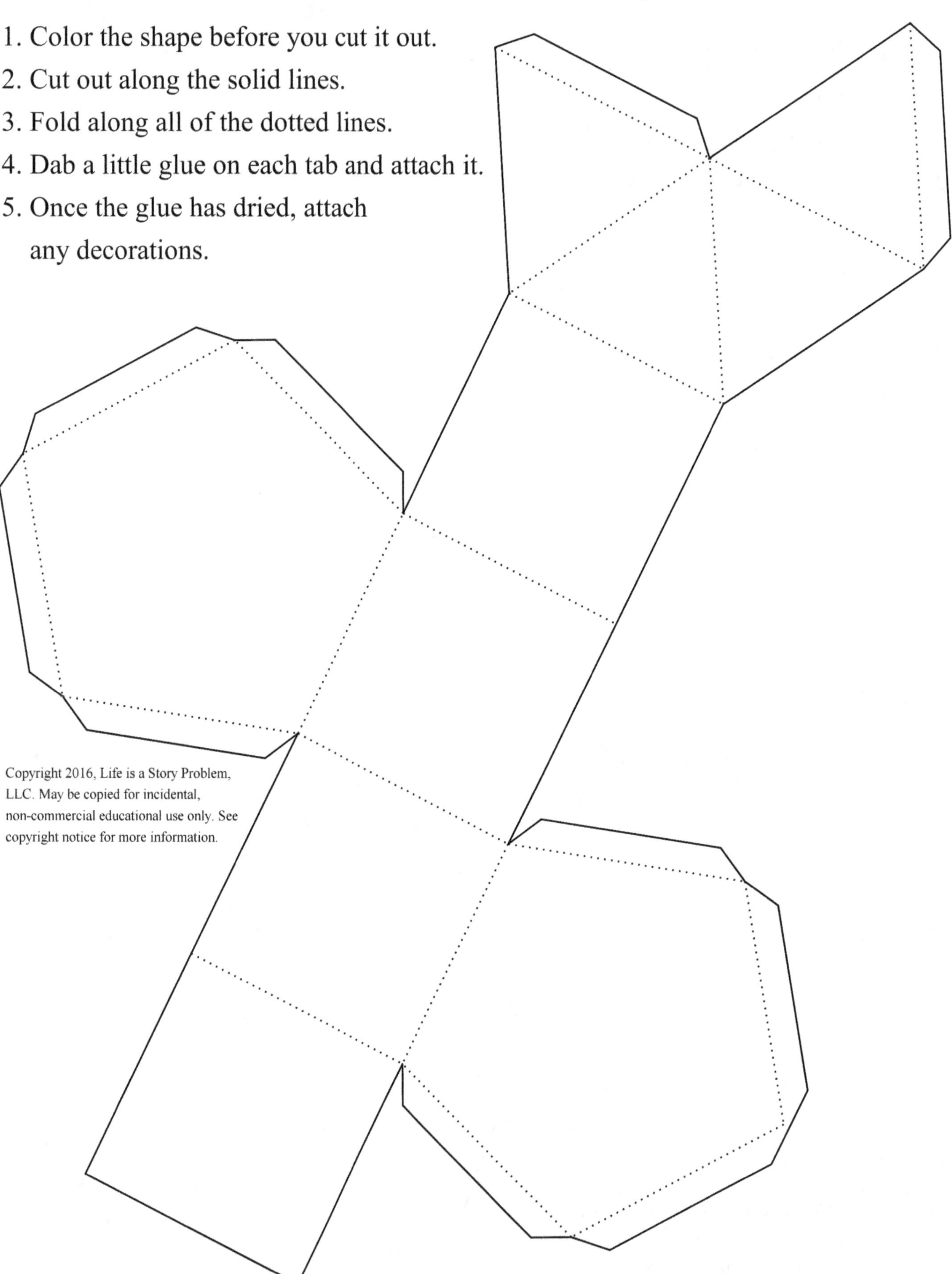

     Geometric Nets Mega Project Book, Tabbed By David E. McAdams

Biaugmented pentagonal prism

1. Color the shape before you cut it out.
2. Cut out along the solid lines.
3. Fold along all of the dotted lines.
4. Dab a little glue on each tab and attach it.
5. Once the glue has dried,
   attach any decorations.

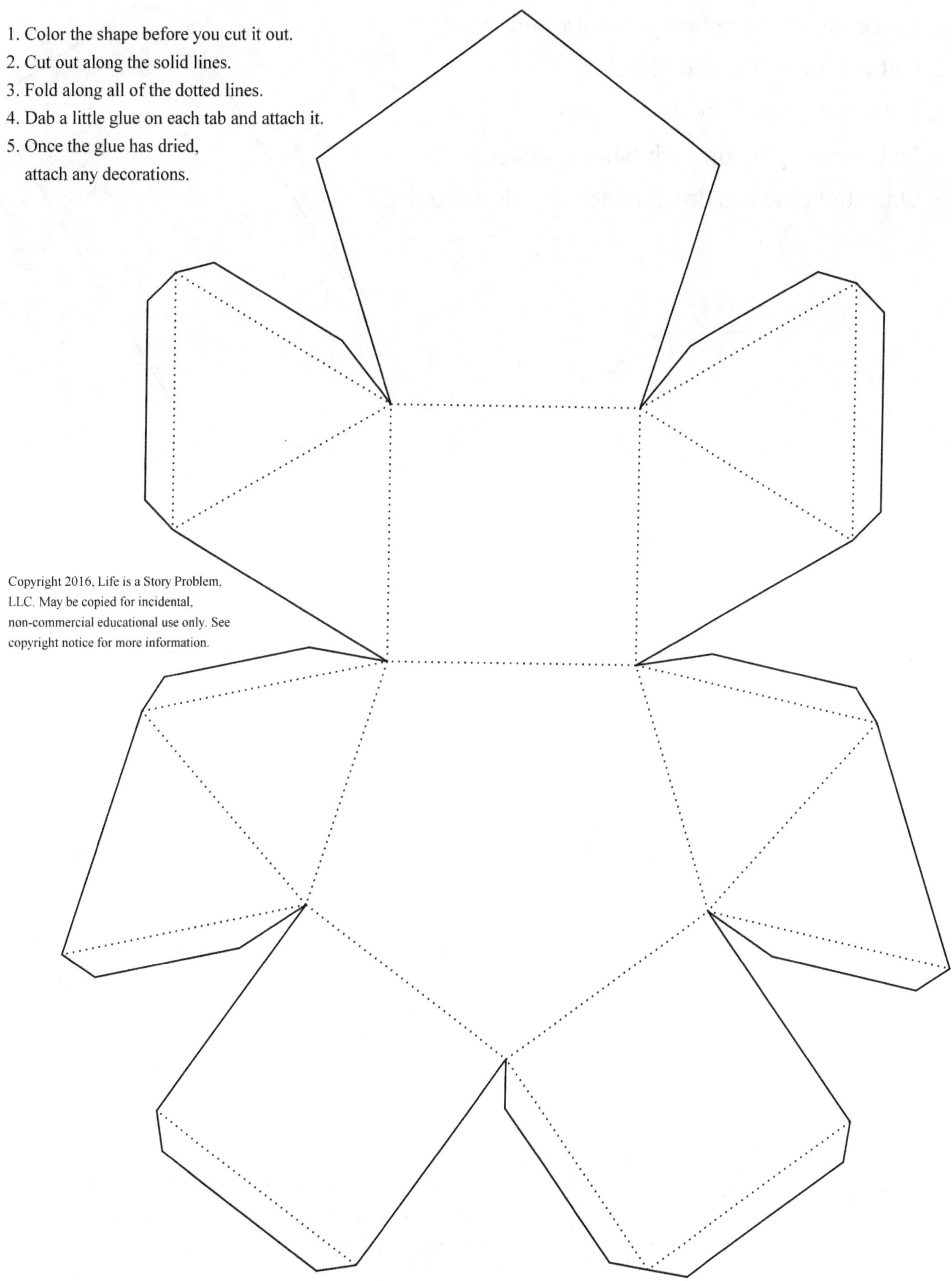

Augmented hexagonal prism

1. Color the shape before you cut it out.

2. Cut out along the solid lines.

3. Fold along all of the dotted lines.

4. Dab a little glue on each tab and attach it.

5. Once the glue has dried, attach any decorations.

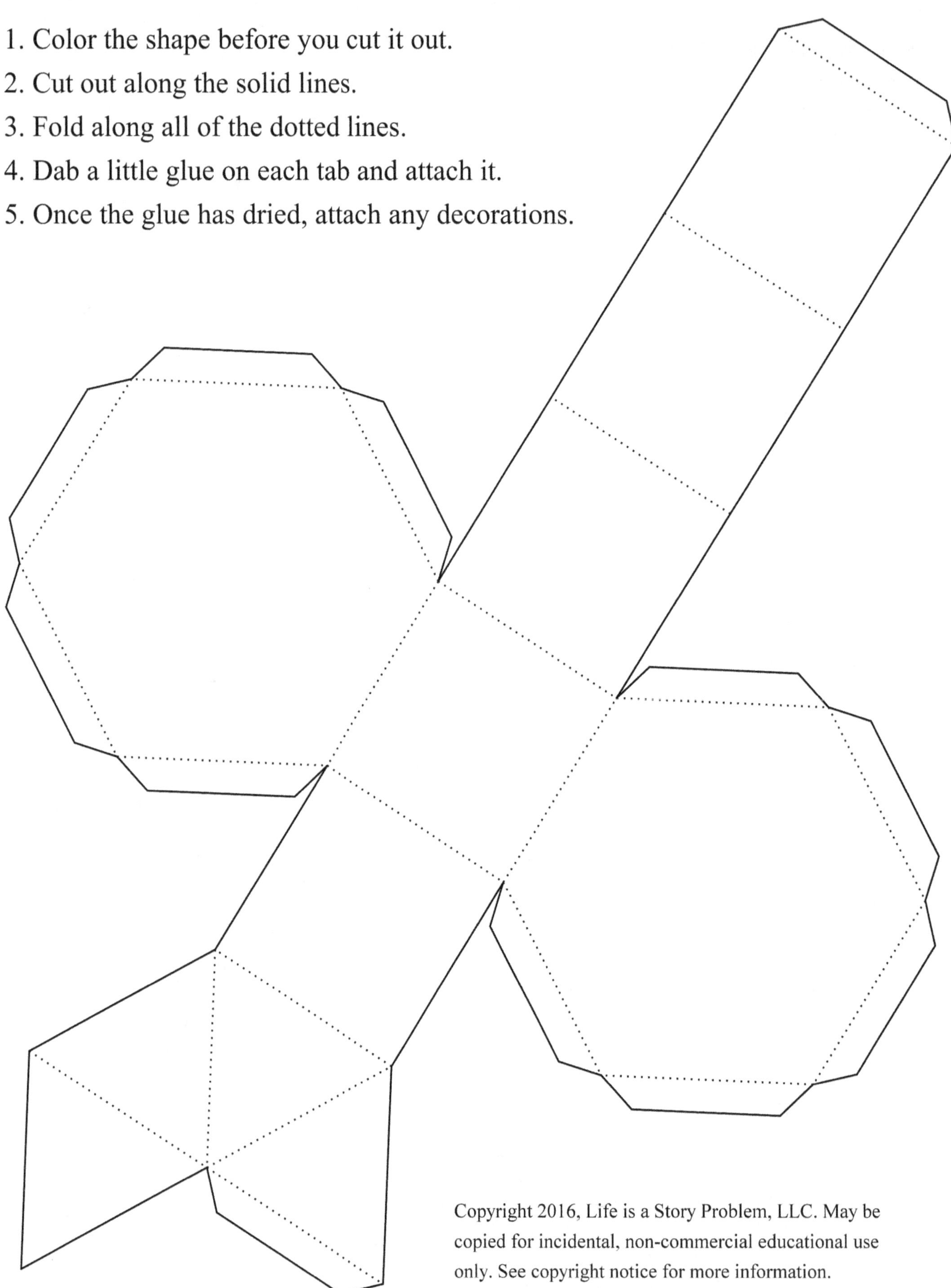

     Geometric Nets Mega Project Book, Tabbed By David E. McAdams

Parabiaugmented hexagonal prism

1. Color the shape before you cut it out.

2. Cut out along the solid lines.

3. Fold along all of the dotted lines.

4. Dab a little glue on each tab and attach it.

5. Once the glue has dried, attach
   any decorations.

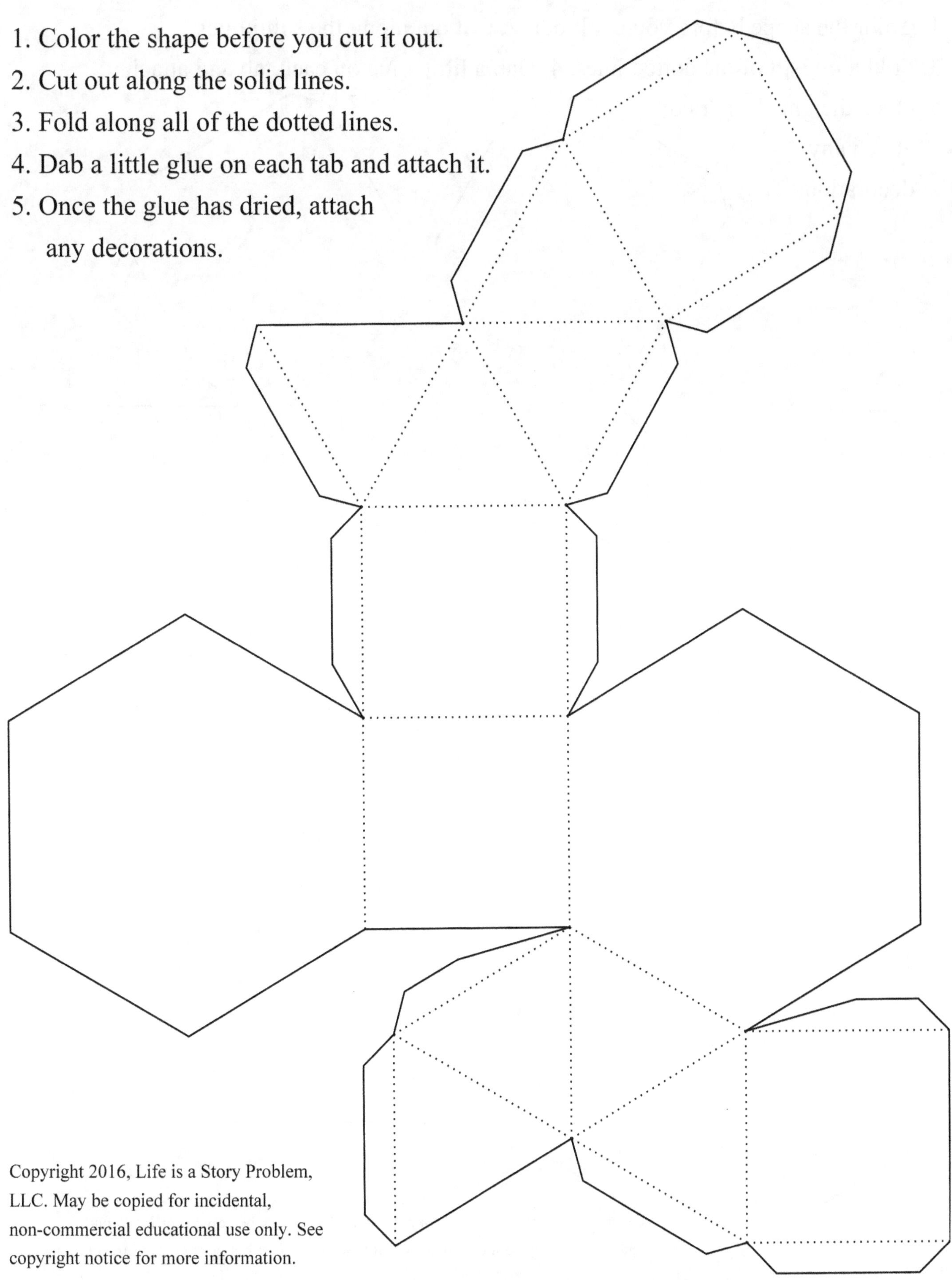

Metabiaugmented hexagonal prism

1. Color the shape before you cut it out. 2. Cut out along the solid lines.

3. Fold along all of the dotted lines. 4. Dab a little glue on each tab and attach it.

5. Once the glue has dried, attach any decorations.

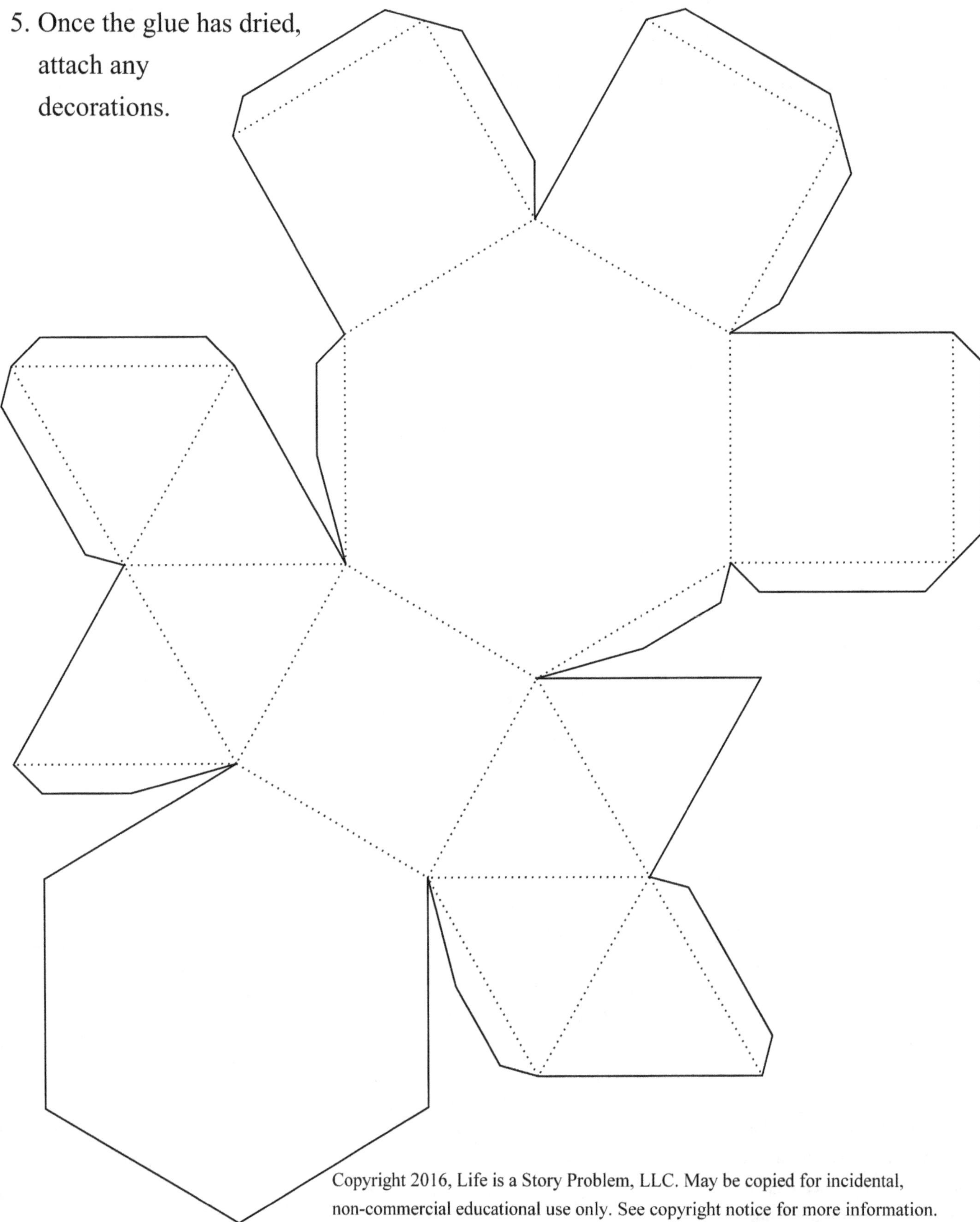

 Geometric Nets Mega Project Book, Tabbed By David E. McAdams

Triaugmented hexagonal prism

1. Color the shape before you cut it out.
2. Cut out along the solid lines.
3. Fold along all of the dotted lines.
4. Dab a little glue on each tab and attach it.
5. Once the glue has dried, attach any
   decorations.

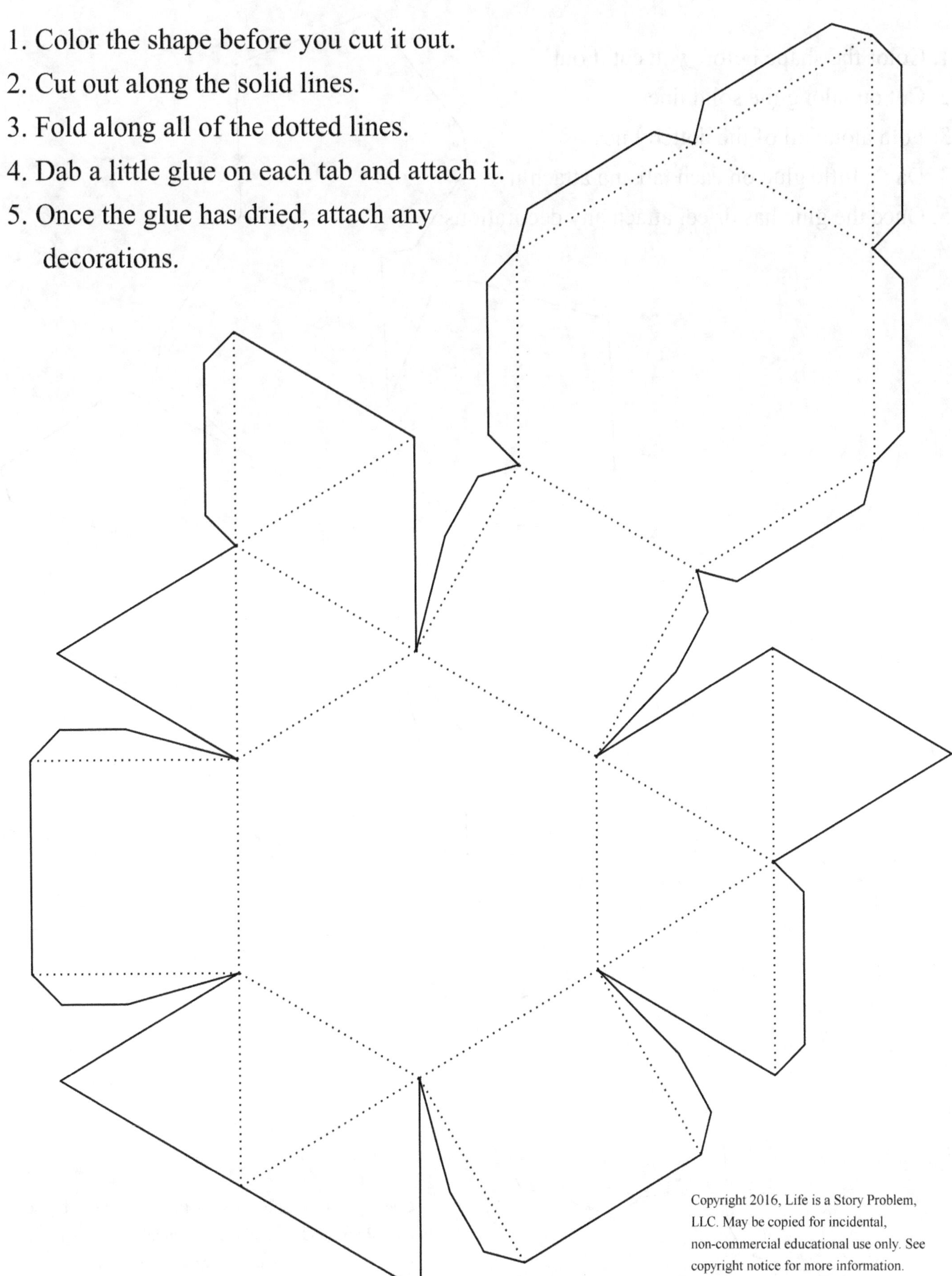

Augmented dodecahedron

1. Color the shape before you cut it out.

2. Cut out along the solid lines.

3. Fold along all of the dotted lines.

4. Dab a little glue on each tab and attach it.

5. Once the glue has dried, attach any decorations.

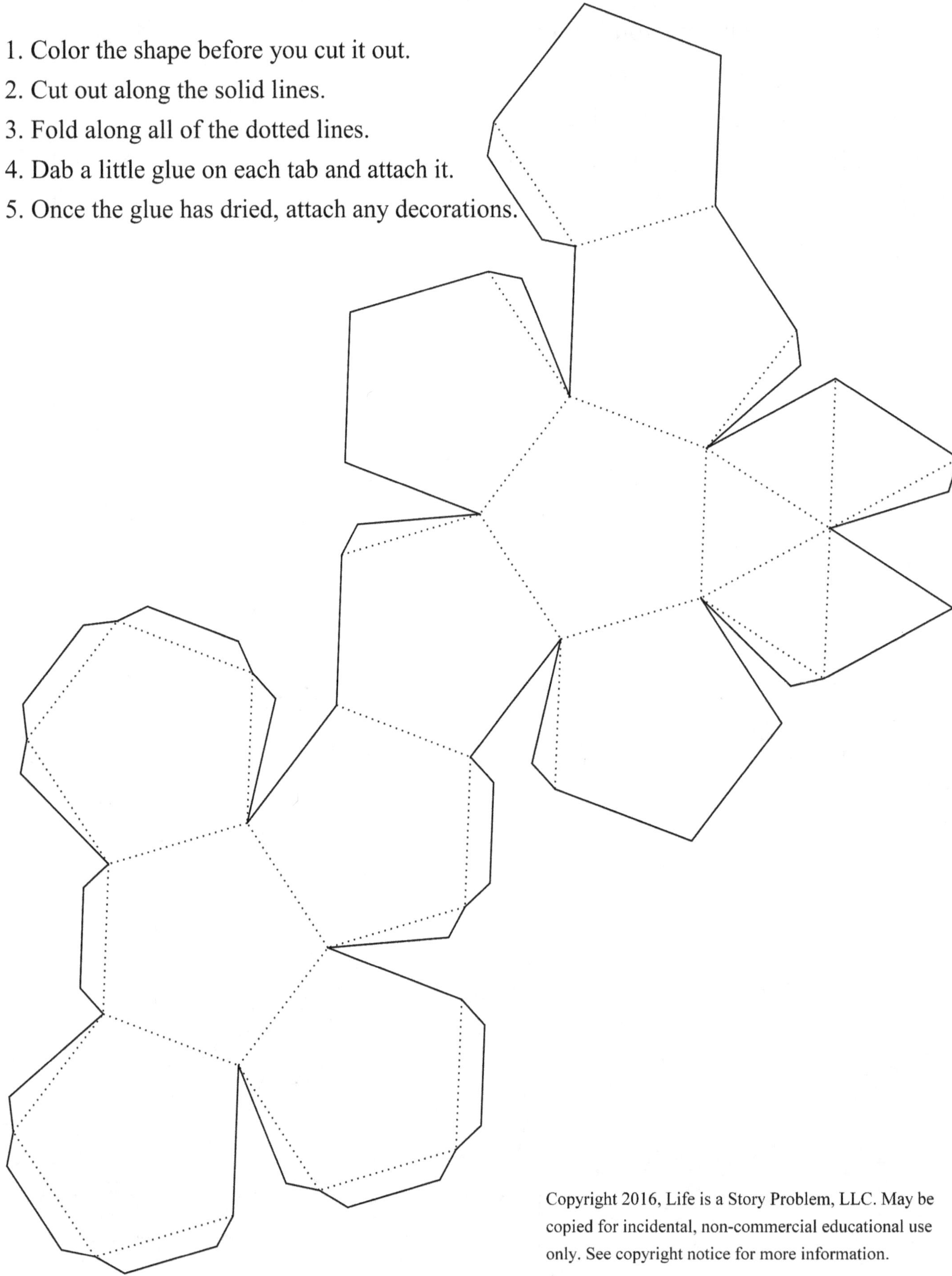

Parabiaugmented dodecahedron

1. Color the shape before you cut it out.
2. Cut out along the solid lines.
3. Fold along all of the dotted lines.
4. Dab a little glue on each tab and attach it.
5. Once the glue has dried, attach any decorations.

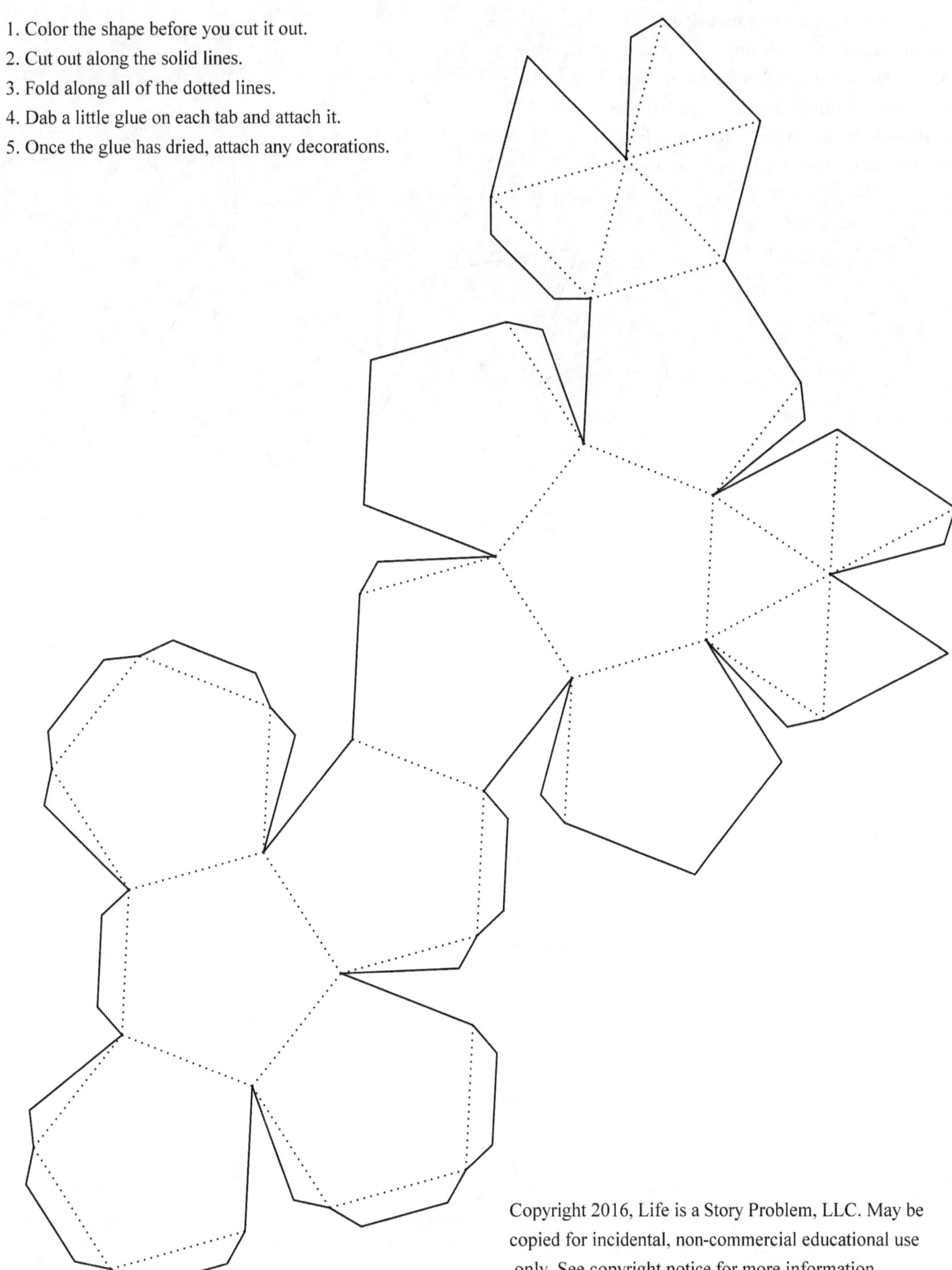

Metabiaugmented dodecahedron

1. Color the shape before you cut it out.
2. Cut out along the solid lines.
3. Fold forward along all of the dotted lines.
4. Fold backward along all of the dashed lines.
5. Dab a little glue on each tab and attach it.
6. Once the glue has dried, attach any decorations.

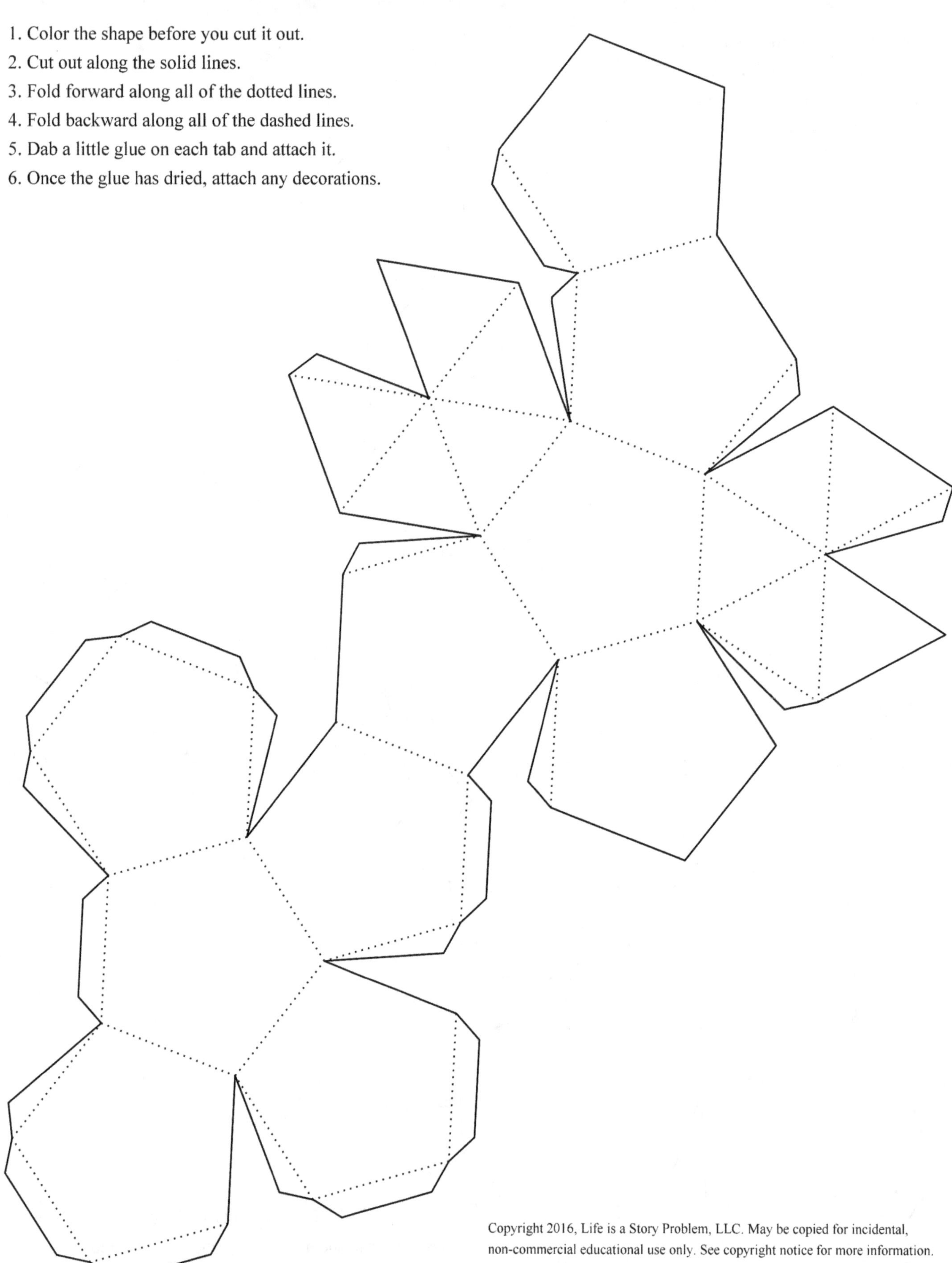

 Geometric Nets Mega Project Book, Tabbed By David E. McAdams

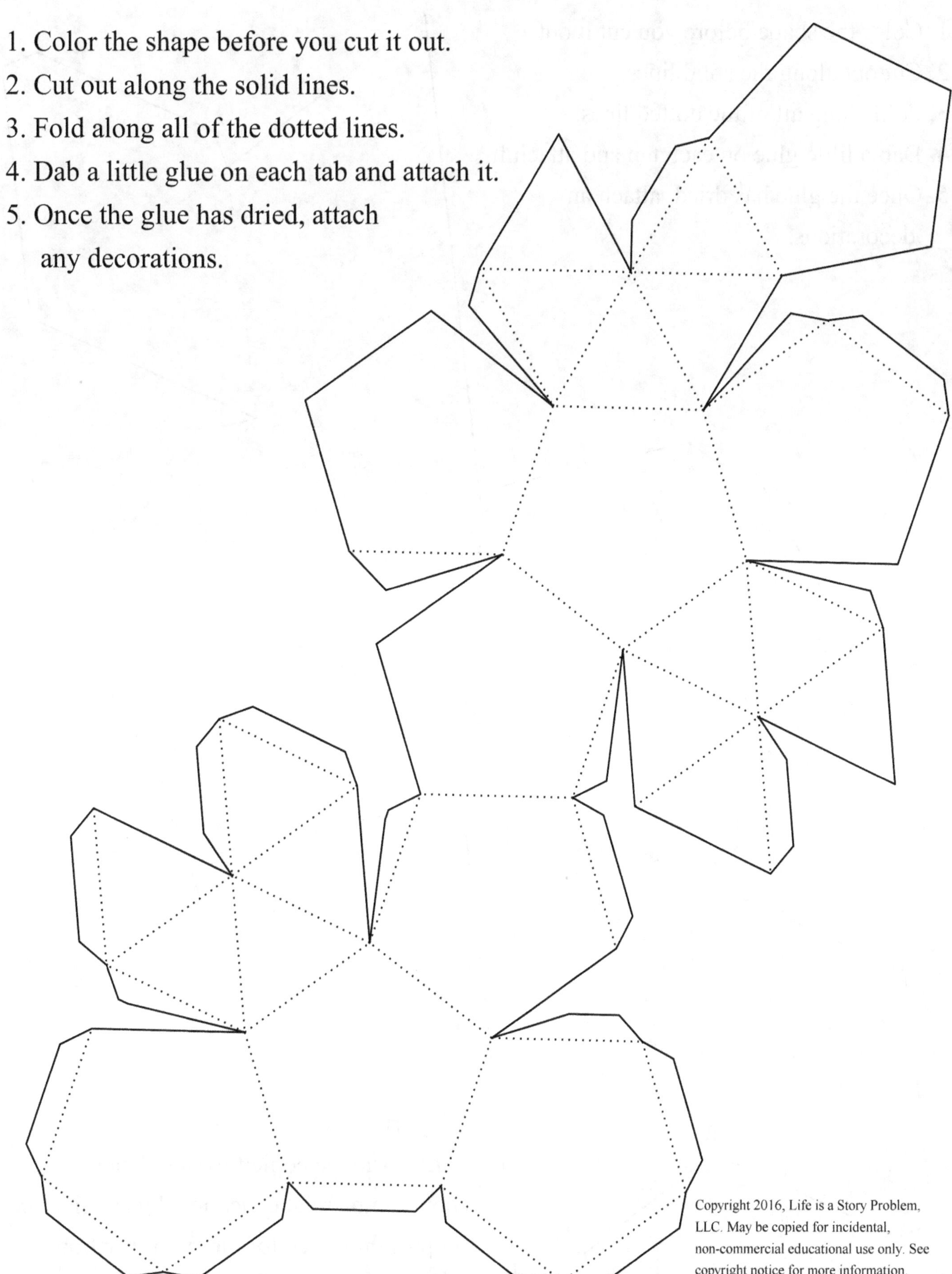

Triaugmented dodecahedron

1. Color the shape before you cut it out.
2. Cut out along the solid lines.
3. Fold along all of the dotted lines.
4. Dab a little glue on each tab and attach it.
5. Once the glue has dried, attach
   any decorations.

Metabidiminished icosahedron

1. Color the shape before you cut it out.
2. Cut out along the solid lines.
3. Fold along all of the dotted lines.
4. Dab a little glue on each tab and attach it.
5. Once the glue has dried, attach any
   decorations.

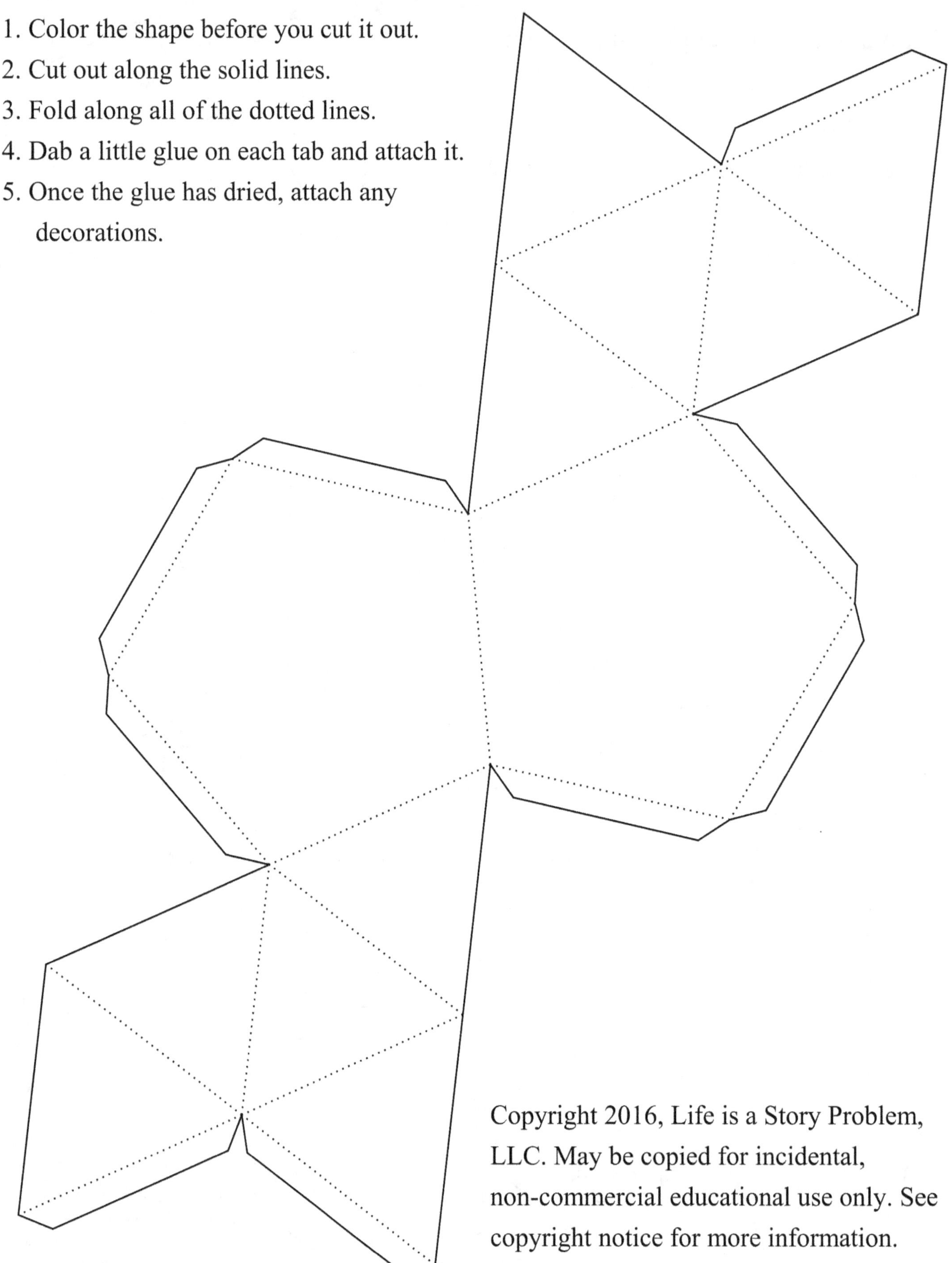

 Geometric Nets Mega Project Book, Tabbed By David E. McAdams

Tridiminished icosahedron

1. Color the shape before you cut it out.
2. Cut out along the solid lines.
3. Fold along all of the dotted lines.
4. Dab a little glue on each tab and attach it.
5. Once the glue has dried, attach any decorations.

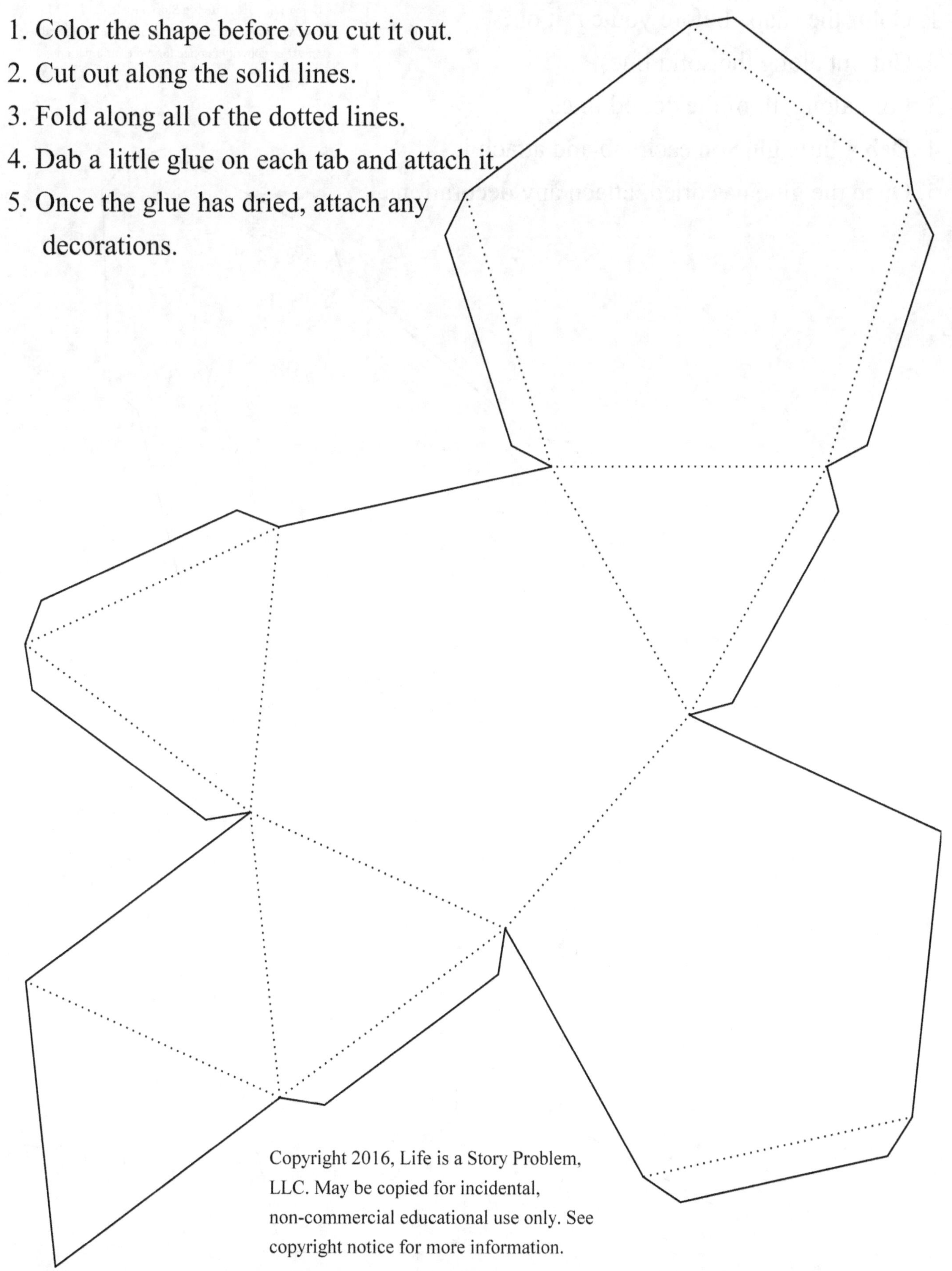

Augmented tridiminished icosahedron

1. Color the shape before you cut it out.

2. Cut out along the solid lines.

3. Fold along all of the dotted lines.

4. Dab a little glue on each tab and attach it.

5. Once the glue has dried, attach any decorations.

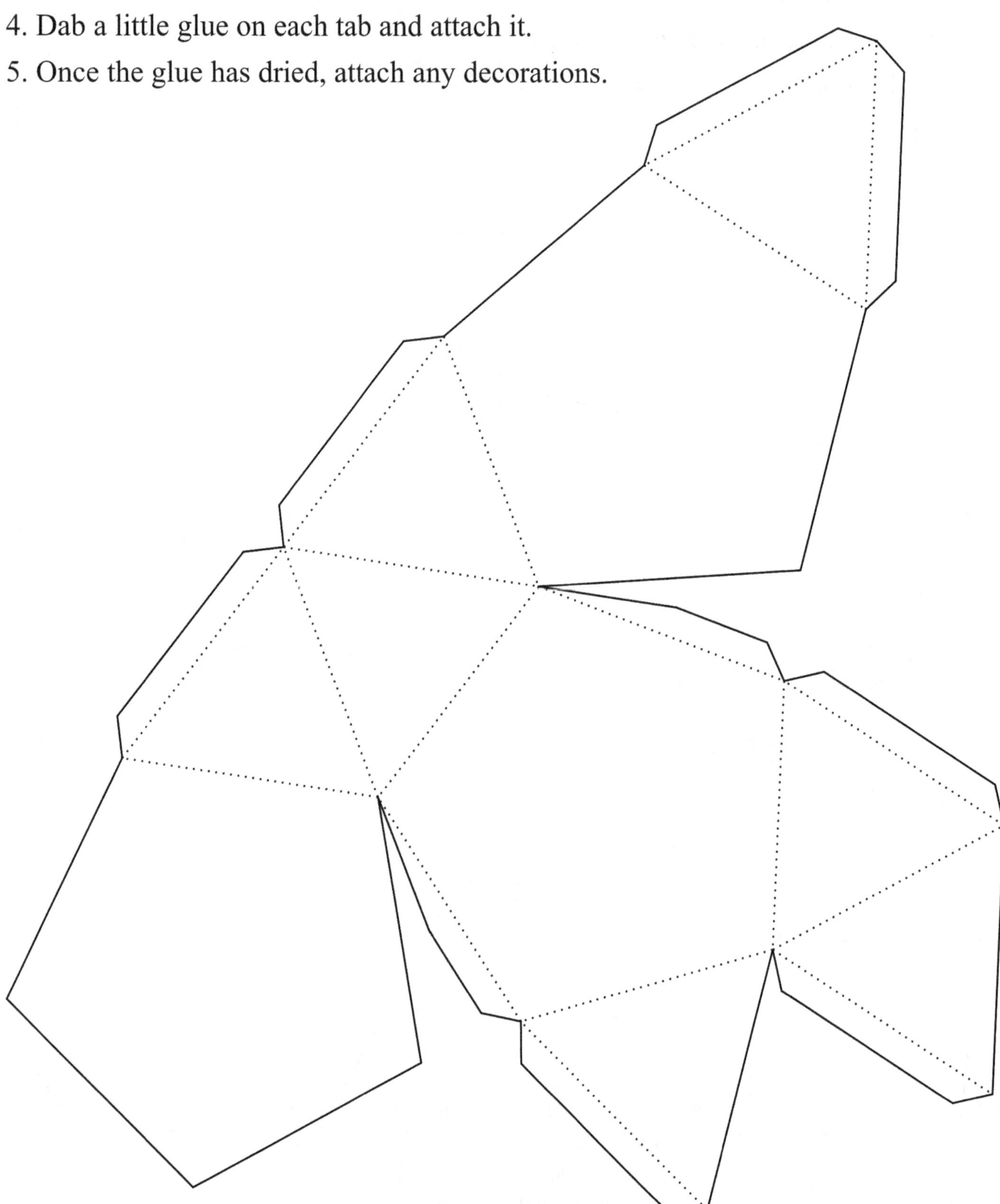

   Geometric Nets Mega Project Book, Tabbed By David E. McAdams

Augmented truncated tetrahedron

1. Color the shape before you cut it out.
2. Cut out along the solid lines.
3. Fold along all of the dotted lines.
4. Dab a little glue on each tab and attach it.
5. Once the glue has dried, attach any decorations.

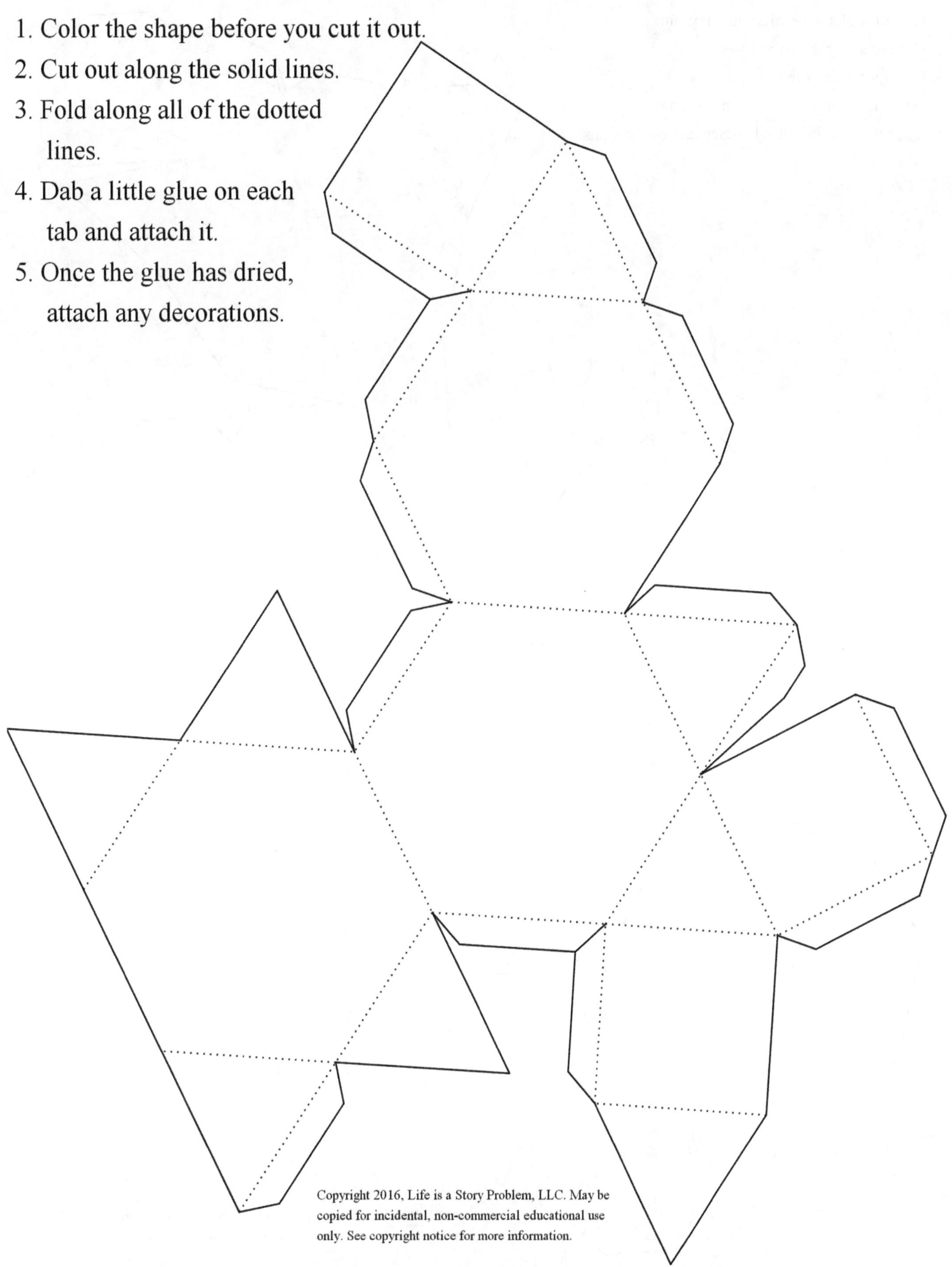

Augmented truncated cube

1. Color the shape before you cut it out.
2. Cut out along the solid lines.
3. Fold along all of the dotted lines.
4. Dab a little glue on each tab and attach it.
5. Once the glue has dried, attach any decorations.

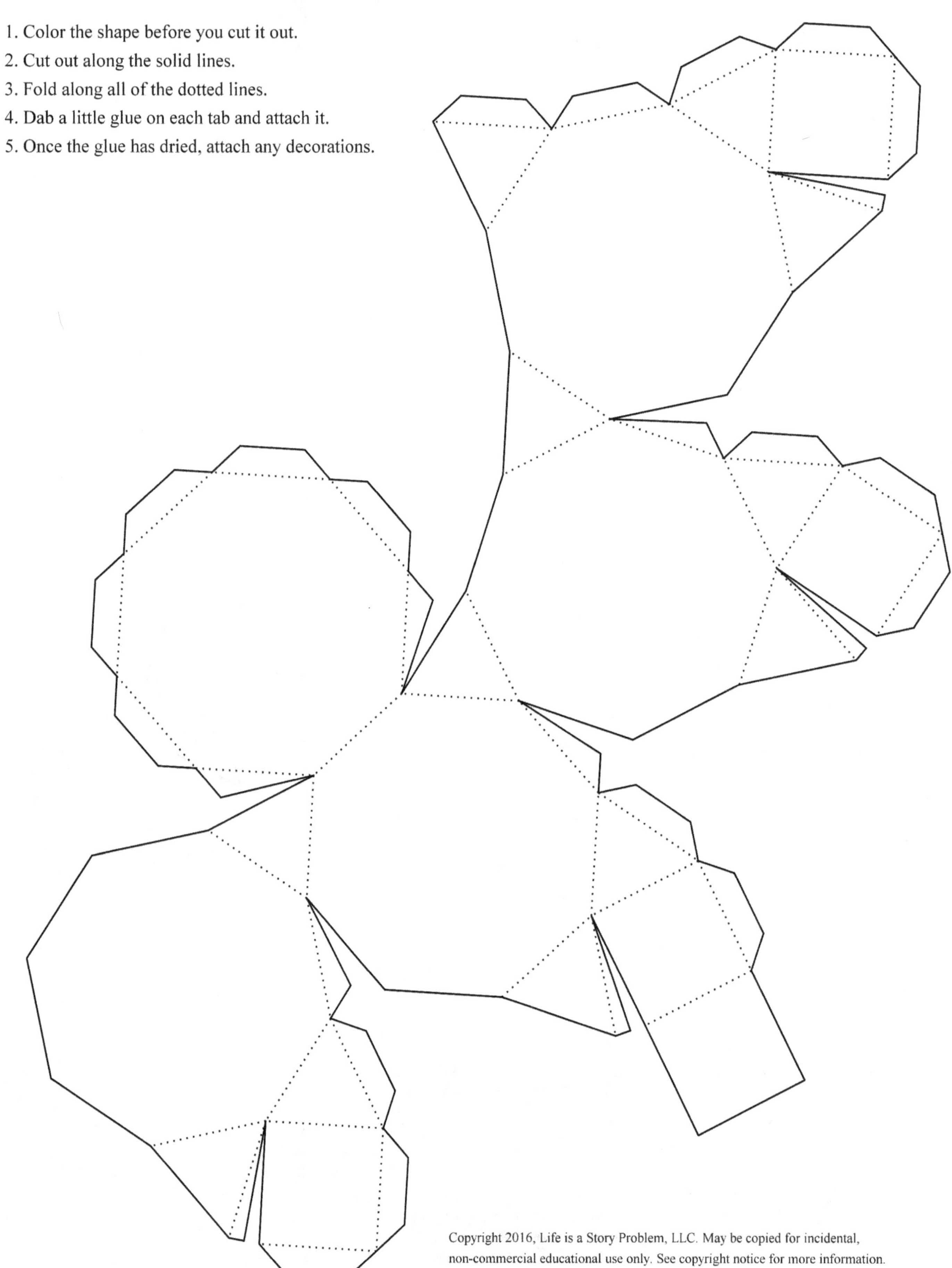

 Geometric Nets Mega Project Book, Tabbed By David E. McAdams

Biaugmented truncated cube

1. Color the shape before you cut it out.
2. Cut out along the solid lines.
3. Fold along all of the dotted lines.
4. Dab a little glue on each tab and attach it.
5. Once the glue has dried, attach
   any decorations.

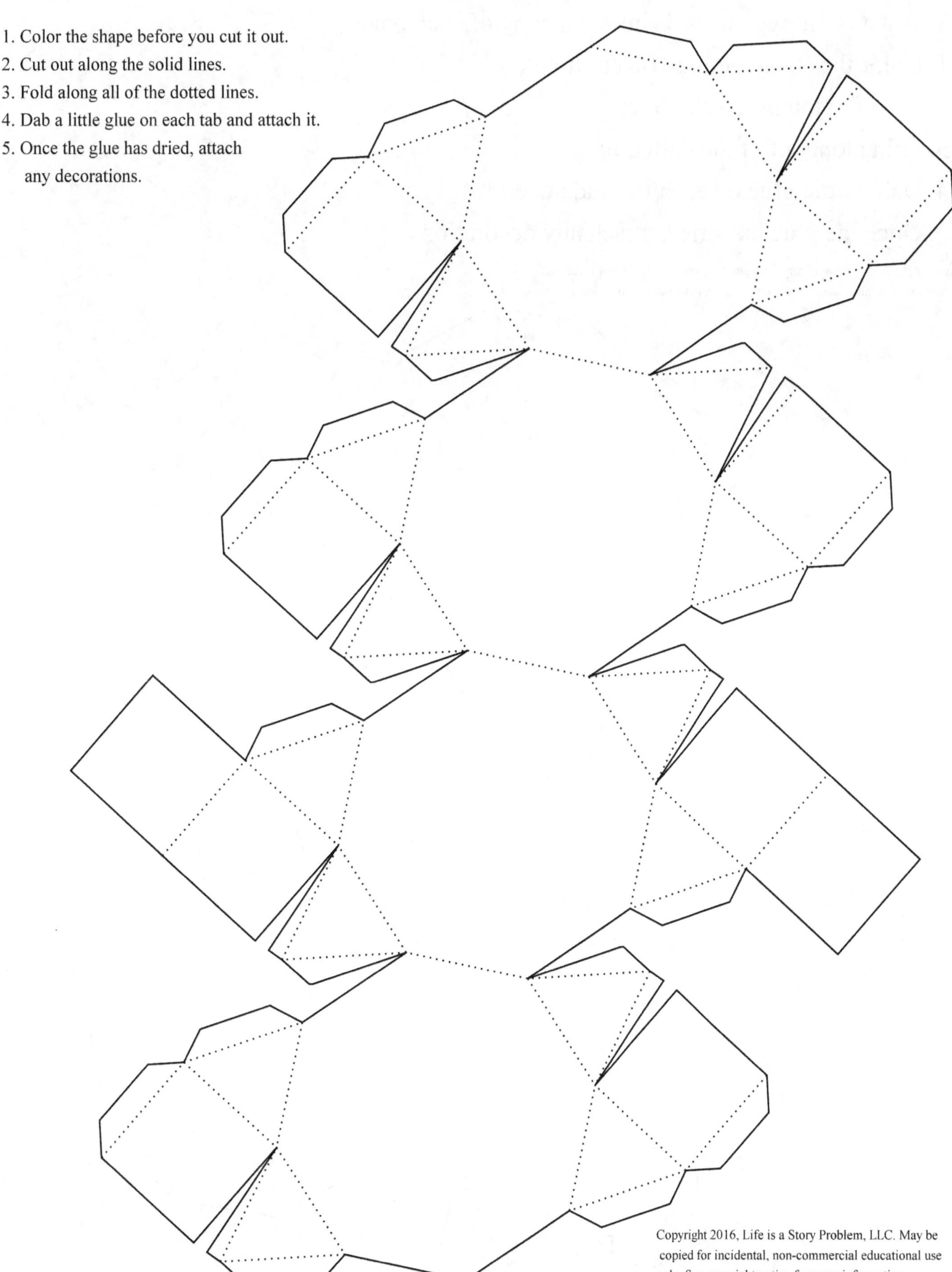

Augmented truncated dodecahedron

This net is on two pages. Print out a copy of each page.

1. Color the shape before you cut it out.

2. Cut out along the solid lines.

3. Fold along all of the dotted lines.

4. Dab a little glue on each tab and attach it.

5. Once the glue has dried, attach any decorations.

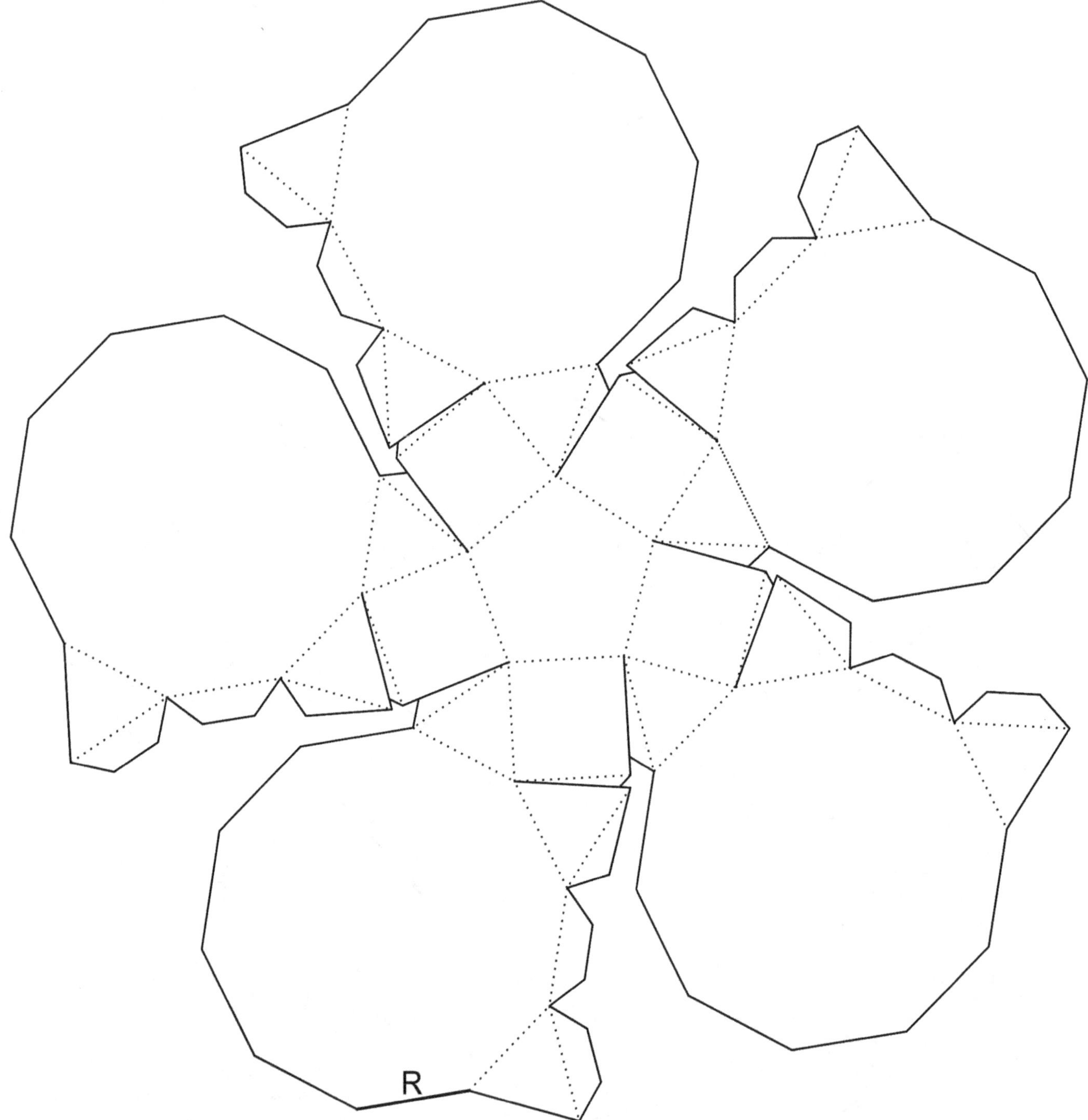

Geometric Nets Mega Project Book, Tabbed By David E. McAdams

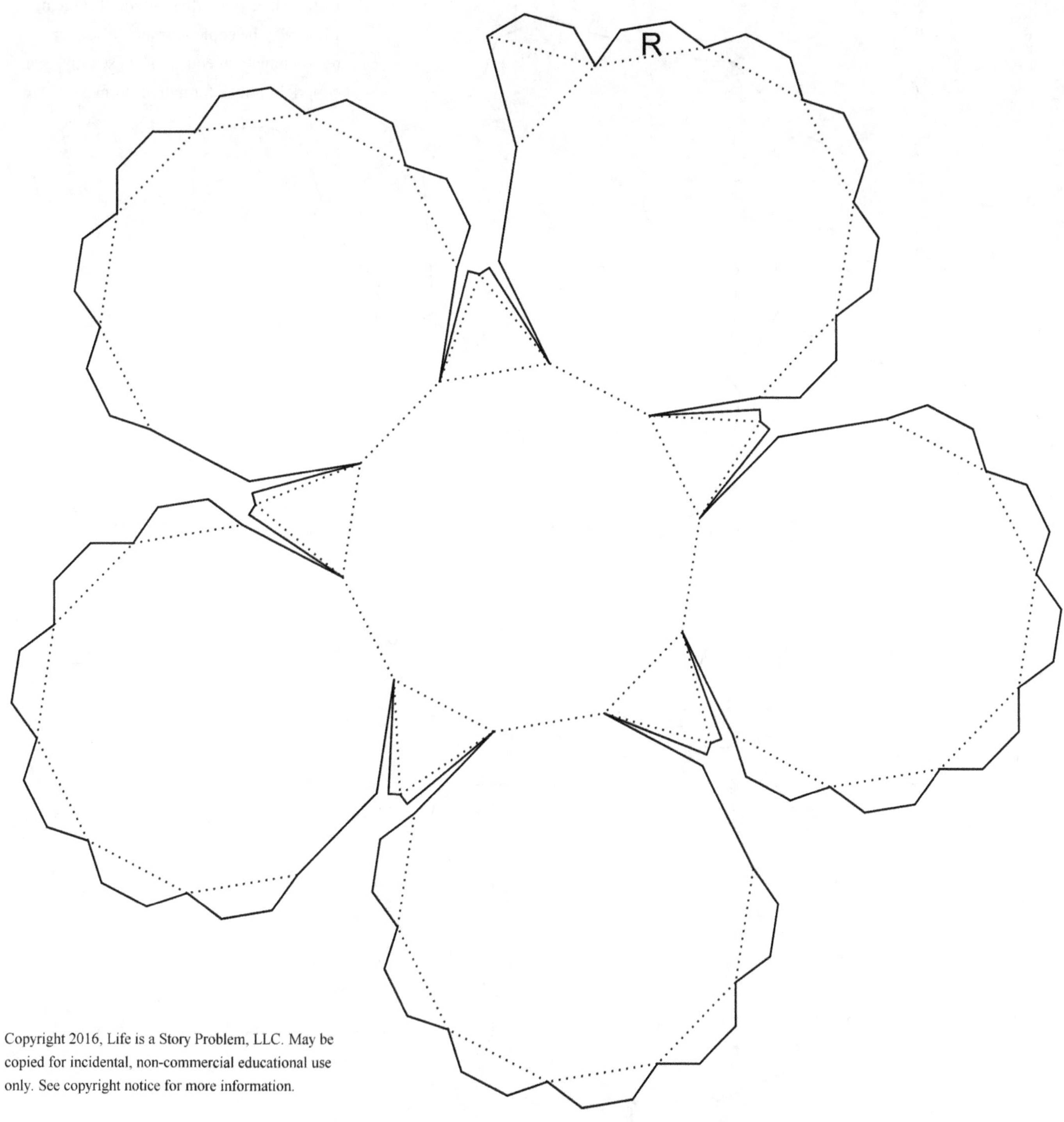
R

Parabiaugmented truncated dodecahedron

This net is on two pages. Print out a copy of each page.

1. Color the shape before you cut it out.

2. Cut out along the solid lines.

3. Fold along all of the dotted lines.

4. Dab a little glue on each tab and attach it.

5. Once the glue has dried, attach any decorations.

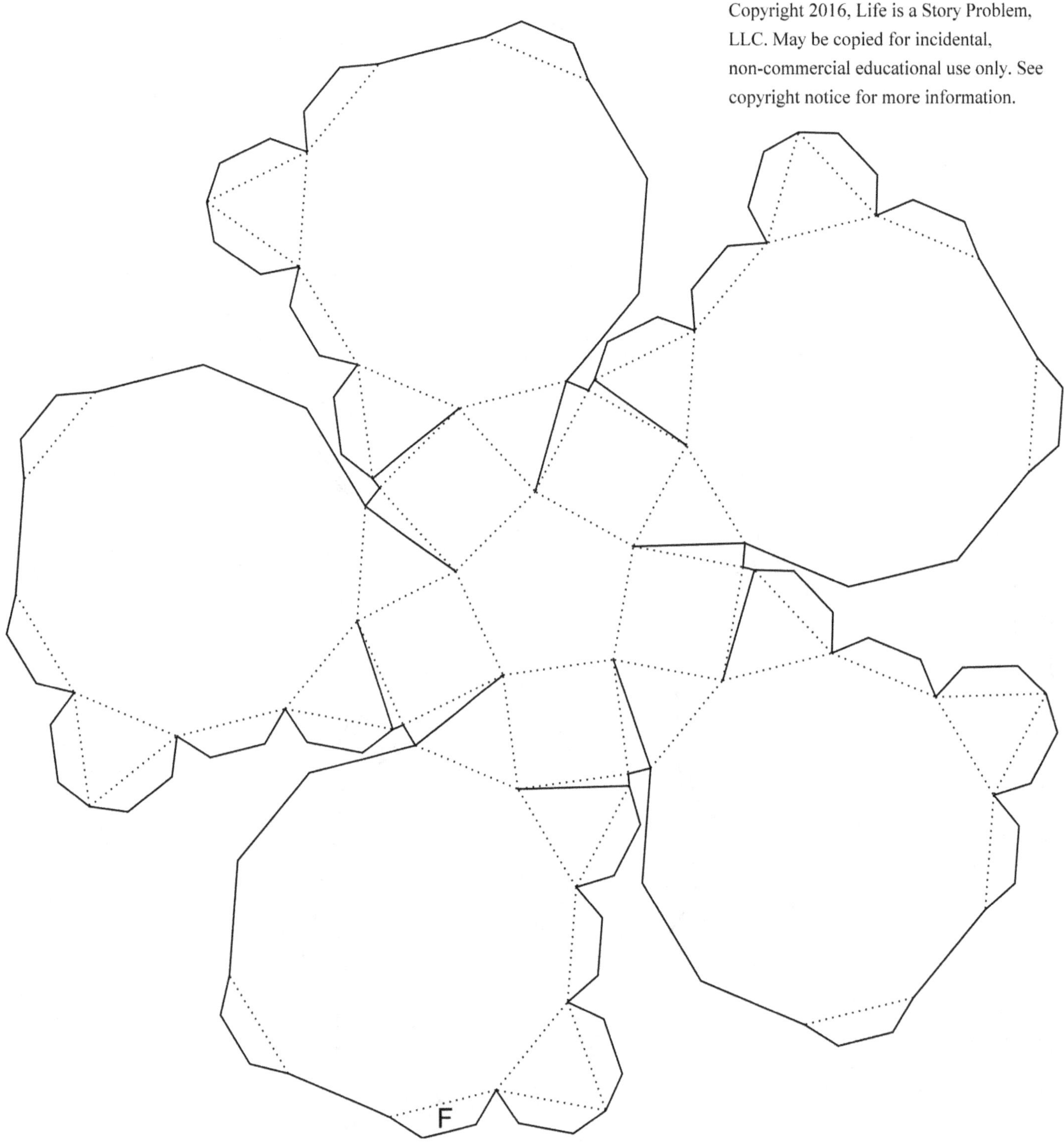

 Geometric Nets Mega Project Book, Tabbed By David E. McAdams

2nd page, Parabiaugmented truncated dodecahedron

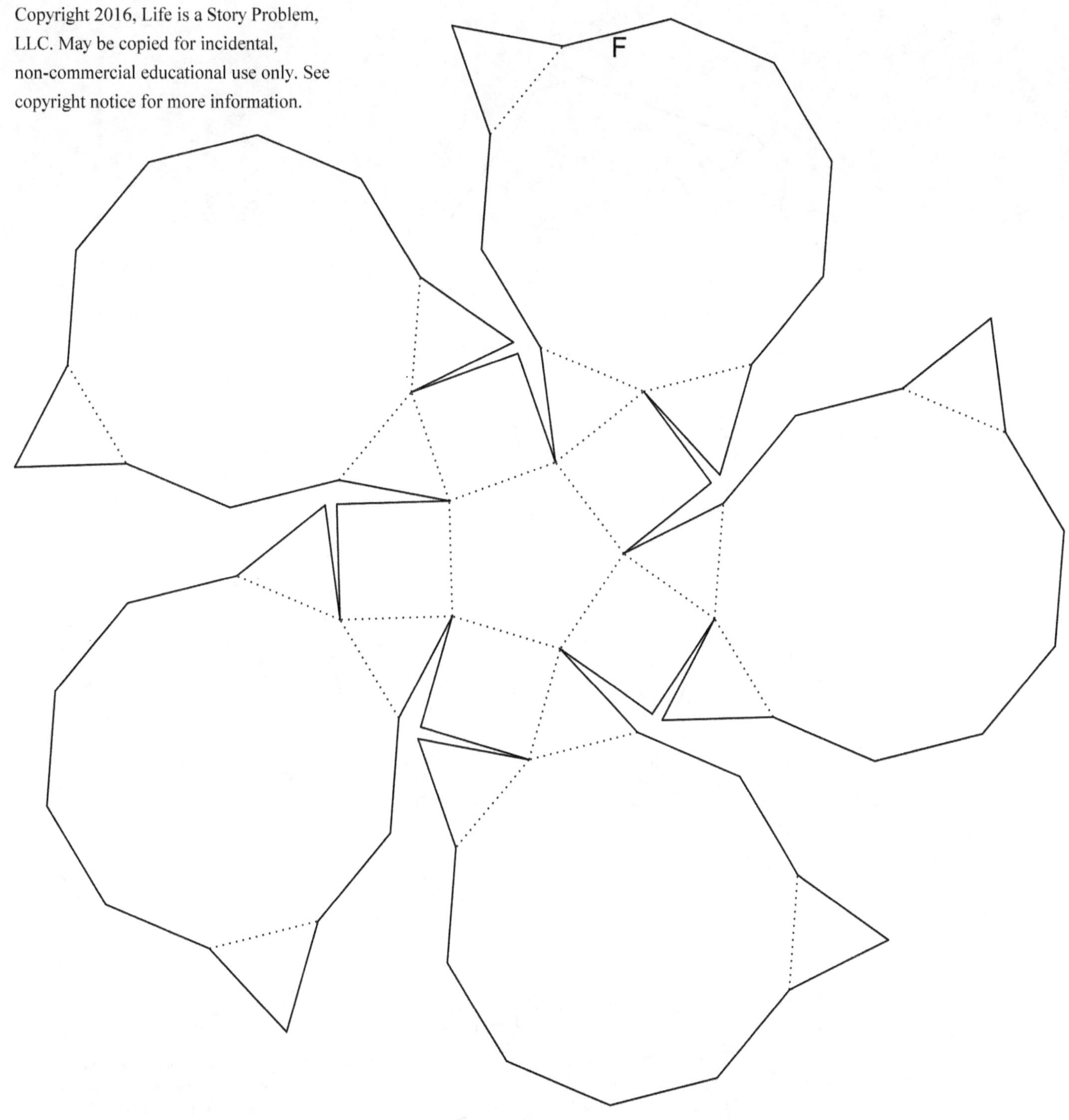

Metabiaugmented truncated dodecahedron

This net is on two pages. Print out a copy of each page.

1. Color the shape before you cut it out.
2. Cut out along the solid lines.
3. Fold along all of the dotted lines.
4. Dab a little glue on each tab and attach it.
5. Once the glue has dried, attach any decorations.

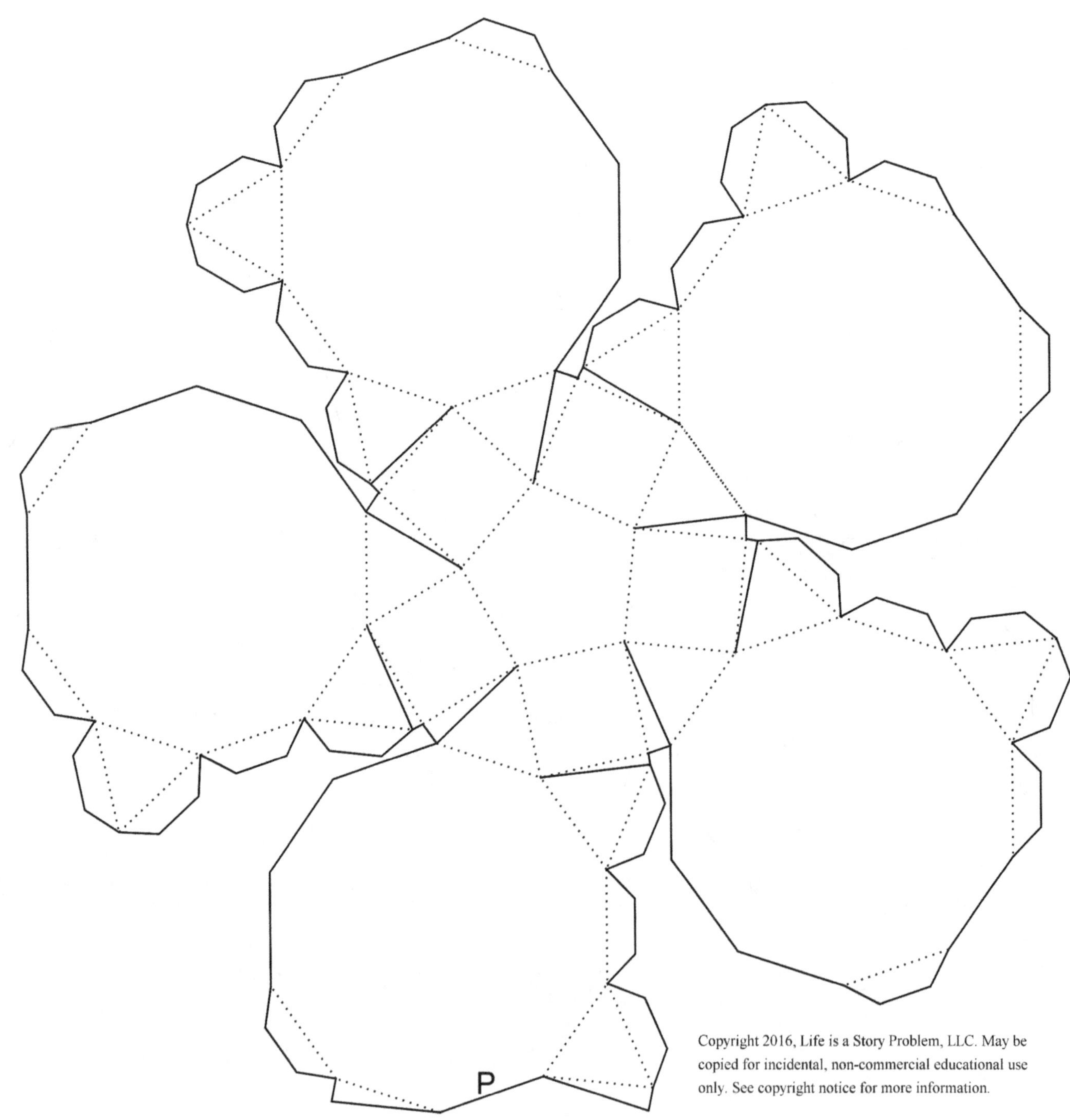

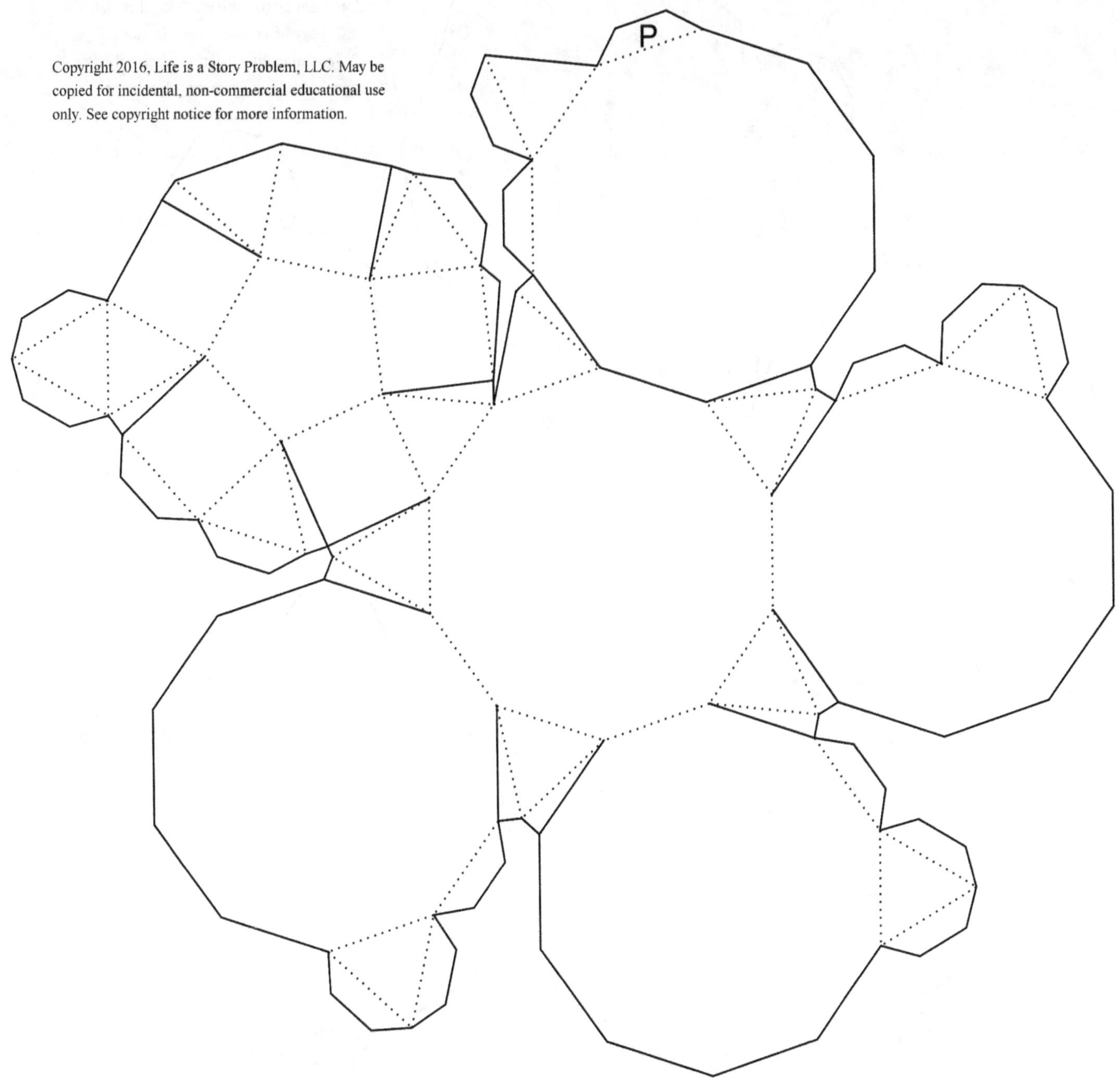
P

Triaugmented truncated dodecahedron

This net is on two pages. Print out a copy of each page.

1. Color the shape before you cut it out.

2. Cut out along the solid lines.

3. Fold along all of the dotted lines.

4. Dab a little glue on each tab and attach it.

5. Once the glue has dried, attach any decorations.

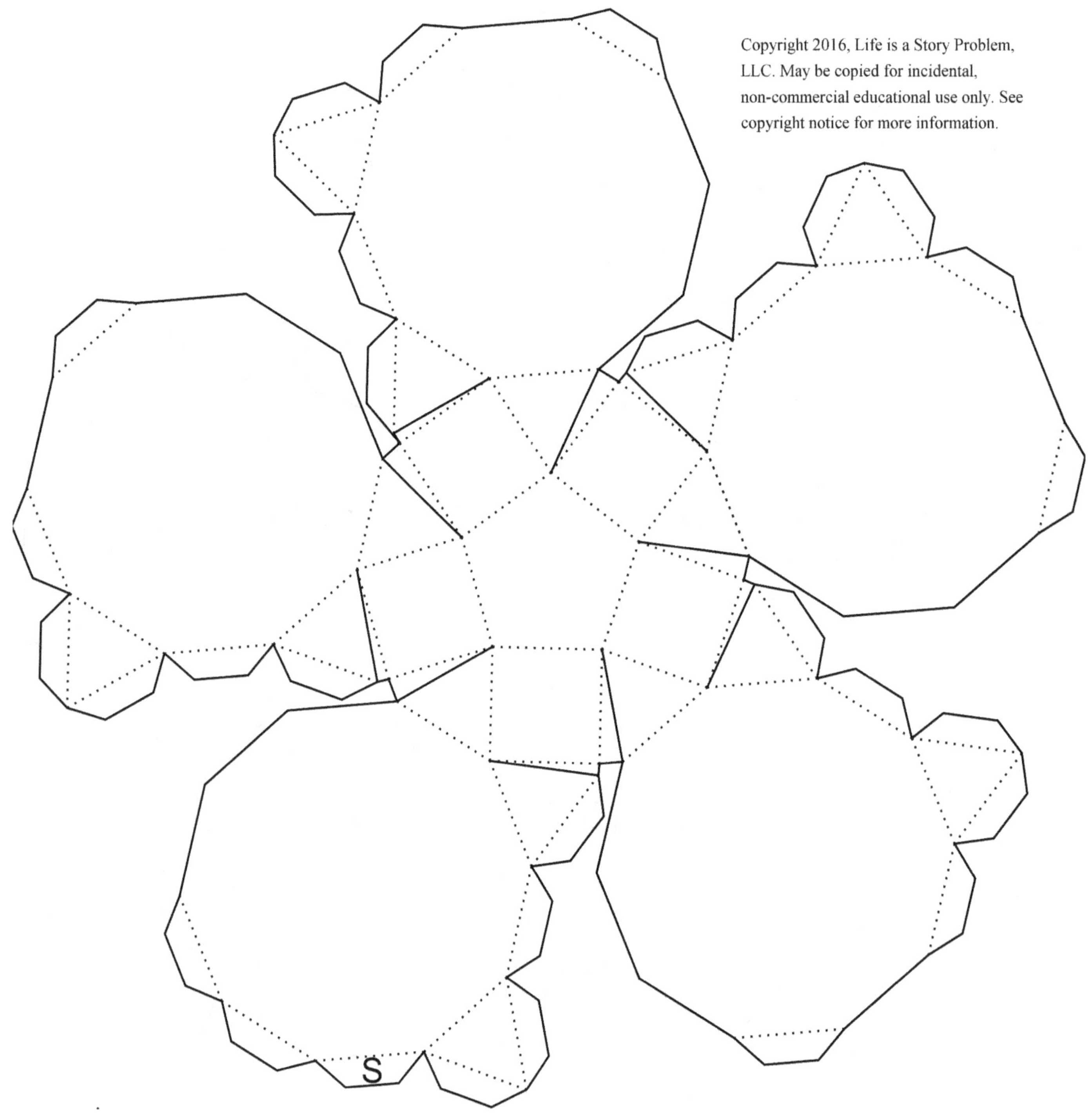

2nd page, Triaugmented truncated dodecahedron

This net is on two pages. Print out a copy
of each page.

1. Color the shape before you cut it out.
2. Cut out along the solid lines.
3. Fold along all of the dotted lines.
4. Dab a little glue on each tab and
   attach it.
5. Once the glue has dried, attach
   any decorations.

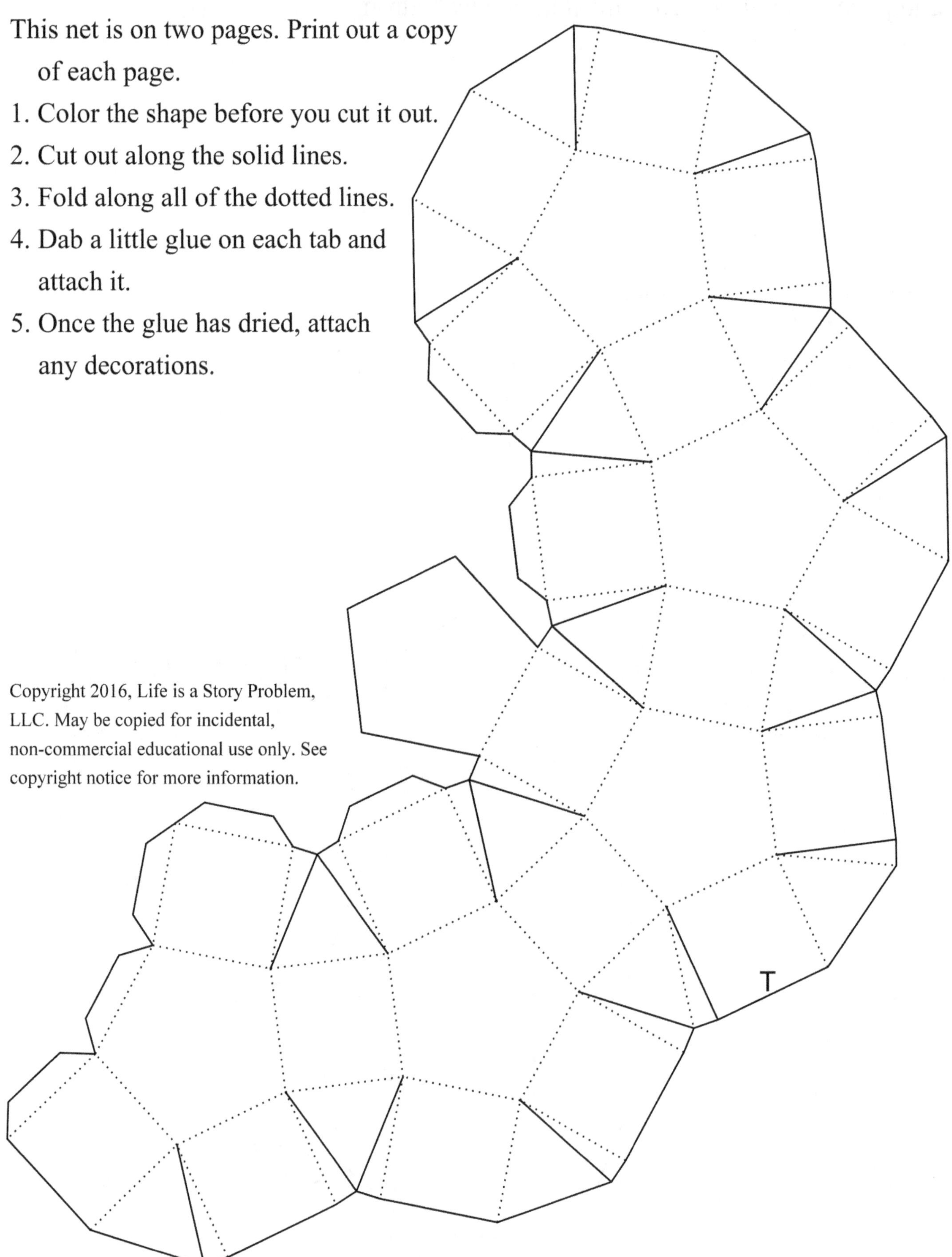

2nd page, Gyrate rhombicosidodecahedron

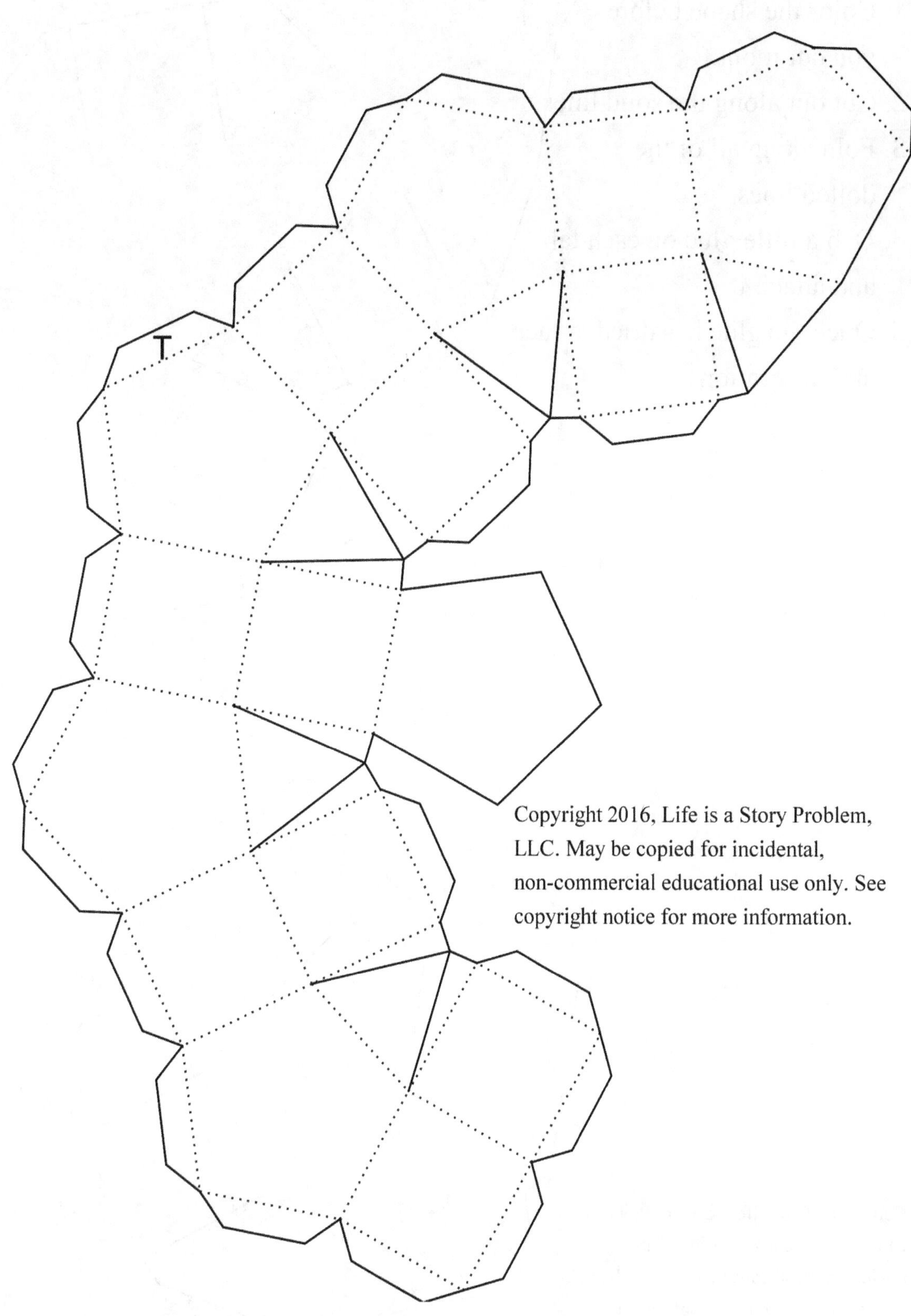

Parabigyrate rhombicosidodecahedron

This net is on two pages. Print out a copy of each page.

1. Color the shape before you cut it out.
2. Cut out along the solid lines.
3. Fold along all of the dotted lines.
4. Dab a little glue on each tab and attach it.
5. Once the glue has dried, attach any decorations.

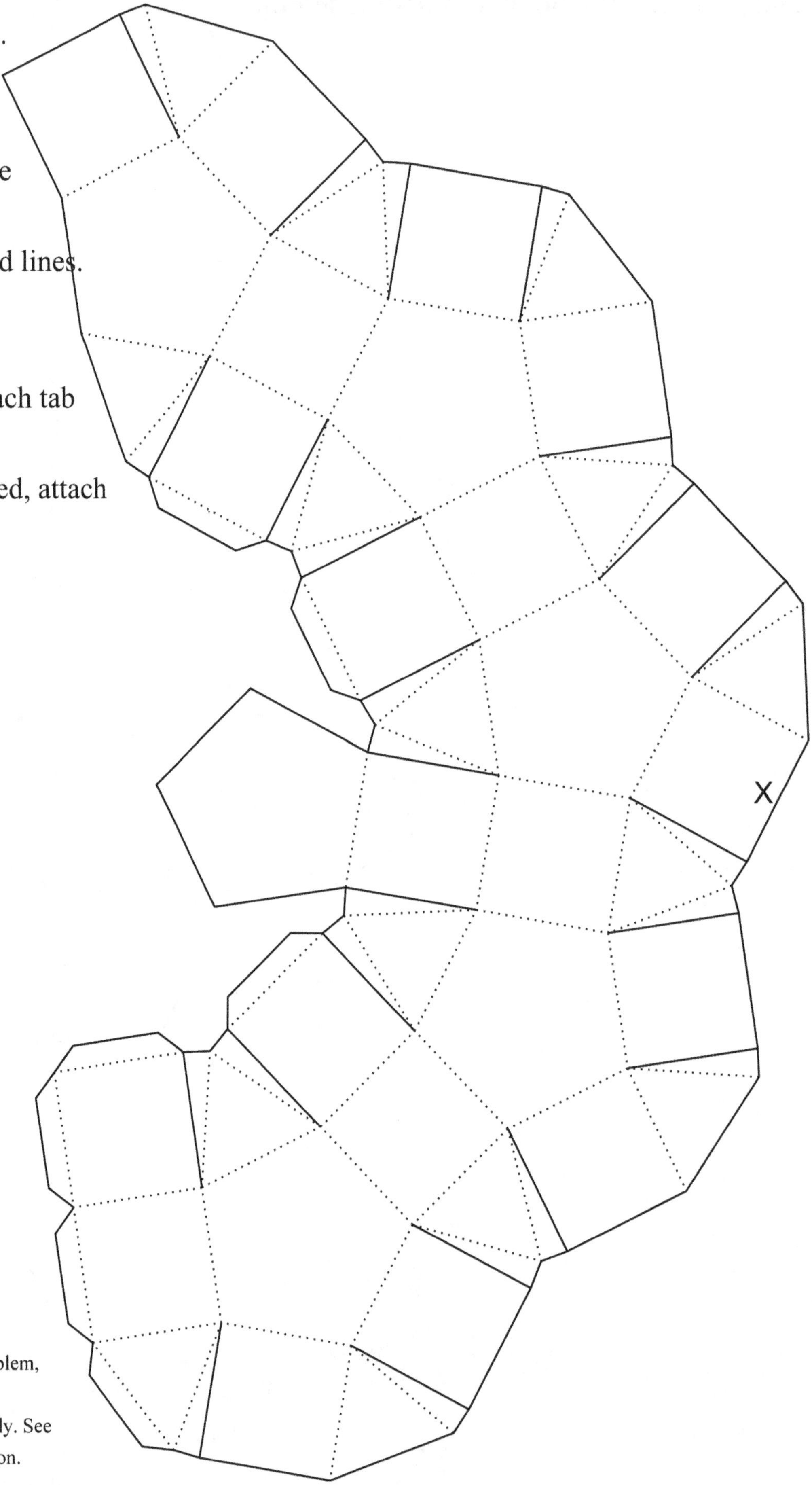

 Geometric Nets Mega Project Book, Tabbed By David E. McAdams

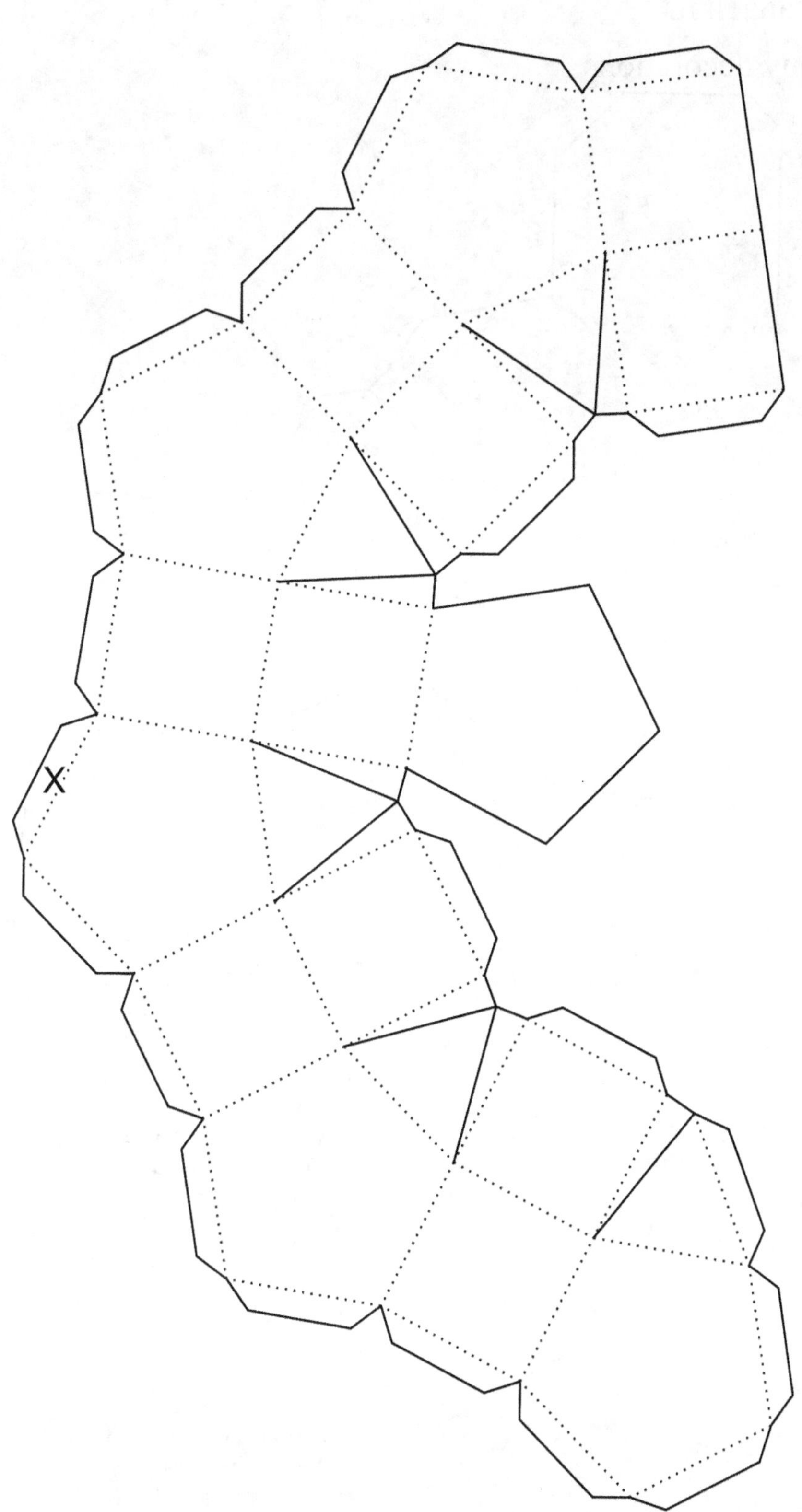

Metabigyrate rhombicosidodecahedron

This net is on three pages. Print out a copy of each page.

1. Color the shape before you cut it out. 2. Cut out along the solid lines.

3. Fold along all of the dotted lines.

4. Dab a little glue on each tab and attach it.

5. Once the glue has dried, attach any decorations.

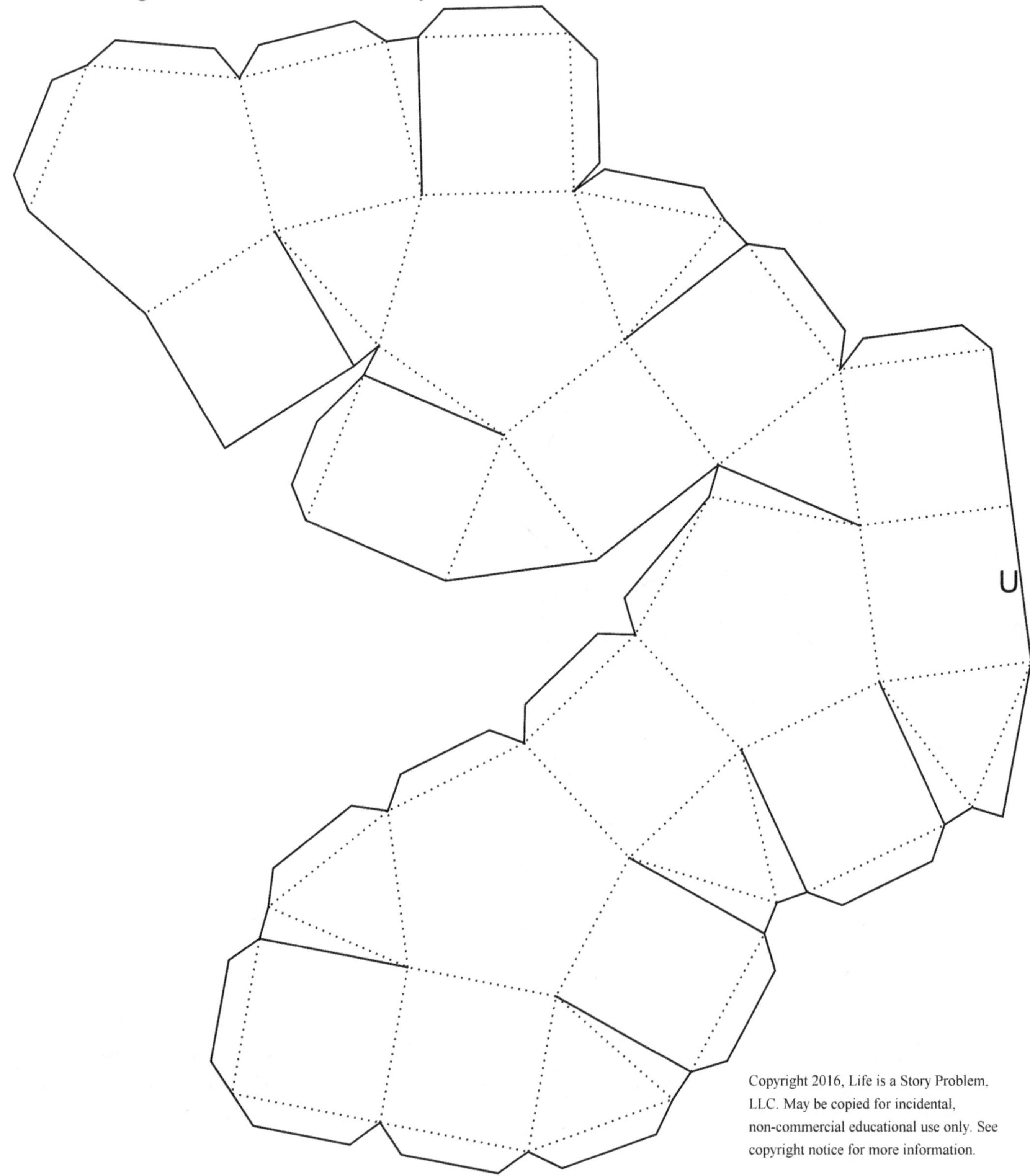

Geometric Nets Mega Project Book, Tabbed By David E. McAdams

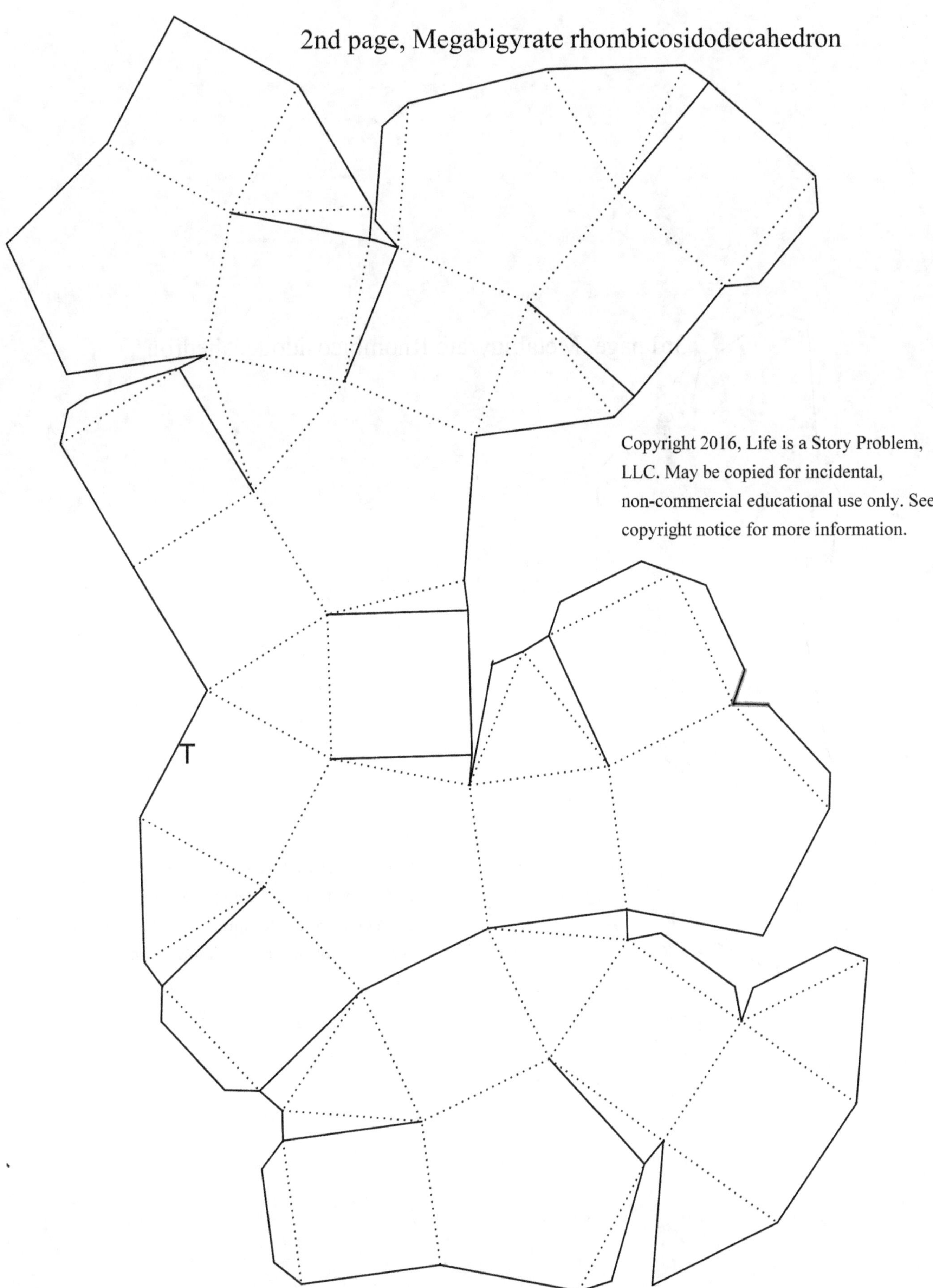

3rd page, Metabigyrate Rhombicosidodecahedron

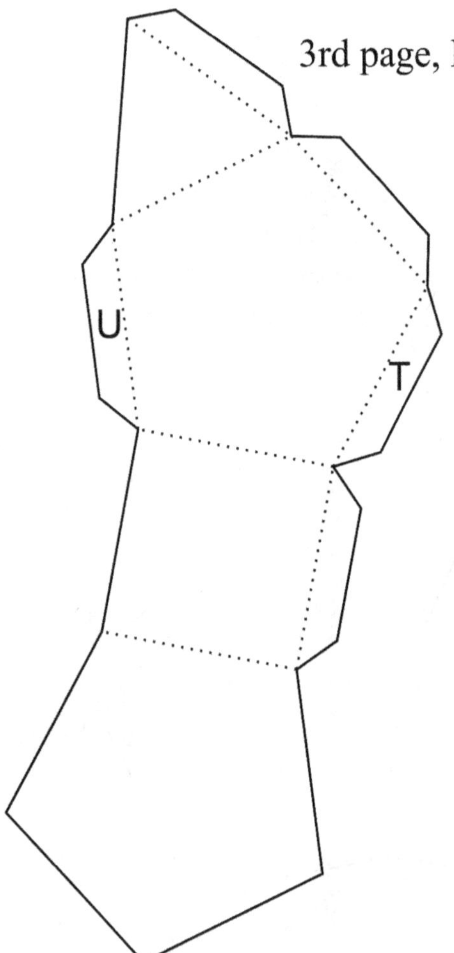

Geometric Nets Mega Project Book, Tabbed By David E. McAdams

Trigyrate rhombicosidodecahedron

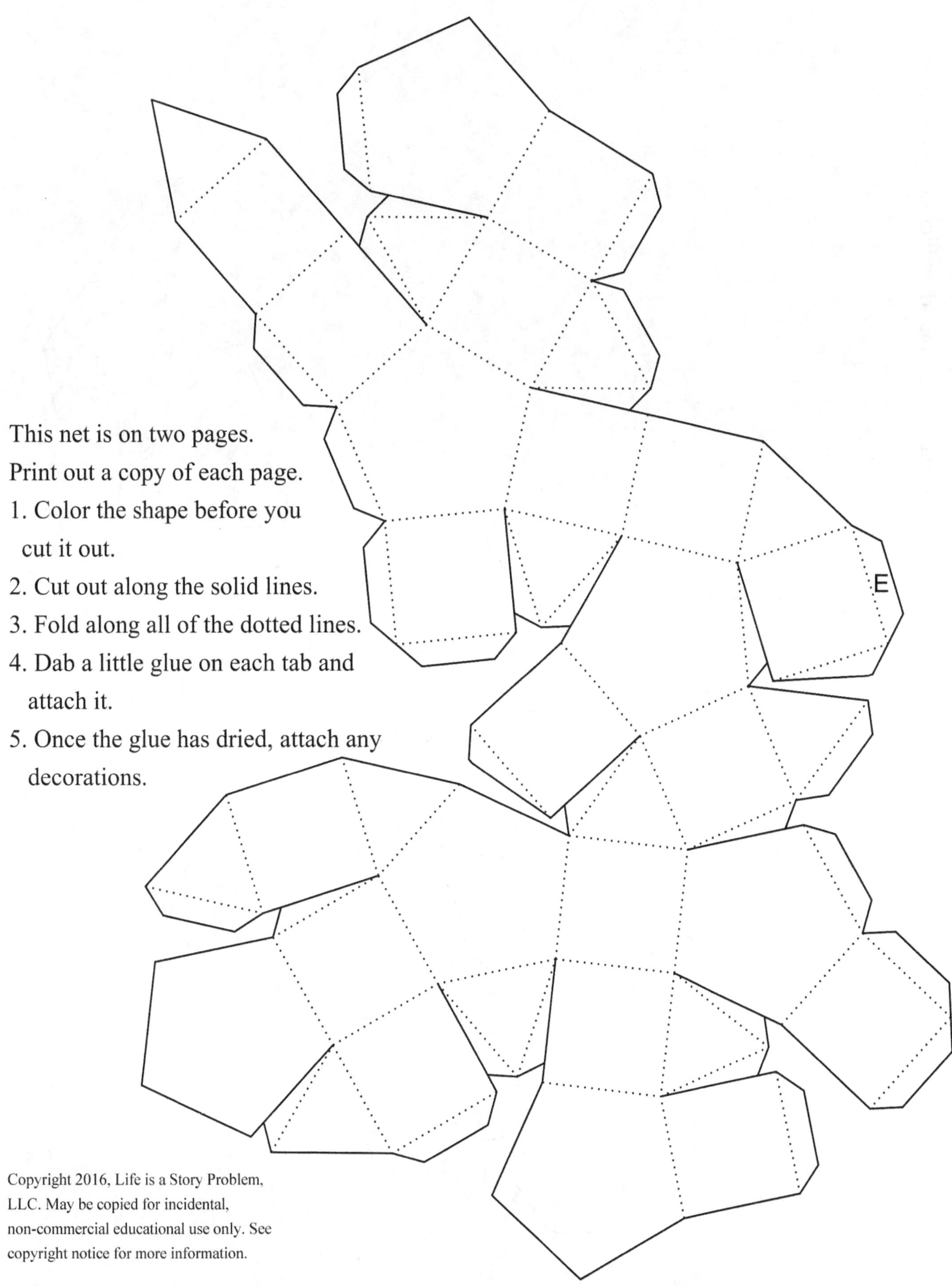

This net is on two pages.
Print out a copy of each page.

1. Color the shape before you cut it out.
2. Cut out along the solid lines.
3. Fold along all of the dotted lines.
4. Dab a little glue on each tab and attach it.
5. Once the glue has dried, attach any decorations.

Geometric Nets Mega Project Book, Tabbed By David E. McAdams

Diminished rhombicosidodecahedron

This net is on two pages. Print out a copy of each page.

1. Color the shape before you cut it out.

2. Cut out along the solid lines.

3. Fold along all of the dotted lines.

4. Dab a little glue on each tab and attach it.

5. Once the glue has dried, attach
   any decorations.

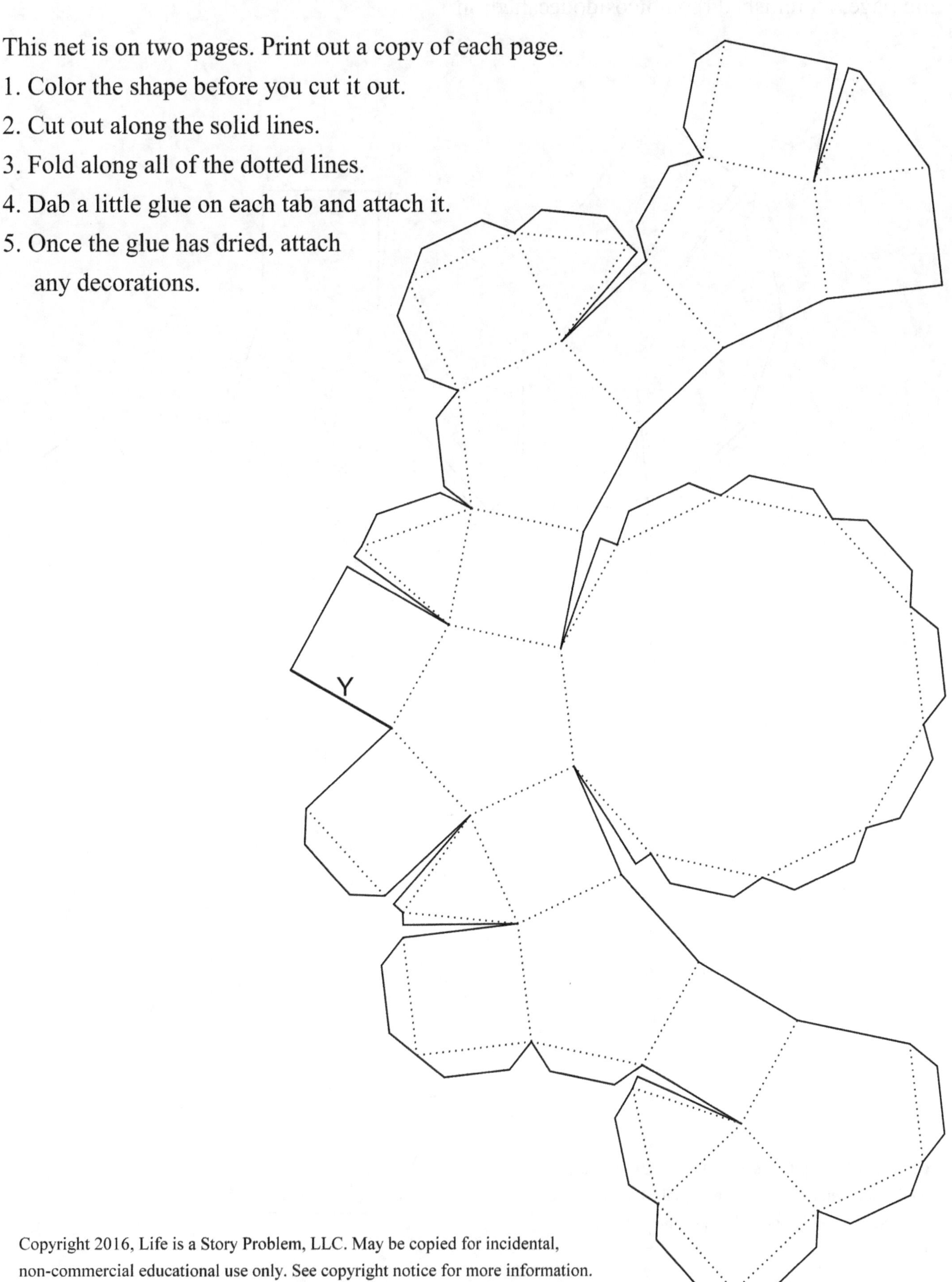

2nd page, Diminished rhombicosidodecahedron.

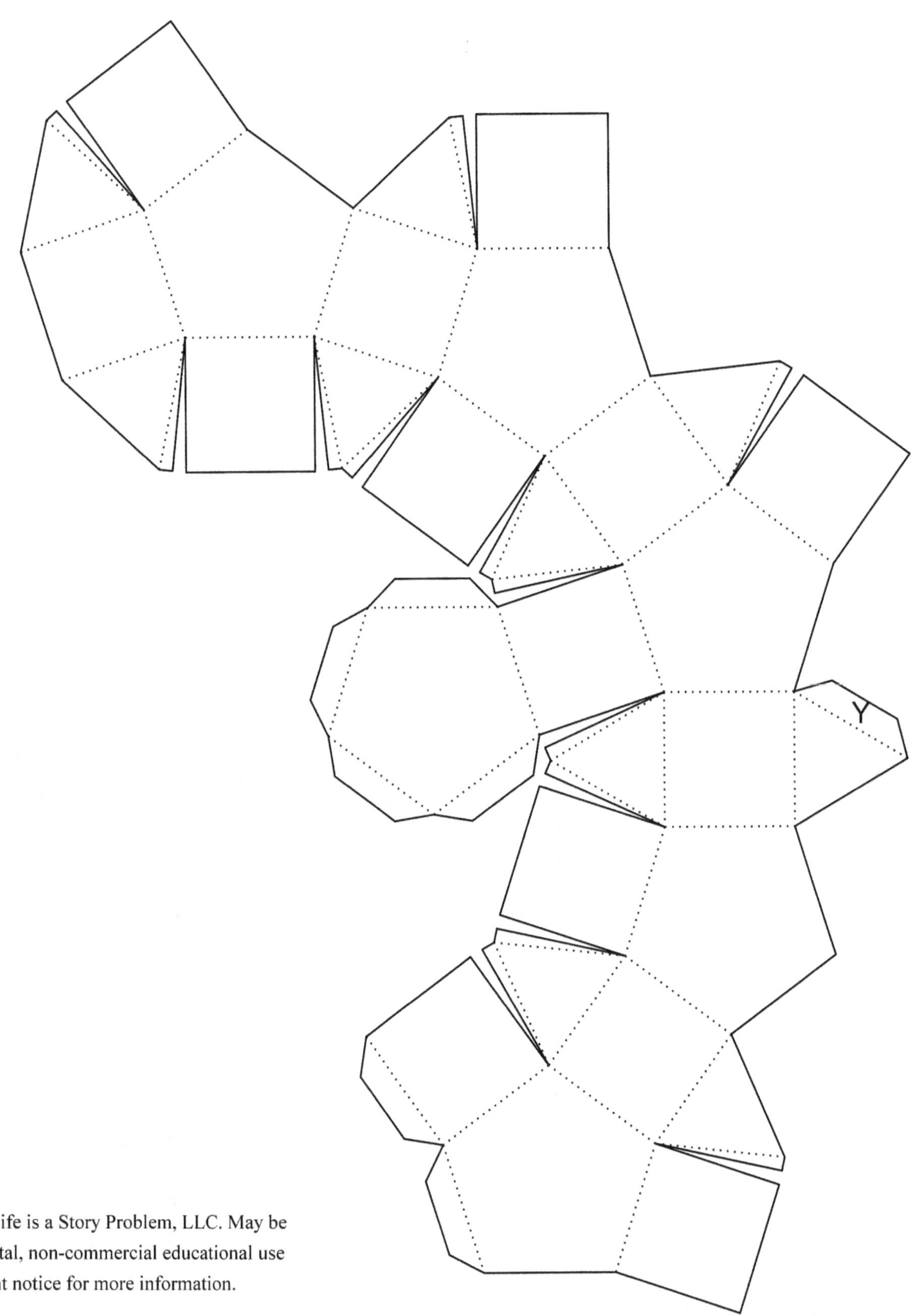

Geometric Nets Mega Project Book, Tabbed By David E. McAdams

Paragyrate diminished rhombicosidodecahedron

This net is on two pages.
Print out a copy of each
page.

1. Color the shape before
   you cut it out.
2. Cut out along the solid lines.
3. Fold along all of the
   dotted lines.
4. Dab a little glue on each tab
   and attach it.
5. Once the glue has dried, attach
   any decorations.

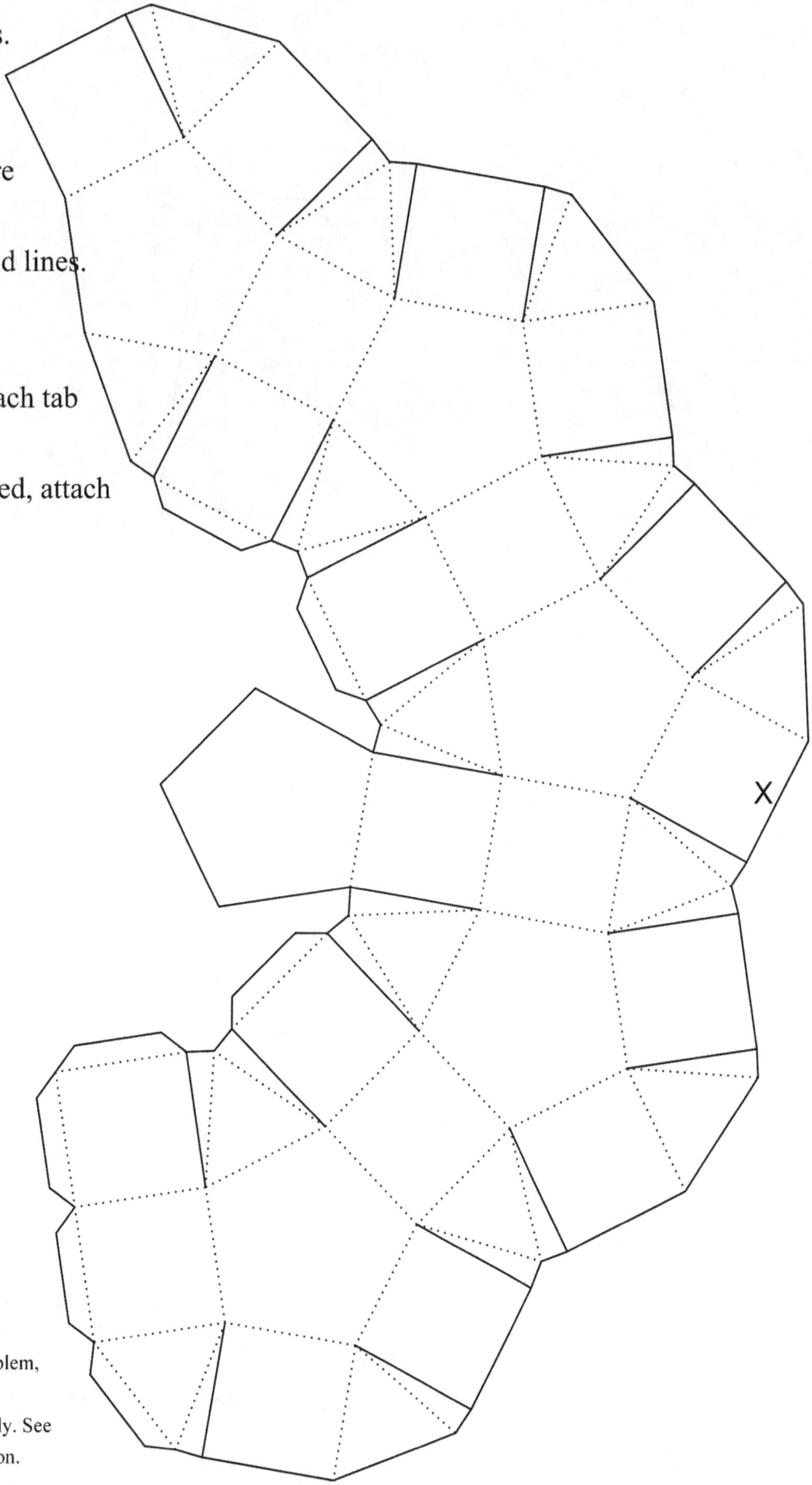

2nd page, Parabigyrate Rhombicosidodecahedron

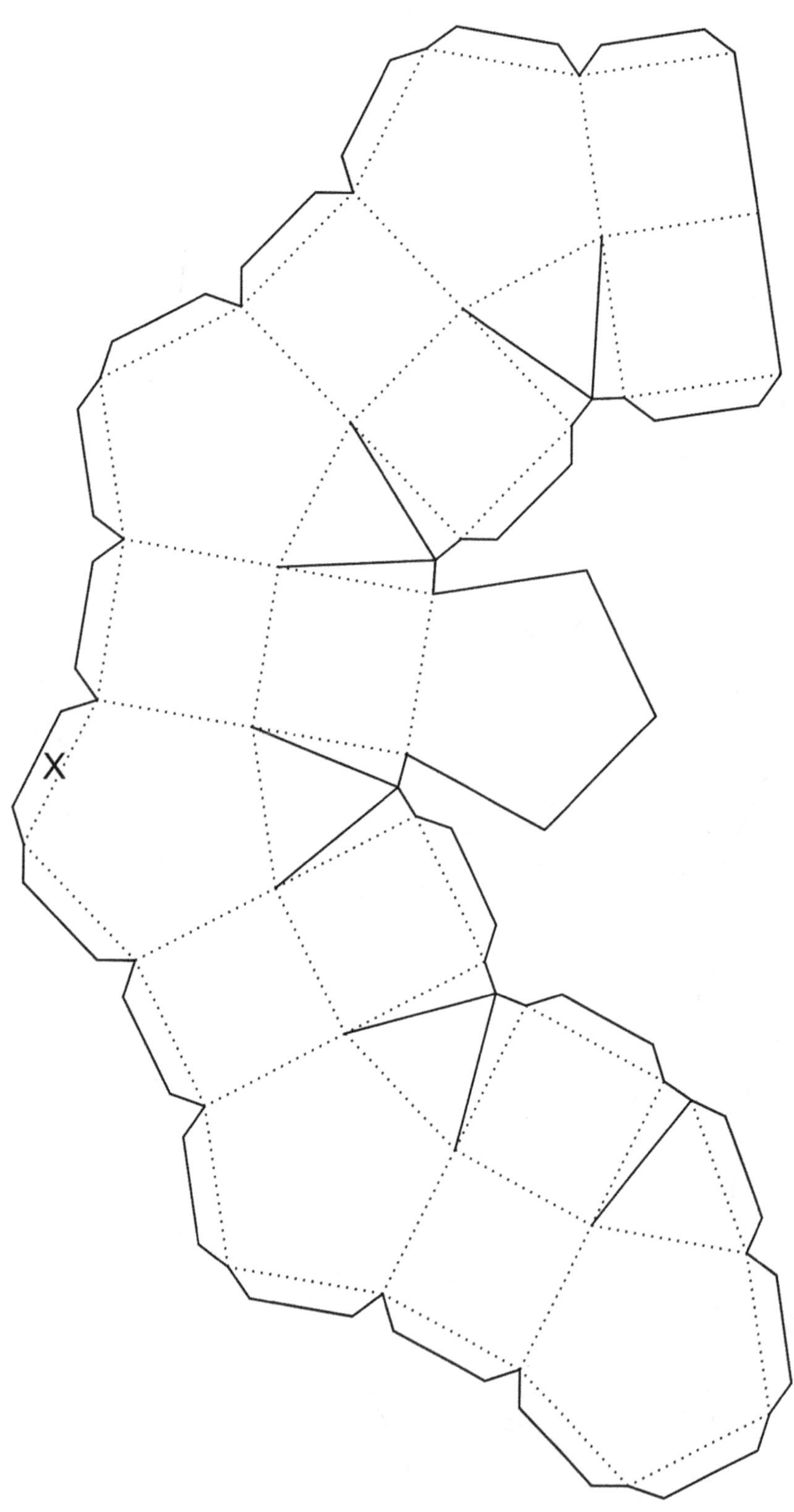

     Geometric Nets Mega Project Book, Tabbed By David E. McAdams

Metagyrate diminished rhombicosidodecahedron

This net is on two pages. Print out a copy of each page.

1. Color the shape before you cut it out.

2. Cut out along the solid lines.

3. Fold along all of the dotted lines.

4. Dab a little glue on each tab and attach it.

5. Once the glue has dried, attach any decorations.

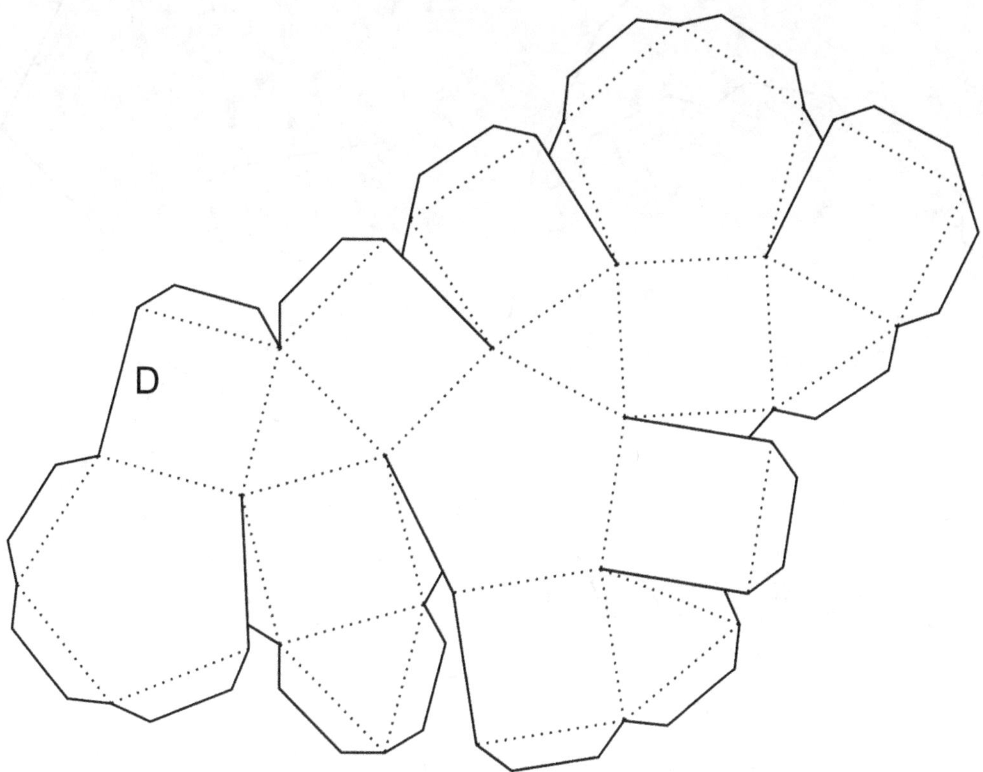

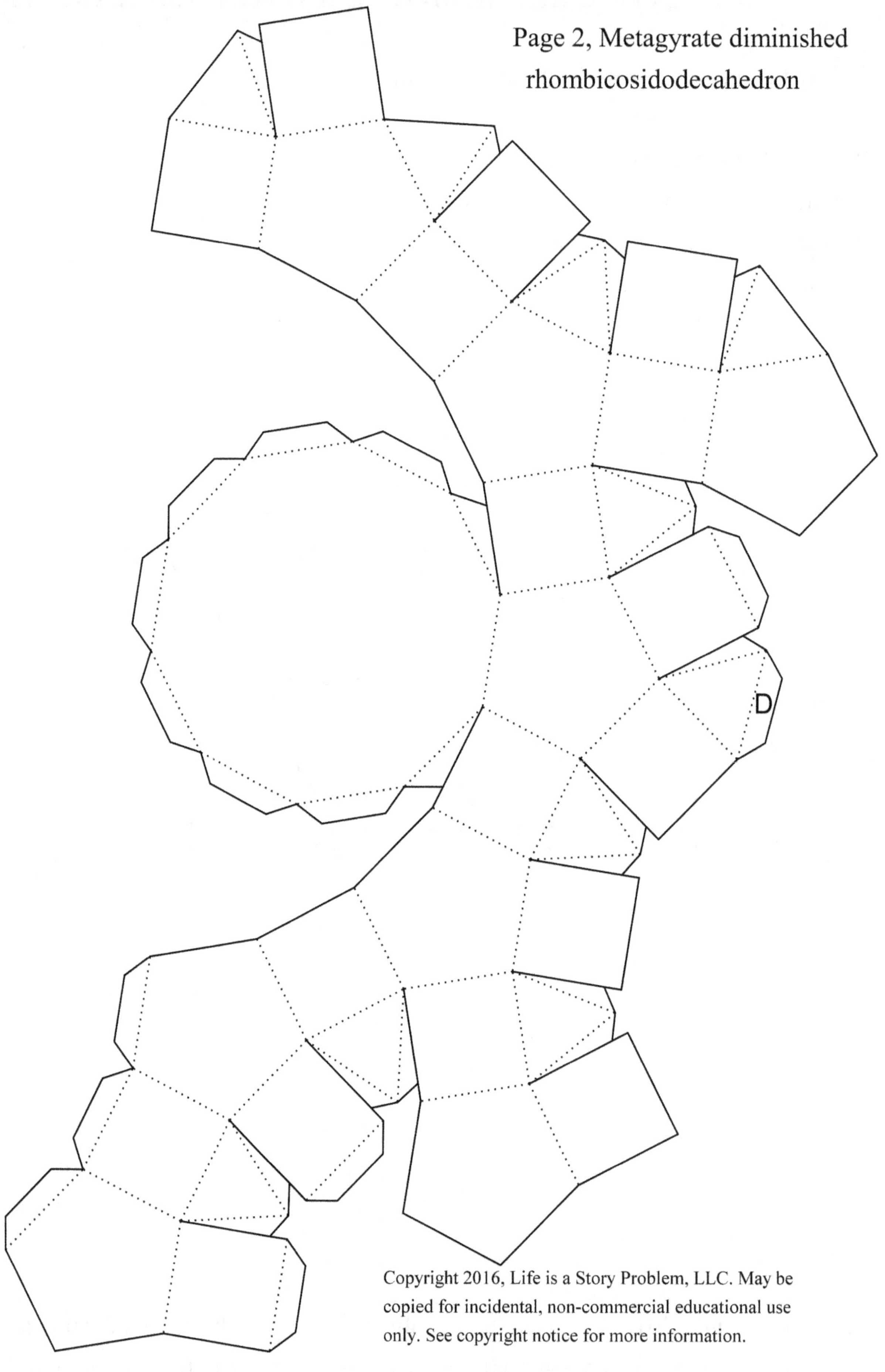

Geometric Nets Mega Project Book, Tabbed By David E. McAdams

Bigyrate diminished rhombicosidodecahedron

1. Color the shape before you cut it out.

2. Cut out along the solid lines.

3. Fold along all of the dotted lines.

4. Dab a little glue on each tab and attach it.

5. Once the glue has dried, attach any decorations.

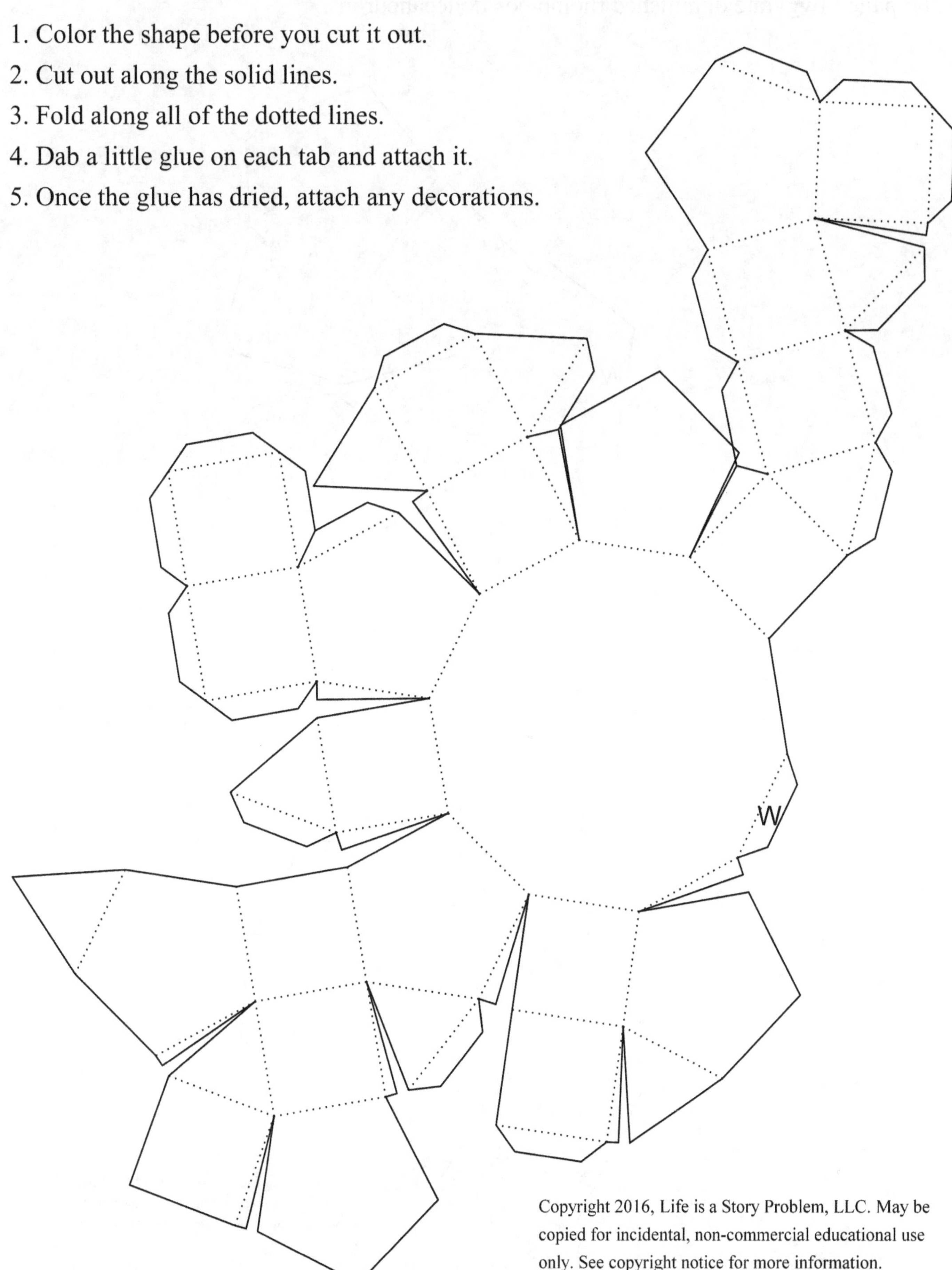

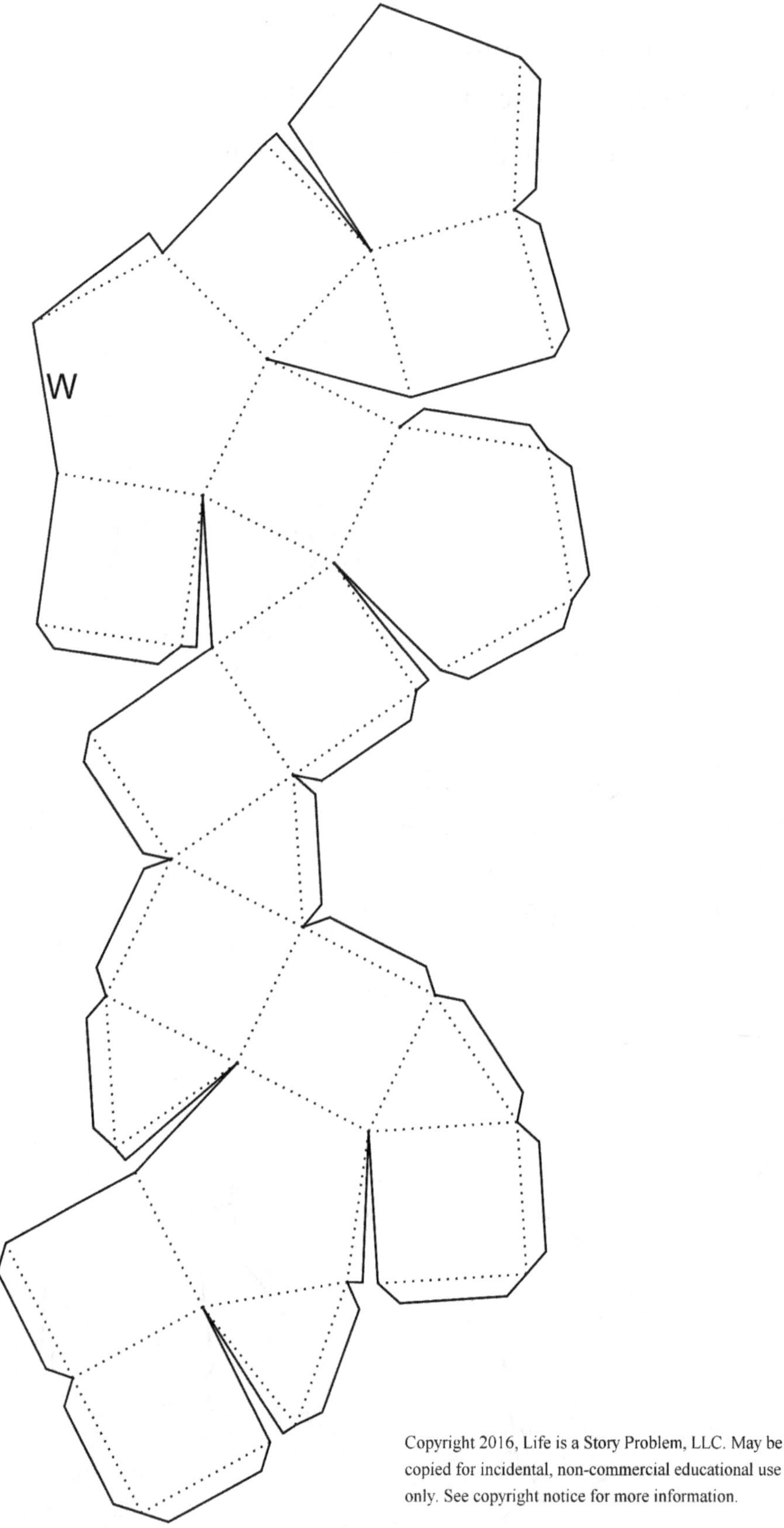

Geometric Nets Mega Project Book, Tabbed By David E. McAdams

Parabidiminished rhombicosidodecahedron

This net is on two pages. Print out a copy of each page.

1. Color the shape before you cut it out.

2. Cut out along the solid lines.

3. Fold along all of the dotted lines.

4. Dab a little glue on each tab and attach it.

5. Once the glue has dried, attach any decorations.

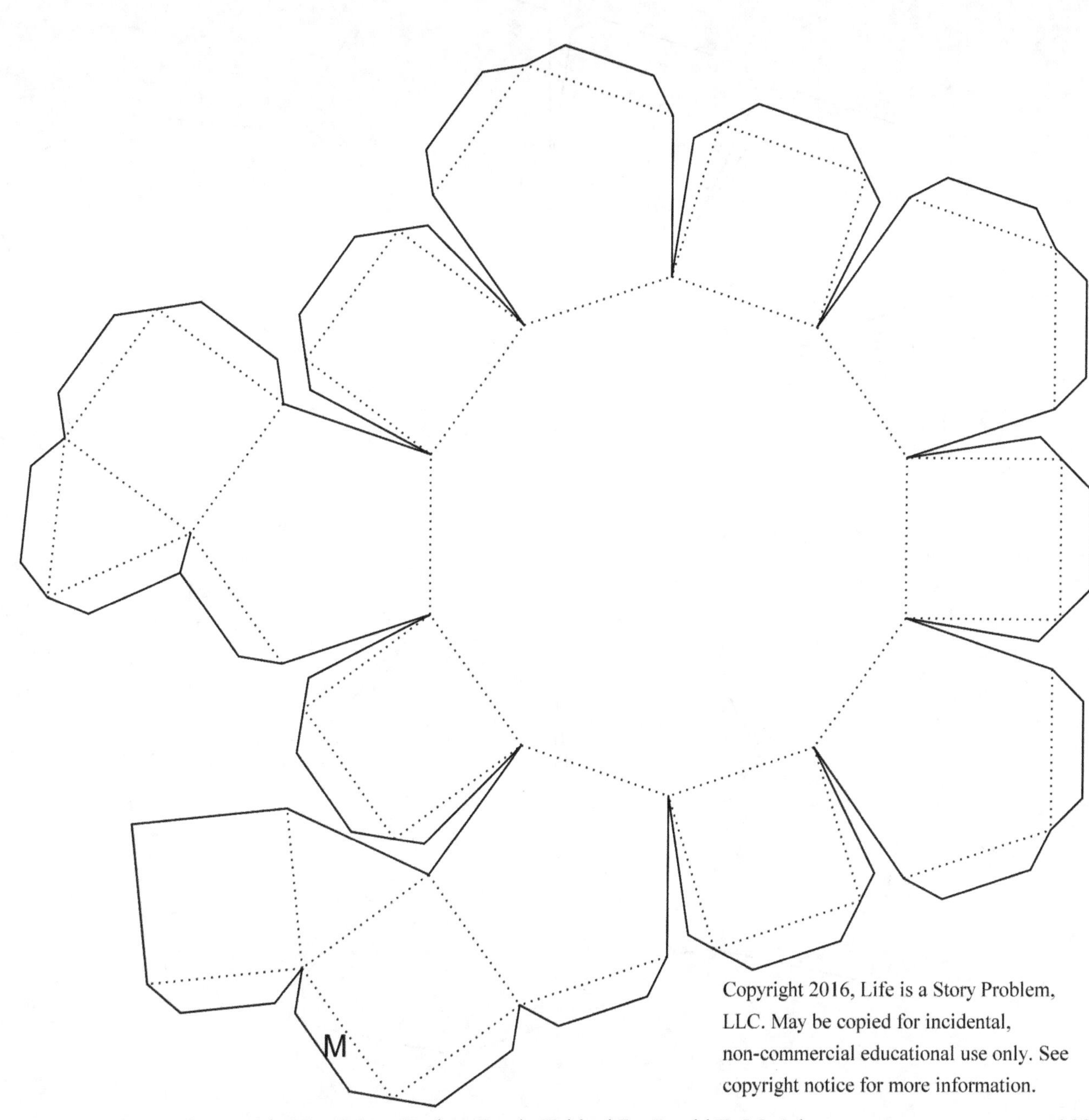

2nd page, Parabidiminished Rhombicosidodecahedron

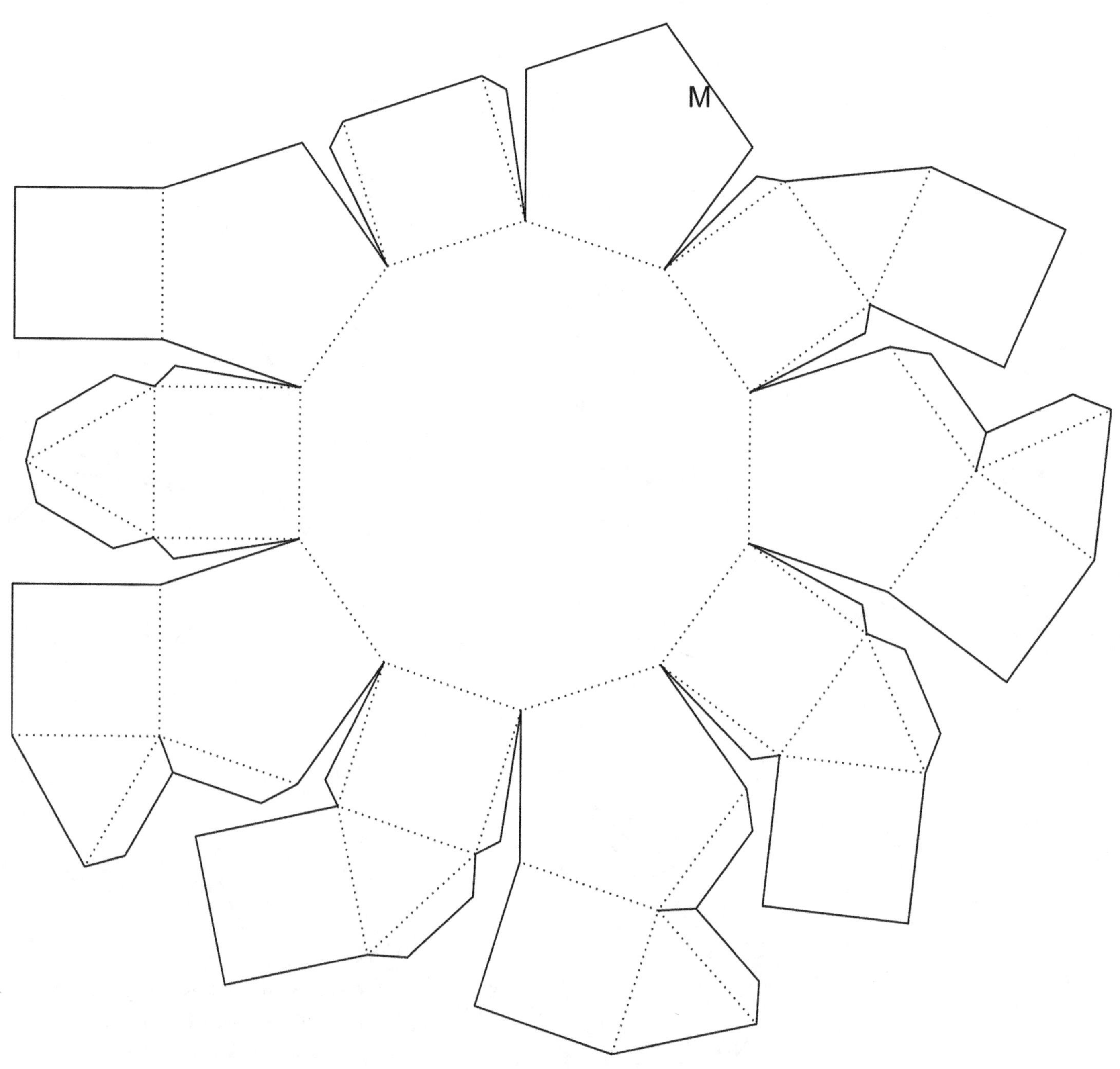

Geometric Nets Mega Project Book, Tabbed By David E. McAdams

Metabidiminished rhombicosidodecahedron

This net is on two pages. Print out a copy of each page.

1. Color the shape before you cut it out.
2. Cut out along the solid lines.
3. Fold along all of the dotted lines.
4. Dab a little glue on each tab and attach it.
5. Once the glue has dried, attach any decorations.

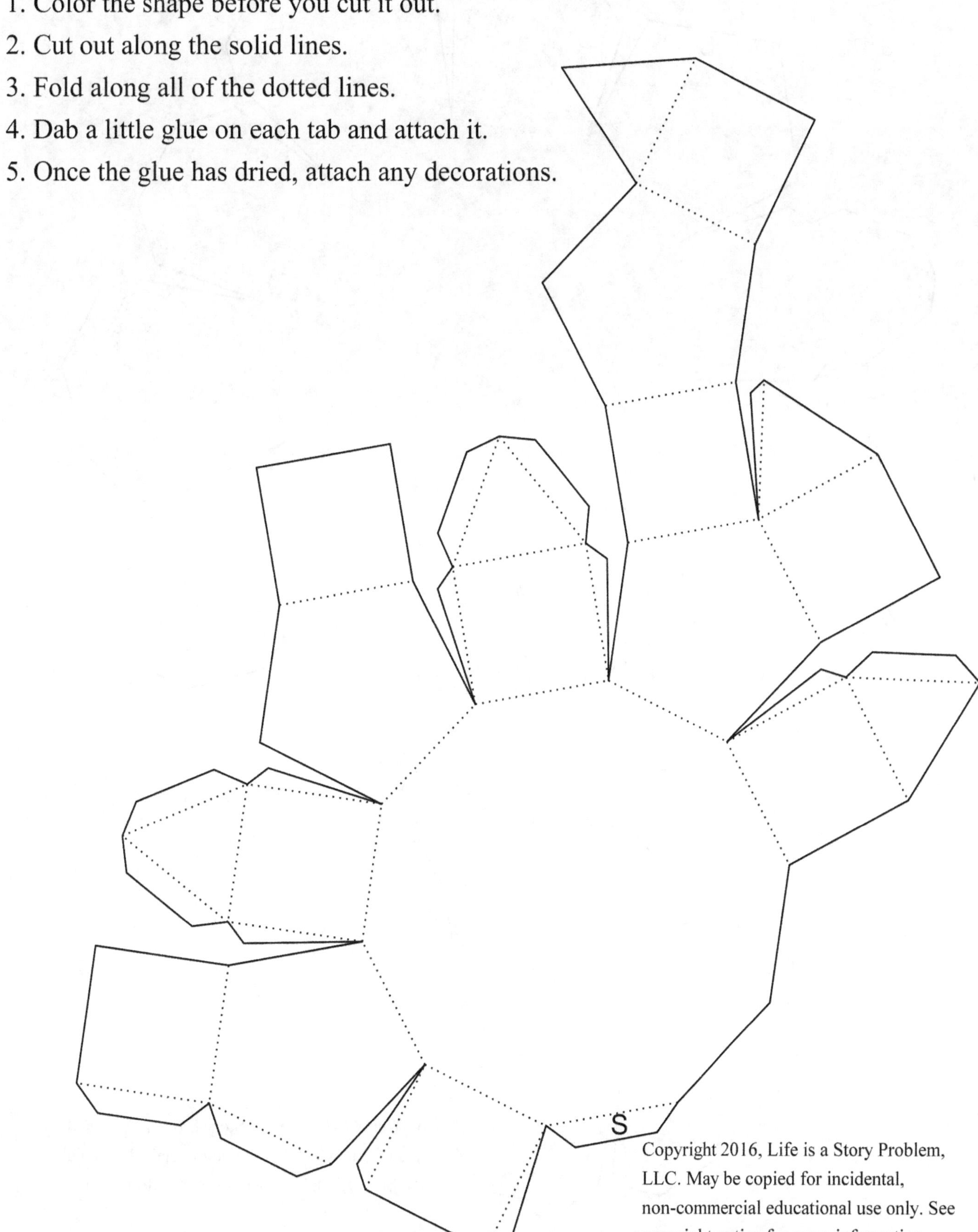

2nd page, Metabidiminished rhombicosidodecahedron

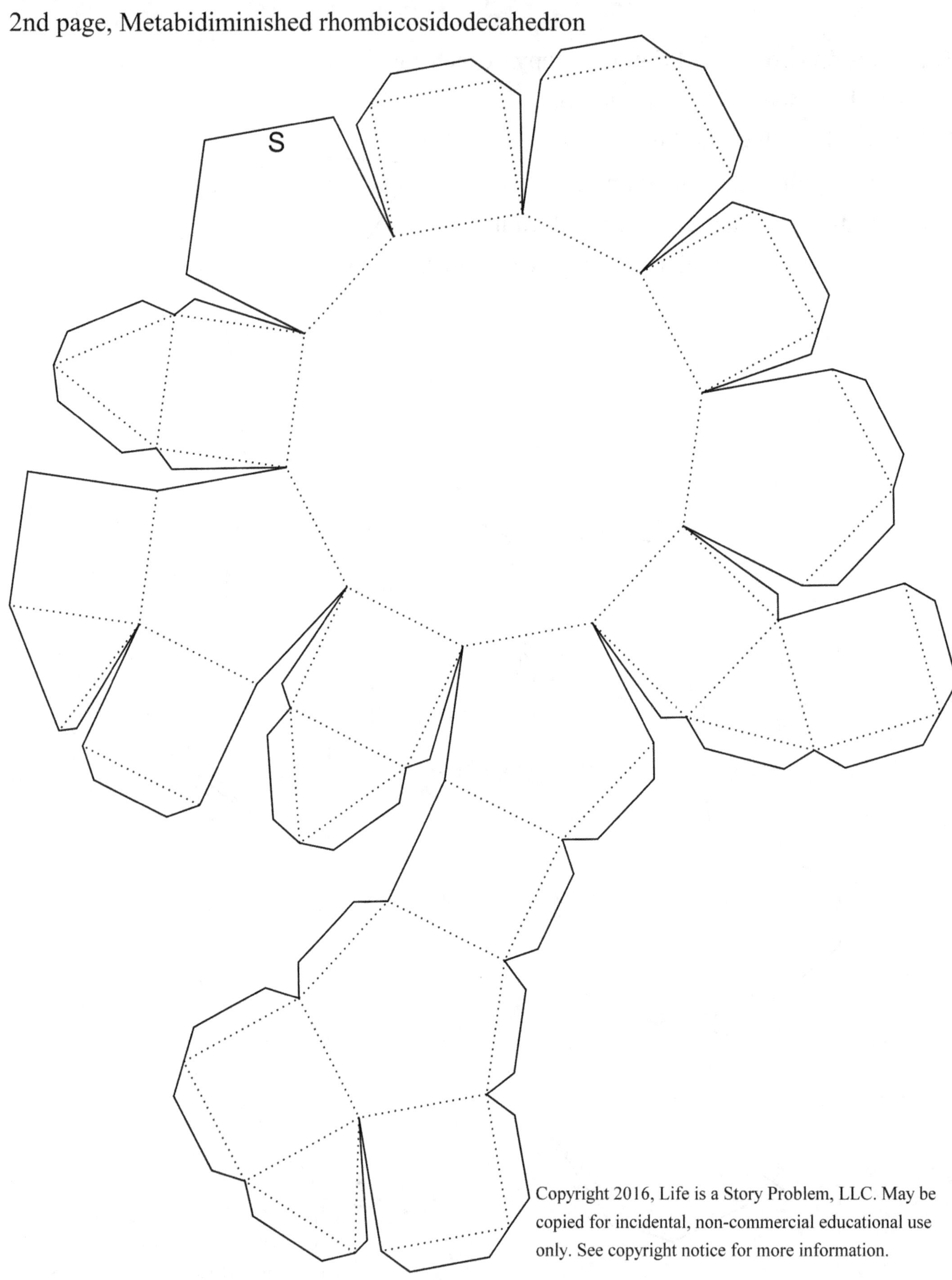

Geometric Nets Mega Project Book, Tabbed By David E. McAdams

Gyrate bidiminished rhombicosidodecahedron

This net is on two pages. Print out a copy of each page.

1. Color the shape before you cut it out.

2. Cut out along the solid lines.

3. Fold along all of the dotted lines.

4. Dab a little glue on each tab and attach it.

5. Once the glue has dried, attach any decorations.

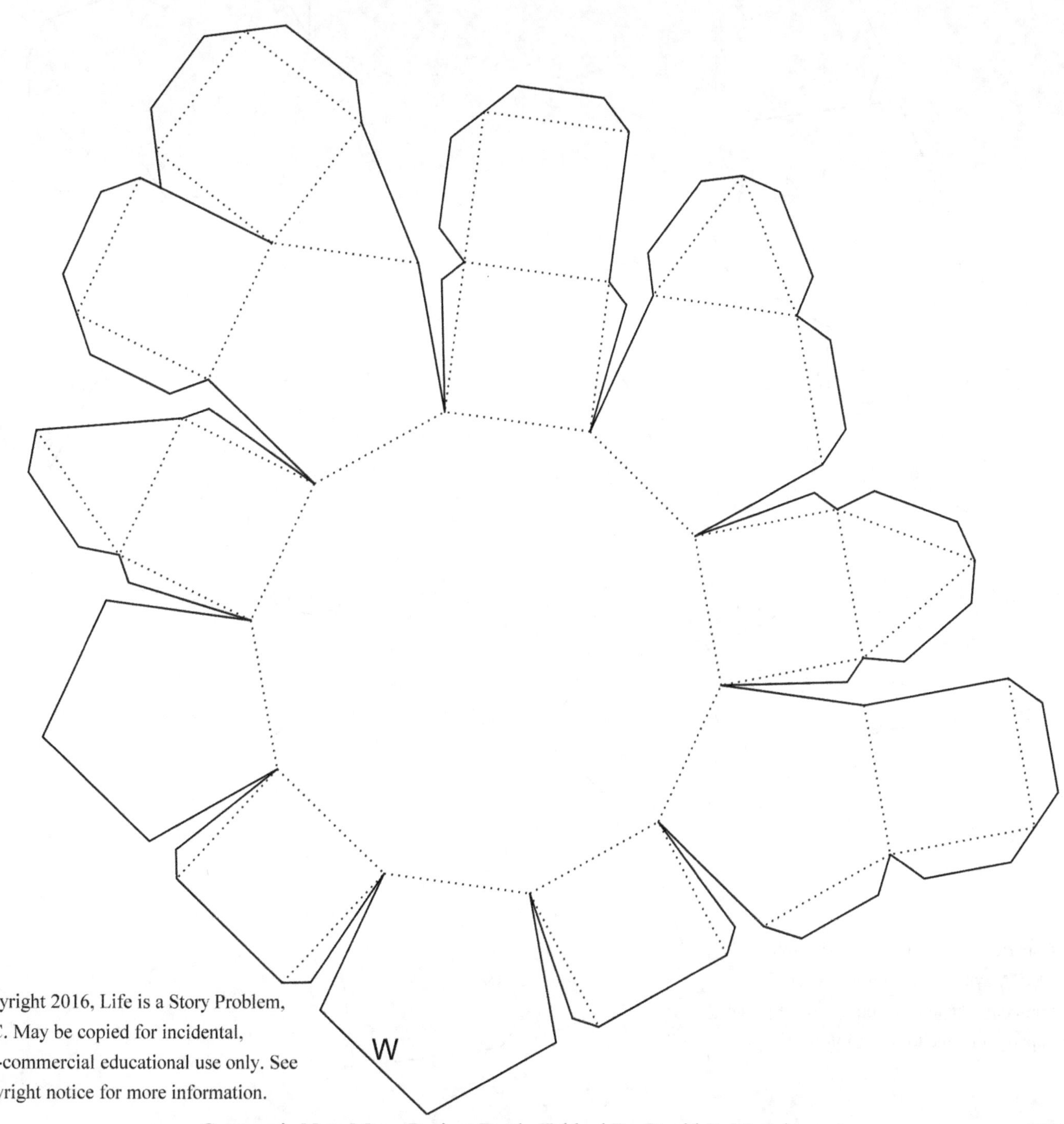

2nd page, Gyrate Bidiminished Rhombicosidodecahedron

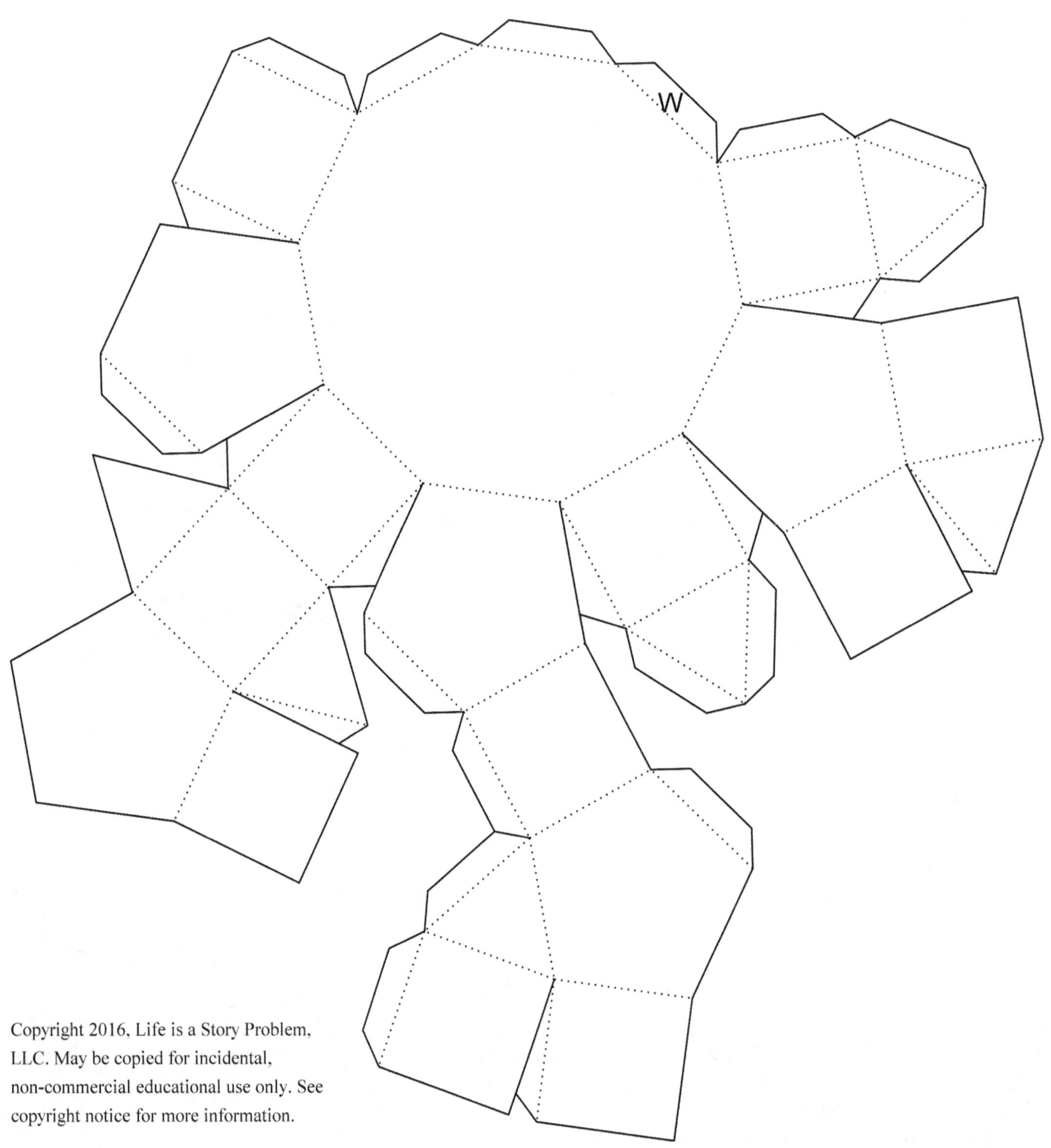

Geometric Nets Mega Project Book, Tabbed By David E. McAdams

Tridiminished rhombicosidodecahedron

This net is on three pages. Print out a copy of each page.

1. Color the shape before you cut it out.

2. Cut out along the solid lines.

3. Fold along all of the dotted lines.

4. Dab a little glue on each tab and attach it.

5. Once the glue has dried, attach any decorations.

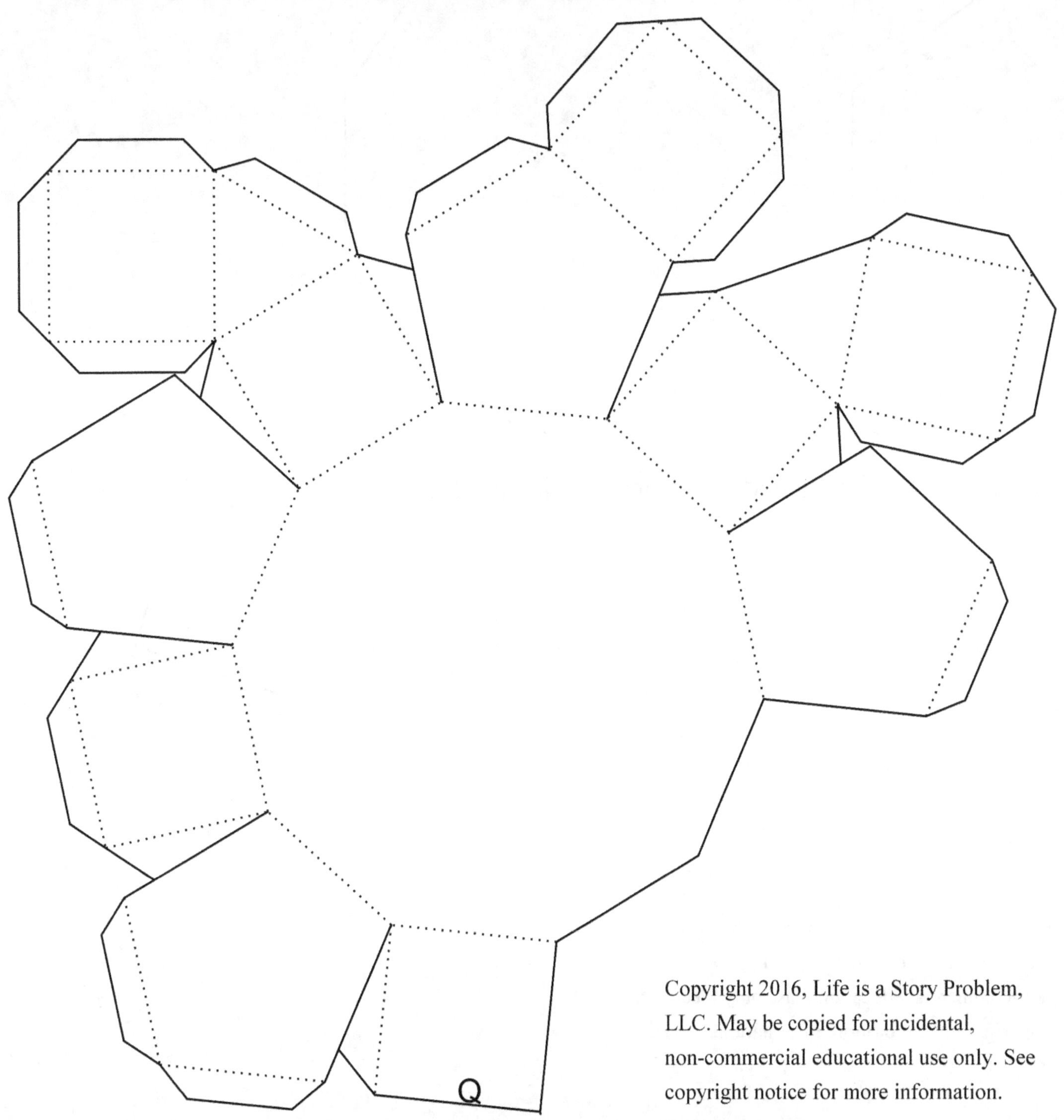

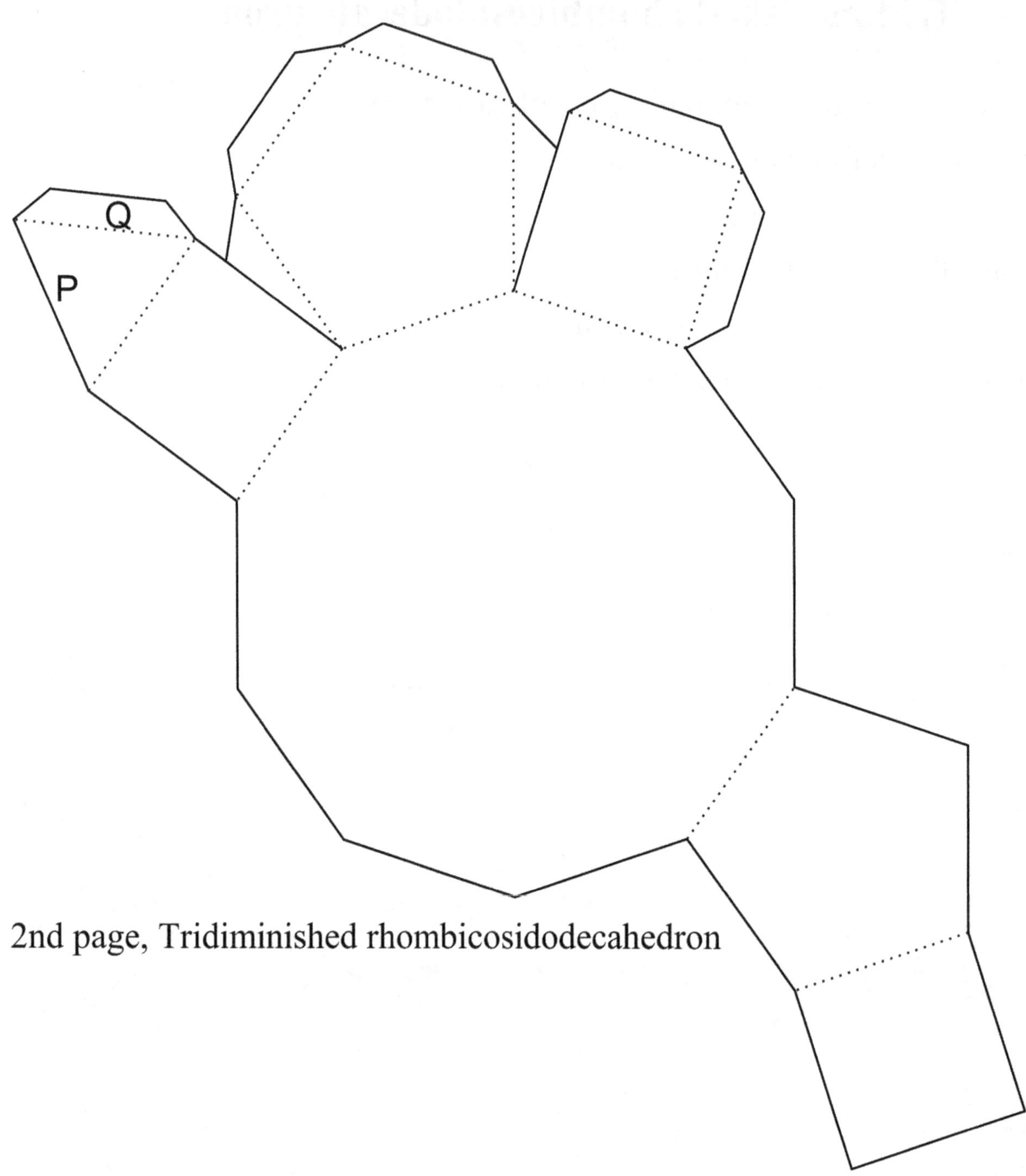

2nd page, Tridiminished rhombicosidodecahedron

Geometric Nets Mega Project Book, Tabbed By David E. McAdams

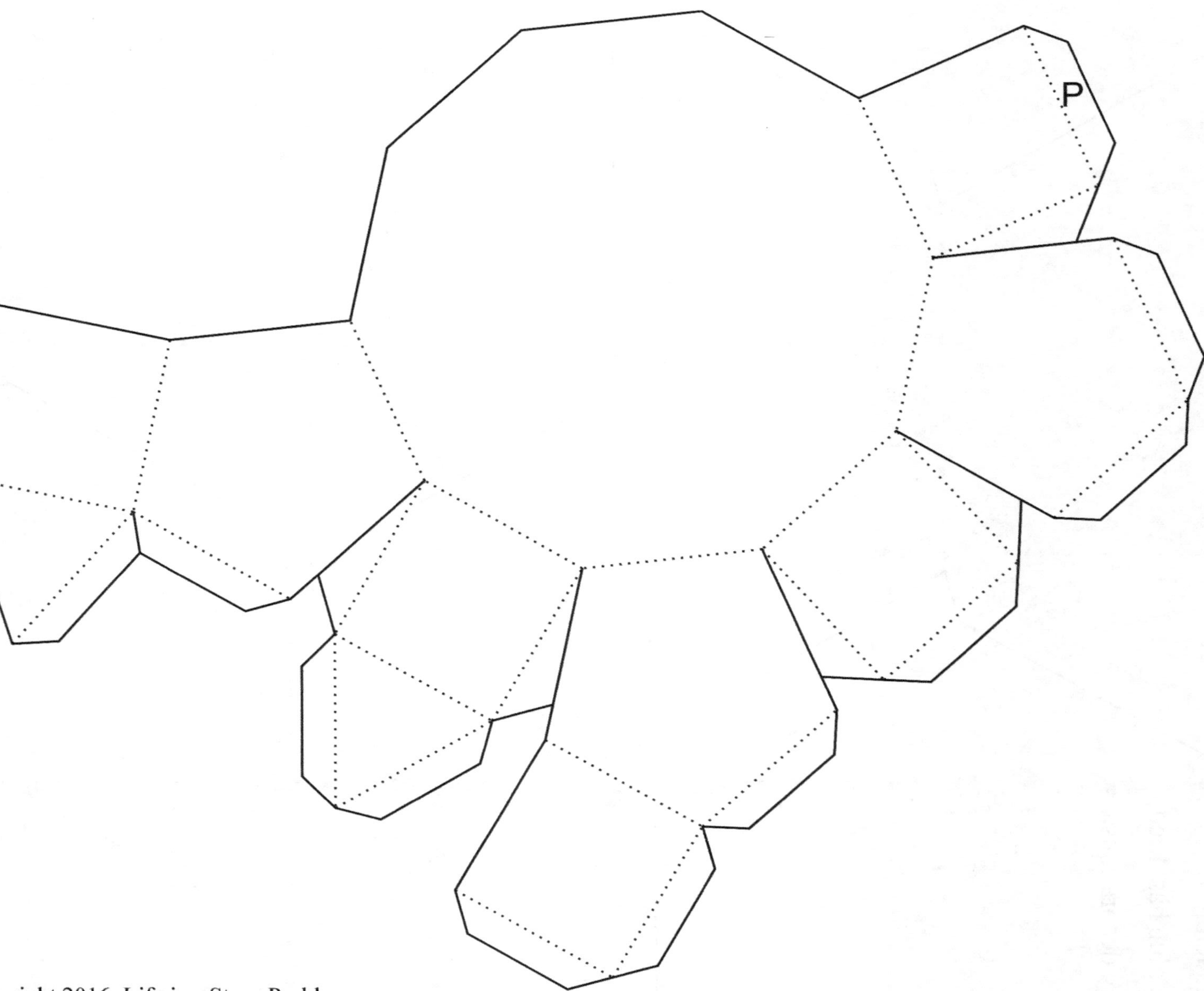

P

Snub disphenoid

1. Color the shape before you cut it out.
2. Cut out along the solid lines.
3. Fold along all of the dotted lines.
4. Dab a little glue on each tab and attach it.
5. Once the glue has dried, attach any decorations.

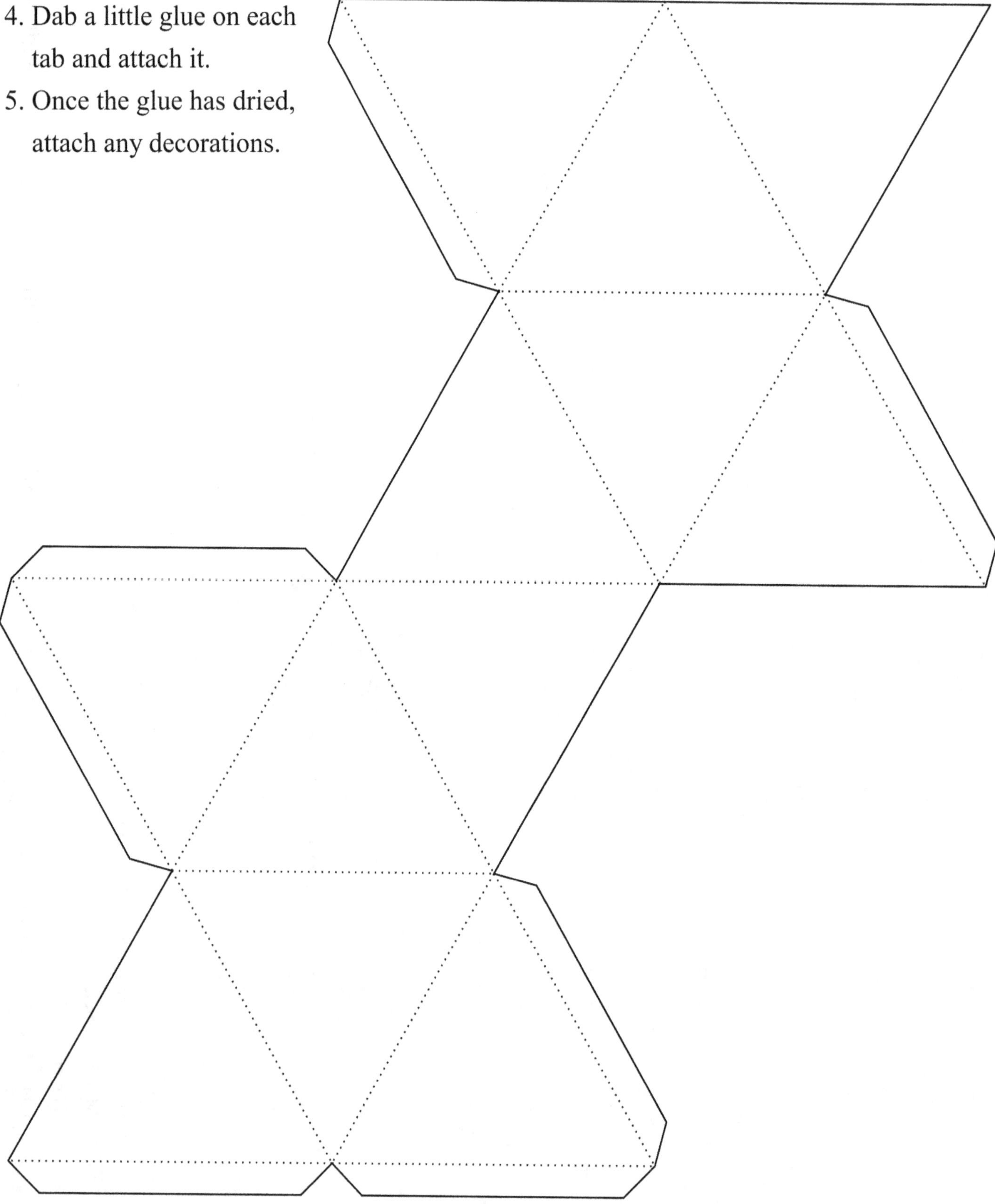

 Geometric Nets Mega Project Book, Tabbed By David E. McAdams

Snub square antiprism

1. Color the shape before you cut it out.
2. Cut out along the solid lines.
3. Fold along all of the dotted lines.
4. Dab a little glue on each tab
   and attach it.
5. Once the glue has dried,
   attach any decorations.

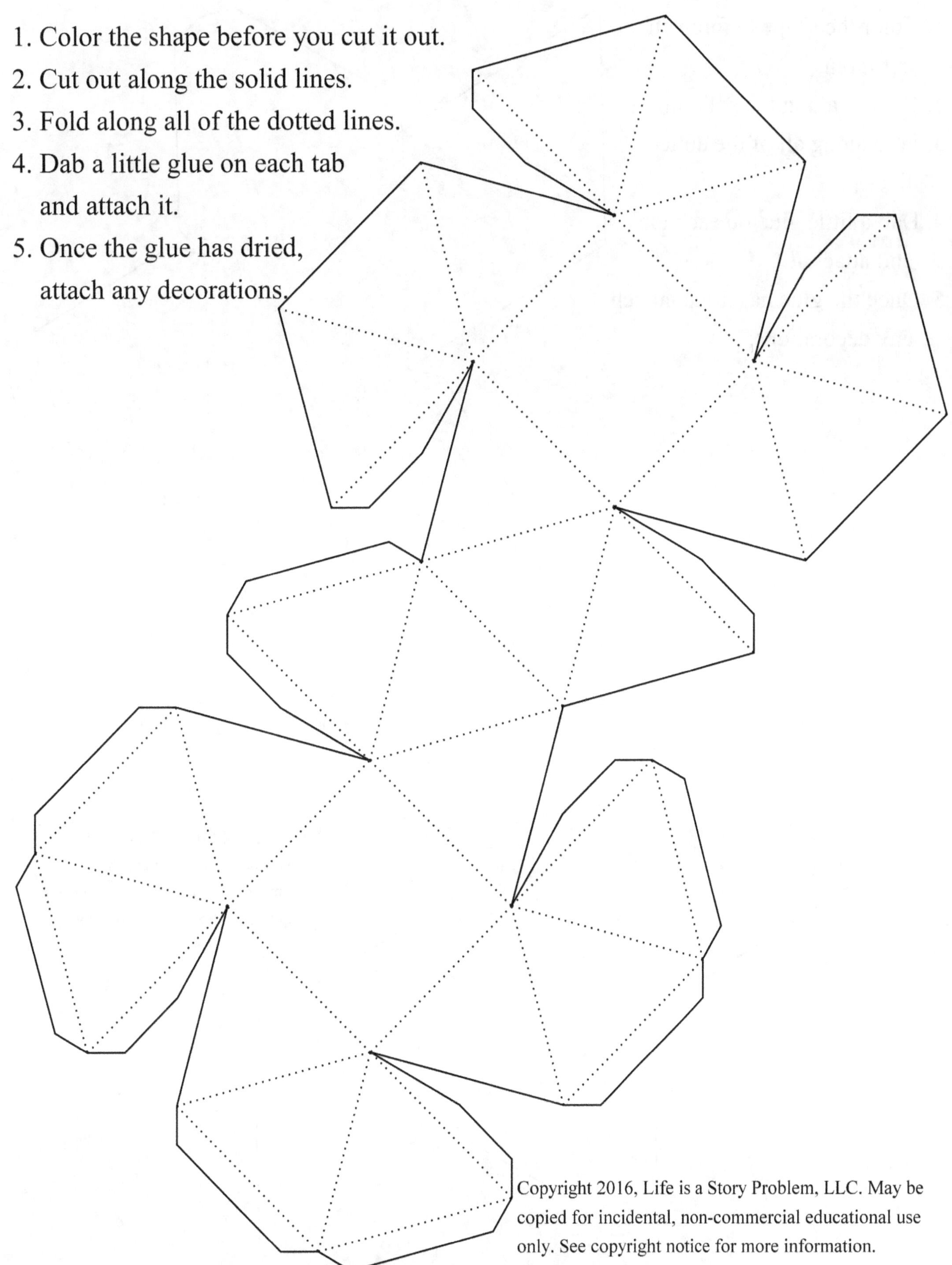

Sphenocorona

1. Color the shape before you cut it out.
2. Cut out along the solid lines.
3. Fold along all of the dotted lines.
4. Dab a little glue on each tab and attach it.
5. Once the glue has dried, attach any decorations.

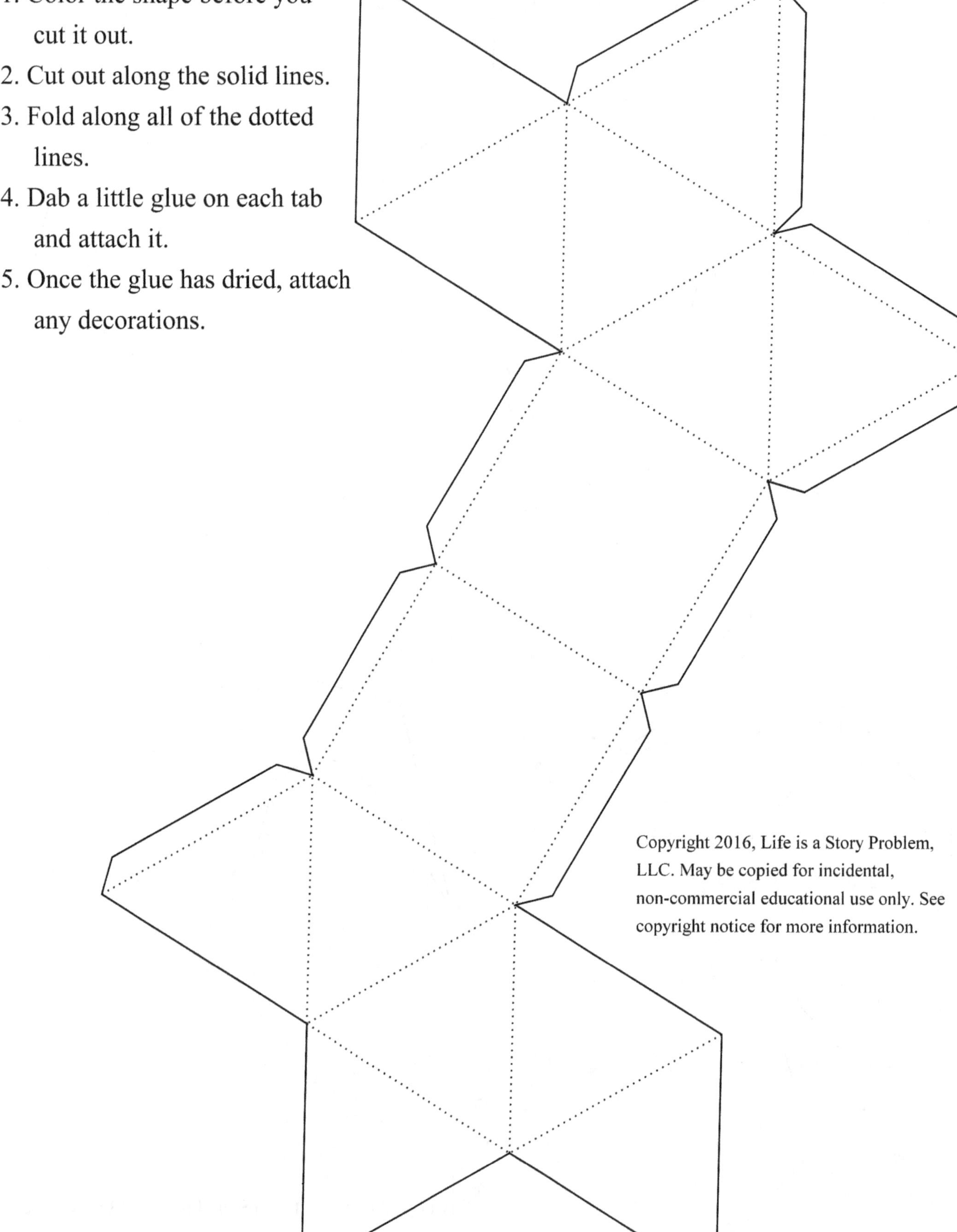

 Geometric Nets Mega Project Book, Tabbed By David E. McAdams

Augmented sphenocorona

1. Color the shape before you cut it out.
2. Cut out along the solid lines.
3. Fold along all of the dotted lines.
4. Dab a little glue on each tab and attach it.
5. Once the glue has dried, attach any
   decorations.

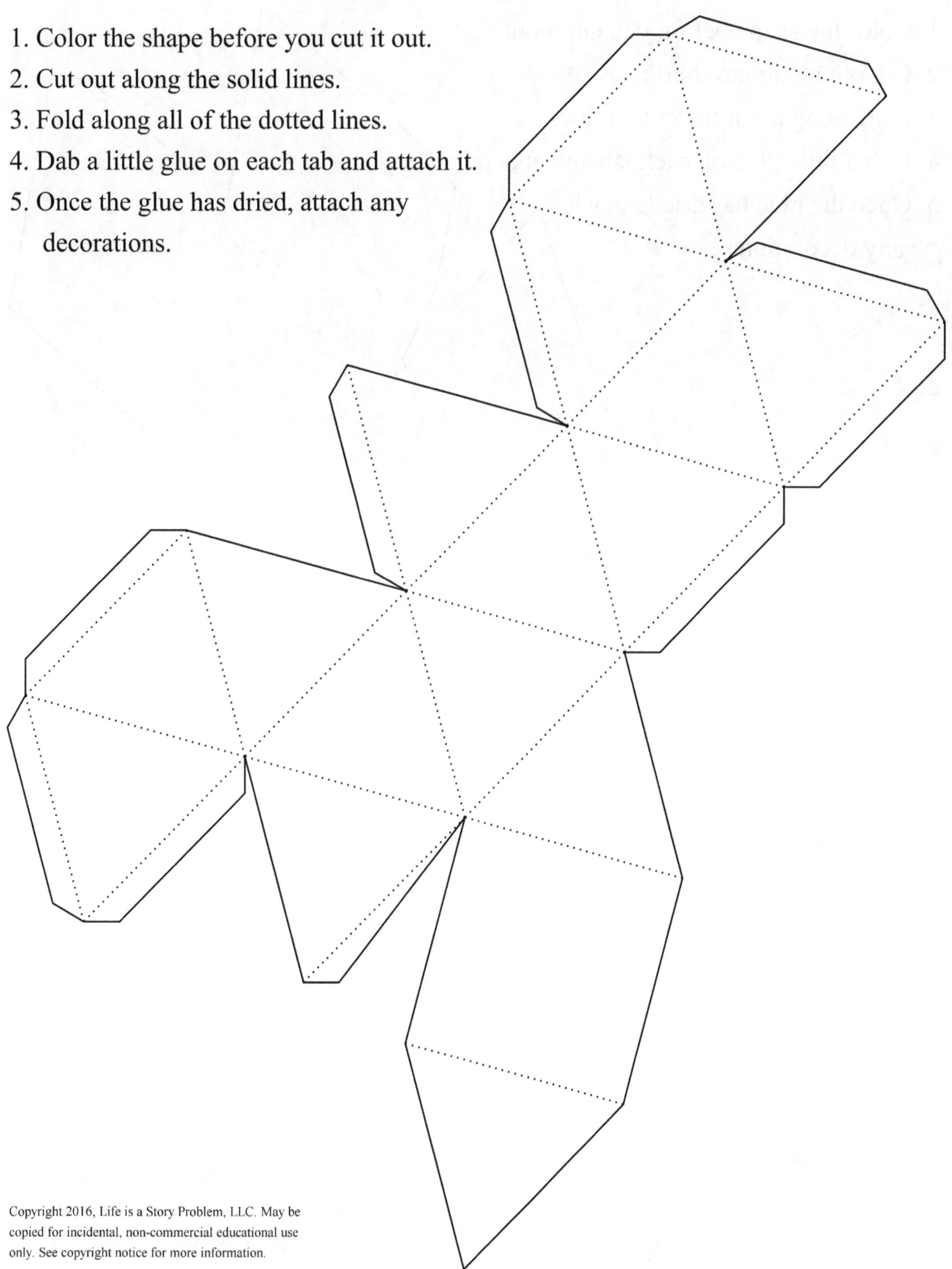

Sphenomegacorona

1. Color the shape before you cut it out.

2. Cut out along the solid lines.

3. Fold along all of the dotted lines.

4. Dab a little glue on each tab and attach it.

5. Once the glue has dried, attach
   any decorations.

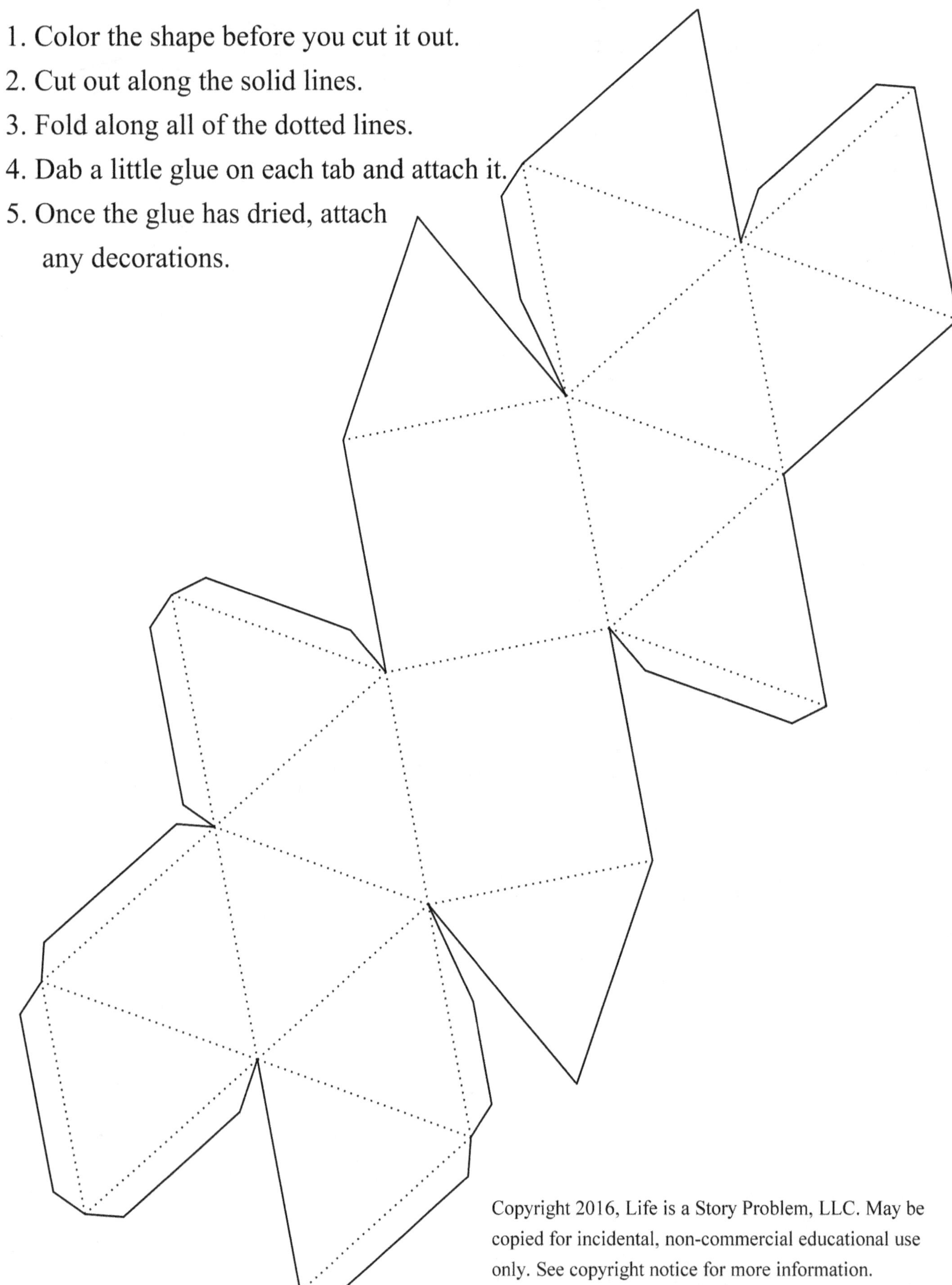

  Geometric Nets Mega Project Book, Tabbed By David E. McAdams

Hebesphenomegacorona

1. Color the shape before you cut it out.
2. Cut out along the solid lines.
3. Fold along all of the dotted lines.
4. Dab a little glue each tab and attach it.
5. Once the glue has dried, attach decorations.

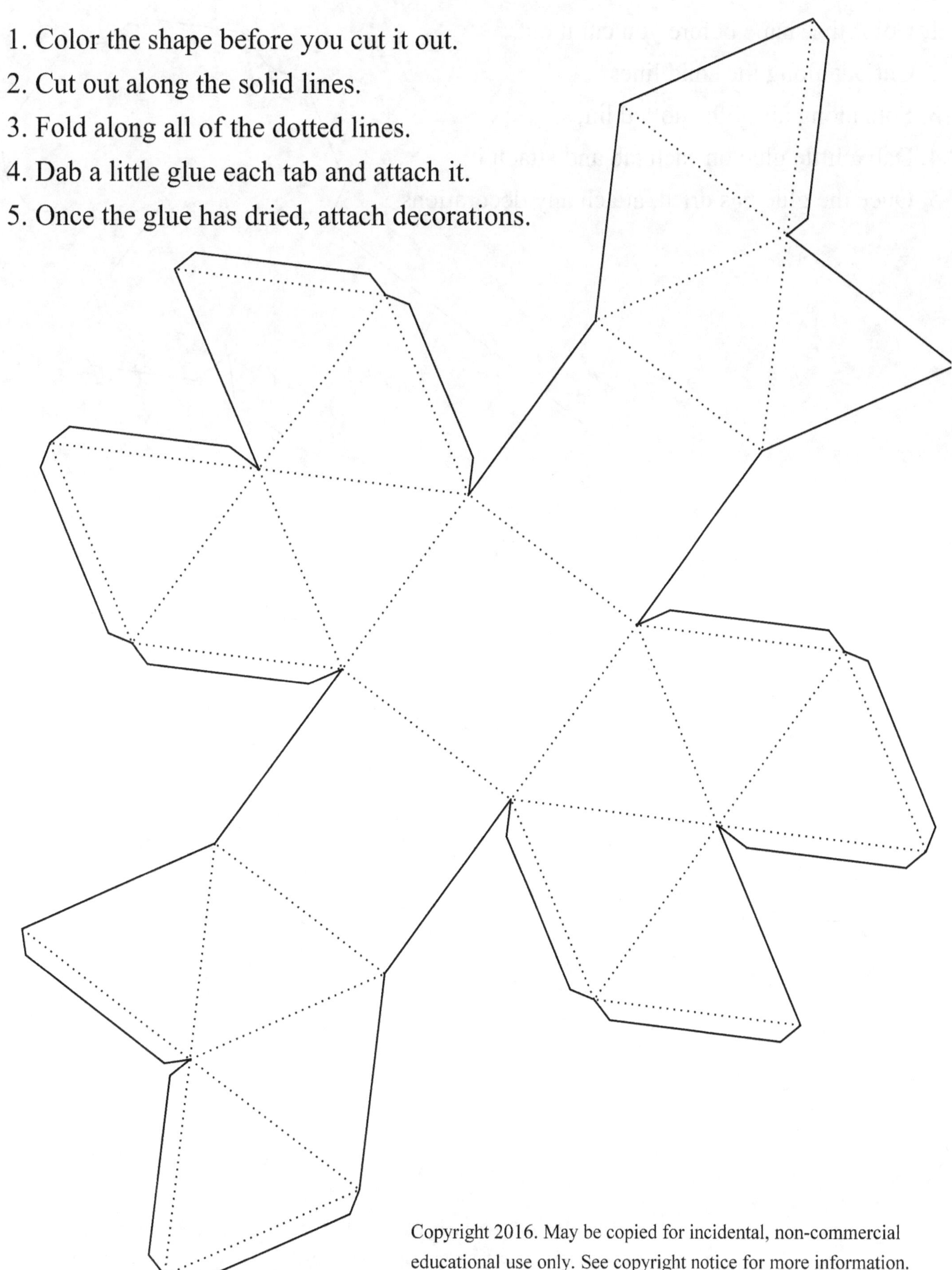

Disphenocingulum

1. Color the shape before you cut it out.
2. Cut out along the solid lines.
3. Fold along all of the dotted lines.
4. Dab a little glue on each tab and attach it.
5. Once the glue has dried, attach any decorations.

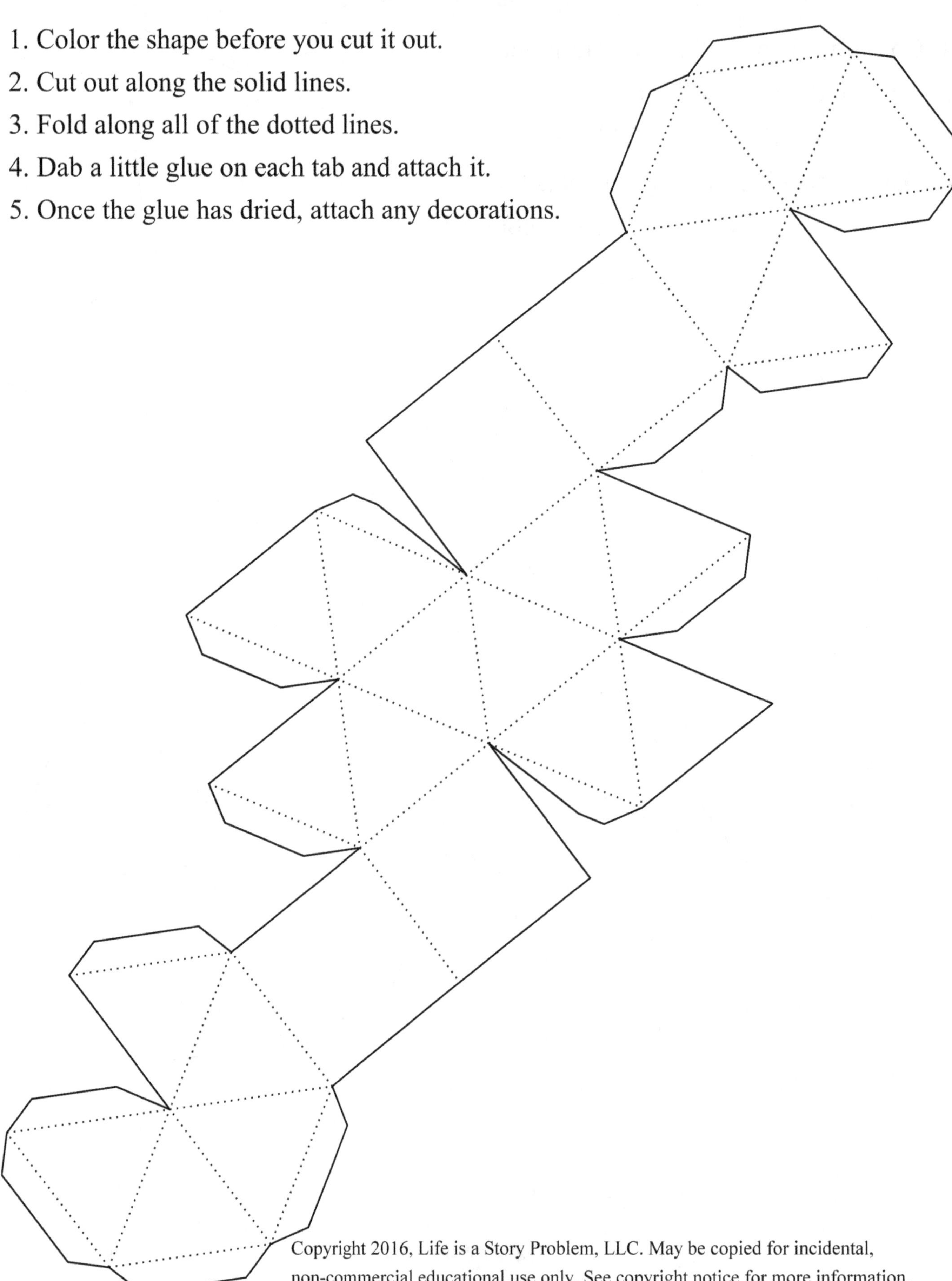

 Geometric Nets Mega Project Book, Tabbed By David E. McAdams

Bilunabirotunda

1. Color the shape before you cut it out.
2. Cut out along the solid lines.
3. Fold along all of the dotted lines.
4. Dab a little glue on each tab and attach it.
5. Once the glue has dried, attach any decorations.

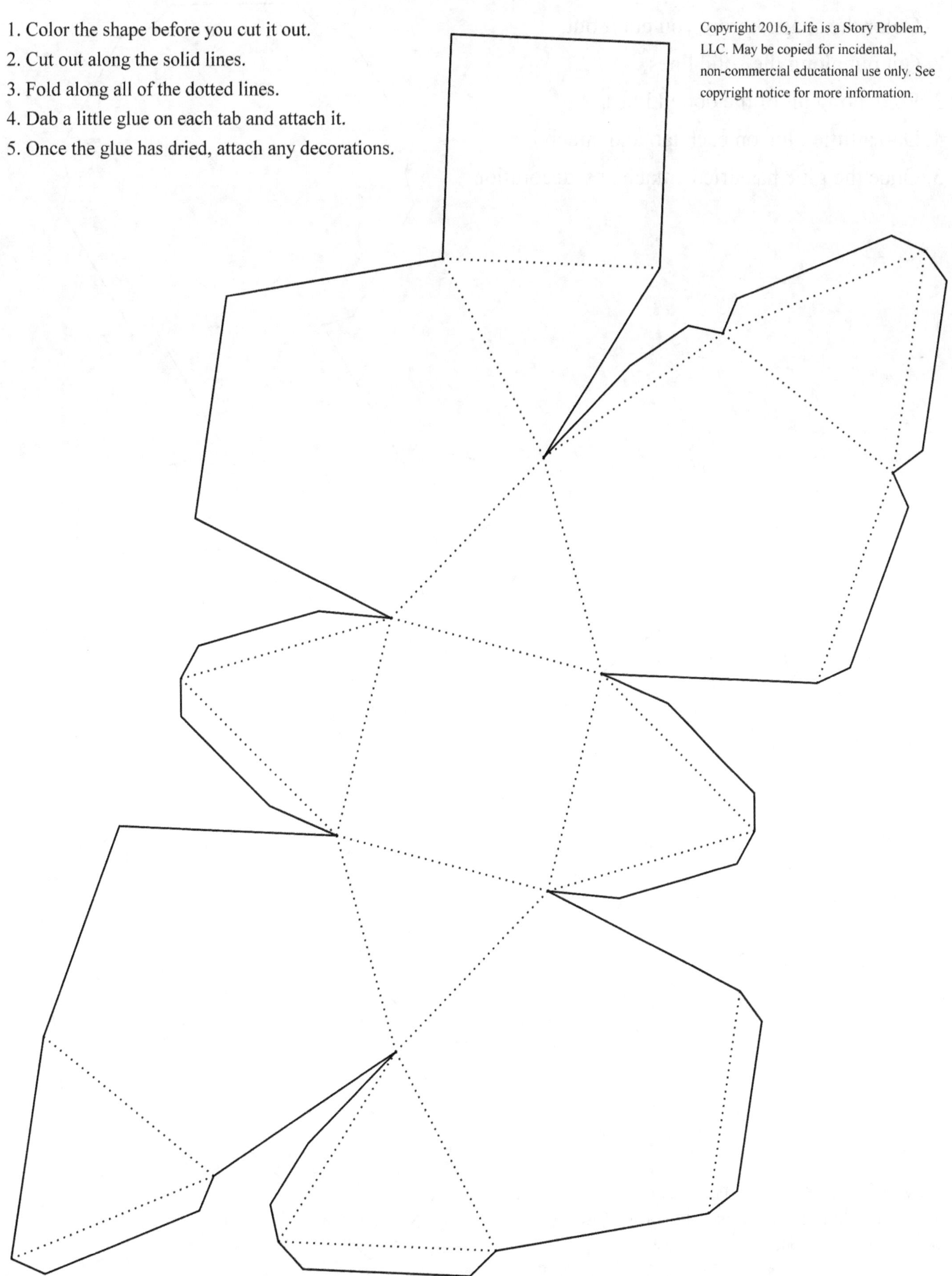

Triangular hebesphenorotunda

1. Color the shape before you cut it out.
2. Cut out along the solid lines.
3. Fold along all of the dotted lines.
4. Dab a little glue on each tab and attach it.
5. Once the glue has dried, attach any decorations.

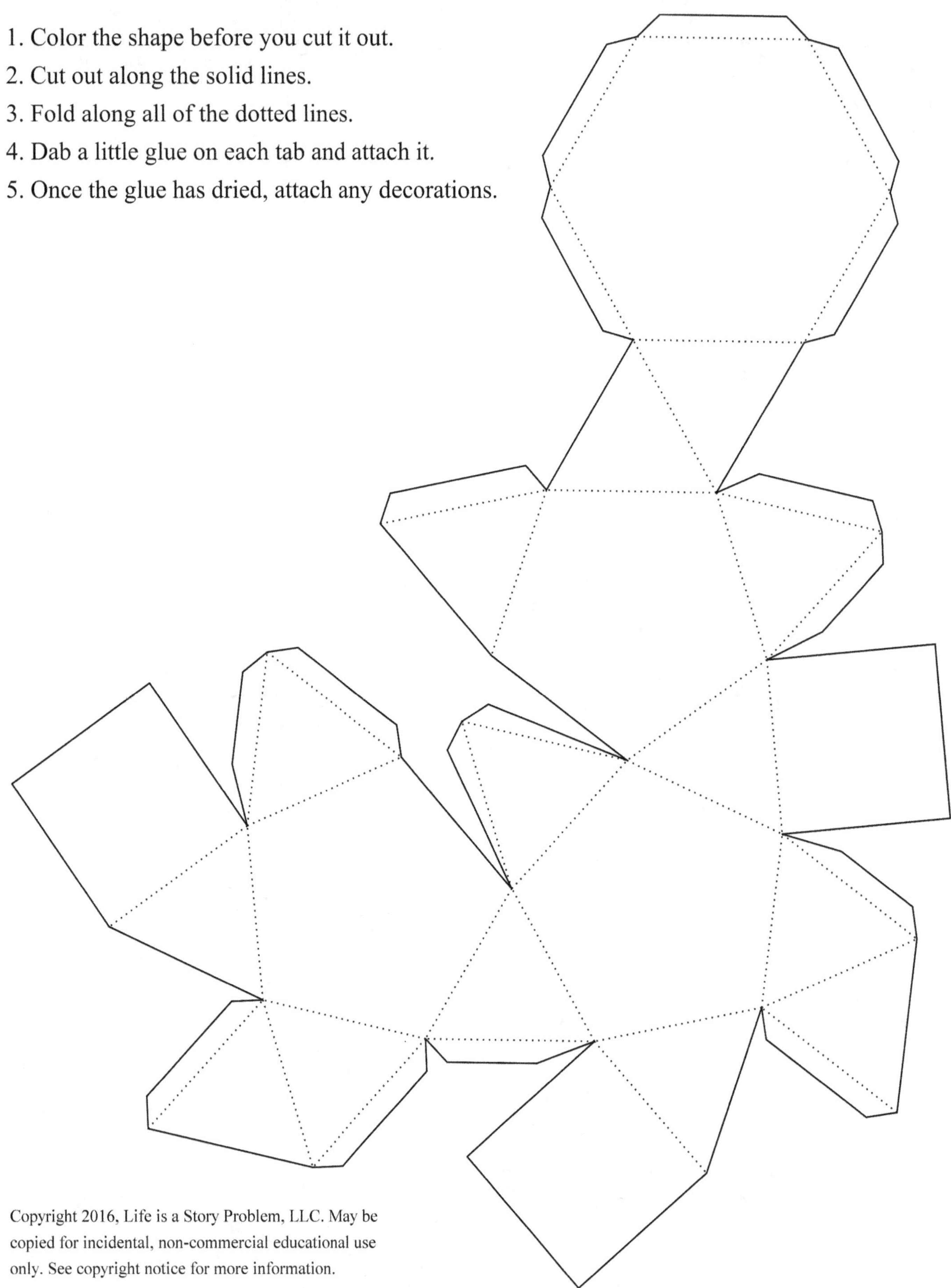

     Geometric Nets Mega Project Book, Tabbed By David E. McAdams

Gyroelongated digonal cupola

1. Color the shape before you cut it out.
2. Cut out along the solid lines.
3. Fold forward along all of the dotted lines.
4. Fold backward along all of the dashed lines.
5. Dab a little glue on each tab and attach it.
6. Once the glue has dried, attach any decorations.

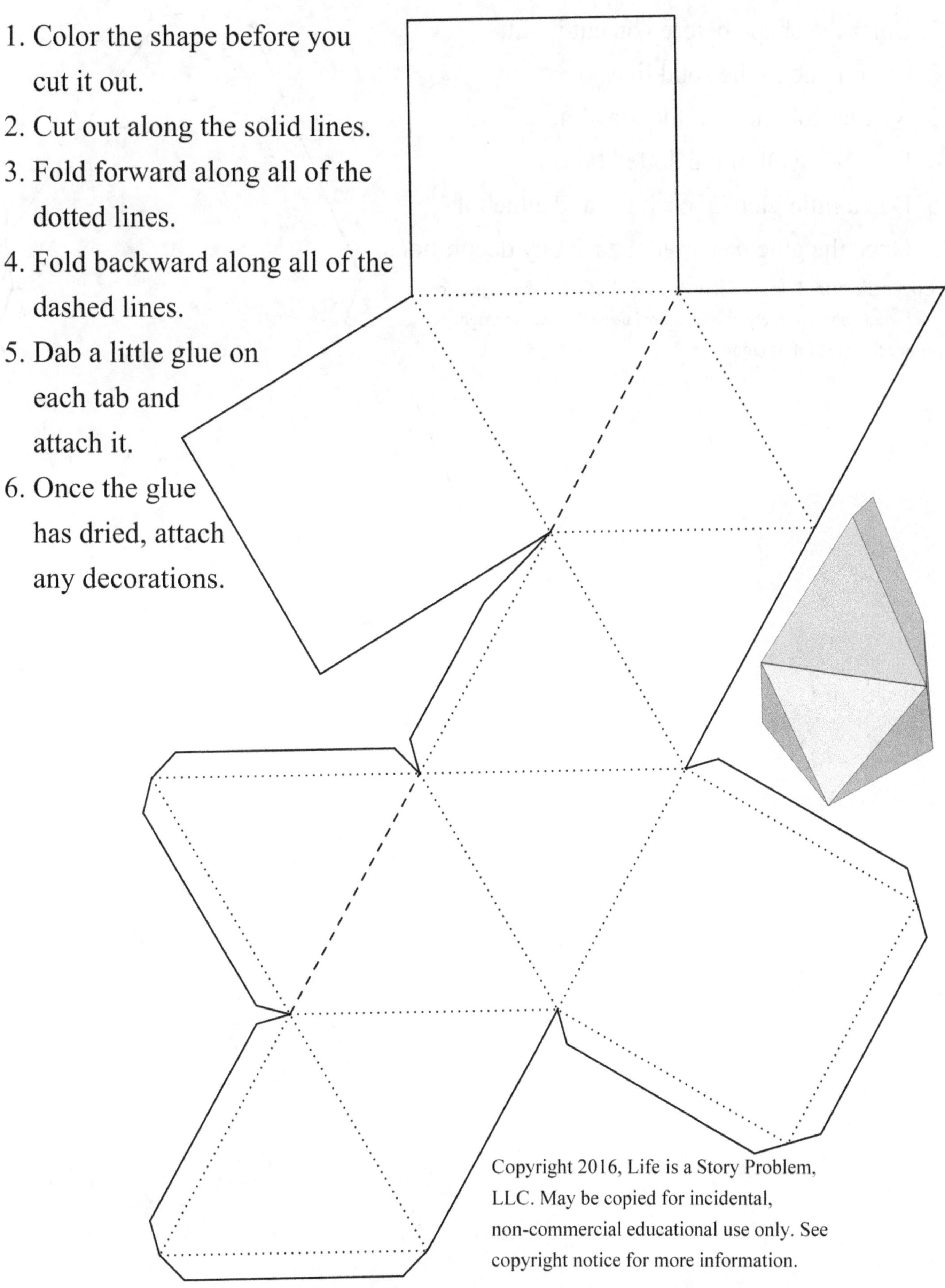

Gyroelongated triangular pyramid

1. Color the shape before you cut it out.

2. Cut out along the solid lines.

3. Neither fold nor cut the gray lines.

4. Fold along all of the dotted lines.

5. Dab a little glue on each tab and attach it.

6. Once the glue has dried, attach any decorations.

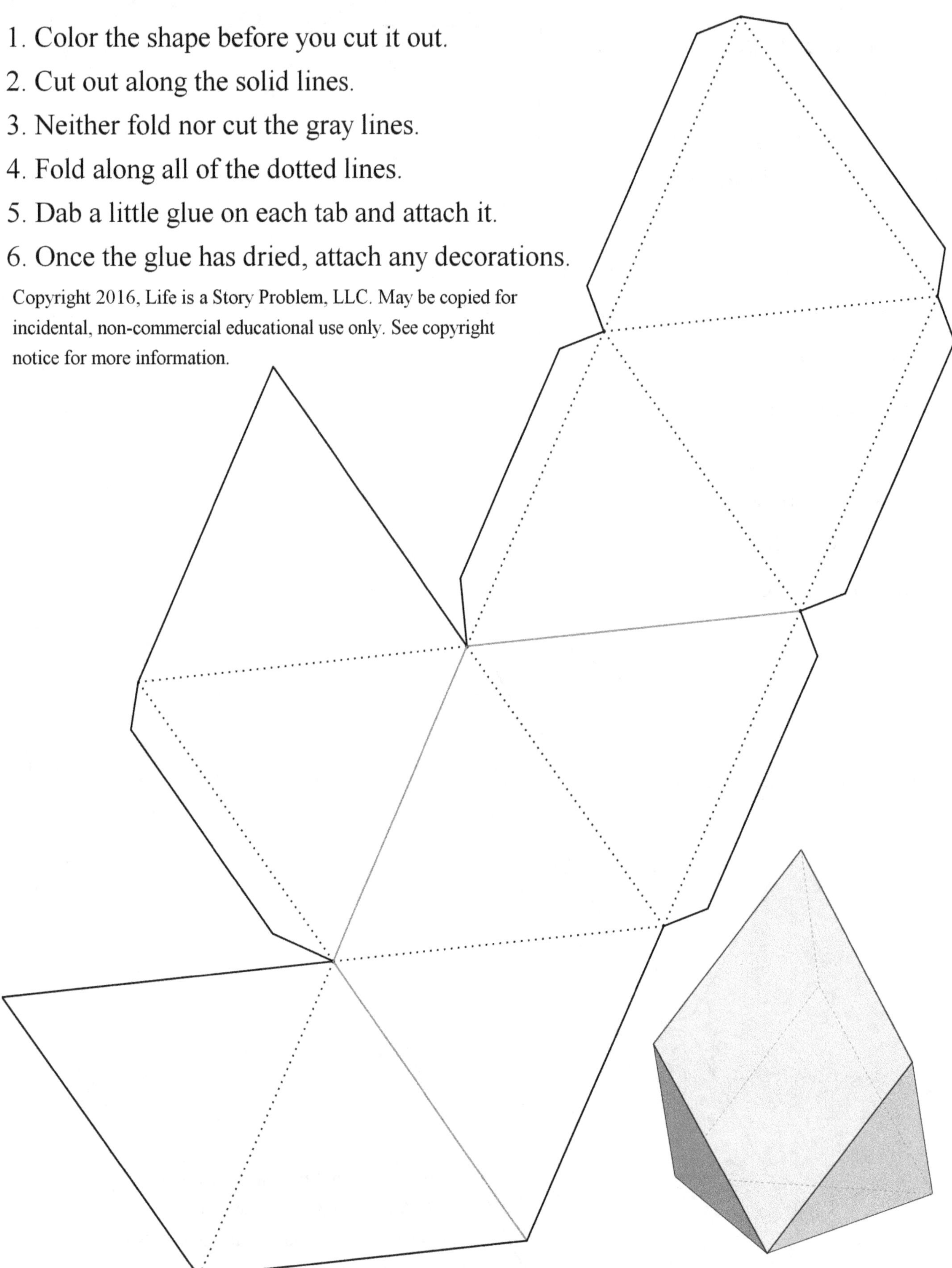

 Geometric Nets Mega Project Book, Tabbed By David E. McAdams

Elongated digonal orthobicupola

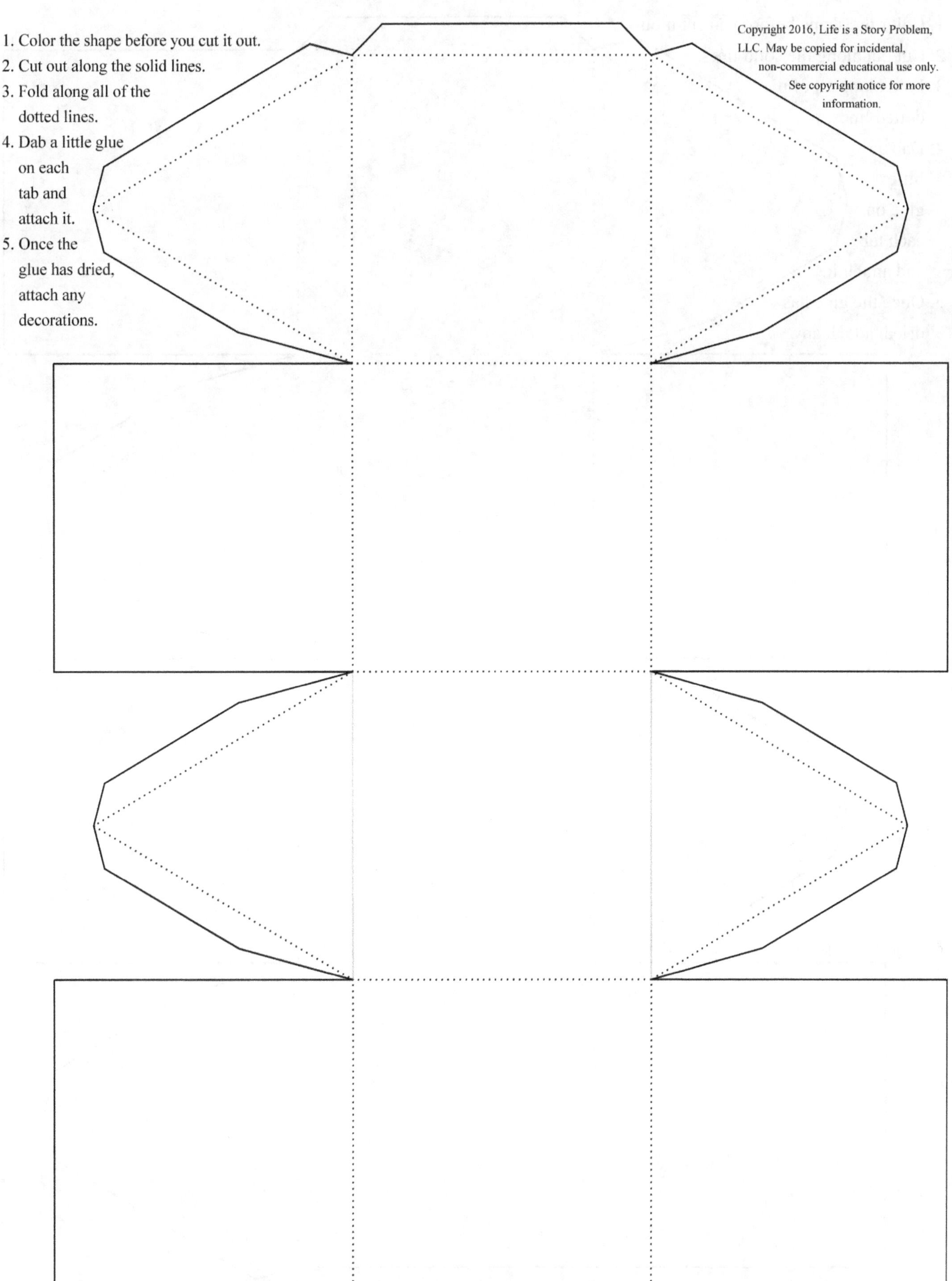

Elongated digonal gyrobicupola

1. Color the shape before you cut it out.
2. Cut out along the solid lines.
3. Fold along all of the dotted lines.
4. Dab a little glue on each tab and attach it.
5. Once the glue has dried, attach any decorations.

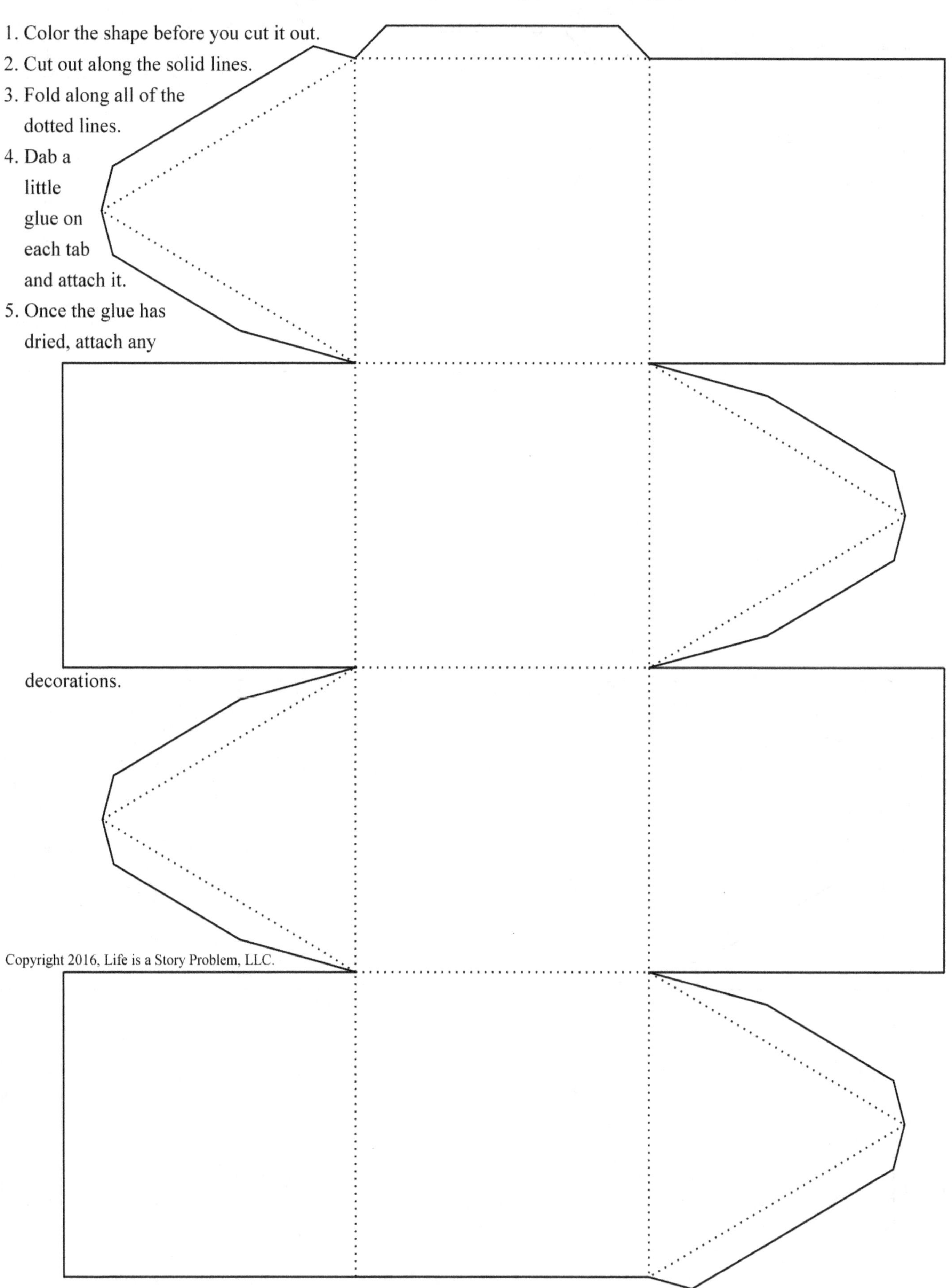

 Geometric Nets Mega Project Book, Tabbed By David E. McAdams

Gyroelongated digonal bicupola

1. Color the shape before you cut it out.
2. Cut out along the solid lines.
3. Fold along all of the dotted lines.
4. Dab a little glue on each tab and attach it.
5. Once the glue has dried, attach any decorations

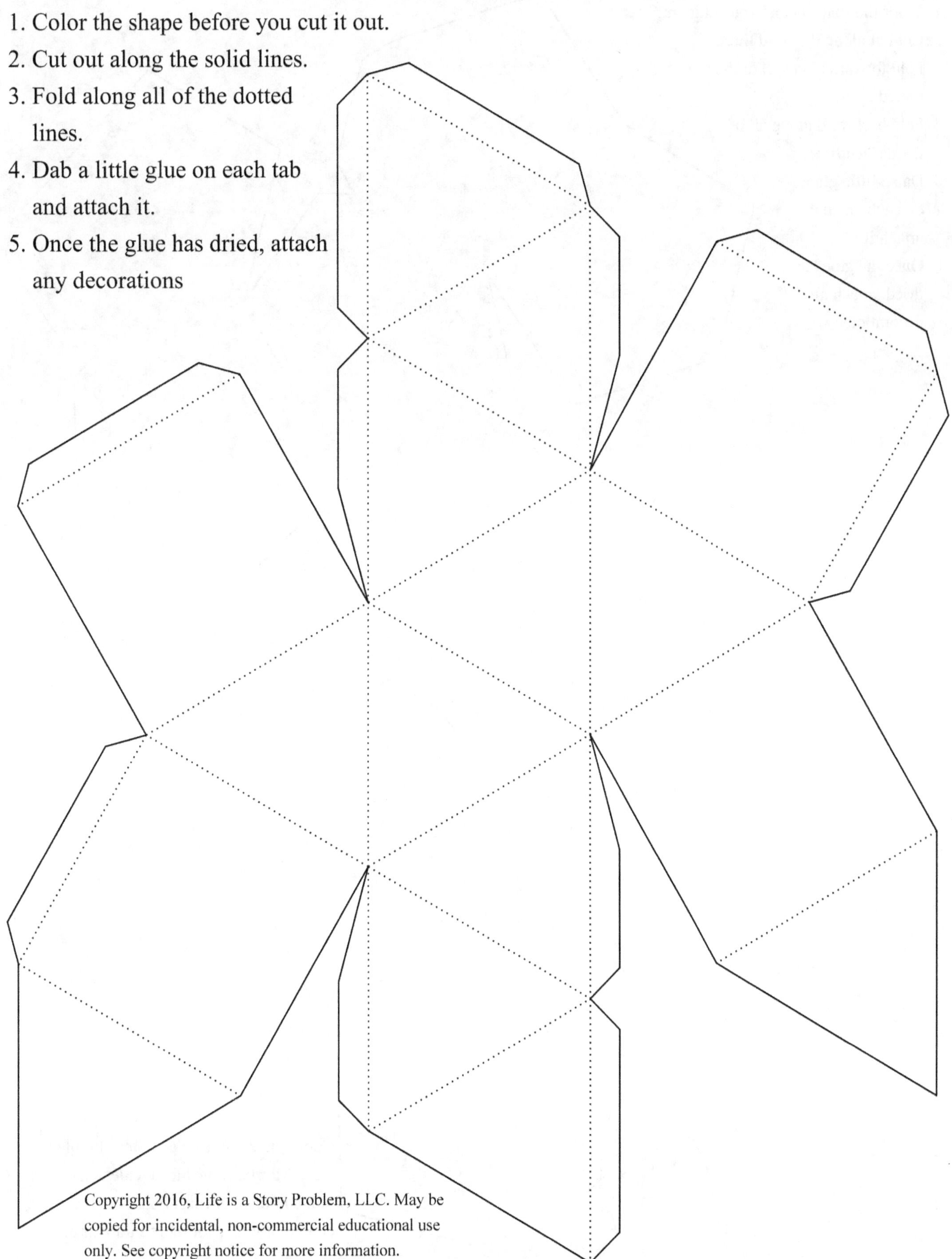

Pyritohedron

1. Color the shape before you cut it out.
2. Cut out along the solid lines.
3. Fold forward along all of the dotted lines.
4. Fold backward along all of the dashed lines.
5. Dab a little glue on each tab and attach it.
6. Once the glue has dried, attach any decorations.

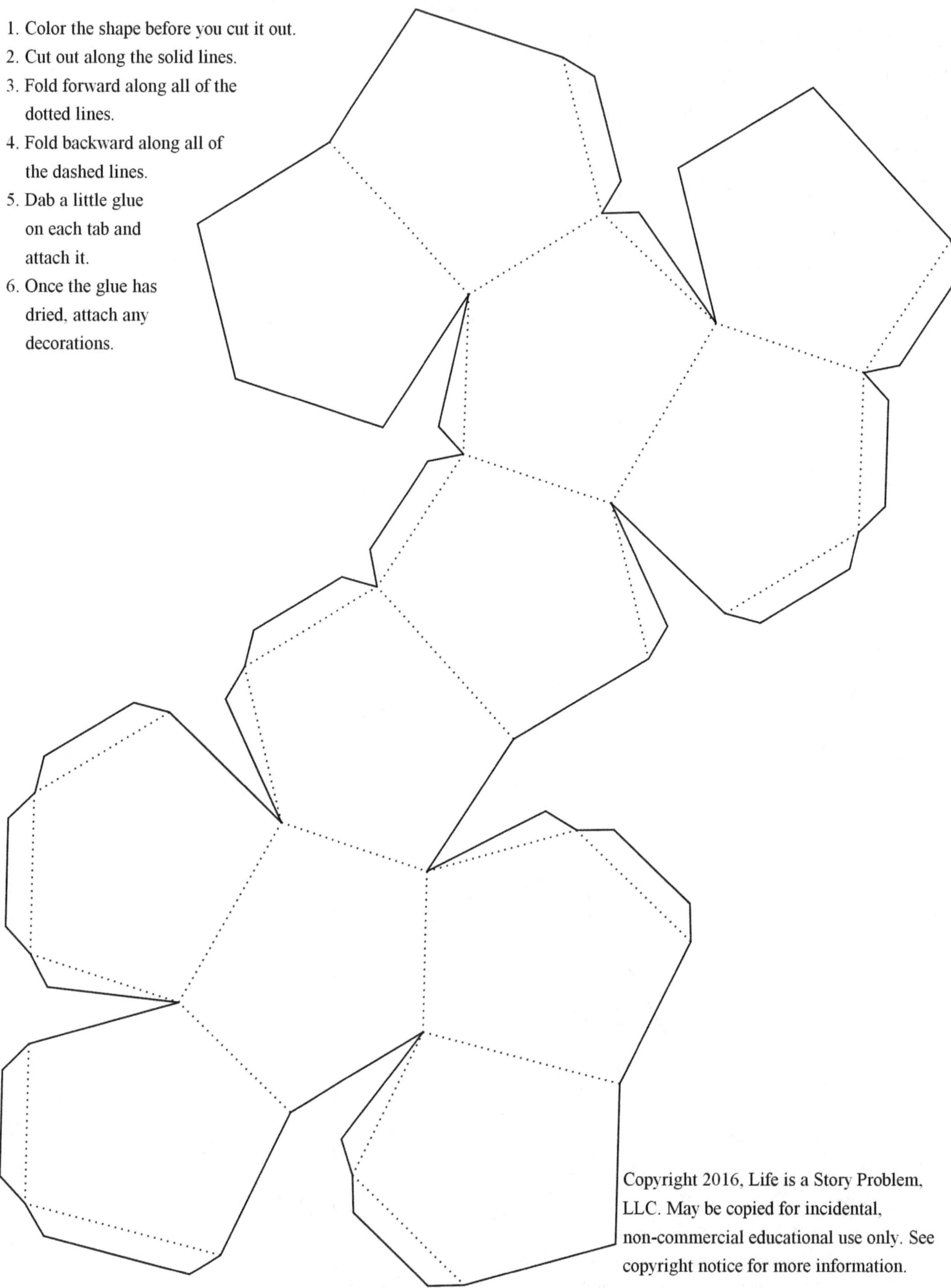

Geometric Nets Mega Project Book, Tabbed By David E. McAdams

Endo-dodecahedron

1. Color the shape before you cut it out.
2. Cut out along the solid lines.
3. Fold forward along all of the dotted lines.
4. Fold backward along all of the dashed lines.
5. Dab a little glue on each tab and attach it.
6. Once the glue has dried, attach any decorations

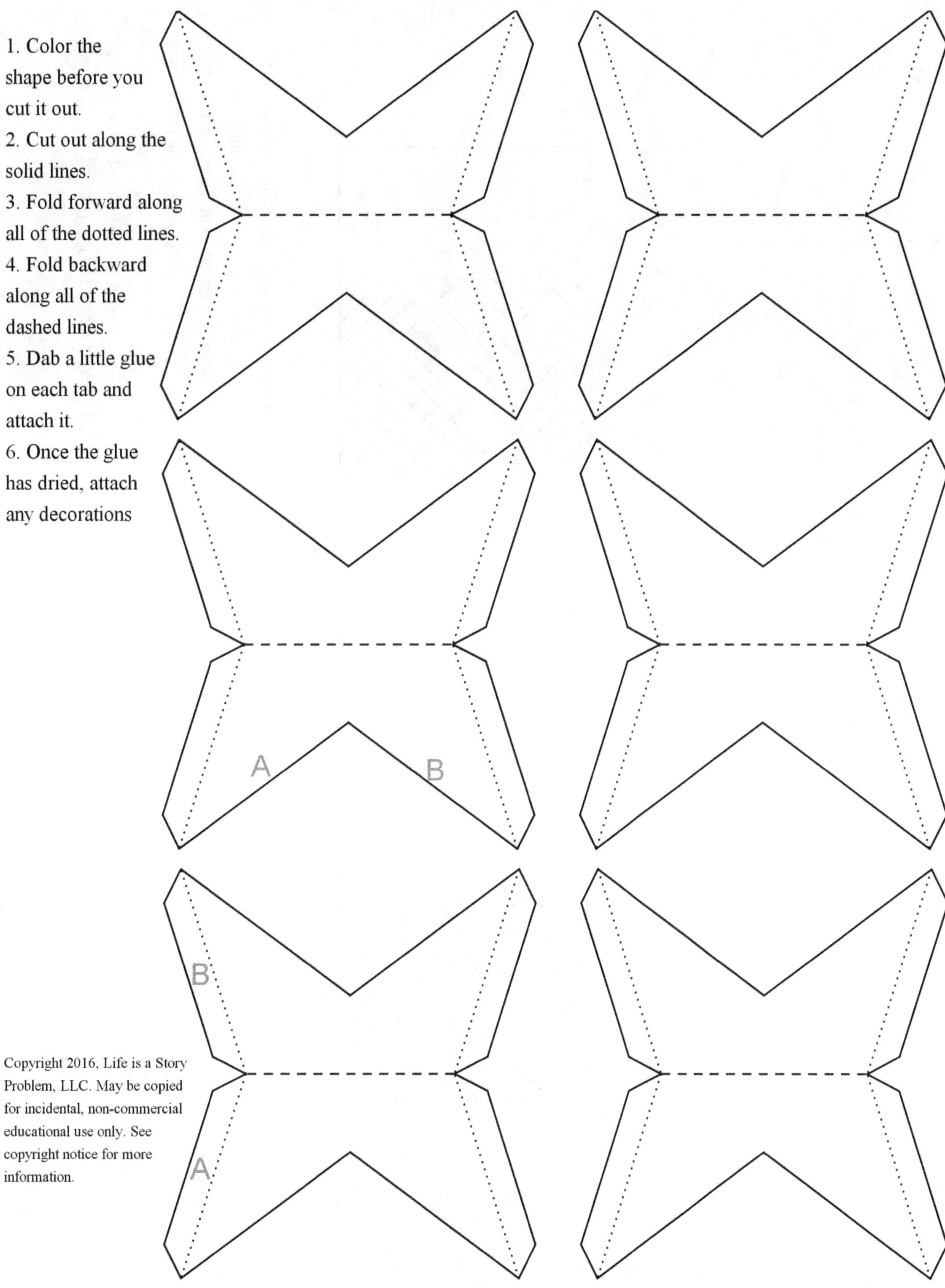

Car

This net is on two pages. Print out a copy of each page.

1. Color the shape before you cut it out.
2. Cut out along the solid lines.
3. Fold forward along all of the dotted lines.
4. Fold backward along all of the dashed lines.
5. Dab a little glue on each tab and attach it.
6. Once the glue has dried, attach any decorations.

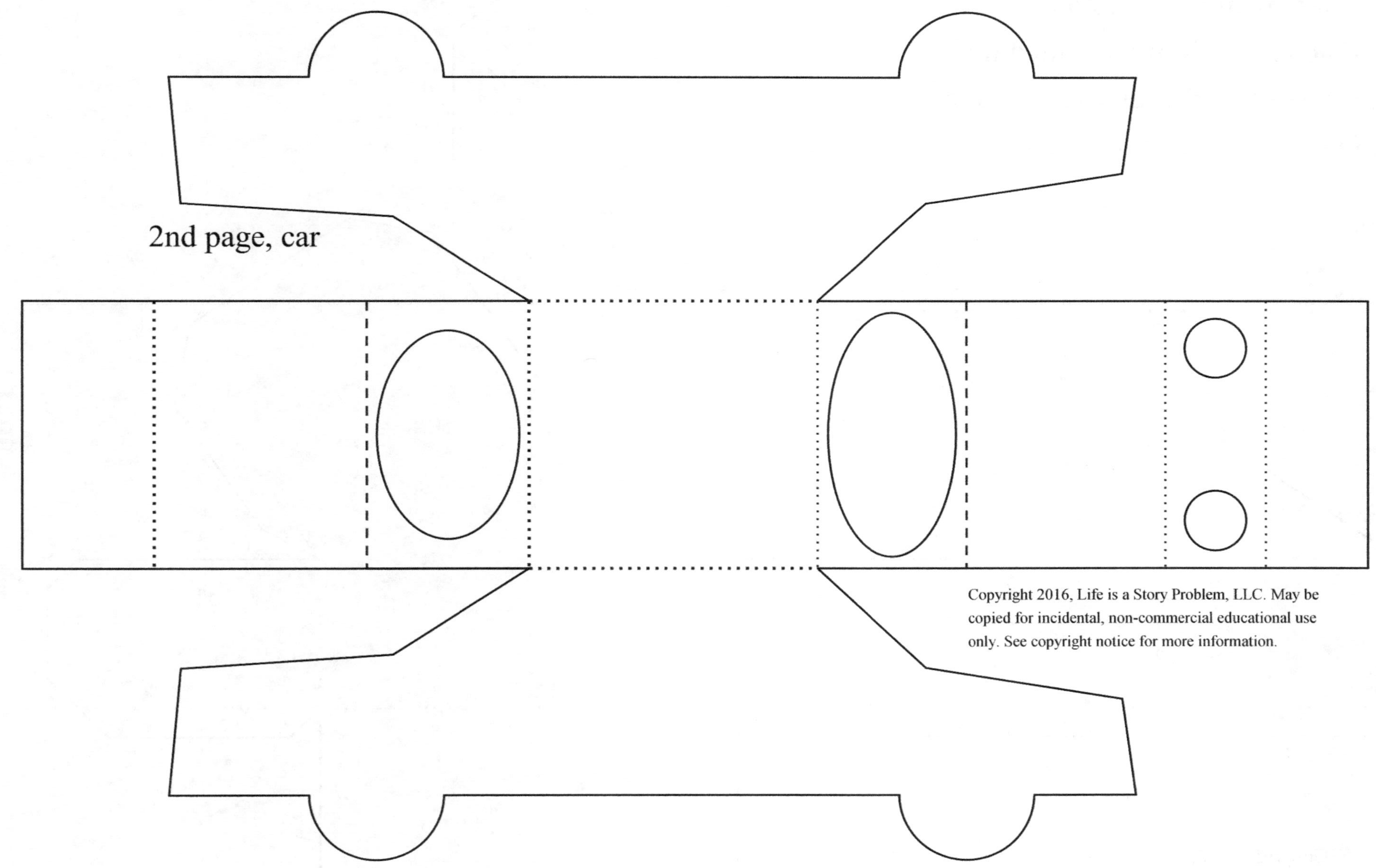

2nd page, car

House

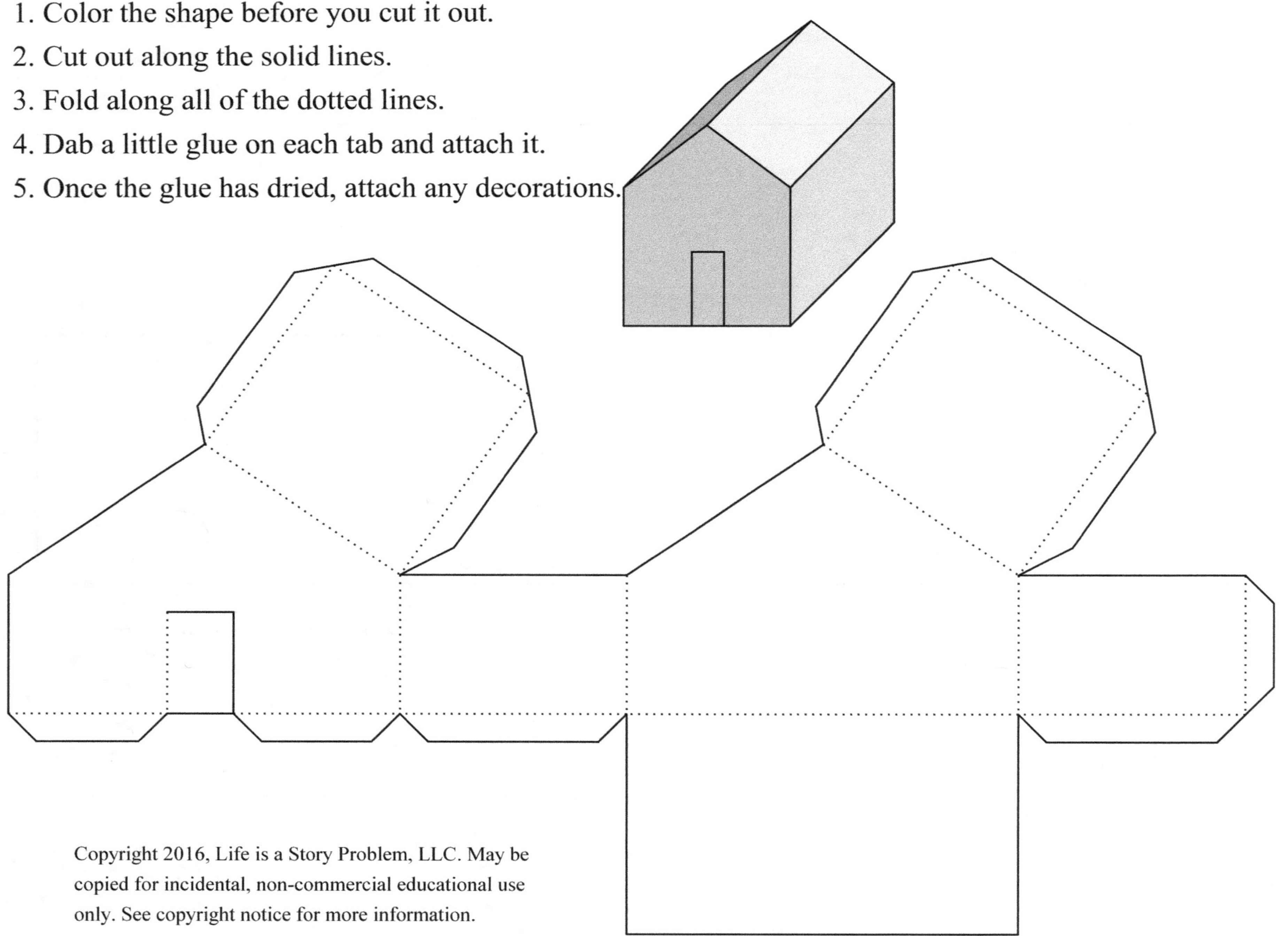

1. Color the shape before you cut it out.
2. Cut out along the solid lines.
3. Fold along all of the dotted lines.
4. Dab a little glue on each tab and attach it.
5. Once the glue has dried, attach any decorations.

Stand – quadrilateral

1. Color the shape before you cut it out.
2. Cut out along the solid lines.
3. Fold along all of the dotted lines.
4. Dab a little glue on each tab and attach it.
5. Once the glue has dried, attach any decorations

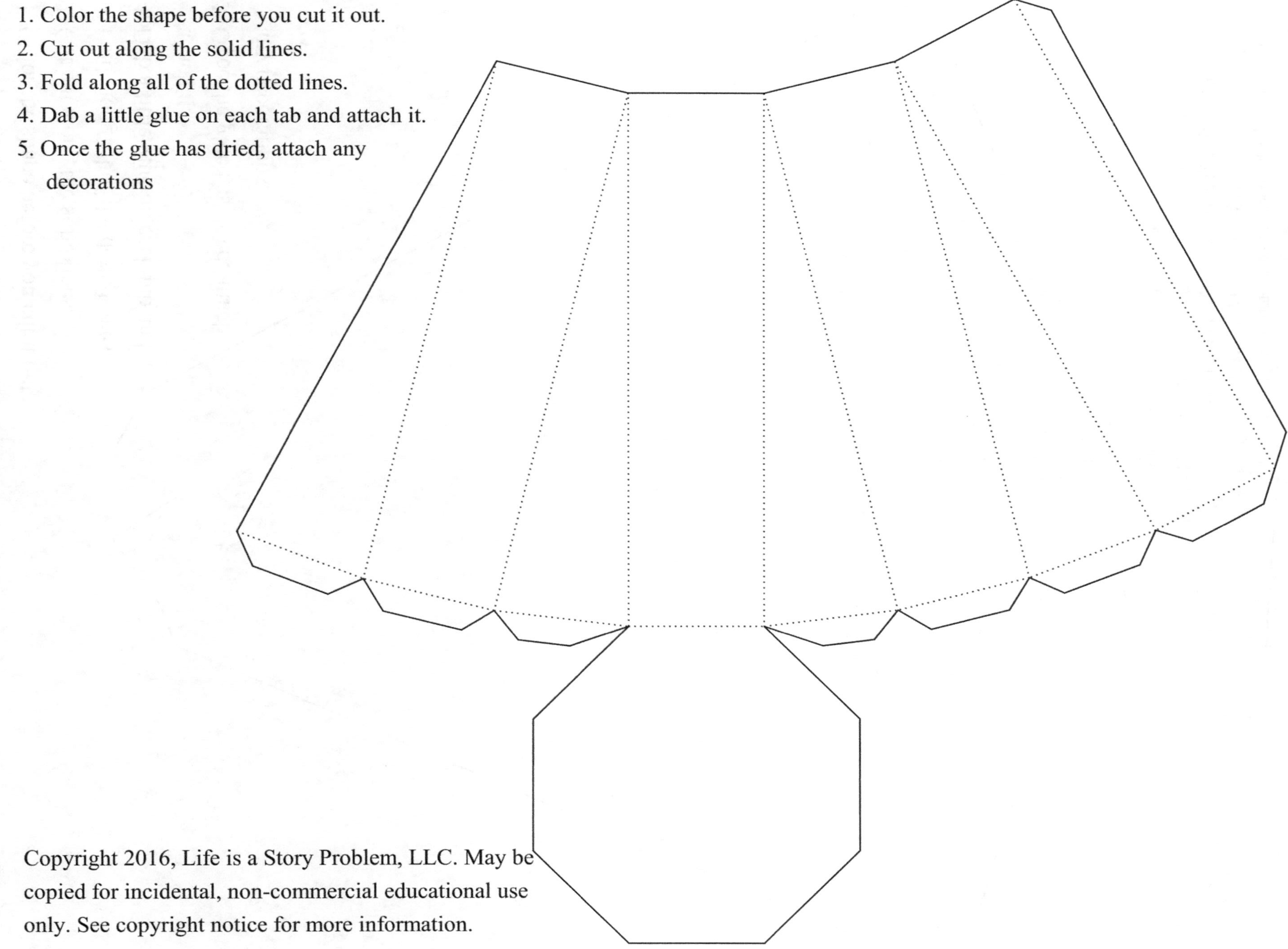

Stand - slotted

1. Color the shape before you cut it out.
2. Cut out along the solid lines.
3. Fold along all of the dotted lines.
4. Dab a little glue on each tab and attach it.
5. Once the glue has dried, attach any decorations.

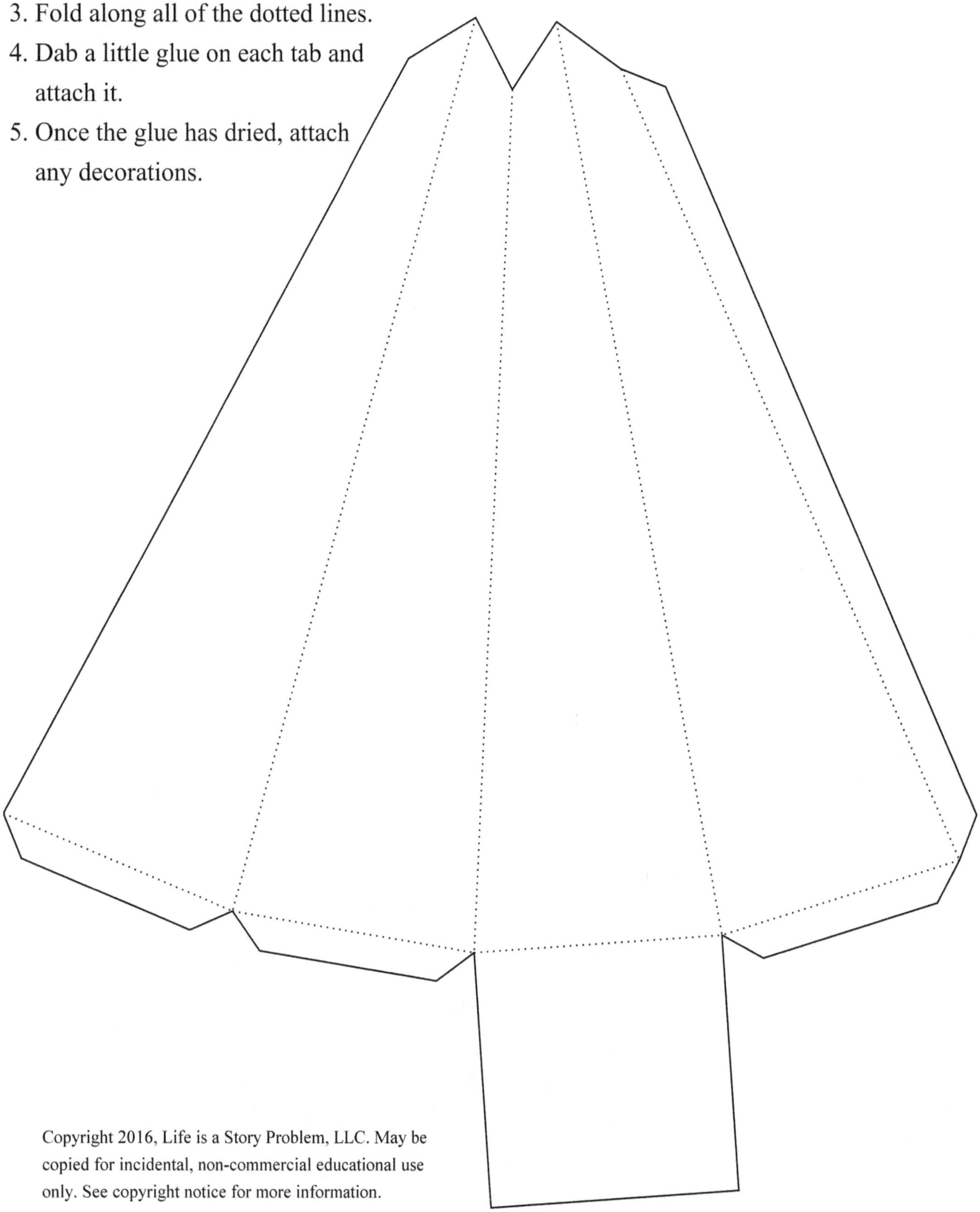

Stand – triangle

1. Color the shape before you cut it out.
2. Cut out along the solid lines.
3. Fold along all of the dotted lines.
4. Dab a little glue on each tab and attach it.
5. Once the glue has dried, attach any decorations.

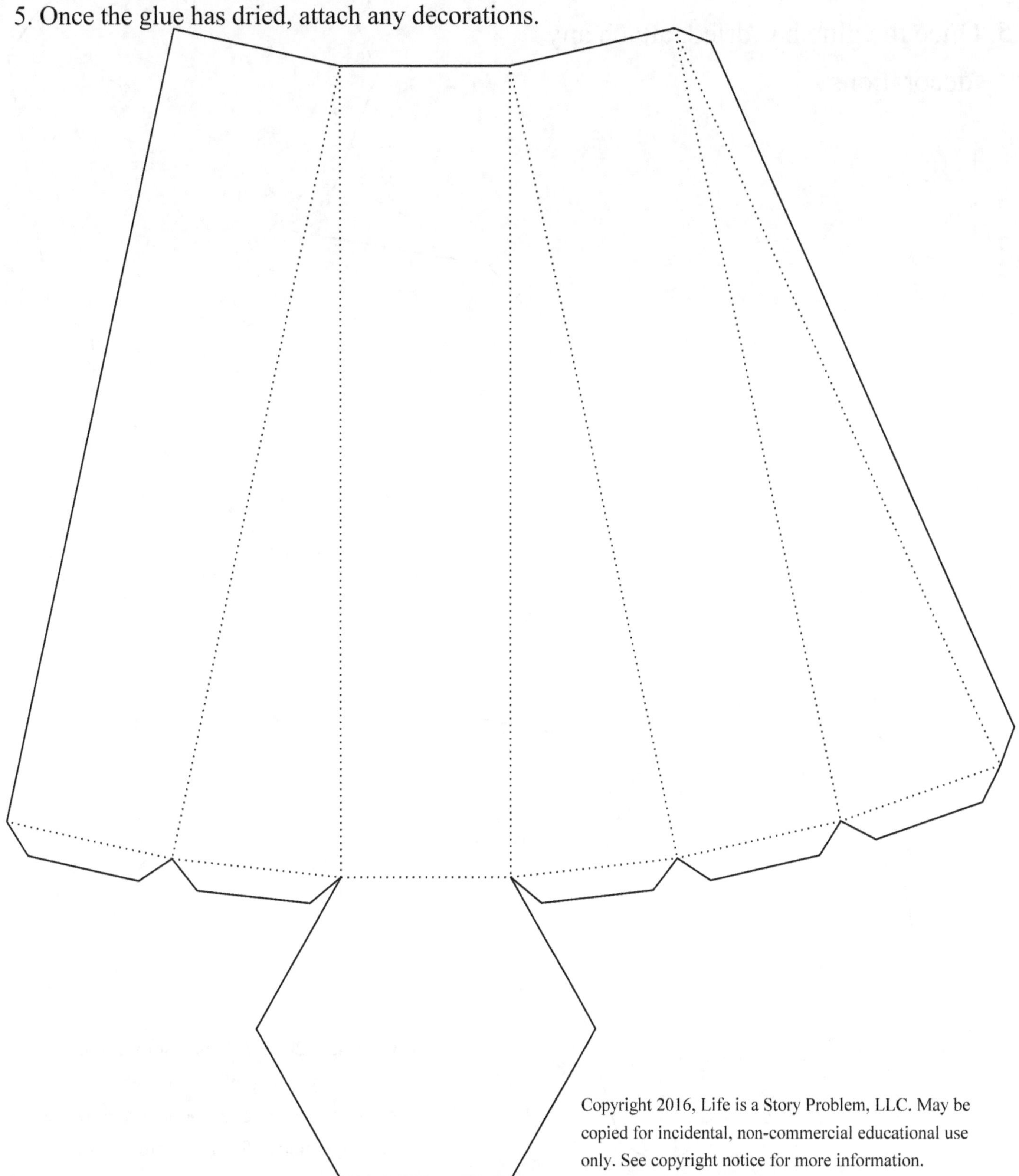

Dodecahedral faceted sphericon

1. Color the shape before you cut it out.
2. Cut out along the solid lines.
3. Fold along all of the dotted lines.
4. Dab a little glue on each tab and attach it.
5. Once the glue has dried, attach any decorations.

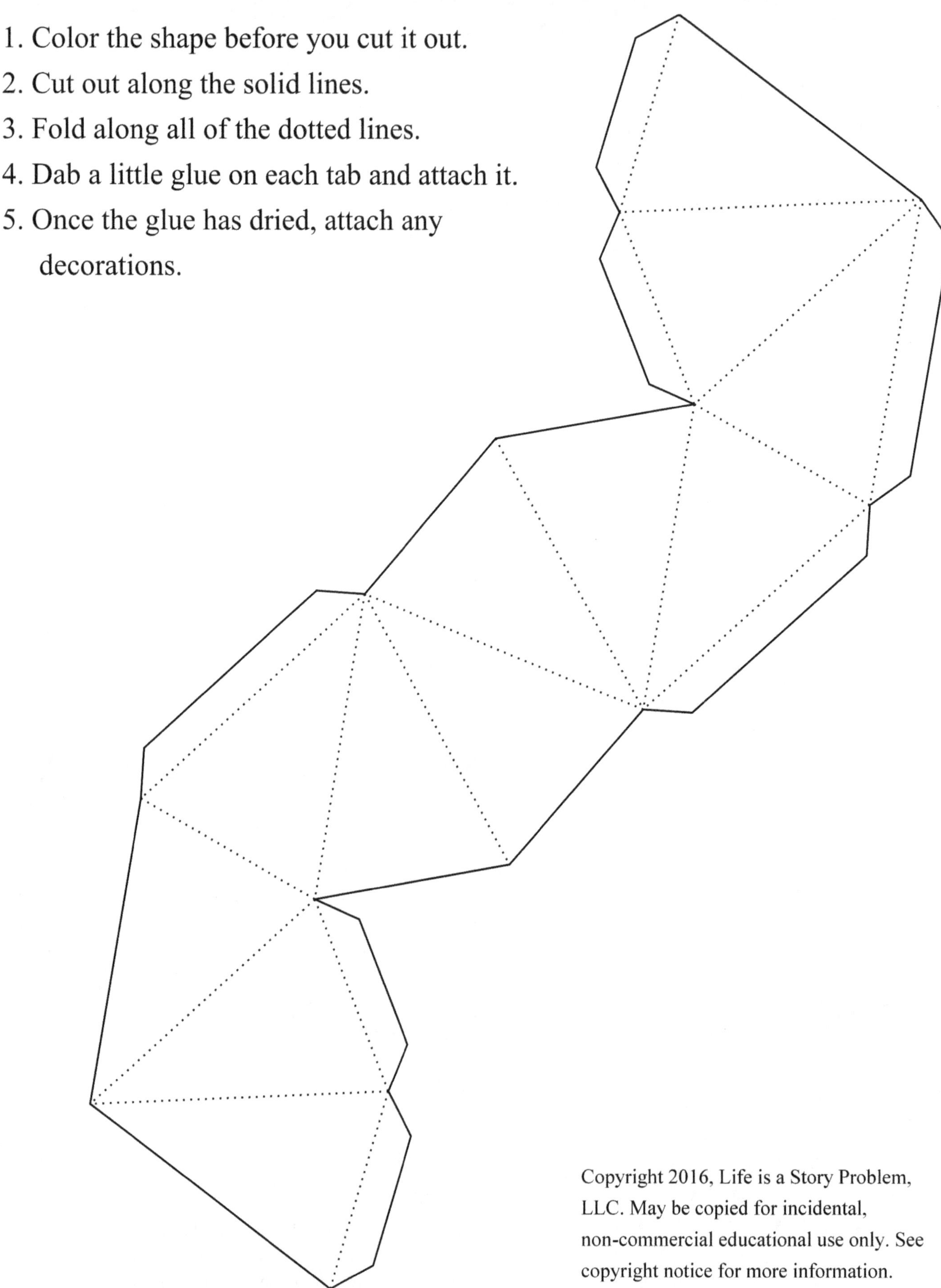

 Geometric Nets Mega Project Book, Tabbed By David E. McAdams

Hexakaidecahedral faceted sphericon

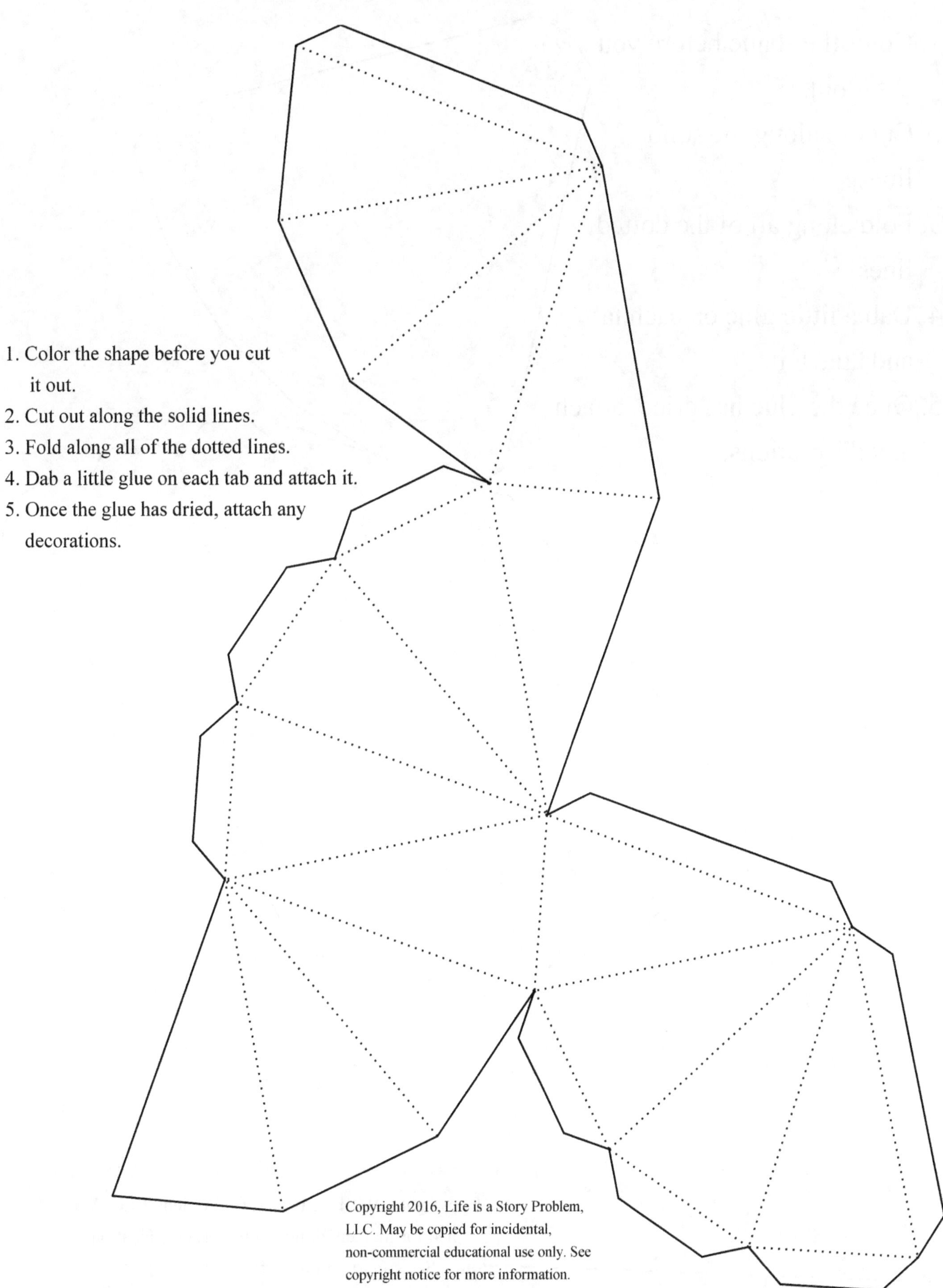

1. Color the shape before you cut
   it out.
2. Cut out along the solid lines.
3. Fold along all of the dotted lines.
4. Dab a little glue on each tab and attach it.
5. Once the glue has dried, attach any
   decorations.

Icosahedral faceted sphericon

1. Color the shape before you cut it out.
2. Cut out along the solid lines.
3. Fold along all of the dotted lines.
4. Dab a little glue on each tab and attach it.
5. Once the glue has dried, attach any decorations.

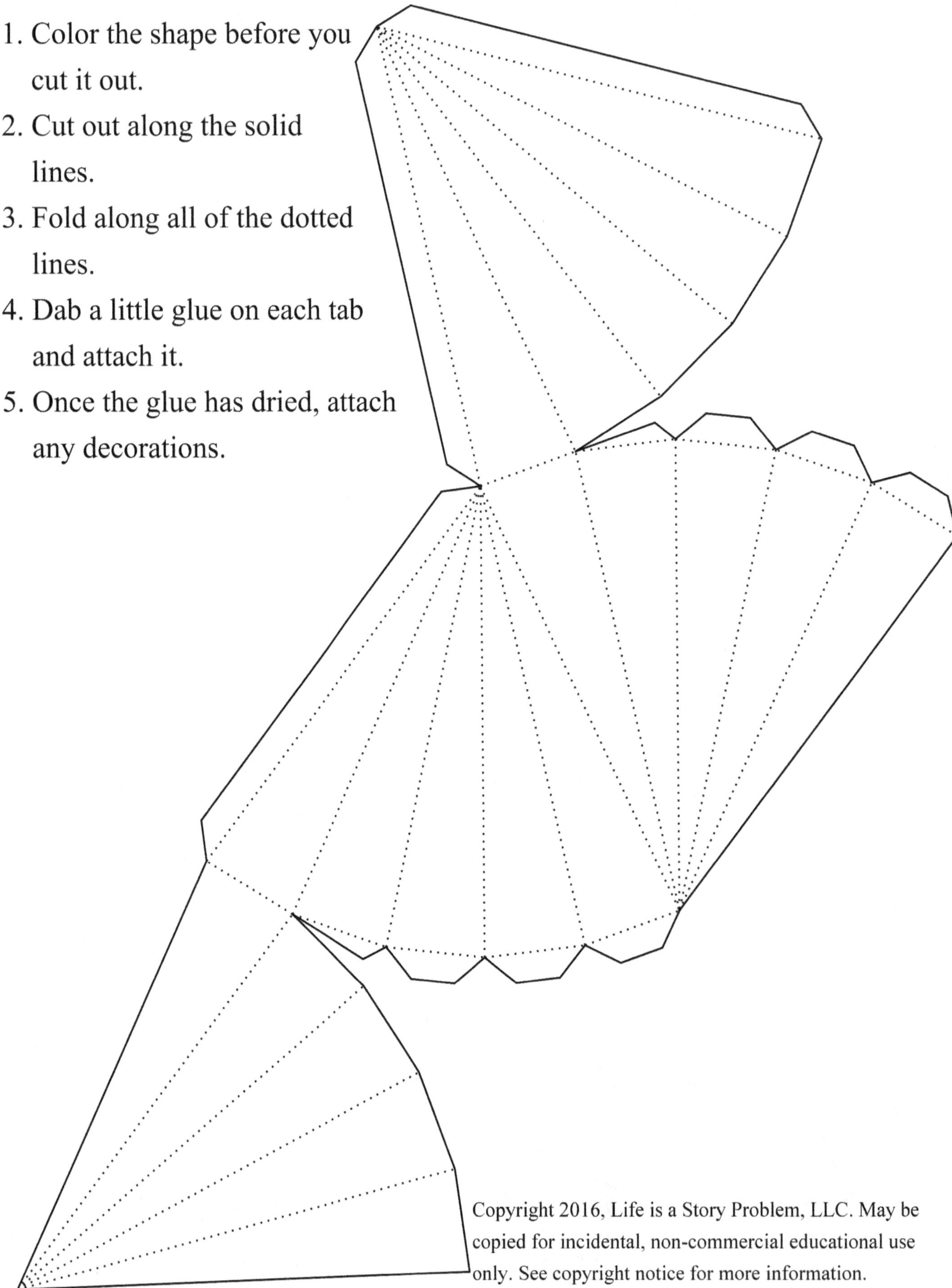

 Geometric Nets Mega Project Book, Tabbed By David E. McAdams

Cylindrical sphericon

1. Color the shape before you cut it out.
2. Cut out along the solid lines.
3. Roll the curved part over the edge of a desk or table.
4. Dab a little glue on each tab and attach it.
5. Once the glue has dried, attach any decorations.

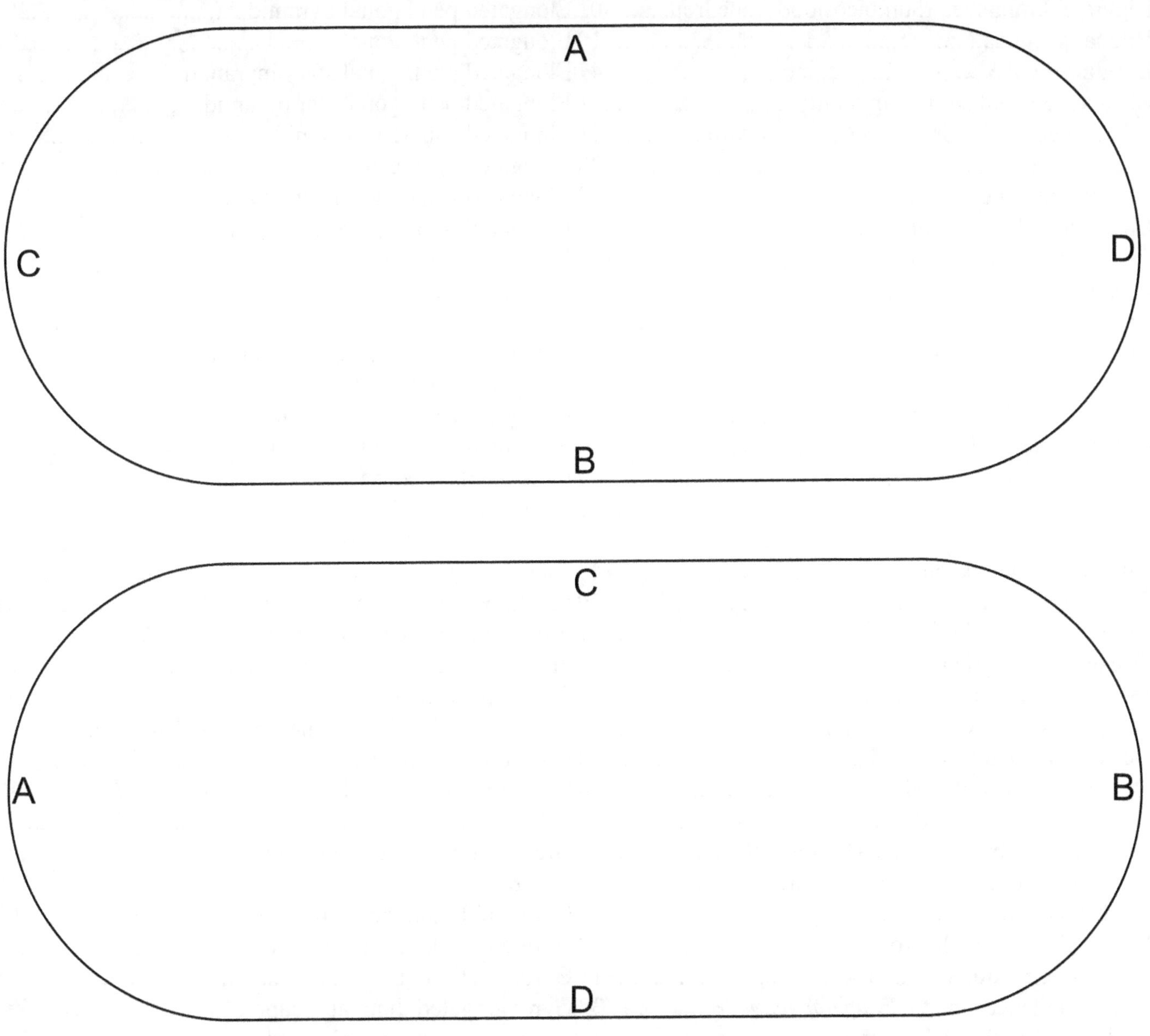

Alphabetical Index

Geometric Nets Mega Project Book, Tabbed By David E. McAdams

Geometric Nets Mega Project Book, Tabbed By David E. McAdams

Keyword Index

Index by Number of Faces